CAUSES AND EFFECTS OF HEAVY METAL POLLUTION

Causes and Effects of Heavy Metal Pollution

Mikel L. Sánchez
Editor

Nova Science Publishers, Inc.
New York

For permission to use material from this book please contact us:
Telephone 631-231-7269; Fax 631-231-8175
Web Site: http://www.novapublishers.com

LIBRARY OF CONGRESS CATALOGING-IN-PUBLICATION DATA

Causes and effects of heavy metal pollution / Mikel L. Sánchez (editor).
p. cm.
Includes bibliographical references and index.
ISBN 978-1-60456-900-1 (hardcover)
1. Heavy metals--Environmental aspects. I. Sánchez, Mikel L.
TD196.M4C38 2008
628.5'2--dc22 *2008030631*

Published by Nova Science Publishers, Inc. ✦ New York

CONTENTS

PREFACE

A heavy metal is a member of an ill-defined subset of elements that exhibit metallic properties, which would mainly include the transition metals, some metalloids, lanthanides, and actinides. Many different definitions have been proposed—some based on density, some on atomic number or atomic weight, and some on chemical properties or toxicity. The term heavy metal has been called "meaningless and misleading" in a IUPAC technical report due to the contradictory definitions and its lack of a "coherent scientific basis". As discussed below, depending on context, heavy metal can include elements lighter than carbon and can exclude some of the heaviest metals. One source defines "heavy metal" as "... common transition metals, such as copper, lead, and zinc. These metals are a cause of environmental pollution (heavy-metal pollution) from a number of sources, including lead in petrol, industrial effluents, and leaching of metal ions from the soil into lakes and rivers by acid rain.

Chapter 1 - Slovenia, a central European state with an area of little more than 20.000 km2, has a more than 500-year of metal mining and smelting tradition. In the previous century, almost all mines and smelting plants were closed, but numerous anomalies were left behind with total area of approximately 70-80 km2, where the concentrations of the heavy metals in soils exceed the critical values, concerning Slovenian legislation. The structure of this paper is as follows: introduction contains the brief description of the historical background of mining and smelting tradition in Slovenia and a description of each site of detailed research from historical perspective. After materials and methods chapter which contains the description of sampling, samples preparation, and data processing, the regional geochemical trends, based on the 60 sampling points in the Slovenian unpolluted areas, is presented together with comparison with urban areas. Further on the detailed description of the geochemical anomalies in the most polluted areas due to metal mining, smelting and iron working follow. The methodology has been the soil and attic dust sampling around past and present smelting plants. The areas, where the pollution has been researched in details, are:

- Mežica valley: Pb and Zn mine with smeltery and ironworks where mining tradition exists since 1665 with the consequence of heavy Pb and Zn pollution, 114 sampling points where soil and attic dust has been sampled; covered area – 101 km^2;
- Celje: 100 years of Zn smelting tradition and 150 years of ironworks left behind heavy Zn and Cd pollution, 99 sampling points where soil and attic dust has been collected, covered area - 92 km^2;

- Jesenice: iron working activities (Jesenice Ironworks company exists from 1937, but small scale smelting is dated back to the middle ages) left behind moderate Cd, Pb, Hg and Zn pollution, 44 sampling locations of soil profiles, covered area – 113 km^2;
- Litija: polymetallic (Pb, Zn, Hg, Ag, Ba) mining with smaller scale smelting activities, dating back to the roman times; moderate As, Mo, Hg, Pb, Sb and Sn pollution, 38 sampling points where soil and attic dust has been collected; covered area – approx. 30 km^2.
- Idrija: world's second largest Hg mine with smelting plant where 500 years of Hg production left heavy Hg pollution, 103 sampling points where soil and attic dust has been collected; covered area – 160 km^2.

A geochemical characteristic of each area is presented with non-parametrical statistical properties (median and average value, range, P25-P75). Bivariate statistics include correlation coefficients between sampling media and multivariate statistics (factor analysis) presents geochemical associations. Scale of the pollution on each area is presented with the help of enrichment factors, calculated on the basis of slovenian background values. Further on maps of spatial distribution of factor scores and of selected chemical elements are made. Discussion contains the brief summary of the research and the short descriptions of most evident anomalies.

Chapter 2 - Natural oxidation of sulphide minerals, exposed to the combined action of oxygen and water, results in the worst environmental problem associated with mining activities, i.e. acid mine drainage (AMD). Waters polluted by AMD are often characterised by low pH, elevated concentrations of iron, sulphates and toxic metals.

Biological remediation options in passive systems (permeable reactive barriers, PRB) usually exploit sulphur production by sulphate reducing bacteria, SRB.

In this report a combined chemical-biological treatment was tested for decontamination of synthetic AMD containing iron, arsenic, copper, manganese and zinc.

Particular attention was paid to the investigation of the mechanisms involved in pollutant removal (chemical precipitation, sorption, bioprecipitation and biosorption) as a fundamental preliminary step for permeable reactive barrier design and long term performance estimation.

Experimental tests were performed both in batch reactors and in a two-column apparatus for sequential treatment by chemical precipitation (first column filled with natural limestone) followed by bioprecipitation/biosorption (second column filled with a natural organic mixture inoculated by sulphate reducing bacteria).

Distinct mechanisms of removal for each metal were identified by combining theoretical data of metal solution chemistry, and results obtained from independent experimental tests: batch and column tests, blank tests using natural organic mixture as biosorbing materials, acid digestions, and selective extractions of metals using solid samples of filling material after column dismantlement.

This analysis allowed isolating metal-specific mechanism of abatement and denoted the relevant contribution of biosorption phenomena in metal removal in biological column. This contribution, generally neglected in biological PRB design with respect to bioprecipitation, should be taken into account in order to avoid misleading estimation of SRB performance and also to better estimate PRB duration.

Chapter 3 - The remediation of mining areas represents a relevant environmental problem in all Europe. The high concentration of heavy metals and the lack of nutrients determines the desertification of wide areas. In Sardinia (Italy) the poor management of Montevecchio – Ingurtosu mining district after mine closure caused the dispersion of high amounts of contaminants by wind and water erosion on wide areas. The wide extension of the contaminated area and the high level of contamination by heavy metals make the application of phytoextraction feasible for this area. The environmental risk related to the presence of heavy metals can be evaluated by determining the bioavailable metal fraction in soil. The Department of Geoengineering and Environmental Technologies of the University of Cagliari made experiments of phytoextraction and assisted phytoextraction both with plants having a high biomass production (Mirabilis jalapa) and with native species (Cistus salvifoliius, Scrophularia canina and Teucrium flavum). Easily biodegradable chelating agents were applied in laboratory experiences (MGDA - methylglycine diacetic acid, S,S-EDDS - [S, S]-ethylenediaminedisuccinic acid, IDSA - iminodisuccinic acid). The ability of the plant species to tolerate and accumulate heavy metals demonstrated the applicability of phytoextraction to the abandoned mining areas remediation.

Chapter 4 - Phytochelatins (PCs) are a group of small, metal-binding peptides that are biosynthesized by higher plants, some fungi and algae in the response to heavy metal exposure. One actual research topic focuses on better understanding the global effect that all elements present in natural environments exert on the PCs production by plants.

In this work, PCs levels were evaluated in the wild plants, chronically exposed to low or moderate levels of heavy metals. The quantification of total PCs in plant extracts was carried out by HPLC with fluorimetric detection, after derivatization of free –SH groups with monobromobimane. Additionally, the distribution of metals in molecular mass (MM) fractions of these same extracts was studied by size exclusion chromatography with on-line UV and ICP-MS detection. All samples were collected in Guanajuato city (Mexico), which has long been an important silver and gold mining area. Among different metals reported in Guanajuato soils, lead, cadmium, copper and silver were selected in this study, because of their capability to induce phytochelatins in plants. The common plants from this region were analyzed, namely: Ricinus communis (castor bean), Tithonia diversifolia (Mexican sunflower) and Opuntia ficus (nopal). The analytical approach involved the ICP-MS analysis of total elements in soil, soil fractions and wild plants and also the evaluation of relationships between PCs, metal levels found in plants/soil and different soil parameters.

In the analysis of plants, PC-2, PC-3 and PC-4 were detected in nopal, PC-2 in castor bean, while in Mexican flower no phytochelatins were found. In further development, the extracts of soil humic substances were obtained and the distribution of metals in molecular mass (MM) fractions was studied by size exclusion chromatography with on-line UV and ICP-MS detection. The soil humic substances (HS) were also assessed.

In search of possible relationship between the parameters measured, the statistical analysis of correlation was performed. The results obtained indicate that the binding of metals to soil HS contributes in lowering their uptake by castor bean plant. On the other hand, the soils collected at nopal roots presented low HS levels and no correlation with metals in plant was found. The results obtained in the sequential extraction of soils and the abundance of sulfide minerals in Guanajuato indicate that the sulfide bound metals were the primary forms of Pb, Cu and Cd in soil adjacent to nopal roots. Owing to their generally poor solubility, rizosphere processes should be important in mobilizing metals and their uptake by nopal.

The authors' results provide further evidence on the role of environmental conditions in the accumulation of heavy metals in relation to PCs production in different plant genotypes. In particular, multi-elemental approach is necessary in studies on PCs induction in actual field situations, where plants (or other organisms) are exposed to a variety of metals and metalloids.

Chapter 5 - Element mobility is of major importance with regard to bioavailability and the potential risk for contamination. Different factors control the ultimate fate of a toxic element in the environment, that is, if it will precipitate or will be adsorbed , or released, transported and taken up by organisms.

The objectives of this work are:

- To evaluate background levels of heavy metals in soils of highly vulnerable area in northern Italy.
- To ascertain metal mobility and possible contamination of some sites, and the related environmental hazard, with special reference to the pollution of the Venice lagoon, which is a unique and delicate ecosystem.
- The Venetian territory is characterized by different pedolandscapes:
- A wide plain formed by alluvial deposits. Most soils here (Entisols, Inceptisols, Alfisols) are cultivated with extensive agriculture;
- Gently ondulating conglomerates, marls and limestones with shallow soils (Entisols and Inceptisols) frequently cultivated with vineyards, or forested;
- Mountain ranges with steep morphology. Forestry and grassland are the main land utilization types on these soils (mostly Inceptisols and Mollisols).

Approximately 900 soil samples from 300 representative soil profiles were analyzed for As, Cd, Co, Cr, Cu, Hg, Mn, Ni, Pb, Zn. Data were statistically processed to find close relationships among elements.

Agriculture soils. The soils examined (more than 200 sites) contain generally high levels of anthropogenic Cu, Zn, Pb, and As.

Forest soils. The heavy metal contents in the soils examined (more than 100 sites) are generally below the target values and depend mainly upon local physico-chemical and geological conditions. Anthropogenic Pb and Cd are concentrated especially in organic layers.

Increasing acidic conditions, redox status, organic matter content and pore solution are the factors responsible for trace elements mobilization within the soil. From the soil, trace elements move to groundwater and to lagoon, where they are concentrated in sediments or transferred to organisms.

The elements with the lowest relative mobility (bioavailability) are Co, Cr, Hg, Pb, As; intermediate elements are Cu, Ni and the most bioavailable are Zn and Cd.

The soils investigated have heavy metal concentrations that are generally within the regulatory guidelines. Exceptions are anthropogenic Cu and Zn in agricultural soils, Cd and Pb at forest sites.

The ecological risk posed by single elements is limited for As and very high for Cd. The cumulative toxic risk indicates a relevant bioaccumulation of trace elements in the lagoon ecosystem.

Chapter 6 - The environmental chemistry of mercury is complex and difficult to predict because it is controlled by a multitude of environmental processes which includes photochemical reactions, chemical oxidation and reduction, microbial transformation, and physiological fractionation. Mercury vapor from natural and anthropogenic sources is released to the atmosphere and distributed globally before being oxidized to a water-soluble form and returned to terrestrial and aquatic systems via deposition. This results in the distribution of mercury contamination to all parts of the earth. Mercury's natural cycle has been significantly disrupted by anthropogenic activities, and atmospheric mercury concentrations have steadily increased since the Industrial Revolution. In aquatic environments, mercury is methylated through a microbially-mediated process primarily involving sulfate-reducing bacteria. Therefore, methylation is strongly influenced by factors that affect these bacterial consortia. Of equal importance are variables that affect the availability of inorganic mercury for uptake by these bacteria. Conditions that favor mercury methylation include low pH, high DOC concentrations, and low redox conditions. However, the factor that asserts the most control is sulfur chemistry and its link to sulfate-reducing bacteria and inorganic mercury availability. Methylmercury is of great concern in aquatic environments because of its ability to bioconcentrate and biomagnify through trophic webs. Human exposure to this neurotoxin ultimately results from consumption of contaminated fish.

Chapter 7 - Development of new analytical techniques and methodologies capable to identify and quantify the composition of complex samples, as the ones related to environmental problems, is an actual tendency. The objectives of this work were: to use X-ray fluorescence technique, with portable system, to identify and quantify the chemical elements present in the water and its concentration in the sampling place, to optimize the preconcentration methodology and to adapt it for field use. The analyses were realized at laboratory and in situ, measuring water samples in natura and pre-concentrated in membranes. It was employed a portable X-ray tube (Ag target, 50 μm Ag filter, 4 W) to excite the samples and a Si-PIN detector (221 eV resolution for 5.9 KeV energy and 25 μm Be window) with standard electronics for acquisition and evaluation of the spectra. The samples were filtered for suspended particulate matter retention. After this, the optimized preconcentration procedure, with APDC precipitation, was applied. The standard reference materials SRM1640 and SRM1643e, prepared in the same conditions of the samples, were analyzed for methodology validation. Samples from several points at Londrina city, Paraná State, Brazil, were analyzed. It was possible to identify and quantify Ca, Ti, Mn, Fe, Cu, Zn and Pb. The equipment performance and robustness were very good and the results satisfactory for in situ analysis employing a portable system. Considering membrane measurements, the system detection limits are below the maximum values established by national and international legislation for drinking water. Besides this, the quantification limit, that is around 0.01 mg L-1 for the majority of the elements, makes viable the methodology application for water analysis. Portable EDXRF has shown to be an useful tool for environmental analysis, as it is a fast, efficient and convenient technique, with potential to substitute high cost and time consuming laboratory methods.

Chapter 8 - Sustainable management of marine bioresources require an ecosystem perspective that includes recognition of natural and anthrapogenic disturbances on supporting food webs and resultant changes in community structure. The marine environment is continuously subjected to chemical pollution, which can have detrimental effect on aquatic organisms living in that environment. Each year several new synthetic chemicals enter the market, many of which are likely to reach and pose impacts on the marine environment. The concept to develop and use biological markers to monitor marine environments is at the base of recent studies. The biological marker reveals the status of environmental health condition by accumulation of metals in their tissues. The migrating organisms may be suitable indicators for a larger zone instead of confined areas. The selection of benthic and sedentary organisms could be ideal to select as bioindicators for notified regions. Among them, primitive multicellular organisms like sponges would be ideal. Sponges are filter feeders and are ideal for assessing the effect of silation and environmental contaminants on the primary food chain. Sponges seem to be filtering a large volume of water and accumulate heavy metals. Heavy metal contamination and sediment deposition had significant influence on the secondary metabolites synthesis of marine sponges. However, accumulation seems to depend on the metal and the species considered. Bacterial endosymbionts isolated from the sponges invariably showed resistance against a battery of heavy metals tested including copper, lead, mercury, cobalt and cadmium. Therefore, the bacteria associated with the sponges can be used as indicators of contamination in marine ecosystem. Seawater bacteria have already been established as biological indicators of contamination. Considering the overall complexity of ecological factors in the marine environment, developing a manageable set of bioindicators is a challenging task. The present report envisages the possibility of developing benthic nematodes as potential bioindicator model for monitoring the heavy metal pollution in the marine environment.

Chapter 9 - With urbanization and economic development in coastal area, metal pollution in coastal environment has been a problem. Estuaries and coastal intertidal zone are important habitats for aquatic and marine life. In the meantime, sediment is a repository of contaminants and records the pollution history. Sediment quality reflects the long-term health status of an estuarine or coastal system and can be evaluated by sediment quality guidelines/criteria, metal enrichment factors and other indicators. In this study, the authors summarize the results mainly from their previous studies in metal pollution in sediments in urban estuarine and coastal systems in the US and China, including New York Harbor and Yangtze River estuary in metropolitan areas and Xiamen Bay and Jiaozhou Bay. The information from this study expands the authors' knowledge in understanding metal pollution in urban coastal systems and assessing environmental quality impacted by industrialization and economic development.

Chapter 10 - Heavy metals are metallic elements with high atomic weights. They tend to accumulate in the food chain and can be toxic and mutagenic. Such elements like mercury, chromium, cadmium, arsenic, and lead, when they are in their ionic and complexed forms, pollute soil, water and even air. Conventional methods of identification of polluted environment are laborious and costly. The presence of contamination is very often difficult to detect. It is even more difficult to evaluate its potential danger to living organisms. In this review, the authors will discuss the use of transgenic plants for the detection of heavy metal pollution and for the evaluation of its potential toxicity and mutagenicity. The greatest advantage of transgenic plants is that they can be made to be more sensitive to a particular pollutant. Plants are an excellent alternative to conventional methods, since they can be planted and grown at the site of pollution. In this chapter, the authors will describe transgenic plants that have already been successfully used for biomonitoring heavy metal pollution, and will also present novel ideas for generating efficient transgenic phytosensors.

Chapter 11 – The paper documents a detailed account of spatial distribution and possible sources of major elements along with heavy metals in fine grained fractions (<63 μm) of core sediments collected from seven sites in Sunderban wetland, northeastern part of Bay of Bengal, India. This work aims to evaluate the fluvio-marine and geochemical processes influencing the trace element distribution and to check the suitability of employing heavy metals data in evaluating biological effects on the basis of sediment quality guidelines. Both Inductively Coupled Plasma Mass Spectrophotometer (ICPMS) and Atomic Absorption Spectrometry (AAS) were employed to determine the elemental concentration in acid-digested sediment samples. Trace element concentrations were available at different core depths with an erratic pattern of distribution. An overall enrichment of majority of the elements has been recorded at the site Kakdwip, located along the main stream of Ganges, and this can be attributed to domestic and industrial effluent discharge, intensive fishing and boating activities coupled with use of antifouling paints. In contrast, the site Canning, located further east in the mudflat of Matla River is characterized by minimum trace element content. An abrupt variations of Mn, V, Cd and U were encountered at the site Jambu Island- an offshore island facing Bay of Bengal. For Cu, Ni and As, a smaller proportion of the samples had exceeded the effects range-low (ER-L) concentrations indicating that the dataset would be suitable for future use in evaluating predictive abilities of sediment quality guidelines

Chapter 12 - This invited paper reviews the status of heavy metal pollution in the paddy soils and rice in Korea, the human health risk assessment for heavy metals, and the remediation approaches to reduce the metal translocation from soil to rice grain. The Soil Environment Conservation Law (SECL) designates the soil pollution standard for As, Cd, Cu, Hg, Cr^{6+}, Pb, Ni, Zn and F, and these are used as both the maximum permissible levels of such metals in agricultural soil. The extensive monitoring for heavy metals in soil and crop revealed that concentrations of metals in paddy soils, without an evident anthropogenic source of contaminants, were mostly below the threshold levels designated by the SECL. Major sources of metal pollution in paddy soils were however related with mining activities. The increased level of Cd and Cu in soil increased the activities of cations (Ca>Mg>K) temporally, decreased the level of exchangeable cations, altered the supply mechanisms and decreased the nutrient buffering capacity of soil. The most widely described effects of metal toxicity in plants were the stunted growth, leaf epinasty and chlorosis. The Korean Government implements various countermeasures to prevent the soil pollution by metals through legislation, monitoring networks, risk assessment and remediation. The potential risk of the adverse effects of metals on human health was assessed based on the human exposure pathways to rice, groundwater and soil in three abandoned mines where metal contents in soil and rice exceeded the safety guidelines. The hazard index (HI) values for As and Cd exceeded 1, representing a potential toxic risk of As and Cd to the human health. The cancer risk for As via the rice and groundwater consumptions exceeded one cancer case in ten thousand. Health risk assessment indicated that a long term exposure to rice grown in the metal contaminated paddy soils could pose a potential health threat. The soil and plant management options have been considered to prevent the heavy metal transfer to rice from the contaminated paddy soils. The soil management options include the uses of soil ameliorants, fertilizers and irrigation control, soil covering/dressing, reversing and soil layer mixing methods. In the plant management options, the 24 rice cultivars were screened to find the accumulating or excluding variety. The Japonica cultivars were considerably low accumulating rice for Cd. These cultivars might be screened to cultivate in the contaminated soil environment to have the metal concentration at a low enough for the safe consumption. The continuous submersion of the soil was interacted better with fertilizer than the intermittent irrigation to retard the Cd uptake by rice. Based on the regulatory criteria of Cd for soil pollution and food safety, the quantity of Cd which should be remediate at most was estimated to be only 0.04% of Cd in the contaminated soils. Is it worthwhile to remove such a small quantity of Cd with effort and budget which may be greater than land price? Are those criteria the risk-based or the concentration-based? At least limiting to rice, we need to devote to the development of protocols for pollution monitoring, risk assessment and remediation to cope with such dilemmas in the paddy soil environment.

In: Causes and Effects of Heavy Metal Pollution
Editor: Mikel L. Sanchez
ISBN 978-1-60456-900-1

Chapter 1

IMPACTS OF THE MINING AND SMELTING ACTIVITIES TO THE ENVIRONMENT – SLOVENIAN CASE STUDIES

Gorazd Žibret* and Robert Šajn†
Geological Survey of Slovenia, Dimičeva ulica 14, SI – 1000 Ljubljana, Slovenia

ABSTRACT

Slovenia, a central European state with an area of little more than 20.000 km2, has a more than 500-year of metal mining and smelting tradition. In the previous century, almost all mines and smelting plants were closed, but numerous anomalies were left behind with total area of approximately 70-80 km2, where the concentrations of the heavy metals in soils exceed the critical values, concerning Slovenian legislation. The structure of this paper is as follows: introduction contains the brief description of the historical background of mining and smelting tradition in Slovenia and a description of each site of detailed research from historical perspective. After materials and methods chapter which contains the description of sampling, samples preparation, and data processing, the regional geochemical trends, based on the 60 sampling points in the Slovenian unpolluted areas, is presented together with comparison with urban areas. Further on the detailed description of the geochemical anomalies in the most polluted areas due to metal mining, smelting and iron working follow. The methodology has been the soil and attic dust sampling around past and present smelting plants. The areas, where the pollution has been researched in details, are:

- Mežica valley: Pb and Zn mine with smeltery and ironworks where mining tradition exists since 1665 with the consequence of heavy Pb and Zn pollution, 114 sampling points where soil and attic dust has been sampled; covered area – 101 km^2;
- Celje: 100 years of Zn smelting tradition and 150 years of ironworks left behind heavy Zn and Cd pollution, 99 sampling points where soil and attic dust has been collected, covered area - 92 km^2;
- Jesenice: iron working activities (Jesenice Ironworks company exists from 1937, but small scale smelting is dated back to the middle ages) left behind moderate Cd, Pb, Hg and Zn pollution, 44 sampling locations of soil profiles, covered area – 113 km^2;

* gorazd.zibret@geo-zs.si; tel.: ++386-1-2809-765.
† robert.sajn@geo-zs.si; tel.: ++386-1-2809-769

- Litija: polymetallic (Pb, Zn, Hg, Ag, Ba) mining with smaller scale smelting activities, dating back to the roman times; moderate As, Mo, Hg, Pb, Sb and Sn pollution, 38 sampling points where soil and attic dust has been collected; covered area – approx. 30 km^2.
- Idrija: world's second largest Hg mine with smelting plant where 500 years of Hg production left heavy Hg pollution, 103 sampling points where soil and attic dust has been collected; covered area – 160 km^2.

A geochemical characteristic of each area is presented with non-parametrical statistical properties (median and average value, range, P25-P75). Bivariate statistics include correlation coefficients between sampling media and multivariate statistics (factor analysis) presents geochemical associations. Scale of the pollution on each area is presented with the help of enrichment factors, calculated on the basis of slovenian background values. Further on maps of spatial distribution of factor scores and of selected chemical elements are made. Discussion contains the brief summary of the research and the short descriptions of most evident anomalies.

1. Introduction

The begining of geochemical investigation on the Geological Survey of Slovenia is dated back to the 1950. Since then all of the analyses have been collected in the database, which now contains more than 9000 entries. The investigations have been focused on the two problems; first: investigation of the regional geochemical trends and properties of the soils, developed on the different types of bedrocks and second: investigation of the anomalies, which are the consequences of past and present mining and smelting activities. Former is the topic of this book chapter.

The brief historical introduction of the metal mining and smelting activities on the area of Slovenia and of the areas of detailed geochemical research is presented in this paper. The next chapter will introduce the reader to the methods of the sampling, analyzing and data interpretation, performed for the geochemical investigation on the Geological survey of Slovenia. Further, data on the brief regional geochemical characteristics and data about geochemical background of the Slovenia will be presented. The main point of this text was to present the influences of the biggest mineral and metal processing plants to the environment in the sense of heavy metal pollution. Five different locations together with multivariate statistical analysis and geochemical maps will be described in more detail. Those locations are: Idrija mercury mine with smeltery (approx. 500 years of operations), Jesenice ironworks (approx. 150 years of iron and steel production), Celje Zn smelting plant and nearby Štore ironworks (100 years of operational time), Mežica Pb-Zn mine with smeltery (450 years of mining and smelting operations) and Litija polymetallic mine with smeltery (500 years of operation).

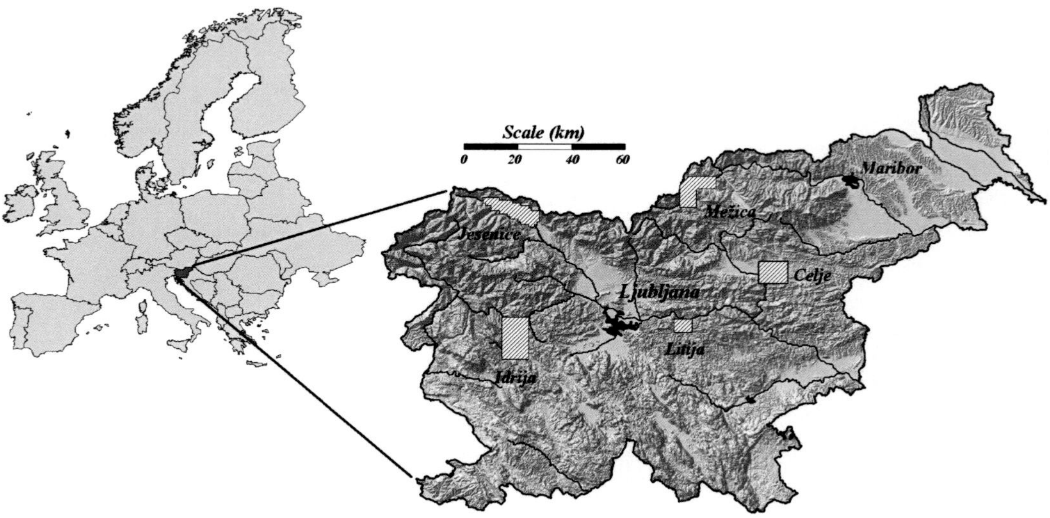

Figure 1. Locations of Slovenia and areas of detailed research.

2. Historical Background

The mining and smelting tradition in Slovenia (Figure 1) has a long history. Some archaeological artifacts, findings of the mining tools on Pohorje region (Tržan, 1989), suggest that mining and metal smelting started in Bronze Age. In the Iron Age period - hallstadt period (800 - 300 B.C.), there have been numerous evidences of iron mining and smelting (findings of molds, tools etc...). The land has been rich in iron ore and mining; smelting and forgery have been wide spread. Also the iron tools and armory from that region has been recognized in the Roman Empire after its quality and this province, called Noricum, has maintained independence and the status hospitum publicum (friends of Rome) for the long period. After annexation of the Noricum and Illyricum provinces to the Roman Empire the beginning of the exploitation of lead and copper ore has taken place.

The exploitation of the biggest ore deposits started in the middle ages. The main branch was the iron smelting. The ore has been collected on the surface and inside small mining shafts. The known bigger mining operations have been in Idrija, Mežica, and Litija. After the fall of feudalism and construction of railways in 1850 the iron smelting activities has increased drastically. Also, the production of other metals, such as Pb, Zn, Hg, Cu and Sb, has gained in its importance. The production reached peak between 1850 and 1900 when taking into account the number of known mining pits. Especially, the Idrija mercury mine had the greatest importance in the Habsburg monarchy. At the peak of mercury production the mine contributed as much as 50% of the monarchy's annual budget.

In the beginning of the 20th century many of operating mines and connected smelters have been closed down due to small ore quantities and its low grade. Only the biggest ones have prevailed. The new, but short-term impulsion to the metal production has been the period of First World War because of the lack of base metals supply. Between both world wars almost all metal production in the territory of Slovenia has been halted, especially after 1929's great depression. Only Mežica Pb-Zn mine was still in operation. Until the end of Second World War the Idrija and Litija mines have been reopened.

After the arrival of communist regime after the end of Second World War the authorities put the great emphasis of on the mineral prospection, but no new large metal deposits has been discovered. The exception was only the opening of Žirovski vrh uranium mine. As the ore processing capacities all over the country have exceeded the mining capacity a lot of metal ore have been imported from other mines in Yugoslavia. In more recent times due to small price of the metals on market and bigger environmental awareness all of the mines and smelters have been closed. Nevertheless larger mines still have the capacity to be reopened because not all of the resources have been exploited.

From the times of Roman empire to present 49 different mines and 25 ore processing plants have been recognized (Figure 2). 4 of them were large (Idrija, Mežica-Topla, Litija and Žirovski vrh). There ware also 33 ironworks nearby the iron ore deposits, 3 of them are still operational (Jesenice, Štore, Ravne na Koroškem).

2.1. Historical and Geographical Description of the Areas of Detailed Geochemical Research

To assure the comparability with other similar mining and smelting sites around the world the historical description of areas of detailed geochemical research is presented. Also where available the data about total production in all operational period has been presented, together with different heavy metal dispersion mechanisms.

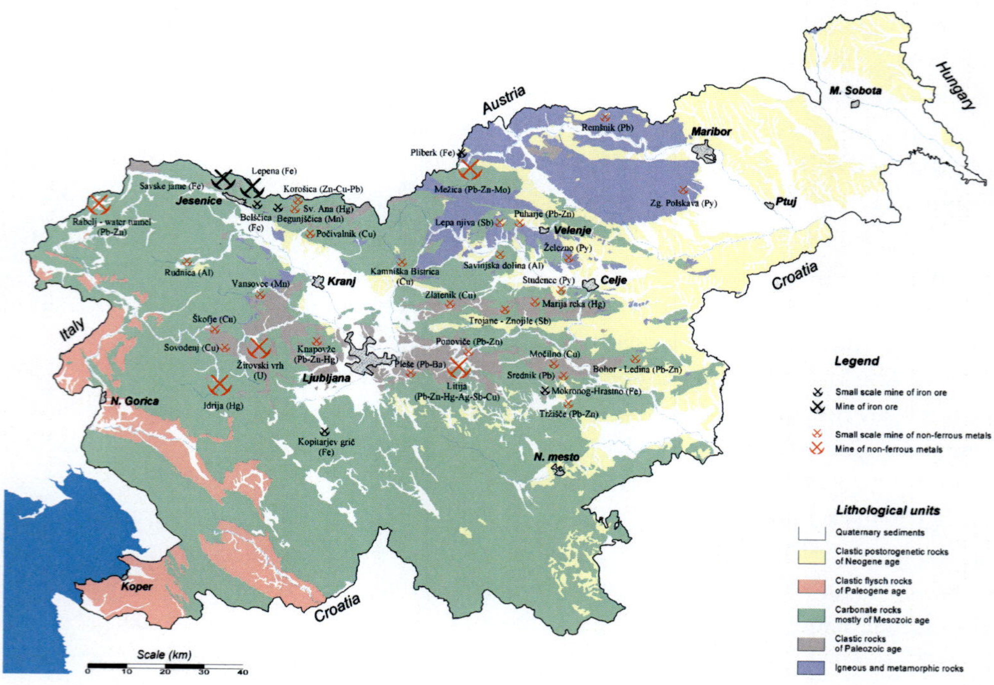

Figure 2. (Continued on next page.)

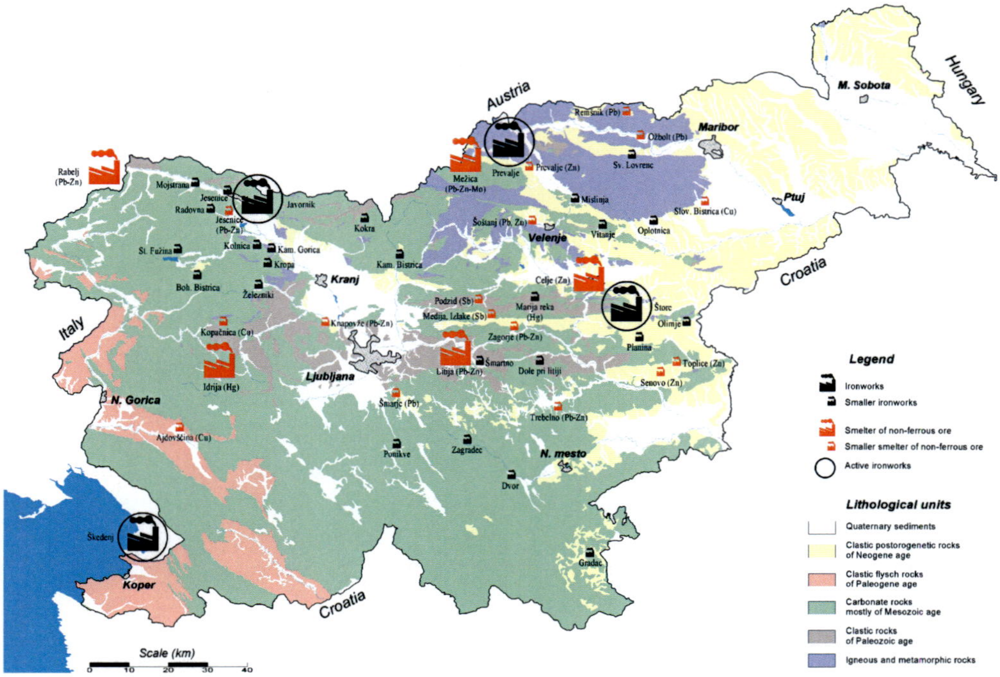

Figure 2. Locations of the past metal mines and smelters (Budkovič et al., 2003).

2.2. Mežica

Historical background of the Mežica mines has been summarized according to the touristic mines promoting web site (http://www.rlv.si/muzej/muzeji/body_m.htm) with additional references where mentioned in text. History of ironworking is summarized according to the successor company's web site (Metal Ravne, http://www.metalravne.com/).

Meža valley lies on the northern part of Slovenia (figure 1), close to the Austrian border. It is cut through the eastern part of Karavanke mountain range. The two settlements (Črna na Koroškem and Mežica) developed inside two widened parts inside overall narrow valley, which is cut through the carbonatic rocks. The Pb-Zn mine with smeltery is situated there. The lower part of the valley is widened and is placed inside metamorphic rocks of alpine foothills. There are the settlements of Ravne and Prevalje where the steel plant is still operational (figure 3).

The Pb mining in that area is dated back to Roman times between Uršlja gora and Peca Mountain. First written document which mention the mining in the Mežica valley is dated back to the 1665, but the smelting activities was reported in the book "De re Metallica" by Agricola in 1556. In that time only the small-scale mining on several different locations is reported. The construction of smeltery in Žerjav village has been in 1746 (Souvent, 1994a). In the year 1809 the Kompoš-Brunner mining company has been established. It has introduced the modern mining machinery which led to the development of the largest Pb-Zn mine in this part of Europe. The production of Pb and Zn since then has never been halted until final closure. The gradual closure of mine begun in 1988 due to severe environmental degradation and low market prices. In year 2000 the mine has been closed finally. The mining operations took area of 64 km^2 in the elevation between 268 and 2060 meters above sea level. The total length of all mining pits is 800 km with 265 mine entrances. Total 19 million tons of ore has been excavated and 1 million tons of Pb and 0.5 million tons of Zn has been produced.

Main ore minerals in Mežica Pb-Zn mines are galena, sphalerite, pyrite, marcasite and in smaller quantities also wulfenite (Pb-Mo oxide), molybdenite (Mo sulfide), smithsonite (Zn carbonate), cerusite (Pb carbonate) and others (Drovenik and Pleničar, 1980).

The iron processing in Mežica valley started in 1620 when the first ironworks has been build in Črna village. In the 18th century the ironworks was operational in Črna, Mežica and Ravne. Further development of iron processing was focused in Prevalje in 1835 and in Ravne after 1899 because of coal mining in the nearby Leše (started in 1818), Holmec and Mežica (Mohorič, 1954). The steel, produced in the Ravne ironworks, was famous because of its quality and the product has been exported all over the world (Spain, Portugal, Egypt, China, Greece, Syria, Turkey, Brazil...). After the First World War the ironworks get into financial troubles. This was the reason that the company became the part of Böhler Company with the head in Vienna and consequently the part of German arms industry. After Second World War the communists nationalized the ironworks and reconstruction and modernization begun. In 1952 the new Siemens-martins furnaces started to operate together with several electric-arc furnaces. Another 40-ton electric-arc furnace has been installed in 1968. In the mid 80's the company reached its peak with more than 6000 employees. After the breakup of Yugoslavia the steel plant got into troubles because of loss of its markets. The company has been split apart into smaller independent units. Today the ironworks is a part of state-owned Industrial Metallurgical Holding.

Main source of heavy metals contamination in the area has been the Pb smelter. In the year 1976 the annual Pb production had been 25.533 tons and it produced 5.812 tons of SO_2 emissions. Daily emissions of dust had been estimated to 500 kg. The dust filters had been installed between years 1968-1978. This reduced atmospheric emissions to 737 tons of SO_2 (15.876 tons of annual Pb production) in year 1991 (Souvent, 1994b). Other minor source of Heavy metals contamination had been the dusting of the mining and smelting waste, placed nearby Žerjav. Also when addressing the heavy metal contamination in this area the 150-year period of iron and steel production in Ravne and Prevalje have to be taken into account.

2.2.1. Heavy Metal Contamination Research on Mežica Area

On the basis of previous research of heavy metal contamination the approximately 100 km^2 large area has been selected for detailed investigation.

It has been divided into 1x1 km big cells and one sampling point per cell has been chosen. In more densely populated areas the additional samples has been taken in the middle of four surrounding sampling points. In total 114 sample sites ware determined (figure 3).

In each sampling site the top layer of soil (0-5 cm) and attic dust has been taken (Šajn, 2006).

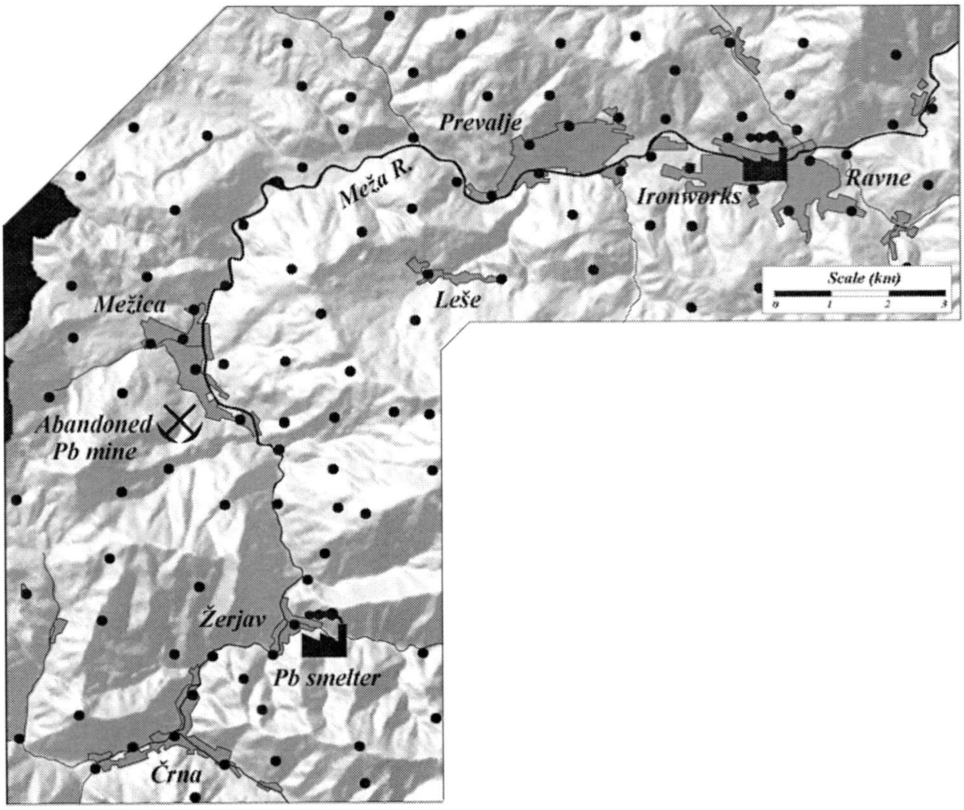

Figure 3. DEM of the Mežica area together with the locations of abandoned mine and smeltery, location of present steel factory (ironworks) and the locations of sampling points of soils and attic dust (Šajn, 2006).

2.3. Celje

Celje is third biggest town in Slovenia with the population of little less than 50.000 inhabitants. It is situated in Celje basin nearby Savinja River in central part of Slovenia (figure 1). In Celje there had been zinc smelting plant, which had been operational for 100 years, between 1874 and 1970. Approximately 2 km east from Celje there is also situated the Štore ironworks.

The plans for Zn production in Celje are dated back to 1873. The first two furnaces started to operate in 1874 in the eastern part of the town. The Zn production extended over time. In 1911 the ore-roasting plant had been upgraded and this date is beginning of sulfuric acid production. Before Second World War the Zn smeltery produced approximately 4.000 tons of Zn annually (Orožen, 1980). After the Second World War the Zn production had been artificially pushing up by the central communist government because of urgent need for foreign currency (approx. 8.000 tons of raw Zn per year). The sulfuric acid production did not follow the Zn production and the environment had been heavily degraded due SO_2 emissions, which ware estimated between 10.000 and 14.000 tons annually on the basis of recovery rate. Meantime the electrolysis method for Zn production had prevailed in developed countries and pyrometalurgical process in Celje was not competitive any more. This was the reason for the shut down of Zn production. At the end of its operational time in 1970 total of 12 furnaces had been installed. In 100-year of Zn production there is estimate that total 580.000 tons of raw zinc had been produced (Žibret, 2007). The Zn ore, needed for the production, had been imported all over the Zn-smeltery operational period.

Another milestone has to be mentioned. After the shutdown of Zn production the titanium dioxide pigment production started in 1973 after the reorganization of Cinkarna Celje chemical factory. The sulfate procedure has been used. The plant had been upgraded in 1989 and in 1994/95. The last upgrade had been made for lowering the titanium dust emissions. In the last years the production of the titanium pigments is growing up and its production is upgraded constantly. Last modernization took place in 2006. Today Cinkarna Celje contributes 1% of the world TiO_2 production (Blagotinšek, 2005).

Historical background of Štore ironworks has been taken from Štore Steel company's web page (http://www.store-steel.si/podZgodovina.asp).

The forgery in nearby Štore has been established in 1845 for the needs for newly opened Vienna - Trieste railway but in 1851 the ironworks started to be operational. In 1894 the ironworks had been modernized with new furnace and employed more than 300 workers. New modernization took place in 1912 with new Siemens-Martin furnaces. In 1937 the Štore Ironworks bought another foundry plant. This made the company the biggest foundry on the territory of present Slovenia with 460 employees. The ironworks grew constantly - in 1950 the company had 1465, in 1964 more than 2000 and in 1973 more than 3000 employees. In 1954 the company started to produce steel for motor cars springs which is also factory's main product today.

In 1970 the factory moved to its present location and in 1978 the Siemens-Martins furnaces ware shut down. In 1984 the company reached its peak with 3675 employees and 140.000 tons of steel produced annually. Due to end of the contract with Fiat car producer the production decreases drastically in 1986 and in 1987 the electric blast furnaces has been shut down. Another shock for the company had been the breakup of Yugoslavia in 1991. The company lost its markets and the steel production has been drastically reduced. Today the Štore steel factory uses scrap iron for steel production inside electric-arc furnaces. The steel is used in car and tool industry. The dust filters had been installed in 2005.

The mayor sources of heavy metals in the air in Celje area ware Zn smelters and Štore steel factory. There are also other possible sources of heavy metals in the atmosphere like coal burning for heating in winter, traffic and other industry in the area but are in minor importance comparing with former two. The report from 1989 (Domitrovič - Uranjek, 1990) claims that Štore steel emits between 364.8 and 641.28 tons of dust annually with the composition 3.6% Pb, 10.7% Zn, 0.054% Cd and 0.012% of As. The analysis of dust, emitted from Štore steel factory in 2001 reports 2.77% of Al_2O_3, 61.12% of Fe_2O_3, 2.23% of FeO, 4.85% of MnO, 9% of Pb, 1.4% of Pb and 0.08% of Cd (Stergar, 2001).

Unfortunately due to highly problematic SO_2 emissions after the World War 2 from Cinkarna Celje no special emphasis had been put to heavy metal contamination from Zn smelters. Exception is only a report from 1972 which quoted another older report, dealing with the measurements of air quality between October 1967 and September 1968 (Planinšek, 1972). Nearby Zn smelting plant of Cinkarna Celje annual average of Pb and Zn in m^3 of air has been 5.4 mg (Pb) and 71.3 mg (Zn).

2.3.1.Heavy Metal Contamination Research in Celje Area

On the basis of previous research of heavy metal contamination of soils (Lobnik et al., 1989) the area of 90 km^2 have been chosen for detailed research with Celje urban zone in the center (Figure 4). The entire area has been covered with the 1x1 km grid. Each grid cell represents one sampling point. In the densely populated Celje - Štore urban zone the sampling density increased that the additional sample in the middle of four samples of basic km grid has been taken. Totally 99 sampling points has been determined.

In each sampling point the sample of top layer of soil (0-5 cm) and attic dust has been taken (Žibret, 2002; Šajn, 2005).

2.4. Jesenice

Jesenice is located on the NW part of Slovenia nearby Austrian border inside the Sava River valley between Karavanke and Julian Alps mountain ranges (figure 1). Historical background is presented according to the Acroni Jesenice steel factory web page (http://www.acroni.si/si/index.php?cat_id=38) with additional references where mentioned.

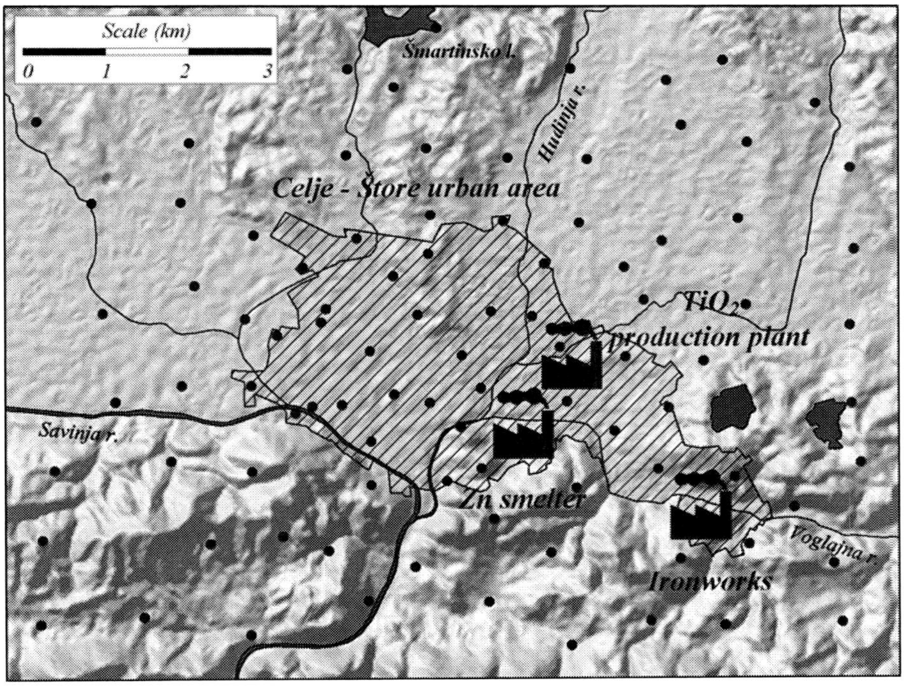

Figure 4. DEM of the Celje area together with the locations of past Zn smeltery, present TiO_2 pigment factory, past and present Štore ironworks and the locations of sampling points of soils and attic dust.

Archaeological evidences indicate that iron smelting dated back to 1000 BC. The first document, which mentions iron smelters in this area, is Otenburg document from 1318. Extensive iron smelting activities are reported from 14th century on. Industrial revolution in 19th century forced small ironworks companies to unify as Karniola industrial company

(Kranjska industrijska družba - KID) between 1869 and 1872. In 1872 there has been a big breakthrough as the process for producing ferromanganese in classical smelting furnaces has been developed which give the KID worldwide reputation. At the end of 19th century the cooperation with the German ironworks extended the iron production widely. Another extension of production dates between 1937 and 1940 when the company produced 100.000 tons of steel annually. Extensions ware also in 1966, 1976 and 1987. After breakup of Yugoslavia the company lost its markets and the production decreased. Since then the production of steel increased drastically again. The company investments lead into 200.000 tons of annual steel production in 1999. Today the Acroni steel factory is second biggest producer of steel sheets in Europe and is a part of state-owned Industrial Metallurgical Holding (as in Ravne case).

The emissions of heavy metals in Jesenice area is mainly because of ironworks. Earlier data (Šipec, 1990) reports daily dust emissions of 48 tons in 1971. Also about 270 tons of ash produced daily should be added. The Jesenice ironworks has been recognized in that time for its red dust emissions. Between 1971 and 1987 the company took partial remediation and modernization which decrease daily dust emissions. The abandoning of the Siemens-Martins furnaces in 1987 decrease the daily emissions to 2 tons of dust and 950 kg of SO_2. In recent days the emissions from Jesenice steel plant are insignificant.

2.4.1 .Heavy Metal Contamination Research in Jesenice Area

The sampling plan in Jesenice area has been focused to the valley and the northern and southern hillsides. The 1.4x1.4 km grid has been made with 44 sampling locations (Figure 5). The covered area was 113 km^2. In each sampling point the pedological profile has been made. Total of 122 samples of soils in different depths has been taken (Šajn et al., 1999).

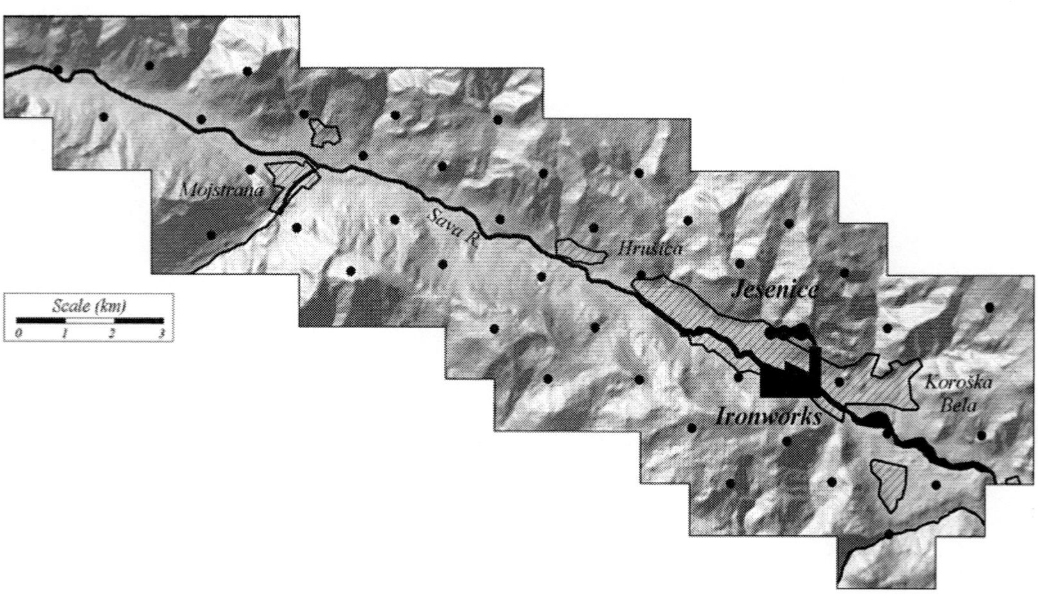

Figure 5. DEM of the Jesenice area together with the locations of past ironworks and the sampling locations of soils.

2.5. Litija

Litija is situated approximately 20 km east from capital of Slovenia - Ljubljana (Figure 1). It is polymetallic mining field with variety of ore minerals, among of them are lead, zinc, mercury, silver and iron sulfides and barite. The minerals are found in veins, which stretch inside approximately 10 km long belt, which categorize it into larger-scale mining field in Slovenia. Litija is also the oldest mining town in Slovenia (Mlakar, 1994).

The mining operations probably begin in the Celtic times but no hard archaeological evidences have been found. First evidence of mining is the findings of slag from roman times (Godec, 1993) and first written evidence of the mining in this area has been finding of the tombstone from 1537 which belonged to the mining perfect. In 16th century there has been higher mining office in Litija. In 1604 the mining operations stopped. In 1689 Valvasor in his book "Die Ehre dess Hertzogthums Crain" writes about big abandoned mines in that area. In 1792 east from the Litija the new iron smeltery has been constructed. The ore has been digging out in local mines (Fabjančič, 1972).

Larger scale mining operations begin in 1838 when the mining company has been established ("Gewerkscahft Littai"; from 1925 it was renamed to "Rudarska združba Litija") which has been operational until 1941. In 1874 the new rich cinnabar vein has been discovered and the excavation of mercury ore has started. At the beginning the ore has been transported to Idrija for processing until 1883 when mercury smeltery has been constructed. It has been reported that cinnabar ore contains up to 15% of mercury but the processed cinnabar ore in 1885 contained on average 2.7% Hg. In this year the Hg production reached its peak - 26.3 tons. The lead production started in 1877 during the deepening of the mines when new galena ore vein has been discovered. Since then lead has been the main product of the mines with the peak annual production in 1884 - 1900 tons. In 1886 the exploatation of silver begun with the peak production in 1890. The mining of cinnabar stopped in 1894 and galena in 1897. Nevertheless the smeltery has been still operational until 1917. The ore has been imported from different mines from Europe and even Africa (Mohorič, 1978). The peak of lead production in smeltery has been in 1914 (3660 tons annually). Also the silver and gold has been extracted (2750 kg of Ag and 2.13 kg of Au in 1915).

Due to many compensation requests because of environmental degradations by local farmers and forestry the smelting plant has been shut down in 1917, but reopened in 1918. It has been shut down again in 1922 because of compensation requests by local beekeepers (Mlakar, 1994). Works has been reopened in 1924 with the lead and silver production. Because of fall of the lead price the smeltery has been shut down again in 1930.

In the year 1941 during the Second World War the mines have been reopened again by the Germans due to great needs for barite. But they have been destroyed by local partisan's guerilla action in 1944 and the complete mining archive has been burned down. Between 1943 and 1944 2299 tons of barite has been excavated.

The mineral prospection has been reopened in 1947 and the mining of the barite started again in 1951. In 1961 the mines have been reorganized as the unit of the Mežica lead mines and the exploatation of the galena ore started again. Because of very poor galena and barite ore, problems with the mining water and the problems with the silicosis by the miners the mining has been shut down in 1965. The official end of the mining operations in Litija is 28th of April 1966 (Fabjančič, 1972).

In all of the several hundred years of the mining operations in the Litija area the estimation is that totally of 50.000 tons of Pb ore, 1000 kg of Ag ore, 42.5 tons of Hg ore and 30.000 tons of barite has been dug out (Drovenik and Pleničar, 1980). When taking into account the imported ore the production in smeltery after 1880 has been estimated to 68.000 t of Pb, 12 t of Ag and 150 t of Hg (Fabjančič, 1972).

The larger mining operations have been in the hill Sitarjevec south from Litija on the area of 600 x 250-300 meters on the altitude between 420 and 171.5 meters above sea level. The length of all mining pits is more than 15 kilometers. Main ore minerals are barite, galena and sphalerite, in smaller quantities also cinnabar and realgar (arsenic sulfide).

The main source of heavy metals in the environment in Litija region are many abandoned mine tailings, heaps of poor ore and slugs from smeltery. Large contribution to the heavy metal contamination has been also the dust emissions from smeltery.

Figure 6 shows the locations of the abandoned smeltery and locations of abandoned mining operations.

2.5.1. Heavy Metal Contamination Research in Litija Area

The sampling of soils and attic dust in Litija area has been conducted on smaller scale. The sampling has been performed on the areas where most extensive pollution is expected. 8 sampling points has been placed in immediate vicinity of past smelting plant and mines, 13 in area around mines and smeltery and 17 points on the wider countryside and up and downstream the Sava River. Total of 38 sampling points has been recognized (figure 6; Jemec, 2006; Jemec and Šajn, 2007). In every sampling location the sample of the top soil (0-5 cm) and attic dust been taken.

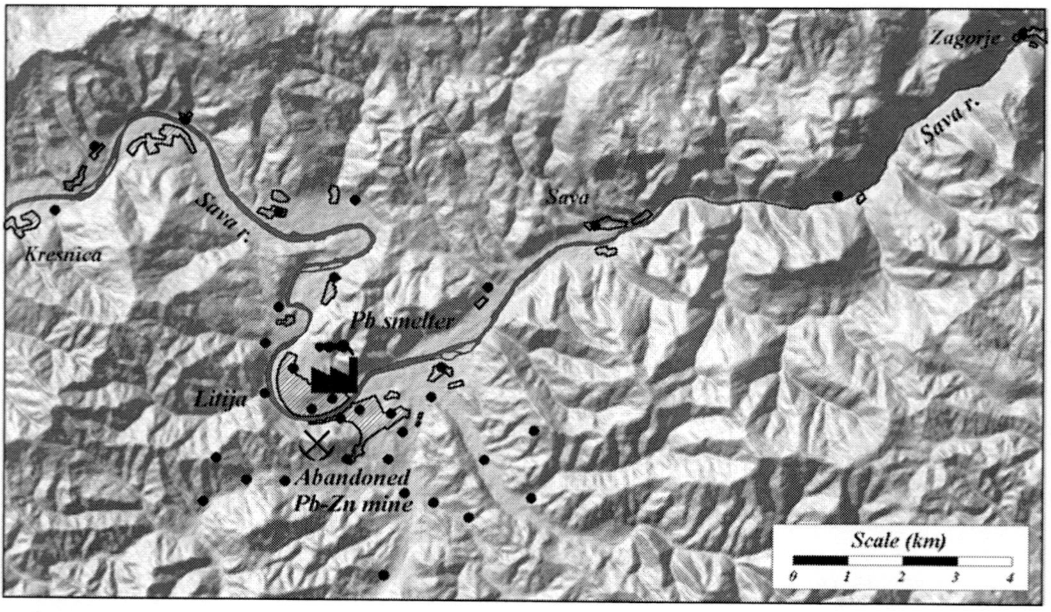

Figure 6. DEM of the Litija area together with the locations of past smeltery, abandoned mines and the locations of sampling points of soils.

2.6. Idrija

Majority of historical background has been summarized according to the official tourist promotional web site by Regional Development Company for Idria-Cerkno region (www.idrija-turizem.si) with additional data where references have been placed.

The town of Idrija is situated approximately 50 km west from the Ljubljana (capital of Slovenia) nearby the Idrijca River. The rivers and streams cut the narrow valleys and gorges through the karstic plateau which lies between 800 and 1000 meters above sea level. The elevation of the Idrija Hg mine and Hg smeltery is 330 meters (figure 1).

The native mercury in Idrija region has been discovered in 1490. The mining and smelting activities by the miners from Friuli region (Italy) and Germanic countries started before 1500. The mining operations extended when the rich cinnabar ore has been discovered in 1508. The ownership of the mines changed in 1575 when they came under the Habsburg rule. Also that was the times when the mines have been modernized and the mining operations have been extended. In the beginning of 17th century the Idrija region became the separate administration unit inside Habsburg monarchy, run by the mining authorities. In the year 1690 the town of Idrija counted 1500 inhabitants, 300 of them ware working in the mines.

In the 18th century Idrija get the town privileges and has been the second largest settlement in the Carniola Region (western part of Slovenia). The mining and smelting activities in that period extended widely. In 1790 the mining company employed 1350 workers and produced between 600 and 700 tons of mercury annually. The cinnabar ore contained on average 18% of Hg. Because of the strategic importance of the mines they have been constantly modernized. With the help of steam power, drilling equipment and improved furnaces the Hg production reached peak in 1913 when 820 tons of Hg has been produced.

Between both world wars and Italian occupation the Hg production decreased and halted after 1945 bombing. After the Second World War the production started again with the annual Hg production between 400 and 500 tons of Hg. The mine has been modernized again. The seventies of 20th century brought a drastically decrease of Hg price on world markets which lead to the decision of progressive mine closure in 1987. The production has been finally closed down in 1995.

In the 500-year period more than 700 km of mining shafts have been dug out. The total Hg production has been estimated to 107.000 tons which put the Idrija Hg mine to the second biggest world mine with the 13% of the total world Hg production right after the Almaden Hg mines in Spain. The main ore mineral is cinnabar combined with the native Hg.

The mining and smelting activities on the other hand caused heavy Hg pollution in this region. Estimates claim that approximately 40,000 tons of Hg has been dissipated in the environment (Mlakar, 1974). The main sources of the pollution before 1652 was small scale smelters, located in the Idrija surrounding. In 1652 the first large smeltery has been constructed nearby the mine. Mercury rich smelting and mining waste has been piling nearby Idrijca River banks which ware eroded by floods. According to the estimates between 20 and 30 kg of Hg ware emitted in the air from the smeltery daily (Kosta et al., 1974, Kavčič, 1974). The other important atmospheric source of Hg has been the mine ventilation shaft. Figure 7 shows the DEM of the Idrija region together with the locations of abandoned smelter.

In Idrija Hg mineralization is connected with lower and middle Triassic hydrothermal activity (Drovenik and Pleničar, 1980). Main mineralization with cinnabar and native mercury occurred in upper Ladinian strata (organic shales, sandstones, quartz conglomerates and anthracite) and in overlying tuffs. Mineralization also occurs as infill and replacement inside fault zones in permo-carboniferous shales up to upper Ladinian beds. Minor ore minerals are pyrite, marcasite and metacinnabar.

2.6.1. Heavy Metal Contamination Research on Idrija Region

The basis for the soil sampling has been the research by Hess (1993), which estimated the dimensions of the Idrija mercury halo. The sampling grid covered 160 km^2 large area which includes the most polluted areas together with the surroundings. For determination of Hg pollution the sample of top soil and attic dust have been taken. The sampling locations ware placed on every kilometer in both x and y directions.

The top layer (0-15 cm) of soils together with the possible organic horizon (A) has been taken. The sample of the attic dust has been taken nearby the site of soil sampling. Together the 103 samples of soils and 103 samples of attic dust have been taken (Gosar and Šajn, 2001; Gosar et al., 2002; figure 7).

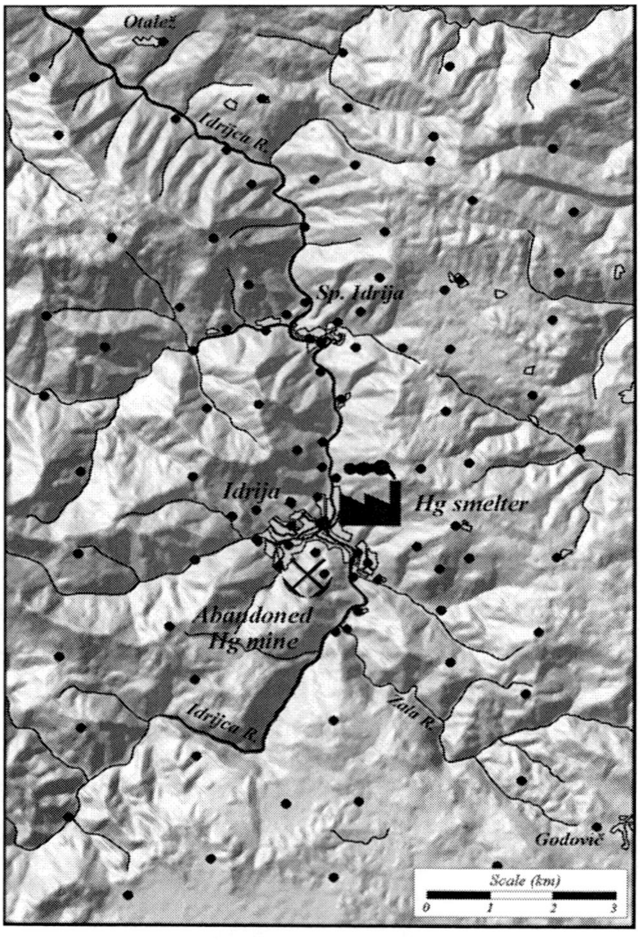

Figure 7. DEM of the Idrija area. Abandoned mine has been located approx. 500 m W from smelter. Black circles represent the locations of sampling points of soils and attic dust.

3. Methods of the Sampling, Sample Preparation, Chemical Analyses and Data Processing

All of the sampling, analyzing and data processing methods are unified to assure comparability of the data from different sampling locations and from different time periods.

When sampling the soils the main emphasis has been put on to sample the soils which have been in situ for longer period of time. In natural and rural areas this has not been the mayor problem but in urban areas this can be very tricky due to many construction operations. The conversation with the owners of the objects has been good indicator of the age of soils in one place. Also the presence of old trees indicates the "age" of the soil. On average the gardens around the houses where no additional soils have been added and parks have been recognized as good sampling points.

In natural areas the grassland soils have been sampled. Also we tried to avoid the possible present and past fields for crops because of the plowing and possible use of fertilizers and fitopharmaceuticals. When possible we also avoid the sampling of soils in forests because the canopy held some pollutants in the air and avoid sampling on steep slopes due to possible gravitational mass movements in the past. The presence of the subsoil B horizon has been a good indicator of the age.

After the good sampling spot has been located the sampling has been performed (figure 8). The sample has been composed of 5 subsamples. The 4 subsamples have been taken around central subsample in circle pattern with the radius of 10 meters. The total weight of composite sample has been approximately 1 kg.

Figure 8. Sampling of the soils.

Pre-analytical sample preparation includes drying on the room temperature (298 K), milling and sieving. The drying has been stopped when no additional weight loss has been observed. During drying process the sample has been stirred several times. If the soil contained a lot of clay or organic component the clots formed ant they have been broken up in the ceramic pot to assure uniform drying. After that the sieving under 2 mm sieve has been performed. Some of the sieved material has been put in the agate pulverizer. The pulverized soil under 0.125 mm represented the material for chemical analyses.

Sampling of attic dust has been performed in the attics of old houses. The condition for attic dust sampling has been that the house is more than 100 years old, desirably more than 150 years. The conversation with the owners of the object helped in determining of its age. Also the usage of wooden nails instead of irons ones and the presence of arches on the ceilings has also been the good age indicator. Another condition was that the house has the original wooden roof-carrying construction and that the attic has not been used for storing hay or crops (figure 9).

After determination of the suitable house and acquisition of the permission of the owners of the object the dust has been brushed from wooden trams with hard brush. Before brushing the top layer of dust with the remains of plants, tiles or other construction materials has been removed. Also when brushing the special emphasis has been put not to collect possible parts of the decomposing trams. The brushing of the trams has been conducted in different parts of the attic until at least 50 grams of dust has been collected.

Pre analytical attic dust preparation included drying on the room temperature and sieving on 0.125 mm sieve to remove all unwanted components (sand, wood particles, fibers, tiles etc...) until only black-colored dry air deposit remains.

In all sampling operations the locations of the sampling spots has been determined with the help of GPS. The position of sampling spot has been mainly determined by the presence of old house and aquisition of permission to sampling the dust. Microlocation of the soil sampling has been determined in the nearby vicinity of the house where dust has been collected. Overall sampling of soils was not so problematic as attic dust sampling. Main problems at attic dust sampling is the presence of suitable houses and aquisition of permission from the owners.

Chemical analyses have been performed in ACME analytical laboratories Vancouver. The inductively coupled plasma mass spectroscopy (ICP-MS) procedure after total 4-acid digestion has been used for determination of the concentration of 41 elements (Ag, Al, As, Au, Ba, Be, Bi, Ca, Cd, Ce, Co, Cr, Cu, Fe, Hf, K, La, Li, Mg, Mn, Mo, Na, Nb, Ni, P, Pb, Rb, S, Sb, Sc, Sn, Sr, Ta, Th, Ti, U, V, W, Y, Zn and Zr). Hg concentration has been determined by the Atomic absorption spectroscopy (AAS) after aqua regia digestion.

Data preparation starts with the removal of the element analyses where more than 30% of the samples have been below the detection limit. In other cases where the sample contains the element concentration below detection limit its concentration has been estimated as 50% of the detection limit. For the presentation of basic statistical properties the non-parametric statistics has been used (median, percentile distribution). The enrichment factors have been calculated according to Slovenian background levels for topsoils.

Bivariate and multivariate statistics has been performed after data normalization by logaritming the data set where the distribution has been log-normal. Bivariate statistics includes calculation of correlation coefficients between different sampling media.

Figure 9. Suitable location for attic dust sampling

For determination of geochemical associations and reduction of the dimensionality the factor analysis has been used. The variables have been standardized to zero mean and unit standard deviation (Reimann et al., 2002) and varimax-raw axis factor rotational method have been used (Statistica 6.1).

The creation of maps of geochemical associations has been made on the basis of calculated factor scores and kriging interpolation method. The percentile distribution of factor scores has been plotted. Separate maps of the selected elements in the area have been made. The basis for scale bars for maps of different heavy metals has also been the percentile distribution (P10 - P25 - P40 - P60 - P75 - P90).

The software used in data manipulation and map making has been Autodesk Autocad 2008, Statsoft Statistica 6.0 and Golden software Surfer 8.0. Digital elevation models used for map making have the resolution of 25 meters.

3.1. Methods of Determining the Regional Geochemical Trends in Slovenia

When addressing the geochemical trends in Slovenia the soils and attic dust has been sampled in the areas where no pollution is expected. The area of Slovenia has been divided into 25x25 km grid which contained 31 "basic" cells. Each basic grid cell has been divided again into 5x5 km subcells. The sampling position in the basic cell has been chosen randomly from 25 subcells. 10 grid cells have been chosen randomly and there additional samples have been taken due to the analysis of variance. Total 41 sampling points has been determined (Šajn, 1999). Later on the additional 19 locations have been sampled where the gaps inside primary sampling plan have been identified. Total number of sampled locations on the unpolluted areas of Slovenia is 60.

In each sampling point the top layer of soil (0-5 cm), bottom layer of soil (20-30 cm) and attic dust has been taken. Figure 10 shows the sampling locations. Samples preparation, chemical analyses, and data interpretation has been performed as described in chapter 3.

Additional sampling has been performed in the larger urban areas (figure 10). Top soil (0-5 cm) and attic dust of old houses (50 - 200 years) has been sampled in 23 locations in the 6 largest towns: Ljubljana (7 sampling locations, 267.000 inhabitants), Maribor (4 sampling locations, 114.000 inhabitants), Celje (4 samples, 45.000 inhabitants), Koper (2 sampling locations, 24.000 inhabitants), Novo Mesto (3 sampling locations, 22.000 inhabitants) and Jesenice (3 sampling locations, 13.000 inhabitants). The number of sampling locations has been determined according to the surface of the urban area. This sampling enables the comparison between heavy metal contamination between urban and rural areas.

4. RESULTS

4.1. Accuracy And Precision of Chemical Analytics and Detection Limits

The detection limits of the analytical laboratory are shown in table 1. Mayor part of the soil and attic dust sampling in the slovenian unpolluted areas and sampling on the Jesenice, Celje, Mežica and Idrija area has been taken before 2002. Other research of the contaminated sites (part of regional geochemical survey of Slovenian unpolluted areas, part of Celje sampling, Litija) has been finished after 2002.

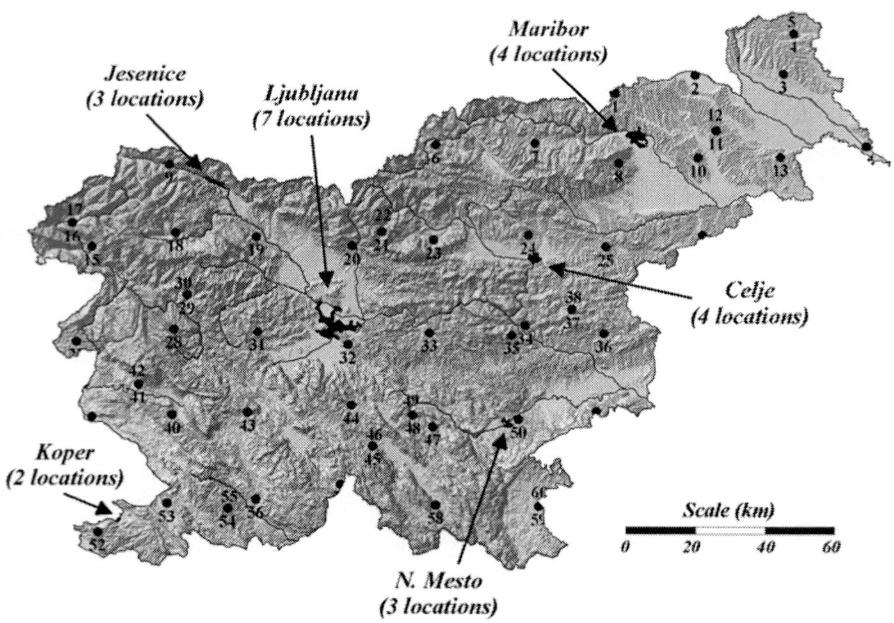

Figure 10. Sampling locations for the determination of regional geochemical trends and geochemical characteristic of urban areas.

Table 1. Detection limits of the ACME analytical laboratories (Vancouver) - 4 acid digestions - of all of the analyzed elements in geochemical research on the Geological survey of Slovenia

	Years	
	1997 - 2002	2002 - present
%		
Al	0.01	0.01
Ca	0.01	0.01
Fe	0.01	0.01
K	0.01	0.01
Mg	0.01	0.01
Na	0.01	0.001
P	0.002	0.001
S*	-	0.01
Ti	0.01	0.001
mg/kg		
Ag*	0.5	0.1
As	5	1
Au*	4	1
Ba	1	1
Be	1	1
Bi	5	0.1
Cd	0.4	0.1
Ce	-	1
Co	2	1
Cr	2	0.1
Cu	2	0.1
Hf	-	0.1
La	2	0.1
Li	-	0.1
Mn	5	1
Mo	2	0.1
Nb	2	0.1
Ni	2	0.1
Pb	5	0.1
Rb	-	0.1
Sb	5	0.1
Sc	1	1
Sn	2	0.1
Sr	1	1
Ta	-	0.1

Table 1. (Continued)

	Years	
	1997 - 2002	**2002 - present**
Th	2	0.1
U	10	0.1
V	2	1
W	4	0.1
Y	2	0.1
Zn	2	1
Zr	2	0.1
μg/kg		
Hg	10	10

The precision and the accuracy have been tested several times. Data are published in articles of Šajn (1999, 2005 and 2006) and are not presented in this book chapter in details. The accuracy of the analytical method has been tested on the basis of the geological standard materials and precission on the basis of double analysis of same samples. Generally accuracy and precission found to be satisfactory for most of the analysed elements. Table 2 shows the elements where deviations (relative percent difference), larger than 15% (for accuracy) and 10% (precission) has been observed in different research.

Table 2. Summary of the accuracy (ACC) and precission (PR) testing of the analytical laboratory; RPD - relative percent difference

Researched area	Publication	ACC (RPD>15%)	PR (RPD>10%)
Slovenia	Šajn, 1999	Co, Nb, P, Y	Nb
Mežica	Šajn, 2006	Ce, La, Nb, Rb, Sb, W	-
Celje	Šajn, 2005	Cd, Cr, La, Mg, P, Pb, Th	Th, Zr

4.2. Geochemical Trends and Background Levels in Slovenia and Geochemical Characteristics of Urban Areas

The non-parametric statistical parameters of chemical analyses of top soil, bottom soil and attic dust in rural non-polluted areas are presented in Table 3. Data of chemical analyses of materials taken on 41 sampling points are taken from Šajn (1999), other have been taken later. The average values can be interpreted as the values of geochemical background in Slovenia.

Table 3. Nonparametric statistical distribution of the elements in the top soil (0-5 cm; N=60), bottom soil (20-30 cm; N=60) and attic dust (N=60) in the unpolluted areas. Mat - material; Dis - distribution (Log = log-normal, N = normal-Gaussian); X, Xg - average value; Md - median value; Min - Max - range; P25 - P75 - quartile distribution. Data rounded on two decimal places

%	Mat	Dis	X,Xg	Md	Min - Max	P25 - P75
Al	top soil	N	6.8	6.9	3.1 - 9.3	6.3 - 7.6
	bottom soil	N	7.4	7.5	3.5 - 11	6.9 - 8.4
	attic dust	Log	3.0	3.0	1.1 - 6.9	2.4 - 4.0
Ca	top soil	Log	0.89	0.68	0.32 - 9.5	0.54 - 1.2
	bottom soil	Log	0.81	0.59	0.22 - 11	0.38 - 1.3
	attic dust	N	7.5	7.6	1.7 - 18	4.9 - 9.7
Fe	top soil	Log	3.5	3.5	2.2 - 4.8	3.1 - 4.1
	bottom soil	Log	4.0	4.1	2.4 - 6.0	3.5 - 4.6
	attic dust	Log	1.9	1.9	0.75 - 4.3	1.5 - 2.5
K	top soil	Log	1.6	1.6	0.84 - 3.0	1.4 - 2.0
	bottom soil	Log	1.7	1.7	0.87 - 3.4	1.5 - 2.1
	attic dust	Log	1.1	1.1	0.45 - 2.0	0.84 - 1.3
Mg	top soil	Log	0.81	0.73	0.46 - 3.1	0.61 - 0.93
	bottom soil	Log	0.98	0.84	0.52 - 5.7	0.72 - 1.1
	attic dust	Log	1.4	1.3	0.36 - 4.8	0.86 - 2.3
Na	top soil	N	0.52	0.52	0.11 - 1.2	0.35 - 0.65
	bottom soil	N	0.58	0.56	0.13 - 1.4	0.41 - 0.74
	attic dust	Log	0.38	0.38	0.17 - 1.1	0.25 - 0.54
P	top soil	Log	0.11	0.11	0.057 - 0.26	0.090 - 0.13
	bottom soil	Log	0.083	0.079	0.039 - 0.34	0.069 - 0.10
	attic dust	Log	0.30	0.27	0.13 - 1.1	0.20 - 0.40
S	top soil	Log	0.062	0.050	0.050 - 0.20	0.050 - 0.10
	bottom soil	Log	0.062	0.050	0.050 - 0.20	0.050 - 0.10
	attic dust	N	2.7	2.5	0.40 - 7.4	1.3 - 3.8
Ti	top soil	Log	0.32	0.33	0.15 - 0.59	0.27 - 0.36
	bottom soil	Log	0.36	0.37	0.18 - 0.64	0.32 - 0.42
	attic dust	Log	0.18	0.18	0.068 - 0.54	0.13 - 0.25
mg/kg						
Ag	top soil	Log	0.070	0.050	0.050 - 0.10	0.050 - 0.10
	bottom soil	Log	0.074	0.100	0.050 - 0.20	0.050 - 0.10
	attic dust	Log	0.30	0.30	0.10 - 3.8	0.20 - 0.50
As	top soil	Log	15	15	6.0 - 37	12 - 20
	bottom soil	Log	16	16	6.0 - 41	12 - 23
	attic dust	Log	12	11	2.0 - 240	9.0 - 16
Ba	top soil	Log	330	360	150 - 690	280 - 410
	bottom soil	Log	370	400	180 - 920	310 - 460
	attic dust	Log	110	120	20 - 730	66 - 200

Table 3. (Continued)

mg/kg	Mat	Dis	X,Xg	Md	Min - Max	P25 - P75
Be	top soil	N	2.3	2.0	0.50 - 5.0	2.0 - 3.0
	bottom soil	Log	2.0	2.0	0.50 - 5.0	1.0 - 3.0
	attic dust	Log	1.0	1.0	0.50 - 2.0	1.0 - 1.0
Bi	top soil	Log	0.40	0.40	0.20 - 0.70	0.30 - 0.50
	bottom soil	N	0.41	0.40	0.20 - 0.70	0.30 - 0.50
	attic dust	Log	0.44	0.40	0.20 - 1.3	0.30 - 0.60
Cd	top soil	Log	0.50	0.45	0.10 - 2.0	0.30 - 0.85
	bottom soil	Log	0.33	0.30	0.10 - 2.2	0.20 - 0.75
	attic dust	Log	1.4	1.3	0.50 - 13	1.0 - 1.9
Ce	top soil	Log	58	60	24 - 110	52 - 71
	bottom soil	Log	63	64	32 - 120	54 - 76
	attic dust	Log	30	31	9.0 - 68	24 - 43
Co	top soil	Log	16	16	5.0 - 37	13 - 21
	bottom soil	Log	19	18	6.0 - 44	15 - 26
	attic dust	Log	6.6	6.7	2.5 - 16	5.4 - 8.6
Cr	top soil	N	90	91	29 - 140	75 - 110
	bottom soil	N	99	98	29 - 170	82 - 120
	attic dust	Log	47	47	23 - 100	36 - 59
Cu	top soil	Log	35	31	17 - 170	24 - 50
	bottom soil	Log	35	33	14 - 150	23 - 51
	attic dust	Log	66	60	19 - 370	37 - 96
Hf	top soil	N	1.4	1.4	0.20 - 2.7	0.95 - 1.7
	bottom soil	N	1.5	1.5	0.20 - 3.1	1.1 - 2.0
	attic dust	N	0.72	0.70	0.40 - 1.3	0.50 - 0.85
La	top soil	Log	31	31	12 - 60	26 - 37
	bottom soil	Log	33	34	16 - 66	28 - 38
	attic dust	Log	15	16	5.5 - 36	11 - 22
Li	top soil	N	50	50	31 - 75	41 - 57
	bottom soil	N	55	55	34 - 83	46 - 64
	attic dust	N	24	24	9.2 - 42	18 - 29
Mn	top soil	Log	1100	1000	380 - 2200	780 - 1500
	bottom soil	Log	1300	1300	420 - 3100	940 - 1800
	attic dust	Log	540	540	260 - 1300	420 - 650
Mo	top soil	Log	0.99	0.80	0.30 - 12	0.60 - 1.2
	bottom soil	Log	0.99	0.75	0.30 - 13	0.60 - 1.2
	attic dust	Log	2.0	1.9	0.70 - 7.4	1.4 - 2.5
Nb	top soil	N	8.8	8.7	3.5 - 16	7.1 - 10
	bottom soil	Log	9.4	9.4	3.9 - 19	8.0 - 12
	attic dust	Log	4.2	4.4	1.4 - 13	3.0 - 5.6
Ni	top soil	Log	48	50	9.2 - 130	31 - 75
	bottom soil	Log	55	56	11 - 160	36 - 90
	attic dust	Log	27	27	13 - 67	21 - 33

mg/kg	Mat	Dis	X,Xg	Md	Min - Max	P25 - P75
Pb	top soil	Log	40	41	20 - 87	32 - 47
	bottom soil	N	42	41	23 - 68	35 - 50
	attic dust	Log	140	120	51 - 1800	97 - 200
Rb	top soil	N	110	110	64 - 140	95 - 120
	bottom soil	N	110	110	60 - 150	100 - 120
	attic dust	Log	54	53	24 - 92	44 - 68
Sb	top soil	Log	1.1	1.1	0.40 - 3.7	0.90 - 1.3
	bottom soil	Log	1.2	1.1	0.40 - 5.7	0.95 - 1.5
	attic dust	Log	2.8	3.0	0.60 - 17	2.2 - 3.8
Sc	top soil	N	12	12	7.0 - 17	11 - 13
	bottom soil	N	13	13	8.0 - 19	11 - 14
	attic dust	Log	5.1	5.0	2.0 - 13	4.0 - 7.0
Sn	top soil	Log	3.3	3.1	1.8 - 33	2.7 - 3.7
	bottom soil	Log	3.4	3.5	1.8 - 6.2	2.8 - 3.8
	attic dust	Log	8.4	8.6	0.50 - 95	4.8 - 14
Sr	top soil	Log	77	77	35 - 170	66 - 90
	bottom soil	Log	83	82	36 - 190	73 - 99
	attic dust	Log	120	130	43 - 320	94 - 150
Ta	top soil	N	0.60	0.60	0.20 - 1.1	0.50 - 0.70
	bottom soil	N	0.68	0.60	0.20 - 1.2	0.50 - 0.80
	attic dust	Log	0.38	0.40	0.10 - 1.3	0.30 - 0.50
Th	top soil	N	11	10	4.7 - 18	8.9 - 12
	bottom soil	N	12	12	5.3 - 18	9.9 - 14
	attic dust	N	5.1	5.0	1.5 - 12	3.8 - 6.8
U	top soil	Log	2.3	2.1	1.2 - 6.1	1.7 - 2.8
	bottom soil	Log	2.5	2.3	1.3 - 6.2	1.9 - 3.3
	attic dust	N	1.9	1.9	0.40 - 3.5	1.5 - 2.2
V	top soil	Log	100	100	41 - 220	86 - 120
	bottom soil	Log	130	130	62 - 250	100 - 160
	attic dust	Log	59	64	20 - 260	42 - 79
W	top soil	Log	1.4	1.4	0.80 - 3.0	1.1 - 1.6
	bottom soil	Log	1.4	1.5	0.70 - 3.9	1.1 - 1.7
	attic dust	Log	1.3	1.4	0.50 - 4.7	1.0 - 1.6
Y	top soil	Log	16	17	7.0 - 44	13 - 21
	bottom soil	Log	16	17	7.4 - 51	13 - 21
	attic dust	Log	8.6	8.7	3.2 - 21	7.0 - 11
Zn	top soil	Log	120	120	75 - 220	110 - 140
	bottom soil	N	120	120	71 - 180	95 - 130
	attic dust	Log	380	320	130 - 1200	230 - 590
Zr	top soil	N	40	39	6.2 - 86	27 - 51
	bottom soil	N	46	45	5.1 - 95	30 - 61
	attic dust	Log	20	21	9.1 - 42	16 - 25
Hg	top soil	Log	0.063	0.065	0.010 - 0.26	0.045 - 0.10
	bottom soil	Log	0.07	0.073	0.025 - 0.32	0.045 - 0.09
	attic dust	Log	1.1	1.0	0.16 - 26	0.54 - 1.

Table 4 shows the analytical results of the samples, taken in the urban areas (Šajn, 1999). The data present the concentration of elements in the top soils (0-5 cm) and attic dust.

Table 4. Nonparametric statistical distribution of the elements in the top soil (0-5 cm; N=23) and attic dust (N=23) in the Slovenian urban areas (Šajn, 1999). Mat - material; Dis - distribution (Log = log-normal, N = normal-Gaussian); X, Xg - average value; Md - median value; Min - Max - range; P25 - P75 - quartile distribution. Data rounded on two decimal places

	Mat	Dis	X, Xg	Md	Min - Max	P25 - P75
%						
Al	top soil	N	4.7	5.0	1.7 - 6.7	4.0 - 5.6
	attic dust	Log	2.7	2.7	1.9 - 4.7	2.2 - 3.3
Ca	top soil	Log	6.7	6.3	3.0 - 22	4.3 - 9.4
	attic dust	Log	9.4	9.5	4.8 - 17	7.6 - 11
Fe	top soil	Log	2.7	2.7	0.99 - 8.9	2.2 - 3.2
	attic dust	Log	2.8	2.4	1.4 - 27	1.9 - 3.5
K	top soil	N	1.0	0.94	0.37 - 1.7	0.87 - 1.2
	attic dust	Log	0.69	0.66	0.46 - 1.8	0.58 - 0.85
Mg	top soil	N	2.4	2.4	0.53 - 4.1	1.5 - 3.3
	attic dust	N	2.1	2.1	0.88 - 3.8	1.3 - 2.6
Na	top soil	N	0.52	0.49	0.20 - 1.1	0.32 - 0.59
	attic dust	Log	0.38	0.34	0.18 - 1.1	0.26 - 0.60
P	top soil	Log	0.19	0.18	0.099 - 0.40	0.14 - 0.25
	attic dust	Log	0.11	0.10	0.053 - 0.49	0.076 - 0.15
S	top soil	Log	0.10	0.10	0.050 - 0.20	0.10 - 0.10
	attic dust	N	5.8	6.8	1.1 - 9.7	3.9 - 7.8
Ti	top soil	Log	0.26	0.27	0.11 - 0.47	0.20 - 0.34
	attic dust	Log	0.17	0.15	0.096 - 0.35	0.12 - 0.27
mg/kg						
Ag	top soil	Log	0.45	0.40	0.20 - 1.7	0.30 - 0.60
	attic dust	Log	0.79	0.80	0.30 - 4.0	0.40 - 1.2
As	top soil	Log	13	12	3.0 - 35	10 - 19
	attic dust	Log	29	24	11 - 170	17 - 56
Ba	top soil	Log	460	440	180 - 2000	350 - 550
	attic dust	Log	47	42	12 - 570	22 - 110
Be	top soil	N	1.7	2.0	1.0 - 3.0	1.0 - 2.0
	attic dust	Log	1.0	1.0	1.0 - 2.0	1.0 - 1.0
Bi	top soil	Log	0.47	0.40	0.20 - 3.3	0.30 - 0.70
	attic dust	Log	0.79	0.60	0.30 - 6.7	0.50 - 1.0
Cd	top soil	Log	1.4	1.2	0.30 - 10	0.90 - 1.6
	attic dust	Log	4.2	3.1	0.50 - 240	1.7 - 5.1
Ce	top soil	N	45	48	15 - 74	37 - 54
	attic dust	Log	23	23	12 - 48	20 - 24

mg/kg	Mat	Dis	X, Xg	Md	Min - Max	P25 - P75
Co	top soil	Log	10	10	6.4 - 19	9.1 - 12
	attic dust	Log	10	9.1	4.7 - 32	7.2 - 13
Cr	top soil	Log	68	60	37 - 310	43 - 86
	attic dust	Log	70	50	36 - 1300	45 - 110
Cu	top soil	Log	74	74	35 - 140	62 - 93
	attic dust	Log	130	110	41 - 870	92 - 190
Hf	top soil	Log	1.1	1.2	0.40 - 2.4	0.80 - 1.4
	attic dust	Log	0.81	0.80	0.50 - 2.0	0.70 - 1.0
La	top soil	N	24	24	8.7 - 40	20 - 28
	attic dust	Log	12	12	4.8 - 24	9.3 - 14
Li	top soil	N	34	35	13 - 51	26 - 39
	attic dust	Log	27	26	18 - 86	21 - 31
Mn	top soil	Log	840	710	380 - 3800	620 - 1100
	attic dust	Log	600	480	230 - 14000	370 – 59
Mo	top soil	Log	1.9	1.8	0.70 - 9.6	1.3 - 2.
	attic dust	Log	5.9	5.7	1.6 - 51	4.0 - 7.
Nb	top soil	N	6.6	6.5	2.4 - 11	5.7 - 7.
	attic dust	Log	4.5	4.1	2.6 - 12	3.3 - 6.
Ni	top soil	Log	34	30	20 - 110	27 - 3
	attic dust	Log	56	49	25 - 520	34 - 77
Pb	top soil	Log	230	230	66 - 820	150 - 320
	attic dust	Log	600	500	190 - 3200	320 - 1000
Rb	top soil	N	66	66	25 - 110	56 - 80
	attic dust	Log	39	38	30 - 61	34 - 44
Sb	top soil	Log	2.6	2.4	1.2 - 5.8	1.9 - 4.0
	attic dust	Log	8.8	9.0	3.3 - 28	5.6 - 13
Sc	top soil	N	8.2	9.0	3.0 - 12	6.0 -
	attic dust	Log	5.0	5.0	3.0 - 9.0	4.0 - 7.0
Sn	top soil	Log	6.8	6.9	3.2 - 14	5.6 - 8.6
	attic dust	Log	6.0	5.7	0.80 - 87	1.9 - 13
Sr	top soil	Log	120	130	71 - 300	100 - 140
	attic dust	Log	160	150	100 - 310	130 - 190
Ta	top soil	N	0.59	0.60	0.20 - 1.0	0.50 - 0.70
	attic dust	Log	0.38	0.40	0.20 - 0.60	0.30 - 0.50
Th	top soil	N	7.6	7.8	2.6 - 11	6.2 - 9.2
	attic dust	Log	3.8	3.7	1.8 - 6.8	3.0 - 4.1
U	top soil	Log	2.9	2.6	1.4 - 6.6	2.2 - 3.6
	attic dust	Log	3.9	3.8	2.0 - 13	3.0 - 5.1
V	top soil	N	74	80	44 - 120	56 - 84
	attic dust	Log	86	92	39 - 190	71 - 100

Table 4. (Continued)

mg/kg	Mat	Dis	X, Xg	Md	Min - Max	P25 - P75
W	top soil	Log	1.6	1.5	0.50 - 5.1	1.2 - 1.9
	attic dust	Log	2.7	3.0	1.0 - 9.8	1.7 - 4.2
Y	top soil	Log	13	13	6.0 - 24	11 - 16
	attic dust	Log	9.2	8.3	5.6 - 17	7.6 - 12
Zn	top soil	Log	460	390	160 - 2300	260 - 650
	attic dust	Log	1600	1300	250 - 28000	750 - 3200
Zr	top soil	Log	34	34	16 - 80	25 - 49
	attic dust	Log	26	24	19 - 60	21 - 30
Hg	top soil	Log	0.55	0.48	0.24 - 3.5	0.32 - 0.95
	attic dust	Log	0.74	0.73	0.21 - 5.7	0.42 - 1.2

Figure 11 shows the correlation coefficients between elements in top soil and attic dust. Logarithmic data from rural and urban areas in Slovenia has been included (N=60) to assure normal distribution.

A result of factor analysis is presented in table 5. Data from both, top soil and attic dust from Slovenian rural areas are included to present the natural geochemical associations. Factor quantity has been determined on the basis of eigenvalues obtained by principal component analysis. 4 factors have been used. Only the characteristic values are presented. Ca, K, Mg, P, S, Ag, Be, Bi, Ce, Hf, Li, Rb, Ta, U, V and W have been excluded from the factor analysis because of lack of significant relations with other chemical elements.

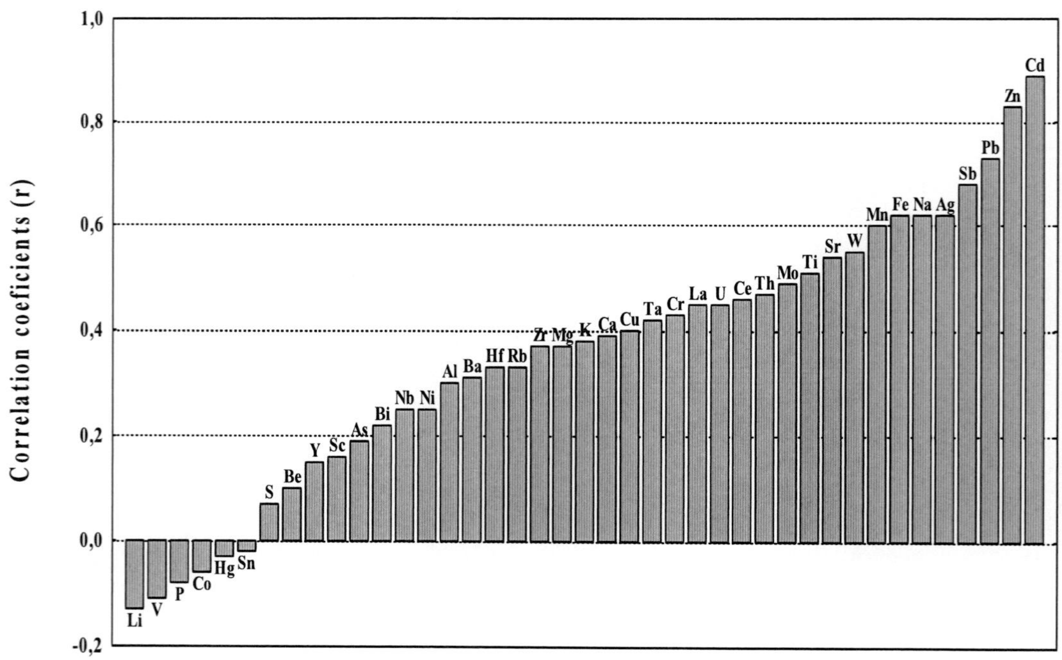

Figure 11. Correlation coefficients between elements concentrations in the top soil (0-5 cm) and attic dust (N=60) based on logarithmic data. Data from Slovenian rural areas has been used.

Table 5. Characteristic factor loadings (abs(r^2)>0.5) obtained from the geochemical data of top soil and attic dust in Slovenian rural areas (N=120). F1-F4 - characteristic factor loadings for each factor; Comm - explained variance of the element concentrations by factor analysis (%), Var - total variance explained by each factor (%)

	F1	F2	F3	F4	Comm
Th	0.90				92.6
Nb	0.88				85.5
Ti	0.85				83.0
La	0.84				85.1
Al	0.81				91.4
Sc	0.81				89.8
Y	0.75				63.5
V	0.71				67.6
Zr	0.70				66.7
Co	0.67		0.56		85.0
Pb		0.90			89.3
Zn		0.87			83.6
Cd		0.85			75.6
Mo		0.81			68.7
Sb		0.77			80.6
Sn		0.75			67.4
As		0.68			69.5
Cu		0.66			69.0
Hg		0.56			60.6
Fe	0.60		0.65		79.9
Ni			0.88		86.9
Cr			0.83		83.9
Mn			0.62		64.9
Sr		0.63		0.63	84.9
Na				0.78	65.6
Ba				0.66	69.1
Var	31.9	25.0	13.3	7.1	77.3

Enrichment factors (figure 12) have been calculated on the basis of comparison with the average values of elements in the top soil in Slovenian rural areas. To obtain clarity the average enrichment factors are presented. The elements have been grouped on the basis of factor analysis. Enrichment of elements in soils in rural areas on figure 12 is 1.

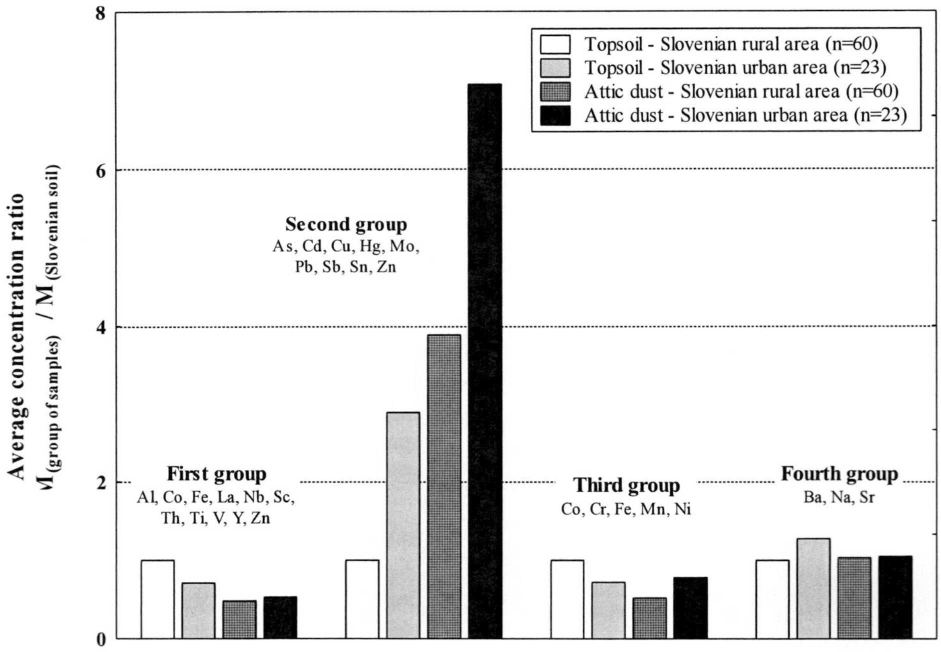

Figure 12. Average enrichment factors of elements in different media comparing to top soils in Slovenian rural areas. The elements are grouped on the basis of factor analysis (table 5).

Spatial distribution of factor scores (figures from 13 to 16) shows the regional natural geochemical trends in Slovenia. Separate maps for Al, Cd, Pb and Zn are presented in figures 17 to 20.

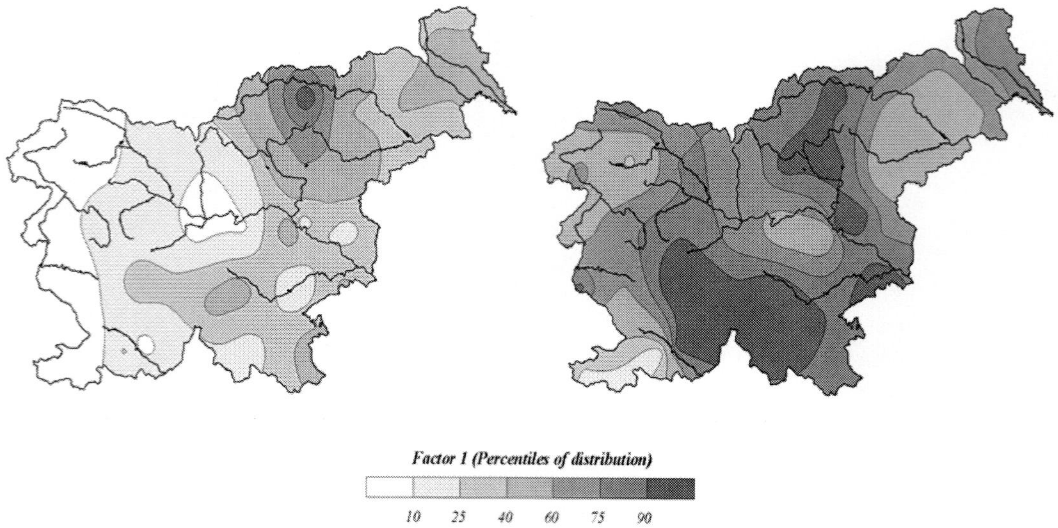

Figure 13. Spatial distribution of factor 1 scores in attic dust (left) and soil (right; Šajn, 1999 with additional data) in unpolluted areas of Slovenia. Characteristic elements for factor 1 are Th, Nb, Ti, La, Al, Sc, Y, V, Zr, Co and Fe. Listed elements are ordered on the basis of the strengths of the connectivity with factor 1 (from 0.90 to 0.60).

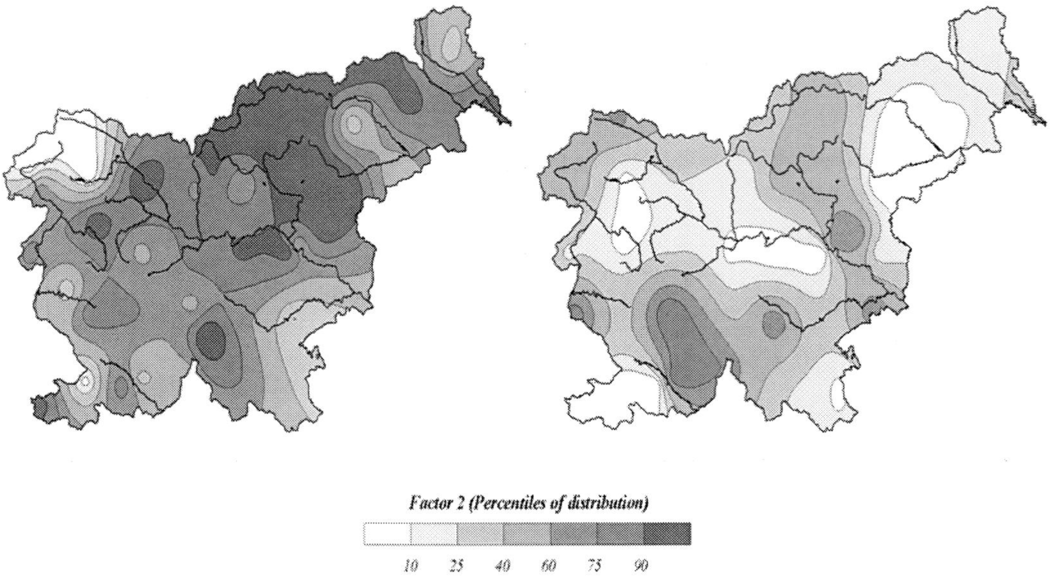

Figure 14. Spatial distribution of factor 2 scores in attic dust (left) and soil (right; Šajn, 1999 with additional data) in unpolluted areas of Slovenia. Characteristic elements for factor 2 are Pb, Zn, Cd, Mo, Sb, Sn, As, Cu, Sr and Hg. Listed elements are ordered on the basis of the strengths of the connectivity with factor 2 (from 0.90 to 0.56).

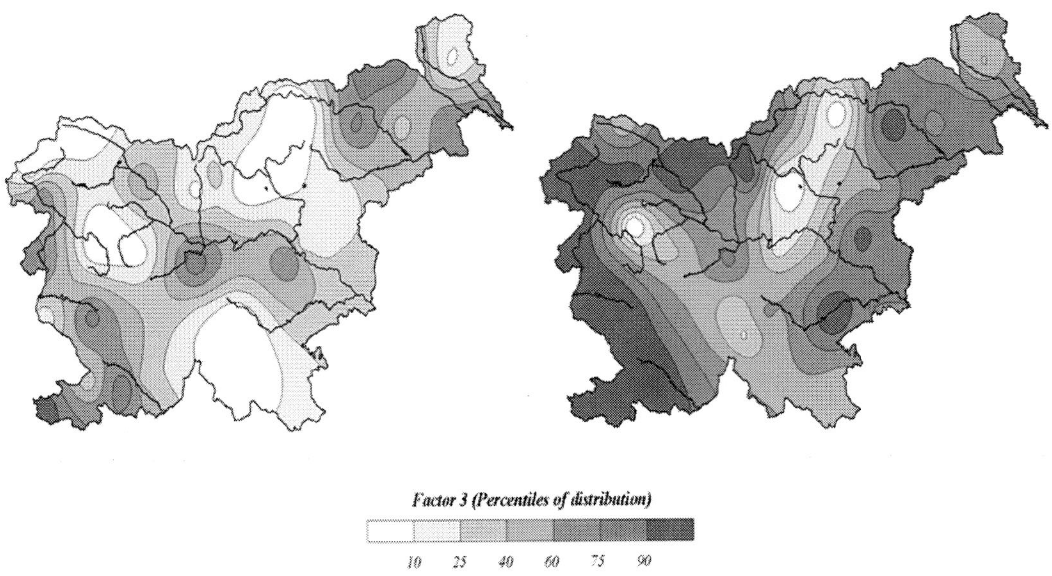

Figure 15. Spatial distribution of factor 3 scores in attic dust (left) and soil (right; Šajn, 1999 with additional data) in unpolluted areas of Slovenia. Characteristic elements for factor 3 are Ni, Cr, Fe, Mn and Co. Listed elements are ordered on the basis of the strength of the connectivity with factor 3 (from 0.88 to 0.56).

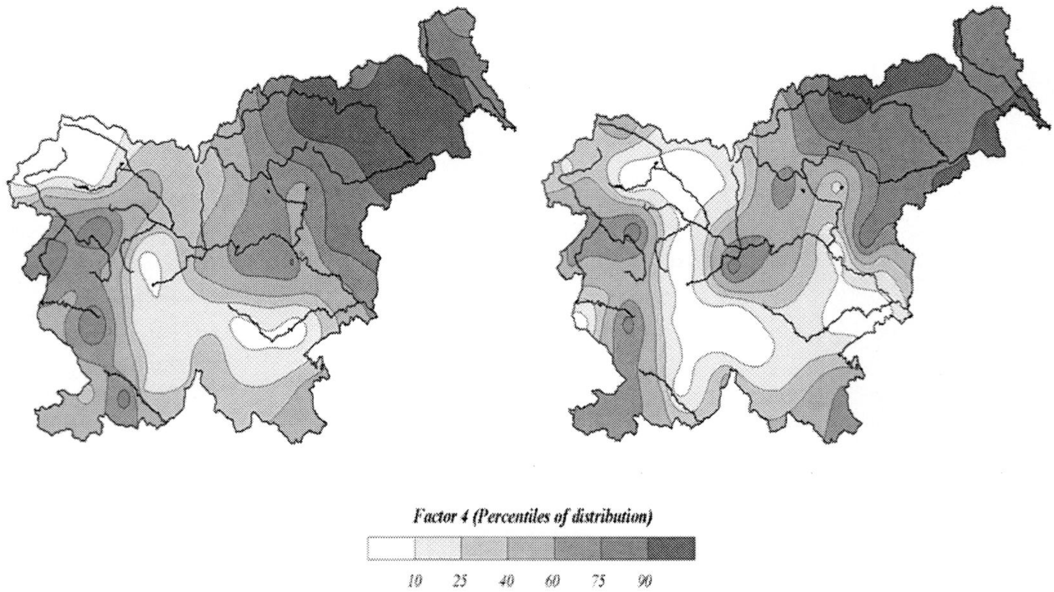

Figure 16. Spatial distribution of factor 4 scores in attic dust (left) and soil (right; Šajn, 1999 with additional data) in unpolluted areas of Slovenia. Characteristic elements for factor 4 are Sr, Na and Ba. Listed elements are ordered on the basis of the strengths of the connectivity with factor 3 (from 0.78 to 0.63).

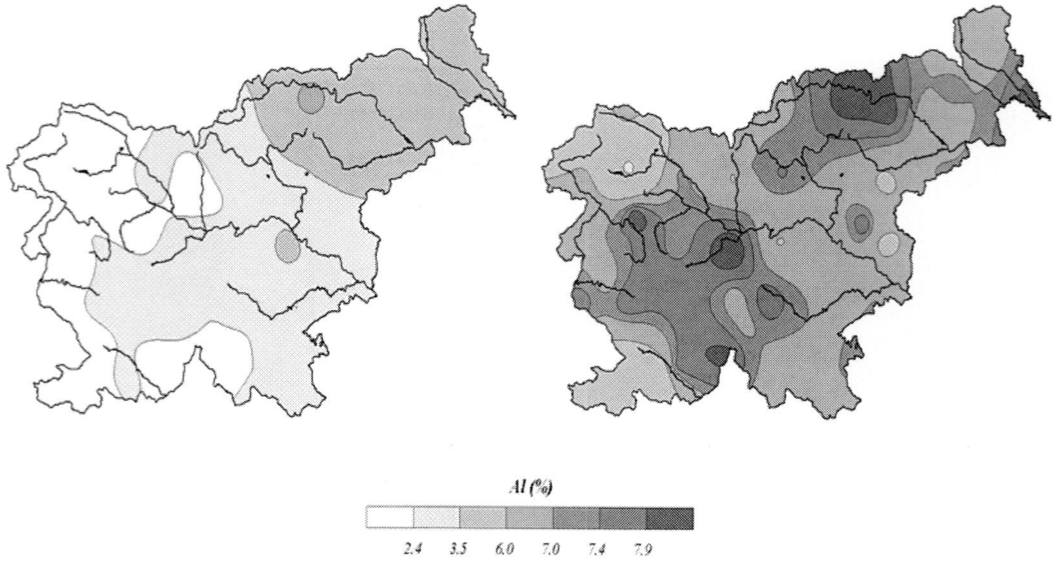

Figure 17. Spatial distribution of Al in attic dust (left) and soil (right; Šajn, 1999 with additional data) in unpolluted areas of Slovenia.

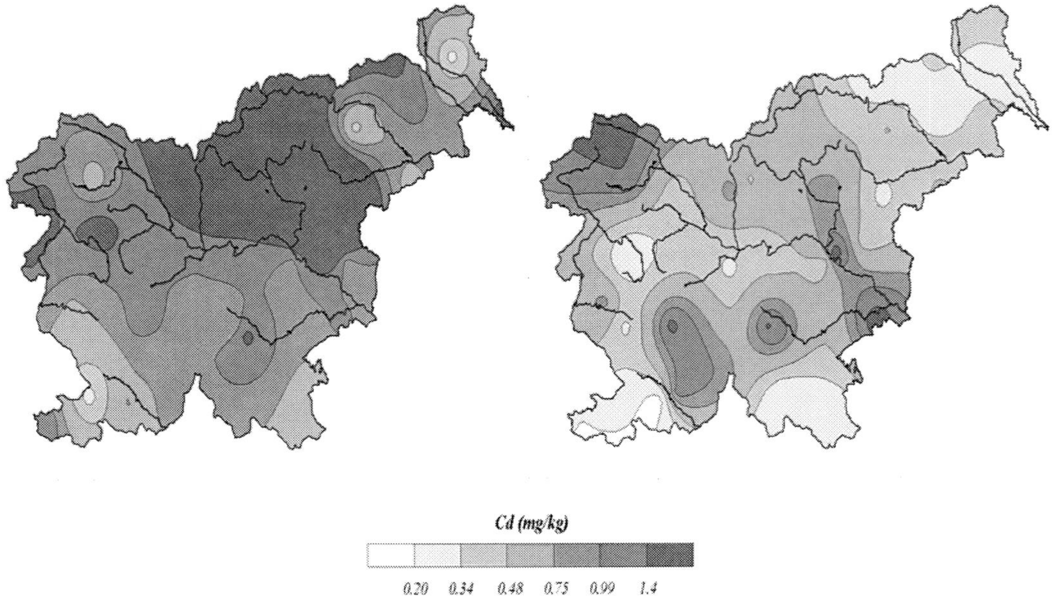

Figure 18. Spatial distribution of Cd in attic dust (left) and soil (right; Šajn, 1999 with additional data) in unpolluted areas of Slovenia.

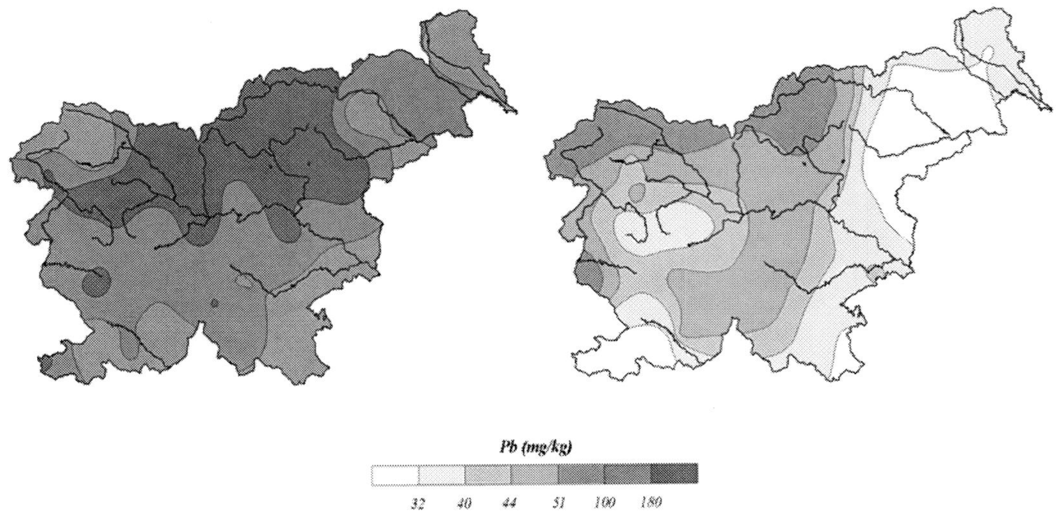

Figure 19. Spatial distribution of Pb in attic dust (left) and soil (right; Šajn, 1999 with additional data) in unpolluted areas of Slovenia.

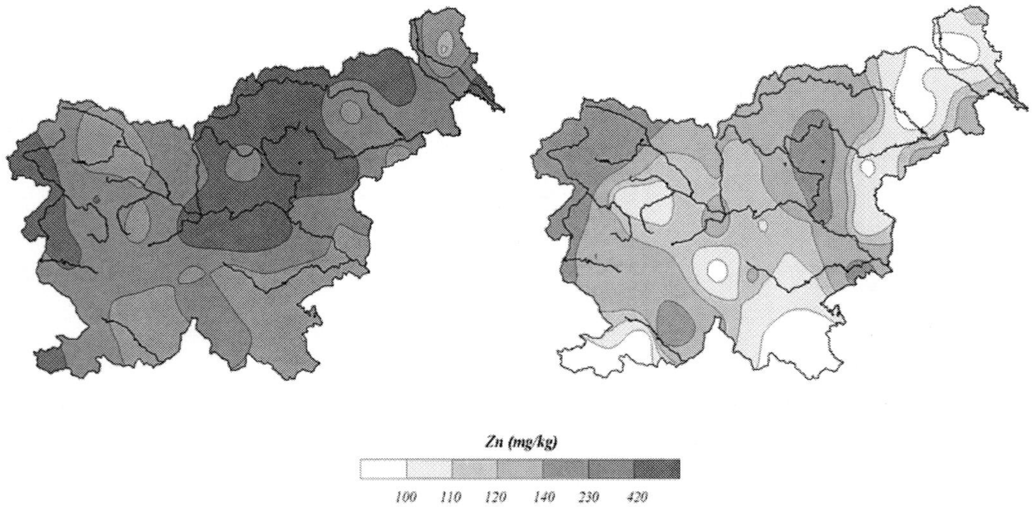

Figure 20. Spatial distribution of Zn in attic dust (left) and soil (right; Šajn, 1999 with additional data) in unpolluted areas of Slovenia

4.3. Heavy Metal Contamination in the Mežica Region Due to Pb-Zn Mining, Smelting and Ironworks Operations

Statistical analysis contains data from 114 sampling locations of top soil and attic dust. 6 sampling sites waer situated on Quaternary deposit, 7 on Miocene sandstone and marl, 45 on Triassic limestone and dolomite, 4 on Permian shale and sandstone and 52 on lower Paleozoic metamorphic and igneous rocks. In total 228 samples of soils (0-5 cm) and attic dust has been taken. Table 6 shows the basic statistics.

Table 6. Nonparametric statistical distribution of the elements in the top soil (0-5 cm; N=114) and attic dust (N=114) in the Mežica-Ravne area (some of the data has already been presented in work of Šajn, 2006). Mat - material; Dis - distribution (Log = log-normal, N = normal-Gaussian); X, Xg - average value; Md - median value; Min - Max - range; P25 - P75 - quartile distribution. Data rounded on two decimal places

	Mat	Dis	X_Xg	Md	Min - Max	P25 - P75
%						
Al	top soil	N	6.3	6.6	1.1 - 11	5.5 - 7.6
Al	attic dust	Log	2.5	2.6	1.0 - 5.6	1.9 - 3.4
Ca	top soil	Log	2.5	2.9	0.17 - 16	1.1 - 7.2
Ca	attic dust	N	6.4	6.3	1.3 - 12	4.4 - 8.4
Fe	top soil	N	3.5	3.7	0.46 - 7.1	2.8 - 4.4
Fe	attic dust	Log	3.6	3.3	1.1 - 20	2.4 - 5.0
K	top soil	N	1.7	1.8	0.12 - 3.8	1.3 - 2.1
K	attic dust	Log	1.1	1.1	0.46 - 2.7	0.90 - 1.4
Mg	top soil	Log	1.9	1.6	0.41 - 8.6	1.1 - 3.3
Mg	attic dust	Log	1.4	1.5	0.23 - 5.0	0.94 - 2.1

	Mat	Dis	X_Xg	Md	Min - Max	P25 - P75
%						
Na	top soil	N	0.63	0.62	0.050 - 1.7	0.34 - 0.88
Na	attic dust	Log	0.35	0.35	0.054 - 1.4	0.27 - 0.47
P	top soil	Log	0.18	0.17	0.076 - 0.55	0.13 - 0.22
P	attic dust	Log	0.28	0.29	0.057 - 0.93	0.22 - 0.38
S	top soil	Log	0.076	0.080	0.020 - 0.35	0.050 - 0.11
S	attic dust	N	3.8	3.7	0.31 - 8.0	2.8 - 5.0
Ti	top soil	Log	0.27	0.28	0.054 - 1.2	0.22 - 0.35
Ti	attic dust	Log	0.12	0.12	0.058 - 0.28	0.10 - 0.15
mg/kg						
Ag	top soil	Log	0.18	0.10	0.10 - 13	0.10 - 0.30
Ag	attic dust	Log	0.97	0.95	0.20 - 23	0.60 - 1.3
As	top soil	Log	17	17	7.0 - 390	13 - 22
As	attic dust	Log	32	31	6.0 - 370	22 - 43
Ba	top soil	N	470	460	38 - 1400	330 - 620
Ba	attic dust	N	110	93	11 - 440	24 - 170
Cd	top soil	Log	2.5	2.0	0.40 - 71	1.1 - 5.3
Cd	attic dust	Log	12	10	2.1 - 280	6.6 - 19
Ce	top soil	N	57	60	11 - 90	50 - 69
Ce	attic dust	N	29	29	11 - 52	22 - 37
Co	top soil	N	15	15	2.0 - 32	11 - 18
Co	attic dust	Log	9.7	9.0	4.0 - 52	7.0 - 13
Cr	top soil	Log	89	95	23 - 830	71 - 120
Cr	attic dust	Log	150	110	37 - 4700	65 - 280
Cu	top soil	Log	42	43	11 - 550	28 - 58
Cu	attic dust	Log	120	110	38 - 740	77 - 160
La	top soil	N	32	34	5.0 - 56	27 - 38
La	attic dust	N	17	16	6.0 - 34	12 - 21
Li	top soil	Log	43	46	6.0 - 240	35 - 60
Li	attic dust	Log	19	21	7.0 - 76	14 - 26
Mn	top soil	Log	1000	1000	180 - 3200	760 - 1400
Mn	attic dust	Log	800	740	320 - 3000	500 - 1300
Mo	top soil	Log	3.3	2.8	0.70 - 290	1.7 - 5.1
Mo	attic dust	Log	15	13	2.4 - 290	7.5 - 30
Nb	top soil	Log	7.6	8.1	1.9 - 22	5.3 - 10
Nb	attic dust	Log	4.0	4.0	1.9 - 9.7	3.0 - 5.3
Ni	top soil	Log	38	43	7.0 - 170	30 - 50
Ni	attic dust	Log	59	50	15 - 830	29 - 110
Pb	top soil	Log	350	300	56 - 27000	130 - 630
Pb	attic dust	Log	2400	1800	220 - 25000	1200 - 4800
Rb	top soil	N	100	110	10 - 180	87 - 120
Rb	attic dust	Log	54	54	28 - 110	43 - 68
Sb	top soil	Log	3.3	3.0	1.0 - 1300	2.0 - 4.0
Sb	attic dust	Log	26	21	4.0 - 1700	14 - 43
Sc	top soil	N	12	12	2.0 - 23	10 - 14
Sc	attic dust	Log	5.0	5.0	2.0 - 10	4.0 - 7.0

Table 6. (Continued)

	Mat	Dis	X_Xg	Md	Min - Max	P25 - P75
mg/kg						
Sn	top soil	Log	6.2	5.6	1.7 - 190	4.1 - 8.0
Sn	attic dust	Log	34	29	6.0 - 510	19 - 51
Sr	top soil	Log	89	90	37 - 300	73 - 110
Sr	attic dust	Log	120	120	37 - 370	92 - 150
Th	top soil	N	9.1	10	1.0 - 15	8.0 - 11
Th	attic dust	N	3.8	4.0	2.0 - 7.0	3.0 - 5.0
U	top soil	Log	2.9	3.0	1.0 - 11	2.0 - 3.0
U	attic dust	Log	1.7	2.0	0.50 - 5.0	1.0 - 2.0
V	top soil	N	110	110	21 - 240	87 - 130
V	attic dust	Log	59	60	19 - 260	40 - 86
W	top soil	Log	1.7	1.0	0.50 - 61	1.0 - 2.0
W	attic dust	Log	6.8	6.0	1.0 - 270	2.0 - 22
Y	top soil	Log	12	12	2.5 - 51	9.0 - 15
Y	attic dust	Log	8.5	8.6	2.7 - 20	6.5 - 11
Zn	top soil	Log	450	320	110 - 4200	240 - 780
Zn	attic dust	Log	1300	1200	300 - 10000	750 - 2100
Zr	top soil	Log	33	34	7.1 - 95	25 - 48
Zr	attic dust	Log	23	21	5.2 - 220	16 - 34
Hg	top soil	Log	0.16	0.16	0.020 - 1.2	0.095 - 0.23
Hg	attic dust	Log	0.68	0.62	0.11 - 15	0.26 - 1.2

Bivariate statistics - correlation coefficients between element concentrations in soil and attic dust in Mežica-Ravne region are presented in figure 21.

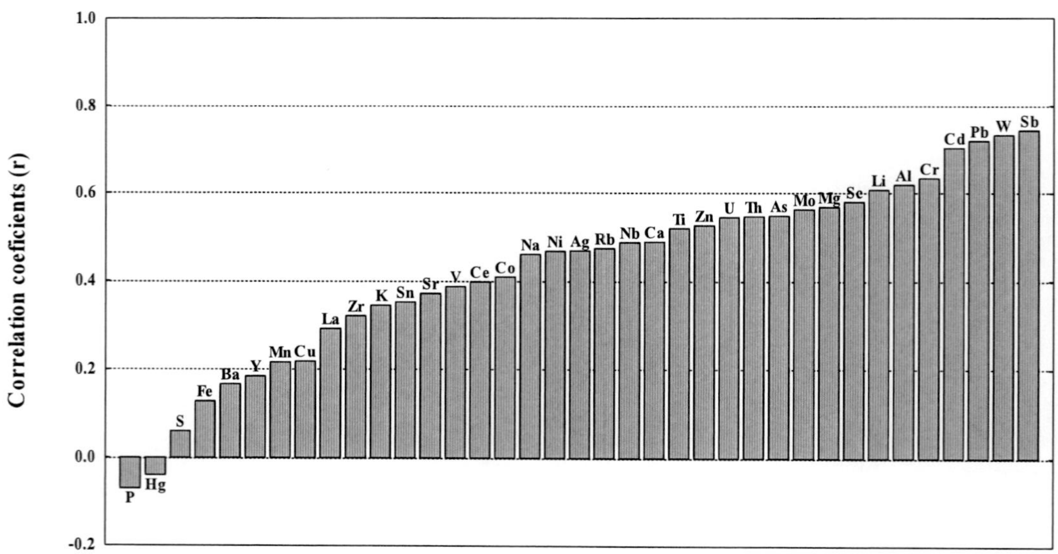

Figure 21. Correlation coefficients between elements concentrations in the top soil (0-5 cm) and attic dust (N=114) based on logarithmic data in the Mežica-Ravne area

In the multivariate factor analysis 32 elements have been retained. P, Sr, U, Y and Zr have been excluded because of lack of significant relations with other chemical elements. Table 7 presents the factor loadings for each of element (Šajn, 2006; renewed and improved). 4 geochemical associations have been found.

Table 7. Characteristic factor loadings (abs(r^2)>0.5) obtained from the geochemical data of top soil and attic dust in the Mežica-Ravne area (N=228). F1-F4 - characteristic factor loadings for each factor; Comm - explained variance of the element concentrations by factor analysis (%), Var - total variance explained by each factor (%). Renewed and improved from analysis, presented in Šajn (2006)

	F1	F2	F3	F4	Comm
Ce	0.91				93.9
La	0.90				89.7
Sc	0.90				94.4
Th	0.89				94.7
Al	0.88				96.2
Ti	0.88				88.7
Li	0.86				86.0
Nb	0.85				82.5
V	0.84				81.6
Rb	0.81				89.0
Na	0.75				68.1
K	0.66			-0.59	81.5
Co	0.65		0.64		88.2
Ba	0.60	-			67.5
S	-0.55	0.67			83.3
Sb		0.89			93.8
As		0.88			78.1
Pb		0.87			93.2
Cd		0.87			94.1
Zn		0.81			85.1
Ag		0.79			85.9
Sn		0.74			82.8
Hg		0.64			57.9
Mo		0.62	0.59		86.1
Cu		0.60	0.59		79.7
Cr			0.93		88.5
Ni			0.90		88.2
W			0.82		76.7
Fe			0.68		71.7
Mn	0.51		0.63		68.3
Mg				0.89	87.7
Ca				0.72	84.7
Var	35.1	25.5	15.3	8.1	84.0

Enrichment factors (figure 22) show the average enrichment of the elements, according to the median values in Slovenian soils in rural unpolluted areas of Slovenia. Two different areas have been separated. First area contains the samples, collected in Ravne surroundings where ironworks exists and second area contains geochemical data of samples in the Mežica surroundings where Pb-Zn mine with smeltery has been situated. Enrichment factors are grouped according to the factor analysis, presented in table 7. Due to clarity average enrichment factors have been used according to the groups of elements obtained by factor analysis.

Figures 23 to 26 show the spatial distribution of factor scores. For each geochemical association separate map have been made. Factor loadings and maps of factor scores indicate that the factor 1 represents natural geochemical association, factor 2 man-made anomalies due to Pb-Zn mining and smelting and factor 3 with pollution, caused by ironworks. Factor 4 is natural geochemical association and is connected with carbonatic rock outcrops. Percentile distributions have been basis for map's color scales.

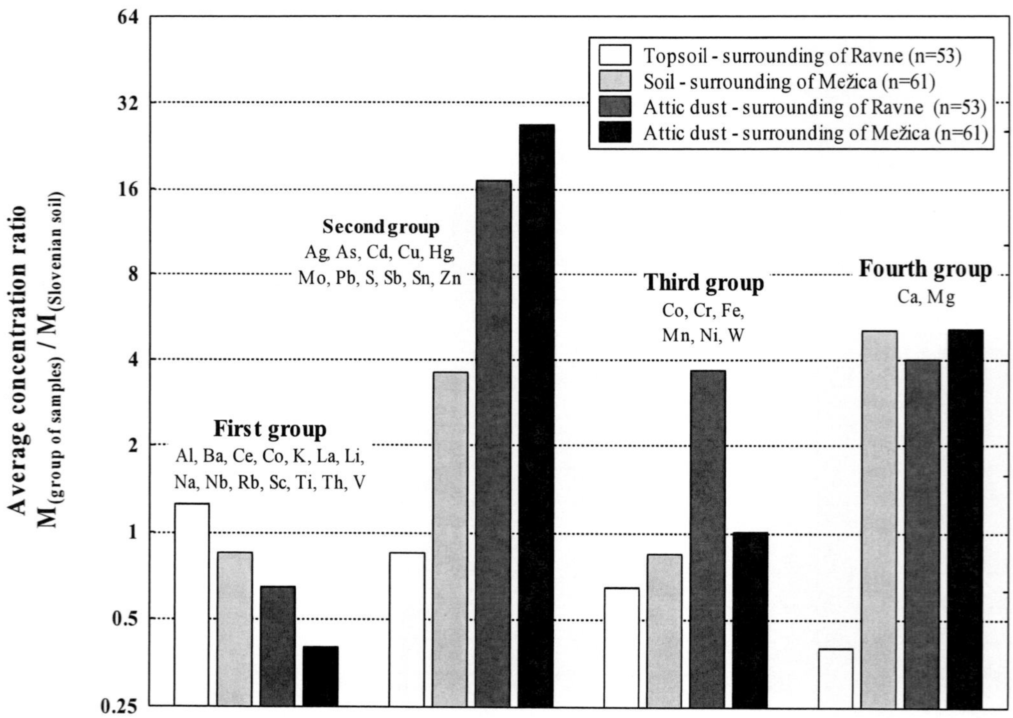

Figure 22. Average enrichment factors of elements in different media comparing to top soils in Slovenian rural areas. The elements are grouped on the basis of factor analysis (table 7). Geochemical data for enrichment factors calculation have been divided into two groups: Ravne area (ironworks) and Mežica area (Pb-Zn mine with smeltery).

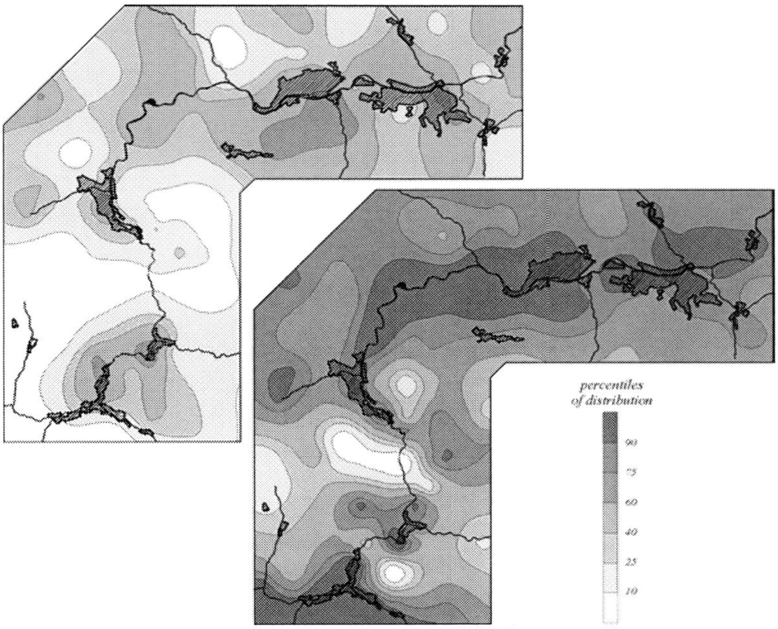

Figure 23: Spatial distribution of factor 1 scores in attic dust (top left) and top soil (bottom right; renewed after Šajn, 2006). Characteristic elements for factor 1 are Ce, La, Sc, Th, Al, Ti, Li, Nb, V, Rb, Na, K, Co, Ba, (-)S and Mn. Listed elements are ordered on the basis of the strengths of the connection with factor 1 (from 0.91 to 0.51). Sign (-) indicates negative correlation.

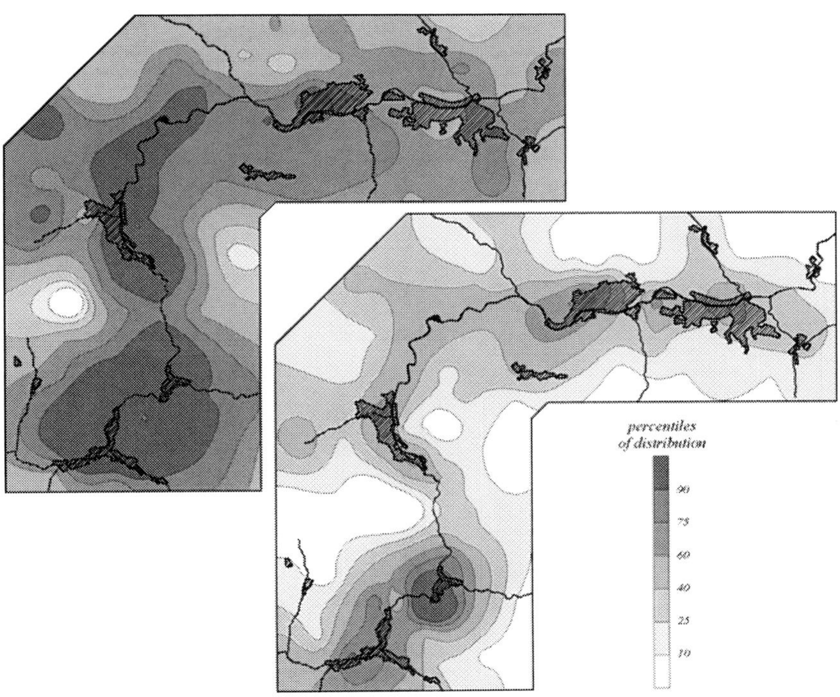

Figure 24. Spatial distribution of factor 2 scores in attic dust (top left) and top soil (bottom right; renewed after Šajn, 2006). Characteristic elements for factor 2 are Sb, As, Pb, Cd, Zn, Ag, Sn, Hg, Mo, Cu and (-)Ba. Listed elements are ordered on the basis of the strengths of the connection with factor 2 (from 0.89 to -0.55). Sign (-) indicates negative correlation

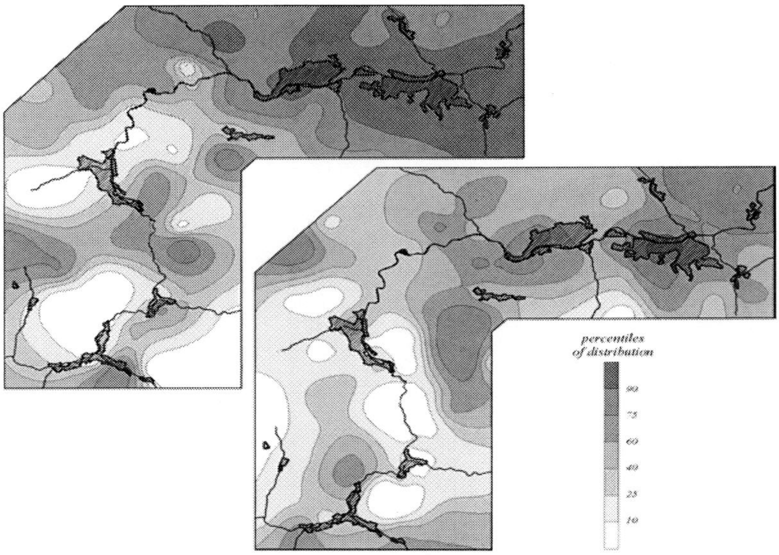

Figure 25. Spatial distribution of factor 3 scores in attic dust (top left) and top soil (bottom right; renewed after Šajn, 2006). Characteristic elements for factor 3 are Cr, Ni, W, Fe, Co, Mn, Cu and Mo. Listed elements are ordered on the basis of the strengths of the connection with factor 3 (from 0.93 to 0.59).

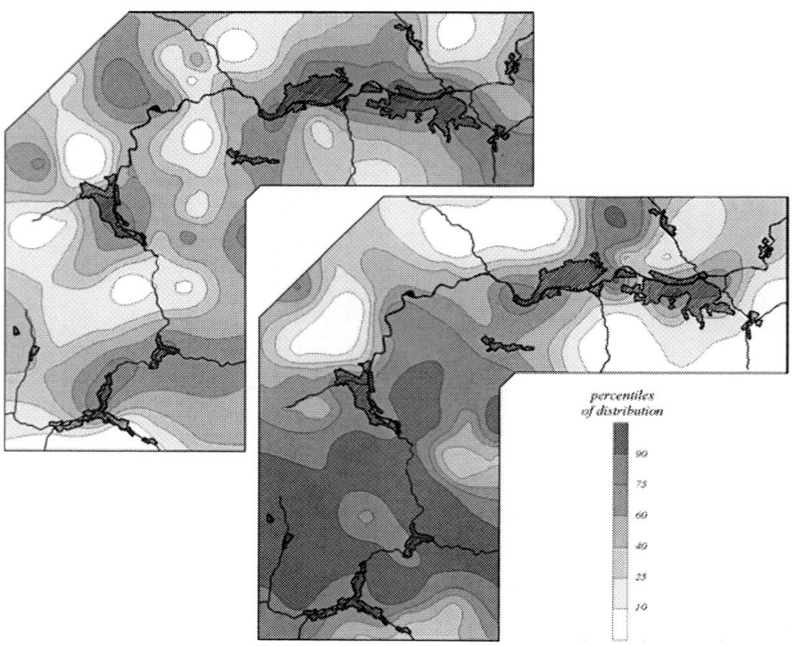

Figure 26. Spatial distribution of factor 4 scores in attic dust (top left) and top soil (bottom right; renewed after Šajn, 2006). Characteristic elements for factor 4 are Mg, Ca and (-)K. Listed elements are ordered on the basis of the strengths of the connection with factor 4 (from 0.89 to -0.59). Sign (-) indicates negative correlation

Figures from 27 to 29 show geochemical maps for different elements, where Al represents natural distribution, Cr anomaly made by ironworks and Pb anomaly made by Pb-Zn smelting and mining operations.

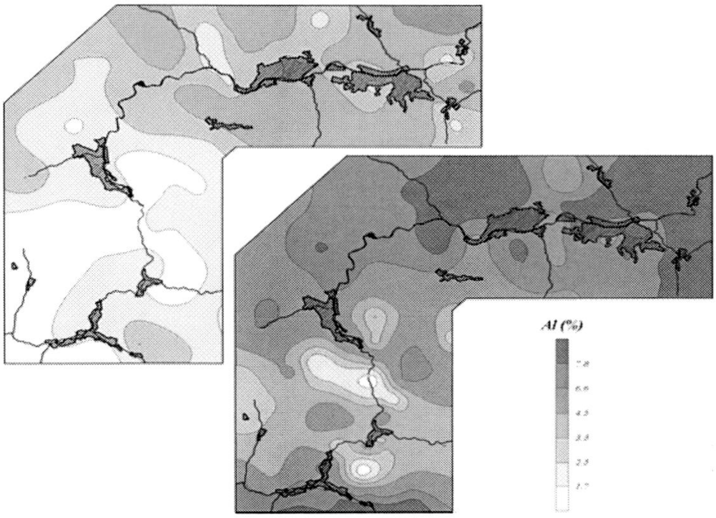

Figure 27: Map of the concentrations of the Al in attic dust (top left) and top soil (bottom right) in the Mežica-Ravne area. Percentile distribution (P10-P25-P40-P60-P75-P90) has been used for contour scaling

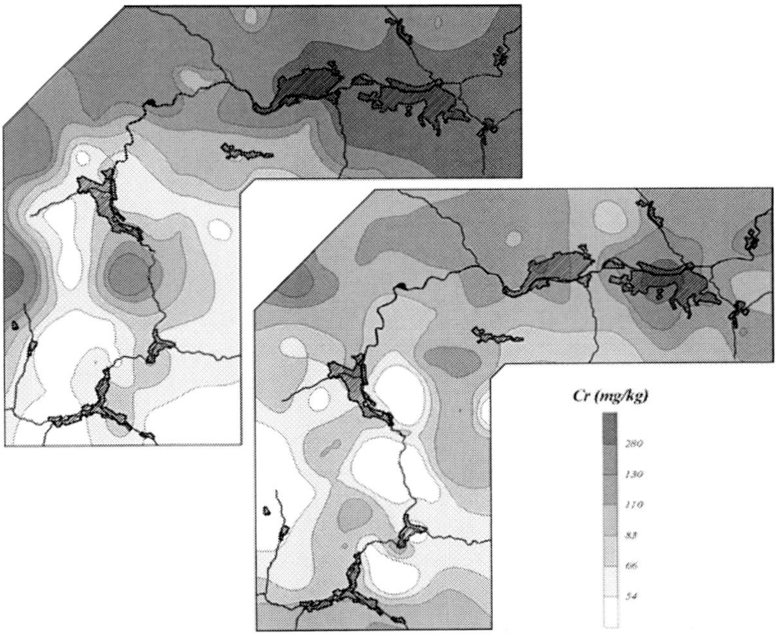

Figure 28. Map of the concentrations of the Cr in attic dust (top left) and top soil (bottom right) in the Mežica-Ravne area. Percentile distribution (P10-P25-P40-P60-P75-P90) has been used for contour scaling.

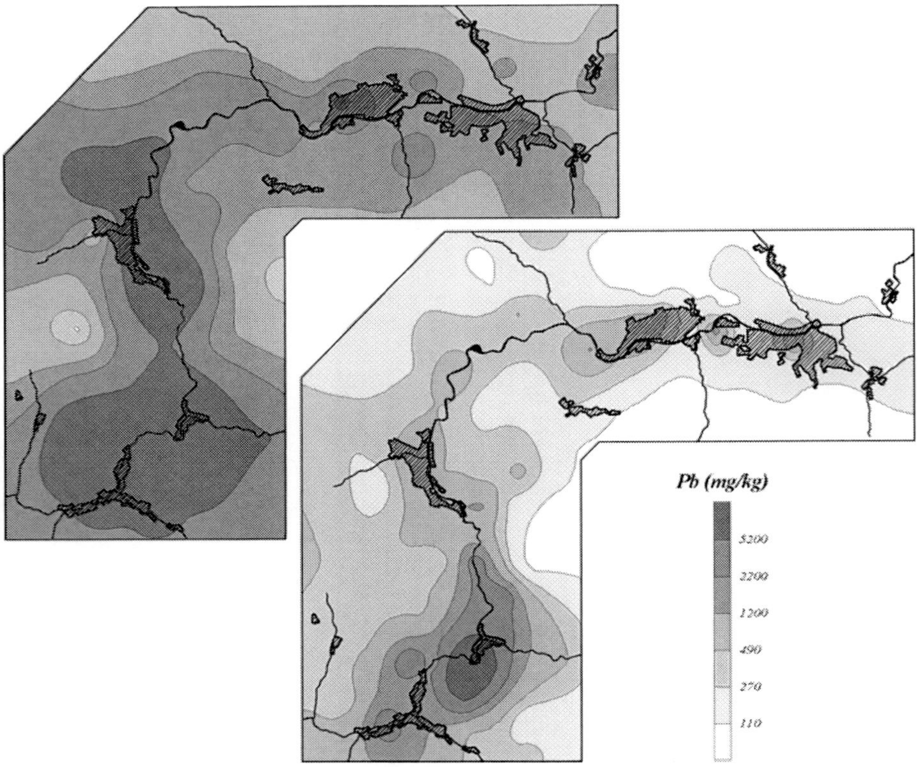

Figure 29. Map of the concentrations of the Pb in attic dust (top left) and top soil (bottom right) in the Mežica-Ravne area. Percentile distribution (P10-P25-P40-P60-P75-P90) has been used for contour scaling.

4.4. Heavy metal contamination in the Celje region due to Zn smelting and ironworks

Preliminary sampling and analysis of soils (0-5 cm) and attic dust have been performed in year 1996 (4 sampling points), but the main sampling (101 sampling points with repetitions; attic dust has not been sampled on all of the locations due to unavailability of old houses) have been performed in year 2000. For grid completion and results verification additional 4 samples of attic dust and 7 samples of top soils have been taken in year 2003. Overall 112 samples of soils and 108 samples of attic dust have been sampled, but in statistical analyses and data presentation only the data where both, soil and attic dust has been collected in same area are included (N=99). Geologically speaking 42 sampling sites has been located on Quaternary alluvial deposits, 26 on Plio-quarternary clays and sands, 9 on Oligocene andesitic tuffs and marine claystones, 9 on Miocene sandstones, marly limestones and conglomerates and 13 on Triassic carbonates, shales and keratophyres (Šajn, 2005). Non-parametric statistics for geochemical composition of top soil and attic dust are presented in table 8.

Table 8. Nonparametric statistical distribution of the elements in the top soil (0-5 cm; N=99) and attic dust (N=99) in the Celje area (Šajn, 2005 & Žibret, 2002; with supplemental data). Dis - distribution (Log = log-normal, N = normal-Gaussian); X, Xg - average value; Md - median value; Min - Max - range; P25 - P75 - quartile distribution. Data rounded on two decimal places.

	Mat	Dis	X, Xg	Md	Min - Max	P75 - P25
%						
Al	top soil	N	6.2	6.2	3.3 - 8.6	5.6 - 6.8
Al	attic dust	Log	3.3	3.3	1.8 - 5.4	2.9 - 4.0
Ca	top soil	Log	1.6	1.3	0.070 - 11	0.83 - 3.2
Ca	attic dust	Log	6.6	6.7	3.5 - 15	5.0 - 8.5
Fe	top soil	Log	3.2	3.2	2.2 - 6.8	2.9 - 3.6
Fe	attic dust	Log	3.8	3.3	1.2 - 15	2.7 - 5.5
K	top soil	Log	1.7	1.7	0.86 - 3.1	1.6 - 1.9
K	attic dust	Log	1.1	1.1	0.62 - 2.2	0.84 - 1.3
Mg	top soil	Log	1.0	0.84	0.41 - 4.4	0.73 - 1.2
Mg	attic dust	Log	1.4	1.3	0.64 - 3.9	1.0 - 1.8
Na	top soil	Log	0.63	0.65	0.23 - 1.4	0.53 - 0.80
Na	attic dust	Log	0.48	0.46	0.30 - 1.8	0.39 - 0.54
P	top soil	Log	0.12	0.12	0.023 - 0.32	0.098 - 0.16
P	attic dust	Log	0.19	0.19	0.056 - 0.72	0.14 - 0.26
S	top soil	Log	0.063	0.060	0.020 - 0.34	0.050 - 0.080
S	attic dust	Log	4.3	4.5	1.7 - 9.2	3.6 - 5.3
Ti	top soil	Log	0.32	0.31	0.19 - 0.51	0.27 - 0.38
Ti	attic dust	Log	0.26	0.25	0.099 - 0.75	0.22 - 0.31
mg/kg						
Ag	top soil	Log	0.23	0.14	0.14 - 3.1	0.14 - 0.30
Ag	attic dust	Log	1.1	1.0	0.20 - 18	0.60 - 1.8
As	top soil	Log	17	16	4.0 - 91	12 - 21
As	attic dust	Log	40	36	6.0 - 480	24 - 59
Ba	top soil	Log	570	530	290 - 1900	460 - 640
Ba	attic dust	Log	110	83	17 - 760	29 - 450
Cd	top soil	Log	3.2	2.9	0.50 - 59	1.6 - 4.8
Cd	attic dust	Log	24	24	2.3 - 460	13 - 40
Ce	top soil	Log	55	55	21 - 100	48 - 61
Ce	attic dust	N	29	29	8.0 - 50	22 - 36
Co	top soil	N	11	11	1.0 - 33	9.0 - 12
Co	attic dust	Log	10	10	4.0 - 41	7.0 - 14
Cr	top soil	N	71	74	22 - 130	56 - 85
Cr	attic dust	Log	74	68	33 - 570	59 - 82
Cu	top soil	Log	44	39	0.70 - 1700	28 - 64
Cu	attic dust	Log	170	160	48 - 1100	100 - 270
La	top soil	Log	33	35	14 - 72	29 - 37
La	attic dust	N	19	19	4.5 - 35	14 - 24
Li	top soil	Log	39	38	24 - 55	35 - 45
Li	attic dust	Log	25	25	16 - 52	22 - 30
Mn	top soil	N	770	720	160 - 1700	640 - 900
Mn	attic dust	Log	630	600	320 - 3600	490 - 750
Mo	top soil	Log	1.5	1.5	0.50 - 6.5	1.1 - 2.0
Mo	attic dust	Log	5.5	5.2	2.1 - 34	4.0 - 7.1
Nb	top soil	N	8.8	8.7	4.0 - 13	7.8 - 10
Nb	attic dust	Log	6.7	6.5	3.8 - 18	5.7 - 7.5

Table 8. (Continued)

mg/kg	Mat	Dis	X, Xg	Md	Min - Max	P75 - P25
Ni	top soil	N	34	33	4.0 - 85	28 - 40
Ni	attic dust	Log	60	57	21 - 370	38 - 81
Pb	top soil	Log	140	120	33 - 1500	74 - 240
Pb	attic dust	Log	720	620	190 - 6900	430 - 1100
Rb	top soil	Log	96	96	48 - 160	85 - 110
Rb	attic dust	Log	56	56	32 - 98	47 - 65
Sb	top soil	Log	1.7	2.0	0.70 - 18	1.0 - 2.5
Sb	attic dust	Log	9.1	9.0	1.0 - 72	6.0 - 14
Sc	top soil	Log	11	11	7.0 - 16	10 - 12
Sc	attic dust	Log	7.0	7.0	4.0 - 10	6.0 - 8.0
Sn	top soil	Log	4.7	4.5	1.2 - 36	3.4 - 5.8
Sn	attic dust	Log	26	22	4.3 - 1800	14 - 38
Sr	top soil	Log	98	98	25 - 220	82 - 120
Sr	attic dust	Log	150	150	100 - 230	140 - 160
Th	top soil	Log	9.9	10	5.0 - 23	9.0 - 11
Th	attic dust	N	4.3	4.0	0.70 - 7.0	3.0 - 6.0
U	top soil	N	4.8	5.0	2.0 - 9.0	4.0 - 5.0
U	attic dust	N	3.9	4.0	0.70 - 8.0	3.0 - 4.0
V	top soil	Log	86	91	37 - 210	71 - 100
V	attic dust	Log	100	100	35 - 260	87 - 120
Y	top soil	Log	13	13	6.6 - 53	10 - 14
Y	attic dust	Log	10	10	5.8 - 19	8.6 - 12
Zn	top soil	Log	600	500	87 - 8600	290 - 980
Zn	attic dust	Log	4600	4400	460 - 56000	2500 - 7700
Zr	top soil	N	38	39	23 - 59	34 - 43
Zr	attic dust	Log	27	27	14 - 78	22 - 30
Hg	top soil	Log	0.15	0.13	0.035 - 1.4	0.095 - 0.22
Hg	attic dust	Log	0.68	0.55	0.11 - 4.6	0.43 - 0.11

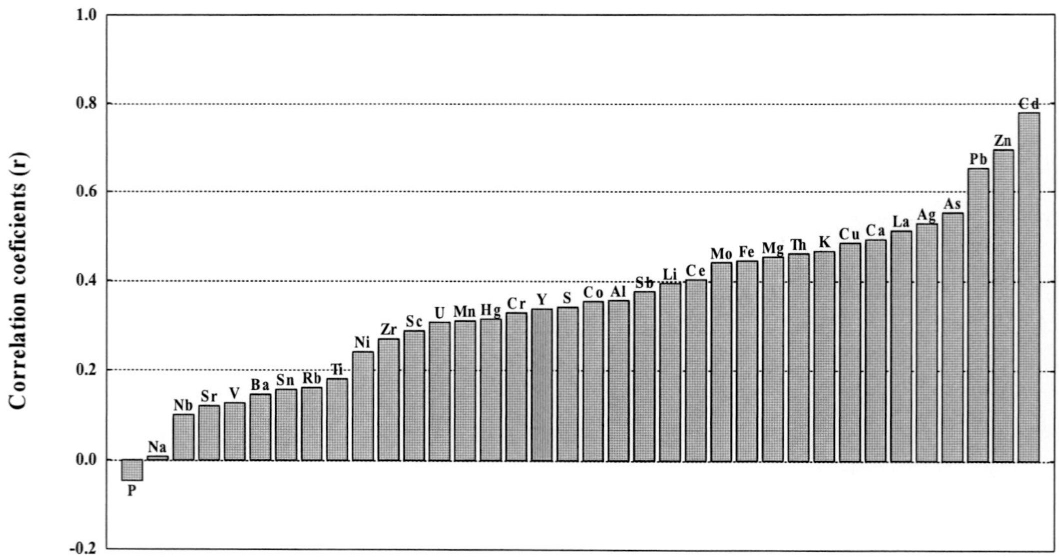

Figure 30. Correlation coefficients between elements concentrations in the top soil (0-5 cm) and attic dust (N=99) based on logarithmic data in the Celje area (Šajn, 2005).

Bivariate statistics have been calculated on the basis of element concentrations in sampling points where both, soil and attic dust have been collected (N=99). Figure 30 shows the correlation coefficients between element concentrations in soil and attic dust.

Multivariate factor analysis has been performed on the basis of logarithmic data to assure normal distribution. The standardization to zero mean and unit standard deviation has been used. Ca, Mg, Na, P, Ba, Sn, Sr, U, V, Y and Zr have been excluded because of lack of significant associations with other elements. Table 9 shows the factor loadings. 4 geochemical associations have been extracted. First factor represents natural geochemical association. All other factors are connected with man-made geochemical anomalies: zinc smelting (F2), ironworks (F3) and recent titanium dioxide production (F4).

Table 9. Characteristic factor loadings (abs(r^2)>0.5) obtained from the geochemical data of top soil and attic dust in Slovenian rural areas (N=198; Šajn, 2005). F1-F4 - characteristic factor loadings for each factor; Comm - explained variance of the element concentrations by factor analysis (%), Var - total variance explained by each factor (%)

	F1	F2	F3	F4	Comm
Rb	0.87				89.5
Ce	0.85				91.2
K	0.84				86.5
Al	0.83				96.0
La	0.82				85.6
Sc	0.81				90.8
Th	0.79	-0.52			89.4
Li	0.70				76.3
As		0.85			87.2
Pb		0.82			91.1
Zn		0.81			91.3
Ag		0.81			85.8
Cd		0.80			88.9
Sb		0.78			84.7
Hg		0.76			77.3
Cu		0.73			80.4
Mo	-0.53	0.64			84.7
S	-0.59	0.64			84.3
Cr			0.85		76.8
Fe			0.78		78.1
Mn			0.76		62.6
Co			0.72		72.1
Ni			0.69		76.1
Ti				0.90	92.4
Nb				0.83	85.9
Var	30.9	30.0	13.8	9.5	84.2

Enrichment factors are calculated according to the average values of the elements concentrations in the Slovenian unpolluted areas. The sampling points in Celje area are divided into two groups. In first group there are data from sampling points, located in the Štore-Celje urban area, where the pollution is most extensive. In second group there is geochemical data from sampling points, located in the surroundings, where pollution is still present but is not so extensive. Figure 31 shows the calculated enrichment factors. Due to clarity average enrichment factors have been used according to the groups of elements obtained by factor analysis.

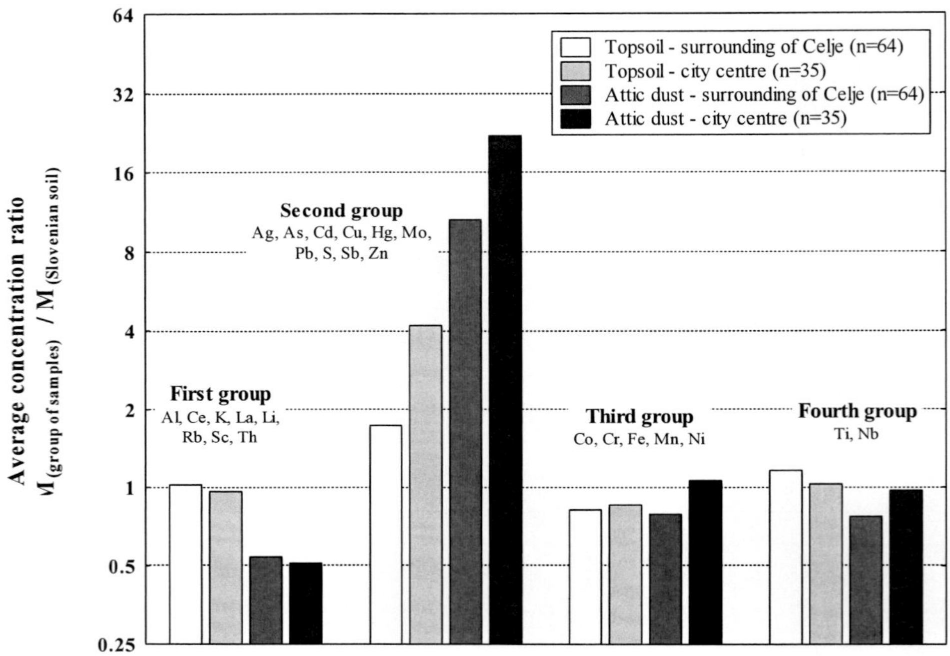

Figure 31. Average enrichment factors of elements in Celje area in different media comparing to top soils in Slovenian rural areas. The elements are grouped on the basis of factor analysis (table 9). Geochemical data for enrichment factors calculation have been divided into two groups: Celje-Štore urban area where pollution is most extensive and surrounding where pollution is still present but not so severe.

Spatial distribution of factor scores (figures from 32 to 35) demonstrates that geochemical association, connected with 1st factor is association of natural distributed elements. Elements, correlated with 2nd factor represent geochemical association, connected with Zn smelting operations as their highest values are around past zinc smelting plants. 3rd group of elements, loaded in 3rd factor are connected with iron smelting activities. Their highest values are in the vicinity of past and present ironworks and last, 4th group of elements are connected with chemical industry (highest values are around present titanium dioxide production plant).

Figures from 36 to 40 show the spatial distribution of different heavy metals, of which Al represents natural pattern. Very distinctive anomaly visible in soil and attic dust, made by past 100-year Zn smelting operations, is presented by the maps of aerial distribution of Zn and Cd. Another two weaker anomalies which are visible only in attic dust but not in soil are presented. Those two anomalies are probably the consequence of Štore ironworks (Cr) and 40-year TiO_2 production by Cinkarna Celje (Ti). All maps are renewed from the work Šajn (2005). Basis for contour scaling has been the percentile distributions (P10-P25-P40-P60-P75-P90).

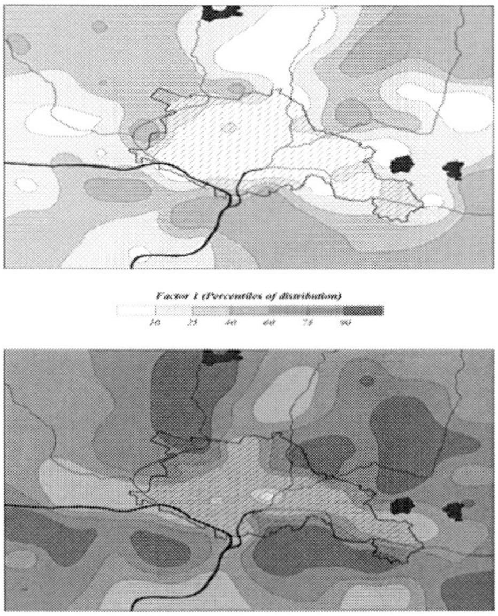

Figure 32. Spatial distribution of factor 1 scores in attic dust (top) and top soil (bottom; Šajn, 2005) in Celje area. Characteristic elements for factor 1 are Rb, Ce, K, Al, La, Sc, Th, Li, (-) S and (-) Mo. Listed elements are ordered on the basis of the strengths of the connection with factor 1 (from 0.87 to -0.53). Sign (-) indicates negative correlation.

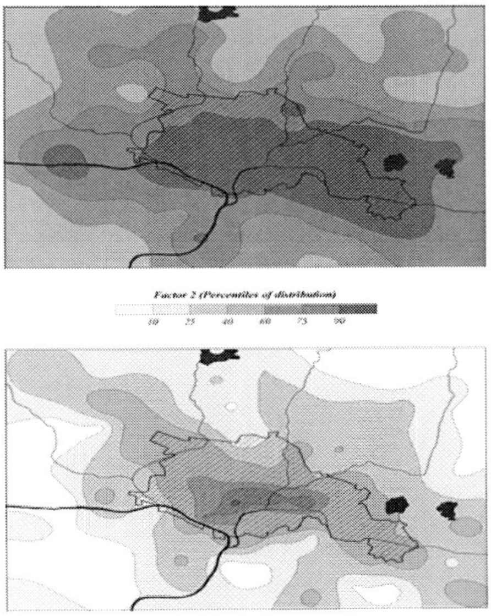

Figure 33: Spatial distribution of factor 2 scores in attic dust (top) and top soil (bottom; Šajn, 2005) in Celje area. Characteristic elements for factor 2 are As, Pb, Zn, Ag, Cd, Sb, Hg, Cu, Mo, S and (-)Th. Listed elements are ordered on the basis of the strengths of the connection with factor 2 (from 0.85 to -0.52). Sign (-) indicates negative correlation.

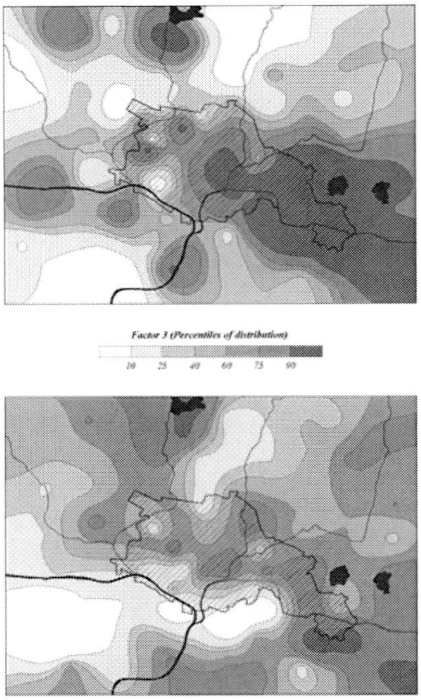

Figure 34. Spatial distribution of factor 3 scores in attic dust (top) and top soil (bottom; Šajn, 2005) in Celje area. Characteristic elements for factor 3 are Cr, Fe, Mn, Co and Ni. Listed elements are ordered on the basis of the strengths of the connection with factor 3 (from 0.85 to 0.69).

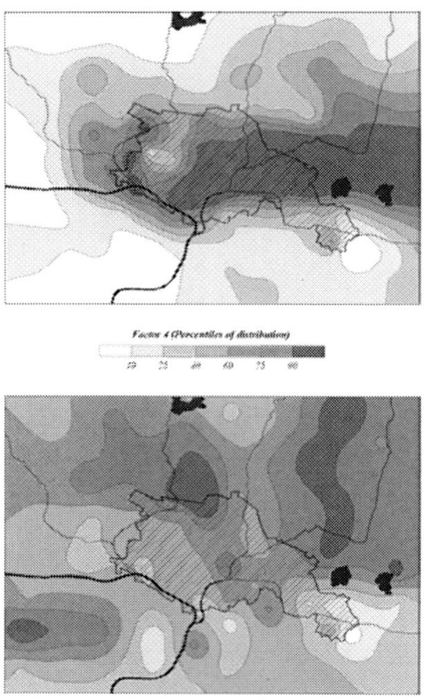

Figure 35. Spatial distribution of factor 4 scores in attic dust (top) and top soil (bottom; Šajn, 2005) in Celje area. Characteristic elements for factor 4 are Ti and Nb. Listed elements are ordered on the basis of the strengths of the connection with factor 4 (from 0.90 to 0.83).

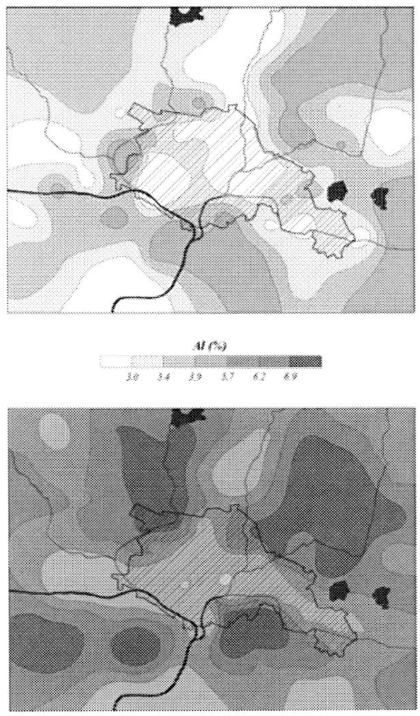

Figure 36. Map of the concentrations of the Al in attic dust (top) and top soil (bottom; Šajn, 2005) in the Celje area. Levels on map are percentile distribution.

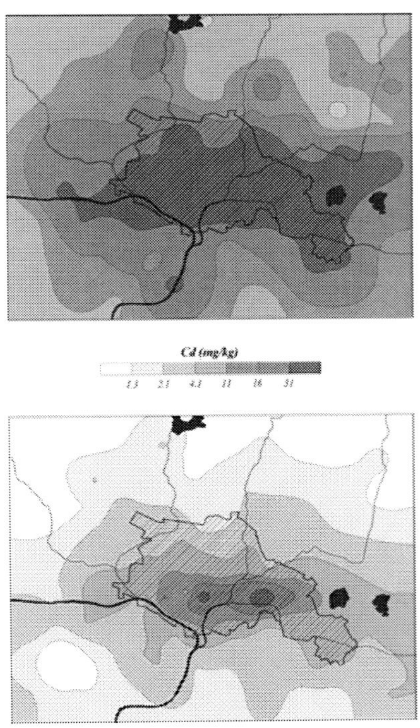

Figure 37. Map of the concentrations of the Cd in attic dust (top) and top soil (bottom; Šajn, 2005) in the Celje area. Levels on map are percentile distribution.

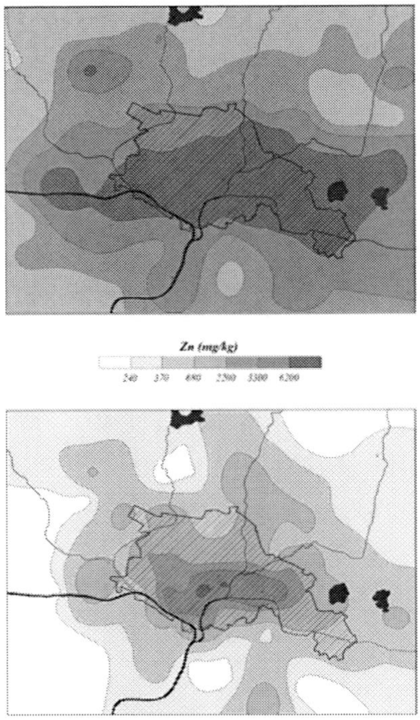

Figure 38. Map of the concentrations of the Zn in attic dust (top) and top soil (bottom) in the Celje area. Levels on map are percentile distribution.

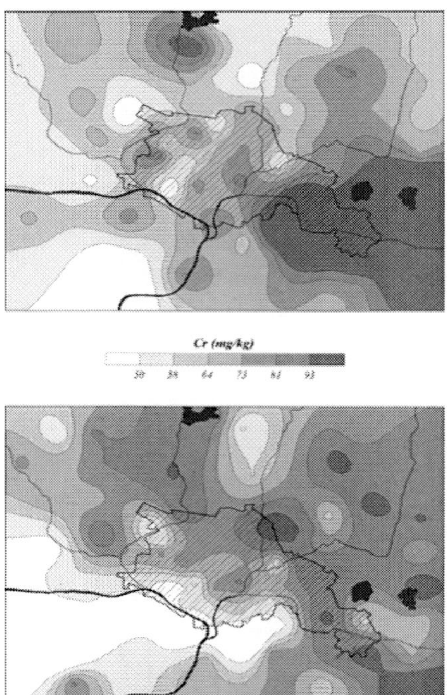

Figure 39. Map of the concentrations of the Cr in attic dust (top) and top soil (bottom; Šajn, 2005) in the Celje area. Levels on map are percentile distribution.

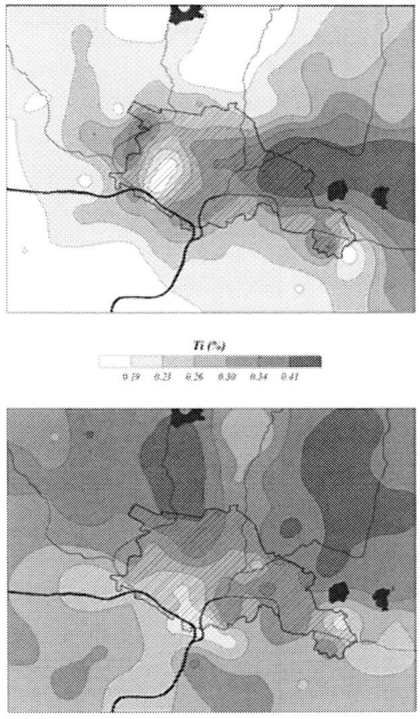

Figure 40. Map of the concentrations of the Ti in attic dust (top) and top soil (bottom; Šajn, 2005) in the Celje area. Levels on map are percentile distribution.

4.5. Heavy Metal Contamination in the Jesenice Region Due to Ironworks

Data for the nonparametric statistical presentation in the Jesenice area has been divided into two groups. First group contains data of the elements concentrations in the top soil (0-5 cm, N=44) where pollution with heavy metals due to ironworks is most clearly visible. Second group of analysis contains geochemical data from different depths from same sampling points where top soil has been collected. Because of smaller vertical mobility of the sedimented airborne particles pollution is not so extensive. Attic dust in Jesenice area has not been sampled. Table 10 shows the nonparametric statistical distribution of the elements for top soil (0-5 cm, N=44) and for the other analyzed samples in different depths (N=78).

Majority of the samples has been taken on carbonate rocks (30 localities), followed by sandstones (6 localities) and alluvial deposits (8 localities). Taking into account pedological soil types 16 sampling profiles has been placed in the rendzinas, 16 cases in calcareous cambisols, 5 cases in dystric cambisols, 2 in ranker, 2 in agricultural soil and 1 in each, fluvisol, luvisol and eutric cambisol (Šajn et al., 1999).

Table 10. Nonparametric statistical distribution of the elements in the top soil (0-5 cm; N=44) and bottom soil (N=78) in the Jesenice area (Šajn et al., 1999). Mat - material; Dis - distribution (Log = log-normal, N = normal-Gaussian); X, Xg - average value; Md - median value; Min - Max - range; P25 - P75 - quartile distribution. Data rounded on two decimal places

	Mat	Dis	X, Xg	Md	Min - Max	P25 - P75
%						
Al	top soil	N	4.1	4.1	0.93 - 7.8	2.4 - 5.2
Al	bottom soil	N	6.2	7.0	0.48 - 11	4.4 - 8.1
Ca	top soil	Log	1.2	1.0	0.19 - 15	0.76 - 2.3
Ca	bottom soil	Log	1.4	1.2	0.080 - 20	0.28 - 9.2
Fe	top soil	N	3.1	2.9	0.88 - 8.3	1.9 - 3.8
Fe	bottom soil	N	3.7	4.0	0.24 - 6.3	2.9 - 4.7
K	top soil	Log	0.56	0.54	0.13 - 2.1	0.36 - 0.93
K	bottom soil	N	1.1	1.1	0.050 - 2.6	0.64 - 1.6
Mg	top soil	Log	0.65	0.46	0.15 - 8.8	0.35 - 0.89
Mg	bottom soil	Log	1.8	1.3	0.24 - 15	0.69 - 6.5
Na	top soil	N	0.22	0.20	0.050 - 0.52	0.10 - 0.31
Na	bottom soil	N	0.36	0.33	0.020 - 0.89	0.14 - 0.56
P	top soil	N	0.070	0.070	0.040 - 0.12	0.057 - 0.076
P	bottom soil	N	0.040	0.040	0.010 - 0.12	0.031 - 0.060
Ti	top soil	N	0.17	0.18	0.040 - 0.35	0.10 - 0.22
Ti	bottom soil	N	0.25	0.26	0.010 - 0.47	0.16 - 0.33
mg/kg						
As	top soil	Log	14	16	4.0 - 76	8.5 - 21
As	bottom soil	N	19	18	4.0 - 61	10 - 25
Ba	top soil	Log	180	180	51 - 610	120 - 260
Ba	bottom soil	N	250	250	10 - 680	140 - 350
Cd	top soil	Log	2.1	2.1	0.70 - 8.7	1.3 - 3.1
Cd	bottom soil	Log	0.88	0.75	0.40 - 4.0	0.40 - 1.5
Co	top soil	Log	8.3	8.0	2.0 - 29	5.5 - 12
Co	bottom soil	N	14	15	3.0 - 29	9.0 - 18
Cr	top soil	Log	50	52	14 - 160	37 - 67
Cr	bottom soil	N	58	60	4.0 - 130	40 - 72
Cu	top soil	Log	28	26	12 - 96	20 - 36
Cu	bottom soil	Log	16	16	3.0 - 45	12 - 24
La	top soil	N	17	18	2.0 - 40	9.0 - 23
La	bottom soil	N	24	25	2.0 - 64	12 - 32
Mn	top soil	Log	780	770	86 - 3400	530 - 1600
Mn	bottom soil	Log	560	590	55 - 1900	340 - 1100
Mo	top soil	Log	2.8	2.0	2.0 - 10	2.0 - 3.5
Mo	bottom soil	Log	2.5	2.0	2.0 - 13	2.0 - 2.0
Nb	top soil	Log	6.3	7.0	2.0 - 26	4.0 - 9.0
Nb	bottom soil	N	12	11	3.0 - 57	7.0 - 15
Ni	top soil	Log	25	27	5.0 - 85	19 - 37
Ni	bottom soil	Log	29	34	4.0 - 110	23 - 43
Pb	top soil	Log	290	290	83 - 1900	170 - 400
Pb	bottom soil	Log	43	44	4.0 - 820	31 - 60
Sc	top soil	N	8.1	8.0	2.0 - 18	5.0 - 10
Sc	bottom soil	N	12	13	1.0 - 29	9.0 - 16
Sr	top soil	Log	84	68	20 - 830	45 - 130
Sr	bottom soil	Log	120	110	37 - 840	70 - 170
Th	top soil	N	5.6	5.5	2.0 - 11	3.0 - 7.0
Th	bottom soil	N	8.4	9.0	2.0 - 17	6.0 - 11
V	top soil	N	55	57	13 - 96	40 - 72
V	bottom soil	N	78	80	6.0 - 130	62 - 100

Table 10. (Continued)

	Mat	Dis	X, Xg	Md	Min - Max	P25 - P75
mg/kg						
W	top soil	Log	13	13	2.0 - 260	6.5 - 22
W	bottom soil	Log	29	28	7.0 - 150	17 - 57
Y	top soil	N	10	10	3.0 - 29	7.0 - 13
Y	bottom soil	Log	13	13	3.0 - 73	8.0 - 18
Zn	top soil	Log	290	290	57 - 1500	180 - 470
Zn	bottom soil	Log	97	110	6.0 - 560	74 - 140
Zr	top soil	N	36	32	10 - 70	24 - 48
Zr	bottom soil	N	47	47	4.0 - 130	36 - 58
Hg	top soil	Log	0.77	0.80	0.35 - 2.5	0.55 - 1.0
Hg	bottom soil	Log	0.24	0.26	0.042 - 1.2	0.16 - 0.39

Factor analysis has been performed on the basis of all available soil geochemical composition data (N=122; Šajn et al., 1999). 4 factors representing 4 geochemical associations have been extracted. Table 11 shows the characteristic factor loadings for different elements. Na, P, As, Co, Mo, Sr, W and Y has been excluded due to lack of connectivity with other elements.

Table 11. Characteristic factor loadings obtained from the geochemical composition of soil profiles (N=122) on 44 localities around Jesenice ironworks (Šajn et al., 1999). F1-F4 - characteristic factor loadings for each factor; Comm - explained variance of the element concentrations by factor analysis (%), Var - total variance explained by each factor (%).

	F1	F2	F3	F4	Kom
Al	0.98				99
Sc	0.95				94
Ti	0.95				93
Th	0.93				90
V	0.90				84
La	0.87				79
Zr	0.79				66
Ba	0.76				60
K	0.72				55
Nb	0.71				54
Zn		0.97			97
Cu		0.92			88
Pb		0.92			88
Hg		0.90			85
Cd		0.88			81
Mn		0.77			62
Cr			0.95		94
Fe			0.93		90
Ni			0.92		88
Ca				0.95	94
Mg				0.94	91
Var	34	23	13	12	82

Spatial distributions of factor scores in Jesenice area for each factor are shown in figures from 41 to 44. Separate maps for Al, Cr and Pb concentrations in top soil and bottom soil has been made and are presented in figures from 45 to 47 where Al represents natural distribution and Cr and Pb man-made pollution.

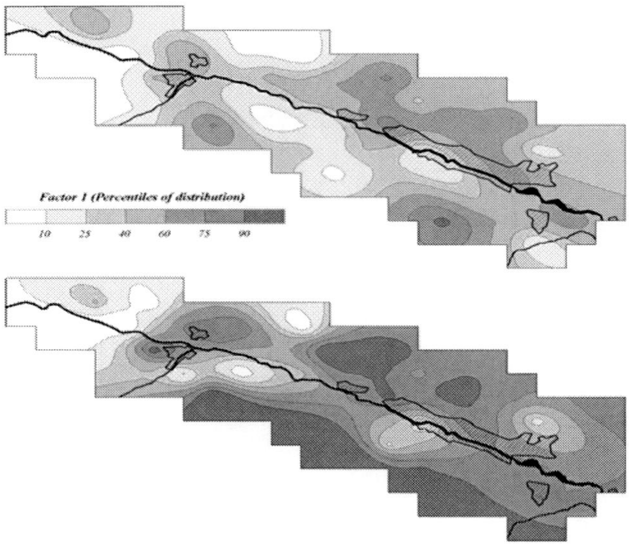

Figure 41. Spatial distribution of factor 1 scores in top soil (up) and bottom soil (bottom) in Jesenice area (renewed from Šajn et al., 1999). Characteristic elements for factor 1 are Al, Sc, Ti, Th, V, La, Zr, Ba, K and Nb. Listed elements are ordered on the basis of the strengths of the connection with factor 1 (from 0.98 to 0.71).

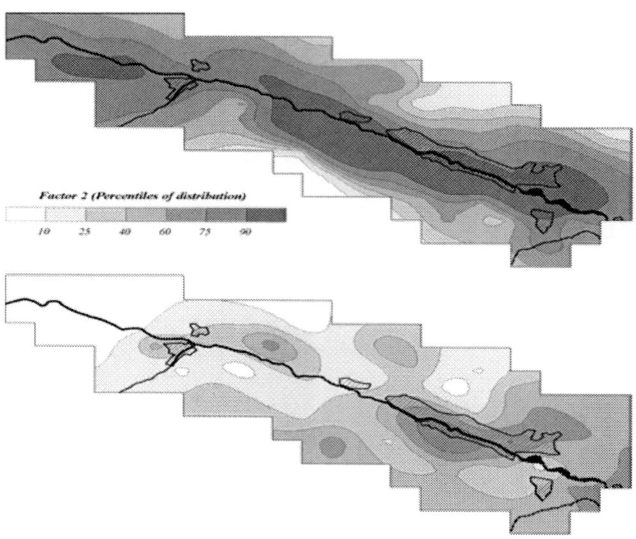

Figure 42. Spatial distribution of factor 2 scores in top soil (up) and bottom soil (bottom) in Jesenice area (renewed from Šajn et al., 1999). Characteristic elements for factor 2 are Zn, Cu, Pb, Hg, Cd and Mn. Listed elements are ordered on the basis of the strengths of the connection with factor 2 (from 0.97 to 0.77).

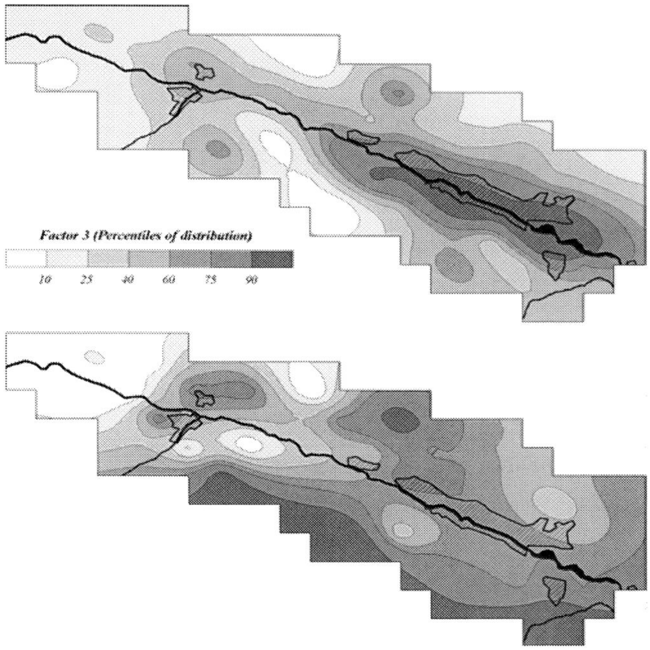

Figure 43: Spatial distribution of factor 3 scores in top soil (up) and bottom soil (bottom) in Jesenice area (renewed from Šajn et al., 1999). Characteristic elements for factor 3 are Cr, Fe and Ni. Listed elements are ordered on the basis of the strengths of the connection with factor 3 (from 0.95 to 0.92).

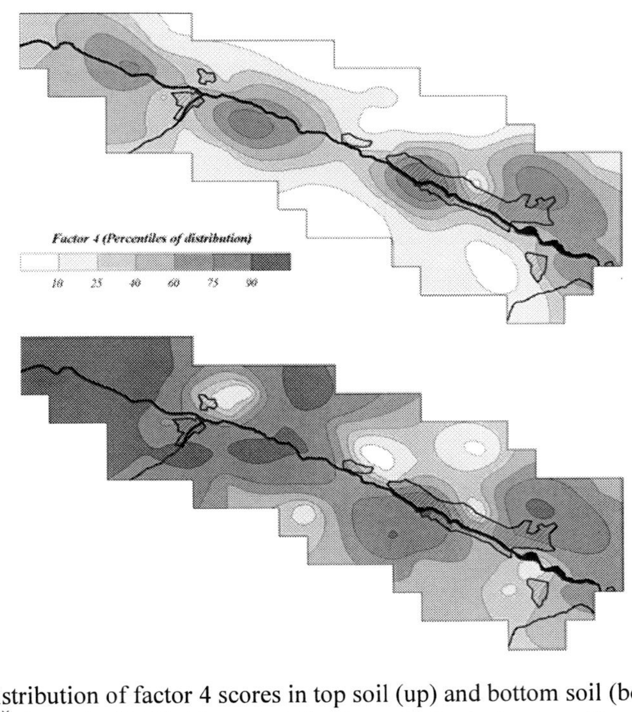

Figure 44. Spatial distribution of factor 4 scores in top soil (up) and bottom soil (bottom) in Jesenice area (renewed from Šajn et al., 1999). Characteristic elements for factor 4 are Ca and Mg. Listed elements are ordered on the basis of the strengths of the connection with factor 4 (from 0.95 to 0.94).

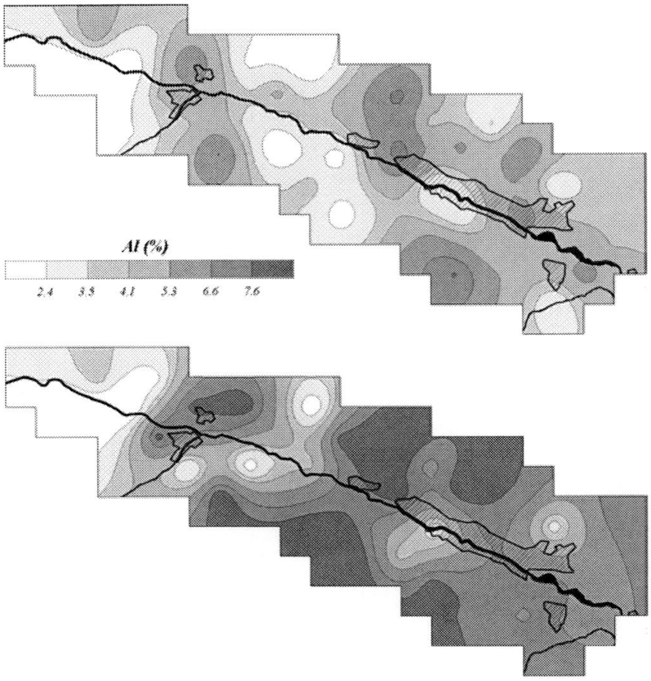

Figure 45. Map of the concentrations of the Al in top soil (up) and bottom soil (bottom) in the Jesenice area. Levels on map are percentile distribution.

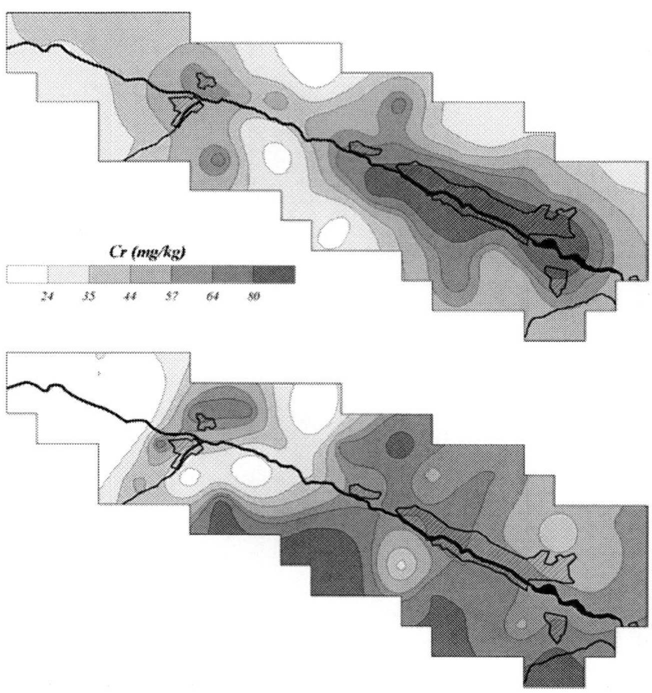

Figure 46. Map of the concentrations of the Cr in top soil (up) and bottom soil (bottom) in the Jesenice area. Levels on map are percentile distribution.

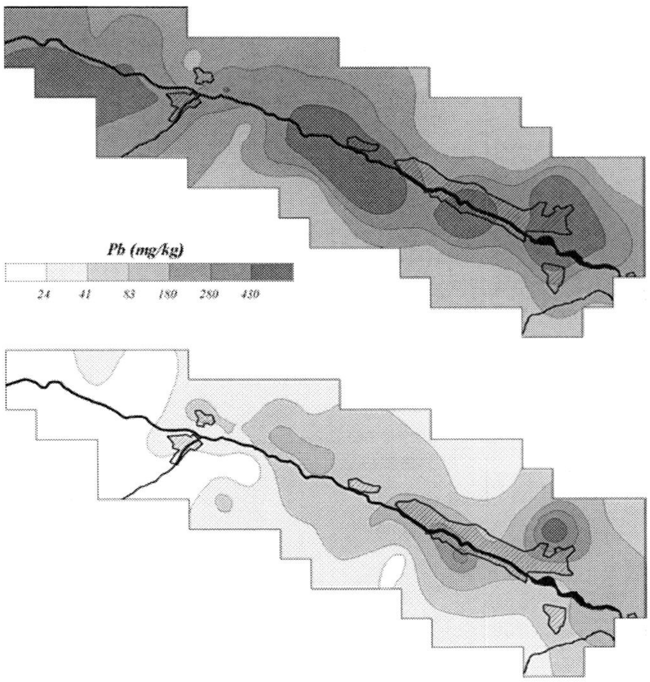

Figure 47. Map of the concentrations of the Pb in top soil (up) and bottom soil (bottom) in the Jesenice area. Levels on map are percentile distribution.

4.6. Heavy Metal Contamination in the Litija Region Due to Polymetallic Mining and Smelting

Nonparametric statistical distributions of element concentrations in the Litija area in top soil and attic dust are presented in table 12. Dataset is taken from the work of Jemec (2006) and Jemec & Šajn (2007).

Table 12. Nonparametric statistical distribution of the elements in the top soil (0-5 cm; N=38) and attic dust (N=38) in the Litija area (Jemec, 2006; Šajn & Jemec, 2007). Mat - material; Dis - distribution (Log = log-normal, N = normal-Gaussian); X, Xg - average value; Md - median value; Min - Max - range; P25 - P75 - quartile distribution. Data rounded on two decimal places

	Mat	Dis	X,Xg	Md	Min - Max	P25 - P75
%						
Al	top soil	Log	5.3	5.3	3.0 - 9.9	4.7 - 6.3
Al	attic dust	Log	2.6	2.6	1.7 - 3.9	2.3 - 3.2
Ca	top soil	Log	2.3	1.8	0.20 - 15	1.1 - 6.0
Ca	attic dust	N	9.6	10	4.9 - 14	7.5 - 11
Fe	top soil	Log	2.6	2.6	1.5 - 4.7	2.2 - 2.9
Fe	attic dust	Log	2.1	2.0	1.2 - 4.2	1.7 - 2.6
K	top soil	Log	1.4	1.5	0.65 - 2.7	1.2 - 1.8
K	attic dust	Log	0.88	0.82	0.53 - 1.4	0.71 - 1.2
Mg	top soil	Log	1.3	1.1	0.35 - 6.7	0.70 - 2.3
Mg	attic dust	Log	1.7	1.6	0.56 - 4.6	1.1 - 2.5

Table 12. (Continued)

	Mat	Dis	X,Xg	Md	Min - Max	P25 - P75
%						
Na	top soil	N	0.42	0.41	0.13 - 0.68	0.29 - 0.54
Na	attic dust	Log	0.28	0.29	0.15 - 0.52	0.23 - 0.34
P	top soil	Log	0.12	0.12	0.059 - 0.32	0.089 - 0.17
P	attic dust	Log	0.20	0.19	0.066 - 0.43	0.16 - 0.28
S	top soil	N	0.19	0.20	0.050 - 0.50	0.050 - 0.30
S	attic dust	N	4.6	4.8	2.1 - 7.8	3.5 - 5.5
Ti	top soil	Log	0.27	0.27	0.15 - 0.47	0.24 - 0.33
Ti	attic dust	Log	0.15	0.15	0.10 - 0.28	0.13 - 0.18
mg/kg						
Ag	top soil	Log	0.13	0.10	0.050 - 1.1	0.10 - 0.20
Ag	attic dust	Log	0.37	0.30	0.20 - 1.5	0.20 - 0.50
As	top soil	Log	13	13	5.0 - 31	10 - 17
As	attic dust	Log	27	25	6.0 - 260	15 - 37
Ba	top soil	Log	550	510	160 - 1600	430 - 780
Ba	attic dust	Log	86	88	29 - 220	63 - 120
Bi	top soil	Log	0.29	0.30	0.20 - 0.70	0.20 - 0.40
Bi	attic dust	Log	0.63	0.60	0.30 - 13	0.40 - 0.70
Cd	top soil	Log	0.67	0.60	0.30 - 2.5	0.50 - 0.90
Cd	attic dust	Log	2.4	2.1	0.80 - 13	1.4 - 4.0
Ce	top soil	Log	57	58	33 - 100	50 - 69
Ce	attic dust	Log	31	30	19 - 54	26 - 38
Co	top soil	Log	10	11	6.0 - 23	8.0 - 12
Co	attic dust	Log	6.0	6.0	3.0 - 12	5.0 - 8.0
Cr	top soil	Log	62	67	37 - 110	51 - 74
Cr	attic dust	Log	52	49	32 - 240	42 - 62
Cu	top soil	Log	33	28	14 - 210	22 - 40
Cu	attic dust	Log	85	78	35 - 290	64 - 100
Hf	top soil	Log	1.1	1.0	0.70 - 2.1	0.90 - 1.2
Hf	attic dust	Log	0.77	0.70	0.40 - 2.4	0.60 - 0.90
La	top soil	Log	27	28	15 - 48	23 - 31
La	attic dust	N	16	14	4.6 - 29	13 - 18
Li	top soil	Log	40	39	25 - 67	34 - 48
Li	attic dust	Log	24	24	17 - 40	19 - 29
Mn	top soil	Log	810	780	400 - 1800	640 - 1000
Mn	attic dust	Log	510	490	300 - 1100	430 - 600
Mo	top soil	Log	0.83	0.80	0.30 - 4.1	0.60 - 1.0
Mo	attic dust	Log	2.7	2.5	1.3 - 8.4	1.7 - 3.4
Nb	top soil	Log	6.5	6.4	3.9 - 11	5.7 - 7.9
Nb	attic dust	Log	3.8	3.7	2.4 - 6.5	3.3 - 4.3
Ni	top soil	Log	22	22	12 - 38	17 - 27
Ni	attic dust	Log	25	24	13 - 57	20 - 31
Pb	top soil	Log	150	150	40 - 5300	79 - 230
Pb	attic dust	Log	450	390	120 - 5500	240 - 670
Rb	top soil	Log	89	93	44 - 180	74 - 110
Rb	attic dust	Log	48	47	32 - 74	40 - 57
Sb	top soil	Log	2.0	1.7	0.90 - 19	1.2 - 2.8
Sb	attic dust	Log	7.2	6.0	2.7 - 52	4.0 - 10
Sc	top soil	Log	8.1	8.0	5.0 - 14	7.0 - 10
Sc	attic dust	Log	4.4	4.0	2.0 - 9.0	4.0 - 5.0
Sn	top soil	Log	4.9	4.3	2.3 - 19	3.6 - 6.4
Sn	attic dust	Log	15	15	5.8 - 71	9.6 - 25
Sr	top soil	Log	84	78	39 - 270	68 - 100
Sr	attic dust	Log	150	150	91 - 350	130 - 170

	Mat	Dis	X,Xg	Md	Min - Max	P25 - P75
mg/kg						
Ta	top soil	N	0.51	0.50	0.30 - 0.80	0.40 - 0.60
Ta	attic dust	N	0.30	0.30	0.20 - 0.50	0.20 - 0.30
Th	top soil	N	9.7	9.9	5.5 - 17	7.7 - 11
Th	attic dust	Log	4.9	4.8	3.1 - 7.6	4.1 - 5.6
U	top soil	Log	2.7	2.7	1.5 - 6.0	2.2 - 3.0
U	attic dust	Log	2.6	2.3	1.6 - 8.1	2.0 - 3.0
V	top soil	Log	68	67	35 - 120	58 - 80
V	attic dust	N	64	65	30 - 110	44 - 84
W	top soil	Log	1.3	1.3	0.70 - 2.6	1.1 - 1.6
W	attic dust	Log	1.8	1.7	0.90 - 7.8	1.3 - 2.0
Y	top soil	Log	10	10	6.3 - 21	8.1 - 12
Y	attic dust	Log	8.3	7.6	4.9 - 21	6.2 - 9.3
Zn	top soil	Log	200	180	76 - 1200	120 - 240
Zn	attic dust	Log	520	450	200 - 3900	290 - 720
Zr	top soil	Log	29	28	19 - 58	26 - 32
Zr	attic dust	Log	21	20	13 - 69	17 - 24
Hg	top soil	Log	0.43	0.35	0.13 - 21	0.21 - 0.63
Hg	attic dust	Log	1.8	1.8	0.36 - 15	1.3 - 2.7

Correlation coefficients between chemical composition of top soil and attic dust are presented in figure 48.

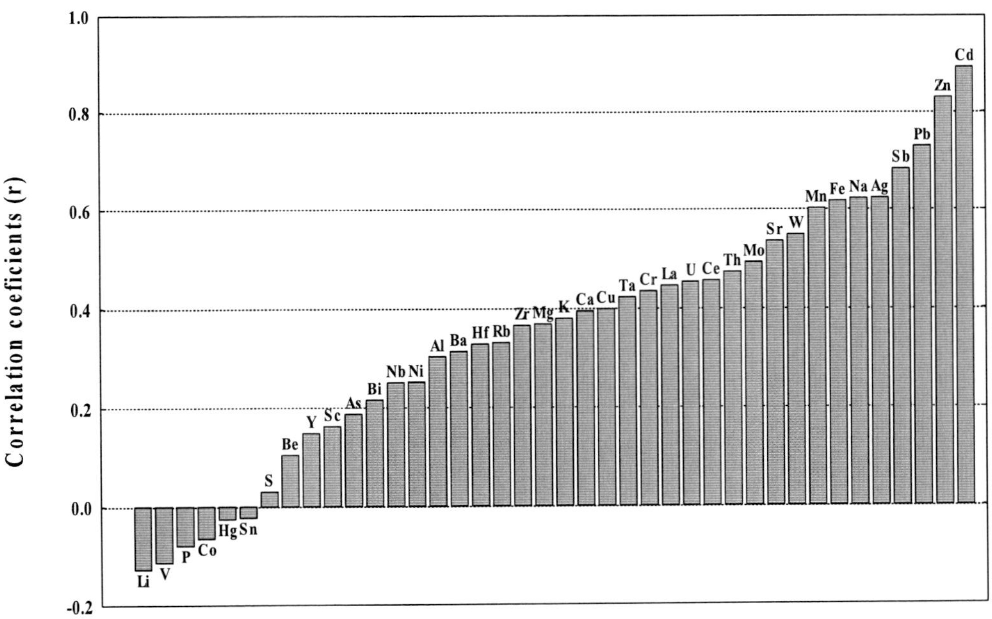

Figure 48. Correlation coefficients between elements concentrations in the top soil (0-5 cm) and attic dust (N=38) based on logarithmic data in the Litija area.

Factor analysis has revealed 4 geochemical associations (table 13). Ca, Fe, Mg, Na, P, S, Ag, Ba, Bi, Cu, Mn, Mo, Rb, Sr, V and W have been excluded. Only 1 association of elements is connected with mining and smelting activities, all other are probably of natural origin.

Table 13. Characteristic factor loadings (abs(r^2)>0.5) obtained from the geochemical data of top soil and attic dust in the Litija area (N=76; Jemec & Šajn, 2007, renewed). F1-F4 - characteristic factor loadings for each factor; Comm - explained variance of the element concentrations by factor analysis (%), Var - total variance explained by each factor (%)

	F1	F2	F3	F4	Comm
Rb	0.92				93.6
Al	0.91				95.9
K	0.90				89.4
Th	0.88				96.0
Ti	0.87				94.2
Li	0.85				84.8
Nb	0.85				93.7
Ce	0.84				89.9
Ta	0.82				84.9
Sc	0.81				88.9
La	0.77				83.5
Co	0.69				79.1
Pb		0.93			90.0
Sb		0.89			93.6
As		0.82			73.8
Hg		0.81			80.5
Cd		0.77			87.2
Zn		0.69			74.2
Sn		0.67			74.0
Zr			0.87		92.6
Y			0.86		80.7
Hf			0.85		89.1
U			0.84		83.0
Ni				0.80	83.6
Cr				0.73	72.5
Var	39.66	22.83	16.42	7.05	86.0

Geochemical data for enrichment factor calculation has been divided into two data sets. Fist set represents samples, collected in the surroundings of the Litija, second set samples, collected in the town of Litija, where smeltery has been operational. Group of elements has been chosen on the basis of factor analysis. Figure 49 shows the average enrichment factors calculated on the basis on the averages of concentration of elements in the Slovenian unpolluted soils.

Sampling plan in Litija area has not been designed in grid pattern and this is the reason that aerial distribution of different factors and element concentration has not been presented.

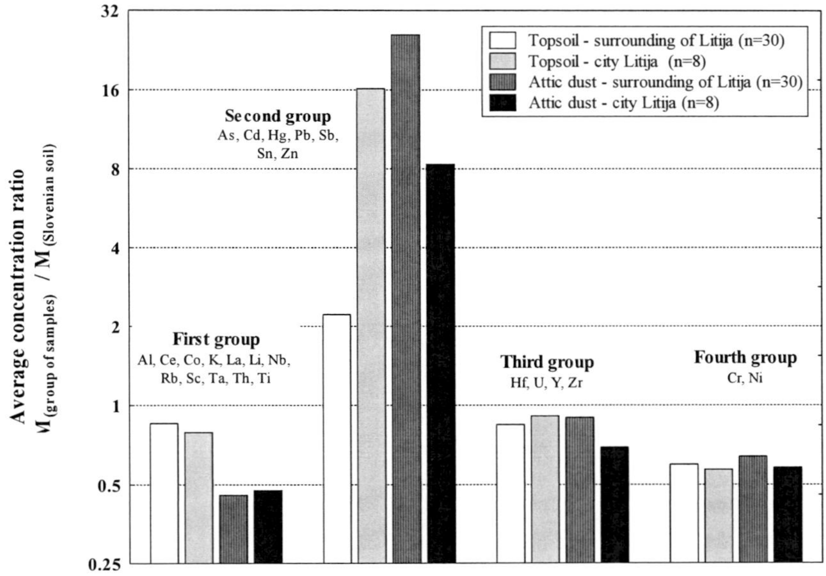

Figure 49. Average enrichment factors of elements in different media comparing to top soils in Slovenian rural areas. The elements are grouped on the basis of factor analysis (table 13). Geochemical data for enrichment factors calculation have been divided into two groups: Litija area where smeltery has been located and surroundings of Litija where impact of the smeltery has not been expected to be so extensive

4.7. Heavy metal contamination in the Idrija region due to Hg mining and smelting

Statistical analysis have been made on the basis of the data from 103 sampling locations. Soil has been sampled to the depth of 15 cm and possible organic horizon has been excluded. Attic dust has been taken in old houses (>100 years old) in the vicinity where soil has been collected. In total 206 samples of soils (0-15 cm) and attic dust has been taken. Table 14 shows the basic nonparametric statistics.

Table 14. Nonparametric statistical distribution of the elements in the top soil (0-15 cm; N=103) and attic dust (N=103) in the Idrija area (Gosar & Šajn, 2001 with supplemental data). Mat - material; Dis - distribution (Log = log-normal, N = normal-Gaussian); X, Xg - average value; Md - median value; Min - Max - range; P25 - P75 - quartile distribution. Data rounded on two decimal places.

	Mat	Dis	X,Xg	Md	Min - Max	P25 - P75
%						
Al	soil	N	6.8	7.0	3.4 - 9.6	6.1 - 7.6
Al	attic dust	Log	2.1	2.1	0.88 - 4.4	1.8 - 2.6
Ca	soil	Log	1.4	1.1	0.090 - 13	0.53 - 3.6
Ca	attic dust	N	7.6	7.6	1.9 - 15	4.8 - 10
Fe	soil	Log	3.5	3.5	1.5 - 7.7	3.0 - 4.2
Fe	attic dust	Log	2.2	2.0	0.85 - 29	1.4 - 3.2

Table 14. (Continued)

	Mat	Dis	X,Xg	Md	Min - Max	P25 - P75
%						
K	soil	Log	2.0	2.0	0.91 - 5.5	1.6 - 2.5
K	attic dust	Log	1.0	0.94	0.47 - 2.5	0.73 - 1.1
Mg	soil	Log	1.6	1.3	0.28 - 6.7	0.92 - 2.6
Mg	attic dust	Log	2.2	2.3	0.45 - 6.8	1.5 - 3.3
Na	soil	Log	0.37	0.39	0.097 - 0.85	0.25 - 0.53
Na	attic dust	Log	0.22	0.22	0.11 - 0.70	0.17 - 0.27
P	soil	Log	0.11	0.11	0.044 - 0.36	0.085 - 0.14
P	attic dust	Log	0.30	0.30	0.078 - 1.4	0.22 - 0.42
S	soil	Log	0.058	0.060	0.010 - 0.16	0.040 - 0.080
S	attic dust	Log	1.9	2.0	0.45 - 9.2	1.3 - 3.2
Ti	soil	N	0.31	0.31	0.13 - 0.55	0.26 - 0.36
Ti	attic dust	Log	0.10	0.10	0.034 - 0.42	0.085 - 0.13
mg/kg						
Ag	soil	Log	0.11	0.10	0.10 - 1.3	0.10 - 0.10
Ag	attic dust	Log	0.29	0.30	0.10 - 3.4	0.20 - 0.50
As	soil	Log	20	20	7.0 - 100	14 - 26
As	attic dust	Log	10	11	1.0 - 140	7.0 - 15
Ba	soil	N	330	330	80 - 840	280 - 400
Ba	attic dust	Log	97	90	18 - 440	68 - 150
Be	soil	Log	2.1	2.0	1.0 - 5.0	2.0 - 3.0
Be	attic dust	N	0.80	1.0	0.50 - 1.5	0.50 - 1.0
Cd	soil	Log	0.53	0.50	0.10 - 4.6	0.30 - 0.90
Cd	attic dust	Log	1.8	1.9	0.40 - 16	1.2 - 2.5
Ce	soil	N	63	63	24 - 130	51 - 75
Ce	attic dust	N	20	19	3.0 - 51	14 - 25
Co	soil	Log	13	14	2.0 - 65	11 - 18
Co	attic dust	Log	5.5	6.0	2.0 - 48	4.0 - 7.0
Cr	soil	Log	73	76	21 - 180	62 - 90
Cr	attic dust	Log	46	47	19 - 320	36 - 57
Cu	soil	Log	31	29	9.0 - 240	25 - 35
Cu	attic dust	Log	79	78	23 - 2000	44 - 120
Hf	soil	Log	1.4	1.0	0.50 - 4.0	1.0 - 2.0
Hf	attic dust	Log	0.69	0.50	0.50 - 2.0	0.50 - 1.0
La	soil	N	33	34	15 - 72	25 - 40
La	attic dust	Log	9.6	10	3.0 - 26	8.0 - 13
Li	soil	Log	47	45	23 - 230	36 - 61
Li	attic dust	N	16	16	6.0 - 28	13 - 20
Mn	soil	Log	990	960	250 - 5800	810 - 1300
Mn	attic dust	Log	500	450	150 - 3000	360 - 620
Mo	soil	Log	1.7	1.6	0.25 - 15	1.0 - 2.7
Mo	attic dust	Log	2.8	2.8	0.60 - 16	1.8 - 4.1
Nb	soil	Log	9.4	9.4	5.8 - 15	8.0 - 11
Nb	attic dust	N	3.8	3.7	1.0 - 7.5	2.9 - 4.3
Ni	soil	Log	33	33	10 - 82	27 - 42
Ni	attic dust	Log	30	28	10 - 660	19 - 40
Pb	soil	Log	64	56	36 - 1200	50 - 68
Pb	attic dust	Log	180	170	46 - 28000	96 - 330
Rb	soil	N	110	110	46 - 160	90 - 130
Rb	attic dust	Log	41	42	17 - 88	34 - 48
Sb	soil	Log	1.7	2.0	0.50 - 4.0	1.0 - 2.0
Sb	attic dust	Log	2.9	3.0	0.50 - 30	2.0 - 5.0
Sc	soil	N	11	11	5.0 - 19	10 - 13
Sc	attic dust	N	4.3	4.0	1.0 - 9.0	4.0 - 5.0
Sn	soil	Log	4.2	4.3	0.80 - 19	3.5 - 5.0
Sn	attic dust	Log	24	23	1.5 - 1100	11 - 52

mg/kg	**Mat**	**Dis**	**X,Xg**	**Md**	**Min - Max**	**P25 - P75**
Sr	soil	Log	75	75	32 - 220	58 - 90
Sr	attic dust	Log	120	110	42 - 320	86 - 160
Th	soil	N	12	12	6.0 - 18	10 - 14
Th	attic dust	Log	2.9	3.0	1.0 - 8.0	2.0 - 4.0
U	soil	Log	5.0	5.0	2.0 - 10	4.0 - 7.0
U	attic dust	Log	2.1	2.0	0.50 - 9.0	1.0 - 3.0
V	soil	Log	110	100	40 - 280	88 - 130
V	attic dust	Log	40	38	15 - 200	29 - 51
Y	soil	Log	17	17	5.4 - 68	13 - 21
Y	attic dust	Log	6.9	7.1	1.9 - 18	5.5 - 8.8
Zn	soil	Log	130	130	38 - 630	110 - 160
Zn	attic dust	Log	710	740	140 - 4700	330 - 1600
Zr	soil	Log	53	52	24 - 120	43 - 64
Zr	attic dust	Log	19	20	6.4 - 54	16 - 23
Hg	soil	Log	5.0	3.2	0.26 - 970	0.85 - 26
Hg	attic dust	Log	23	27	0.58 - 1100	6.1 - 84

Correlation coefficients between element concentrations in soil and attic dust are presented in figure 50.

Multivariate factor analysis (table 15) shows four geochemical groups. K, Na, P, S, Ba, Be, Hf, Mo, Sn, Sr, U, Zr and Hg has been excluded. Note that Hg, which is the only polluter, is not connected with any other element. So factor analysis in this case shows only natural associations.

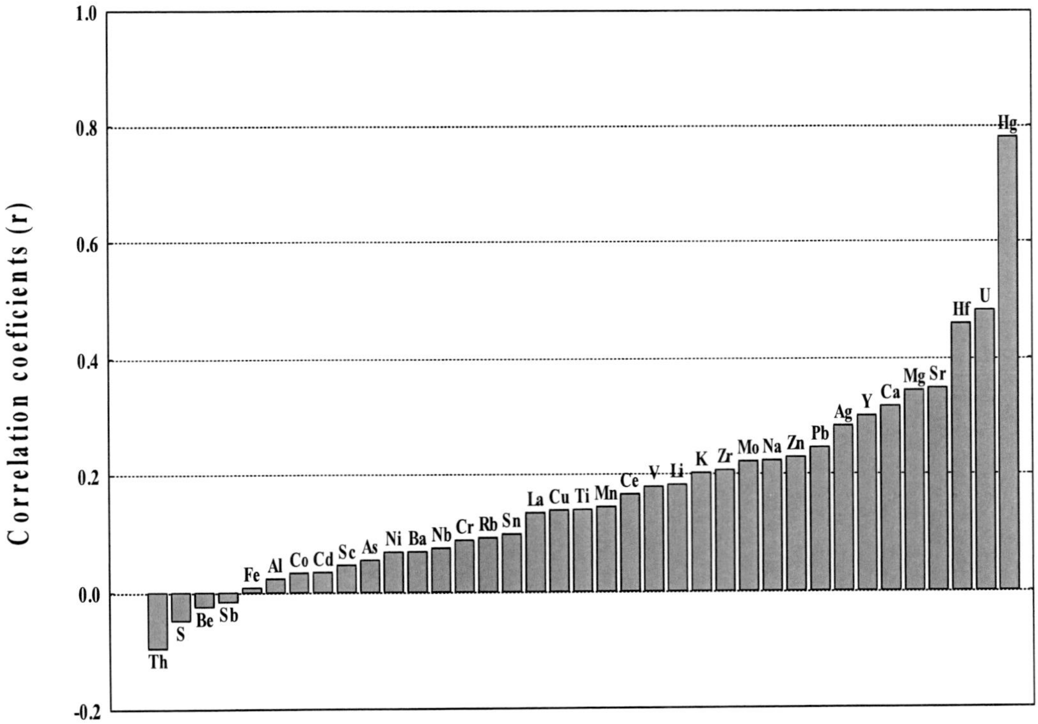

Figure 50. Correlation coefficients between elements concentrations in the soil (0-15 cm) and attic dust (N=103) based on logarithmic data in the Idrija area (Gosar & Šajn, 2001).

Table 15. Characteristic factor loadings (abs(r^2)>0.5) obtained from the geochemical data of top soil and attic dust in the Idrija area (N=206). F1-F4 - characteristic factor loadings for each factor; Comm - explained variance of the element concentrations by factor analysis (%), Var - total variance explained by each factor (%)

	F1	F2	F3	F4	Comm
Ti	0.90				93.1
Nb	0.89				90.3
Sc	0.86				89.6
Al	0.85				95.4
Li	0.84				86.5
V	0.84				82.9
Ce	0.81				89.7
Y	0.81				85.5
Th	0.81				92.0
La	0.80				85.8
Rb	0.72				80.5
Pb		0.86			83.4
Ag		0.80			77.5
Zn		0.73			82.6
Cu		0.68			77.4
Sb		0.66			68.0
Cd		0.57			73.2
Co	0.61		0.68		90.6
Mn	0.57		0.59		71.9
Cr	0.52		0.57		63.5
Fe			0.82		81.5
As			0.78		72.3
Ni			0.77		69.9
Mg				0.88	86.6
Ca				0.77	93.1
Var	39.1	18.6	15.6	9.1	82.5

Scale of environmental pollution with Hg is visible when comparing data with Slovenian background values for soils. Average enrichment factors (figure 51) grew up to 1000, but separate ones can grow up to 17.000 on most polluted areas comparing with Slovenian soils.

Geochemical maps shows the aerial distribution of factor scores (figures from 52 to 55) and aerial distribution of Al (figure 56), which represent natural distribution and aerial distribution of Hg, which represents man-made anomaly (figure 57).

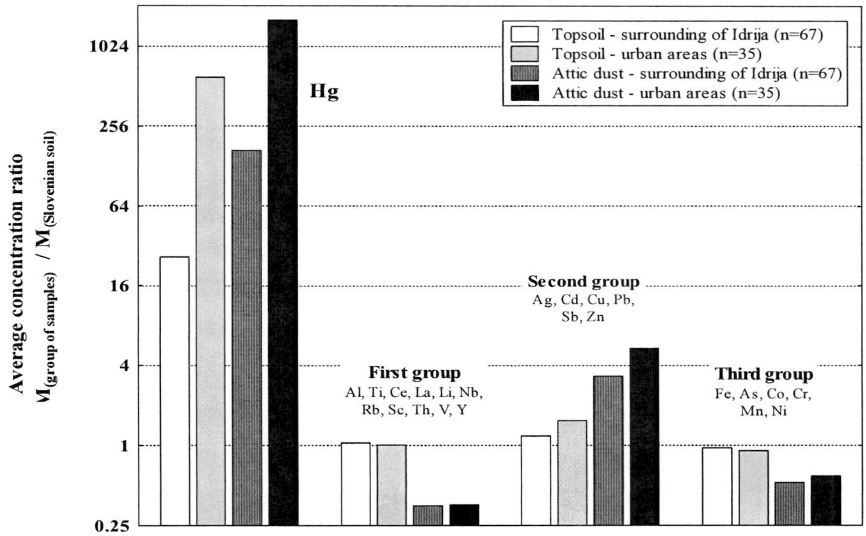

Figure 51. Average enrichment factors of elements in Idrija area and in different media comparing to top soils in Slovenian rural areas. The elements are grouped on the basis of factor analysis (table 15). Geochemical data for enrichment factors calculation have been divided into two groups: Idrija urban area where Hg mines and smeltery has been located and uplands surrounding Idrija.

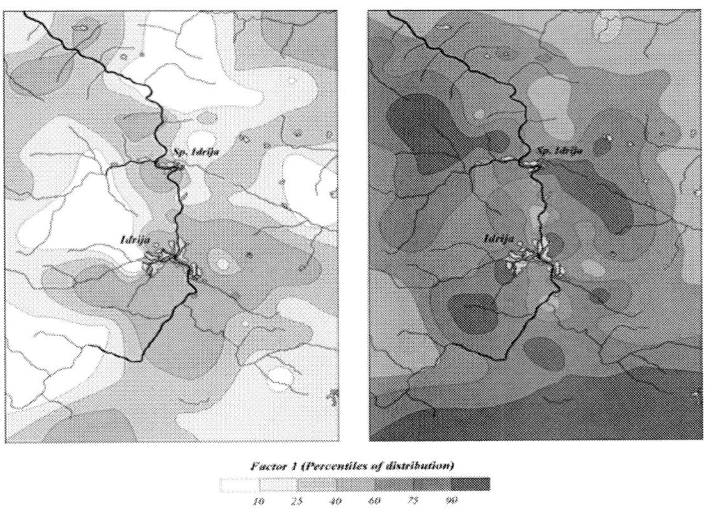

Figure 52. Spatial distribution of factor 1 scores in attic dust (left) and top soil (right) in Idrija area. Characteristic elements for factor 1 are Ti, Nb, Sc, Al, Li, V, Ce, Y, Th, La, Rb, Co, Mn and Cr. Listed elements are ordered on the basis of the strengths of the connection with factor 1 (from 0.90 to 0.52).

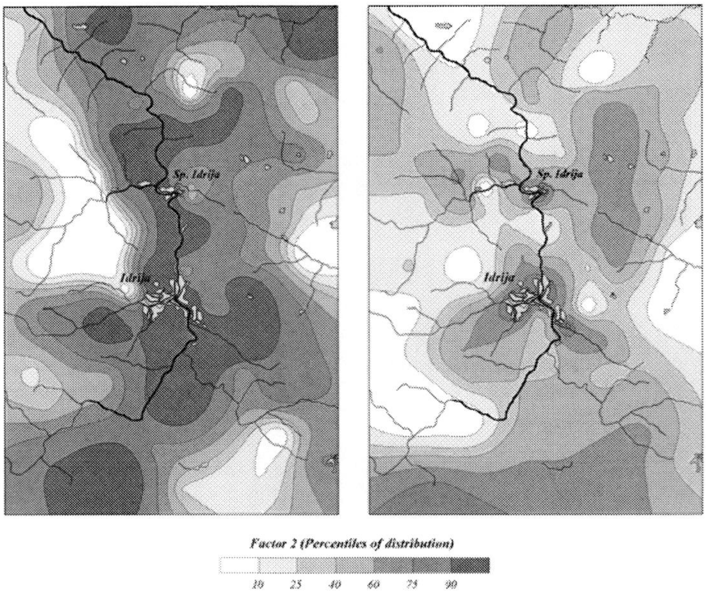

Figure 53. Spatial distribution of factor 2 scores in attic dust (left) and soil (right) in Idrija area. Characteristic elements for factor 2 are Pb, Ag, Zn, Cu, Sb and Cd. Listed elements are ordered on the basis of the strengths of the connection with factor 2 (from 0.86 to 0.57).

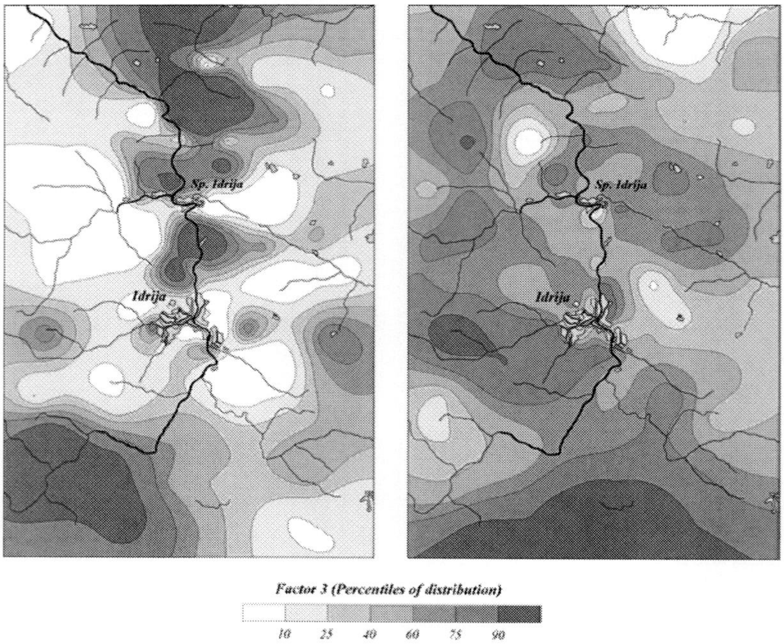

Figure 54. Spatial distribution of factor 3 scores in attic dust (left) and top soil (right) in Idrija area. Characteristic elements for factor 3 are Fe, As, Ni, Co, Mn and Cr. Listed elements are ordered on the basis of the strengths of the connection with factor 3 (from 0.82 to 0.57).

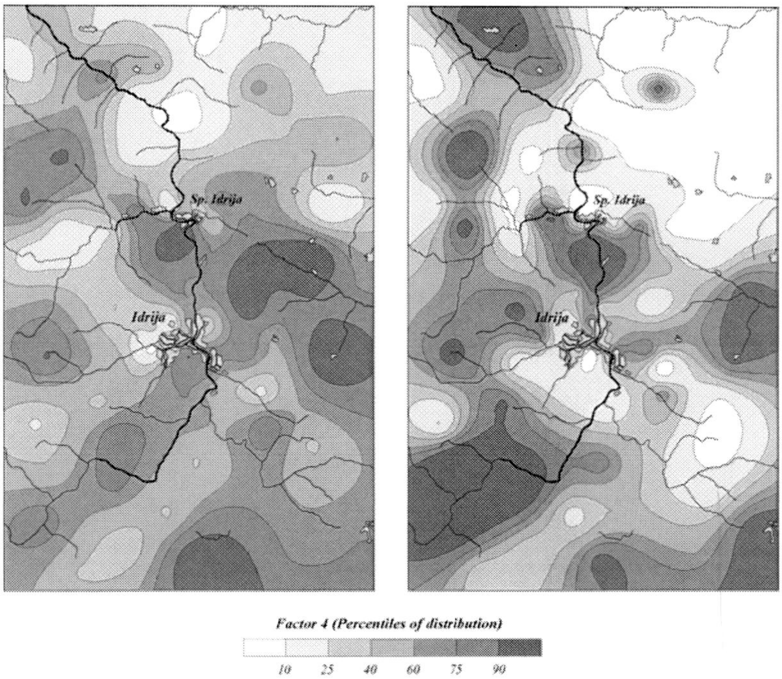

Figure 55: Spatial distribution of factor 4 scores in attic dust (left) and top soil (right) in Idrija area. Characteristic elements for factor 4 are Mg and Ca. Listed elements are ordered on the basis of the strengths of the connection with factor 4 (from 0.88 to 0.77).

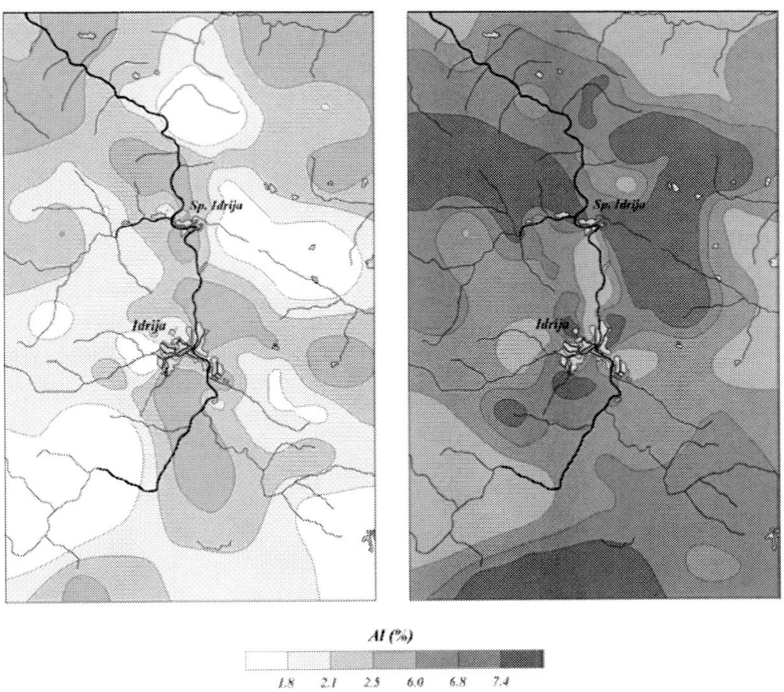

Figure 56. Map of the concentrations of the Al in attic dust (left) and soil (right) in the Idrija area. Levels on map are percentile distribution.

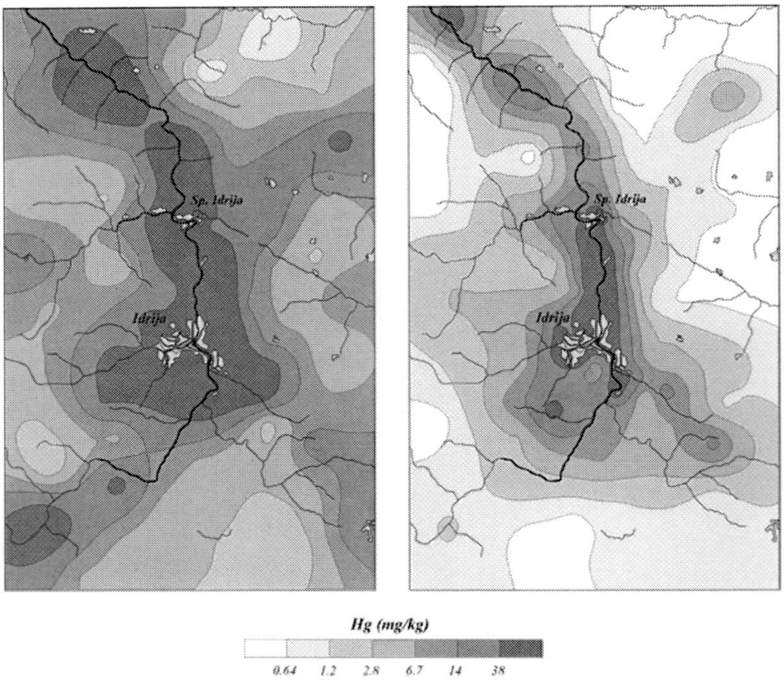

Figure 57. Map of the concentrations of the Hg in attic dust (left) and soil (right) in the Idrija area (Gosar & Šajn, 2001). Levels on map are percentile distribution.

5. Discussion

In last 10 years different geochemical surveys in the Geological survey of Slovenia has been performed by several researchers. They have been focused on the general geochemical composition of soils and attic dust and on the most polluted areas due to metal mining and smelting. Because every author and every research used a bit different methodology for data processing and presentation and also additional sampling in some areas has been performed after the publication of the original results the focus of this paper is the harmonization of data processing and presentation. Also previous articles concerning geochemical research of Slovenian area are successfully dispersed in different Slovene and international journals. Moreover some of the data has not yet been published outside Slovenia. All of this allow direct comparison between natural patterns and different polluted areas in Slovenia, based also on the latest acquired data.

General geochemical survey of Slovenia is based on 60 sampling points where no pollution is expected. Concerning the polluted areas the research has been focused on the areas of dense population (towns) and areas of historical and present smelting and ironworking industry. 5 localities have been investigated: Mežica (Pb-Zn mining, Pb-Zn smeltery and ironworks), Celje (Zn smelter, ironworks, TiO_2 production), Jesenice (ironworks), Litija (Pb-Hg smelting and polymetallic mining) and Idrija (Hg mining and smelting). Several man-made anomalies concerning heavy metal pollution has been determined.

5.1. Comparison of geochemical associations

Factor analysis allows us to group the elements with similar distribution. In all of the research four factors has been used and this number seems to be sufficient for approx. 80% of explained variance. More factors did not yield good results because grouping of elements starts to be unreasonable and also more factors did not contribute much more to explained variance.

In general first, the strongest factor, grouped elements with natural distribution, regardless of possible man-made anomalies. Elements which are most commonly connected with first factor are Al, Co, Ce, La, Nb, Sc, Th and Ti. Also Li and Rb can be appended to the first group. In different areas different litophile elements can be also found to be connected with first factor. Characteristically for this group is low abundance in attic dust comparing to the soils (figures 12, 22, 31, 49 and 51).

Second strongest factor represents chalcophile elements taking into account Goldschmidt classification. Most common elements, connected with factor 2 are Pb, Zn, Cd, Hg, Sb, Cu subordinately also As, Ag and Mo. Aerial distribution of factor 2 scores reveal that most commonly this geochemical association is connected with base metal smelting activities, as in Mežica, Celje and Jesenice case. Where no smelting has been present this factor can possibly be connected with pollution caused population (coal and oil combustion, transportation) as clearly seen in the case of values of the enrichment factors (figure 12) of second group when comparing Slovenian urban and rural areas. A characteristic for these elements is also enrichment in attic dust comparing to the top soil as evident from figures 12, 22, 31 and 51. Litija is an exception (figure 49), possibly because of relative low number of collected samples, irregular sampling plan and low population density.

Third strongest factor represents siderophile elements. Cr (exception, very close to siderophiles), Fe and Ni ware most common representatives, conditionally also Mn and Co. Where the ironworks have been presented the areal distribution of factor scores showed man-made geochemical association, connection with it, as in Mežica, Celje and Jesenice case. Where there is no such activity aerial distribution indicates connectivity with non-carbonatic rock, preferably metamorphic and igneous rocks, tuffs and flysch.

Fourth, weakest factor, groups together different elements in different surveys. In all cases this factor was connected with litophile elements. Most common association was Ca and Mg, possibly connected with poorly developed soils such as rendzinas, situated on limestones and dolomites, quarternary alluvial sediments or sidehill gravel of carbonate origin (Mežica, Jesenice and Litija cases). Ti-Nb association in Celje is most probably connected with man-made anomaly because of 40-year TiO_2 production.

Table 16 summarizes description of geochemical associations

5.2. Main Characteristics of Man-Made Anomalies Because of Metal Mining and Smelting Activities and Ironworks

In the last section a comparison of different man-mad anomalies will be presented. In all cases the "action value" represents the action value, according to the "New Dutchlist" (Swartjes, 1999) and Slovenian legislative values (OG RS, 1996). Where these two values differ it is noted.

Table 16. Most common geochemical associations and their origin

geochemical association no.	1	2	3	4
Goldschmidt's classification	litophile	chalophile	siderophile	litophile
common elements	Th, La, Al, Sc, Nb, Ti, Co, Ce	Pb, Zn, Cd, Hg, Sb, Cu	Ni, Cr, Fe	Ca, Mg
possible other associated elements	Li, Rb	As	Mn, Co	usually alkali and alkaline earth metals
source	natural	population, fossil fuel combustion, smelting	natural, ironworks	natural
occurrence if natural	possible intensive weathering of bedrocks		igneous and metamorphic rocks, tuffs, flysch	rendzinas or other types of undeveloped soils on carbonates, alluvium or sidehill gravel of carbonate origin

Main characteristic for man-made anomaly due to metal smelting is that the atmosphere was the carrier of the heavy metal bearing particles to the place of observation and that the pollution is extensive. Attic dust of old houses on these areas has been sampled for the reason to collect dry atmospheric deposit and compare its composition with the equivalent concentrations in the soils, collected on same location. At elements where the source of pollution in soils has been the sedimentation of airborne particles it is commonly observed a very high correlation (mostly >0.7) between chemical composition of soils and attic dust in comparison to other elements. This is most clearly evident in the case of Hg pollution in Idrija area (figure 50) and also from Mežica case (Pb, figure 21), Celje case (Cd & Zn, figure 30) and Litija case (Cd, Zn, Pb, figure 48).

Other feature can be observed that the concentrations of pollutants in attic dust are approximately 10 times higher than in soil (figure 58). This is also the reason that the enrichment factors of elements, bearers of the pollution, are higher in attic dust than in soils. We can summarize that if in some area correlation coefficient of elements concentration in soil and attic dust is greater than 0.7 this indicates contamination of atmosphere with this particular element. Two other conditions has to be fulfilled before acceptance of such conclusion: high concentration of specific heavy metal in soil in center of pollution and sampling conducted until low concentrations similar to geochemical background levels are reached. Such condition can be observed in Litija and Idrija area. Correlation coeifficient for Hg concentration in soil and attic dust in Litija (low-scale Hg smelting, low Hg pollution; see also figure 60) is close to 0, but in Idrija (high Hg pollution, large-scale smelting operations; figure 60) it is high (0.79).

When comparing the top and bottom soil conclusion can be made that heavy metals are low-mobile materials and that the level of pollution decrease in relation to depth. Top layer (first 5 cm) is most contaminated. This is most evident in Jesenice case (figure 59) where several soil horizonts has been sampled. This case is valid only at the case where the cause of heavy metal pollution was the emissions of particles in atmosphere. That is why similar trends are not evident when comparing top and bottom layer of soil in Slovenian non-polluted areas.

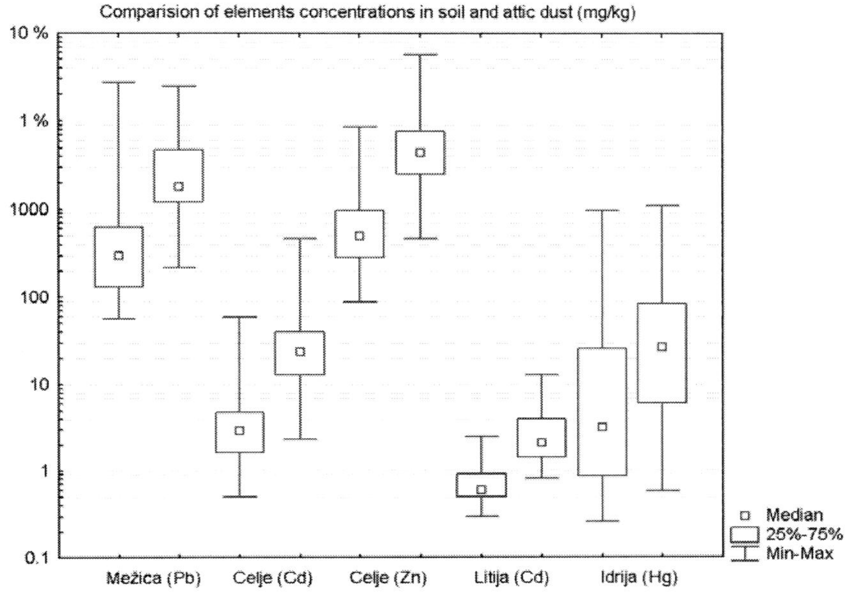

Figure 58. comparison of concentration in soil (left box) and attic dust (right box) of selected heavy metals in the cases of heavy contamination by atmospheric emissions from smelters.

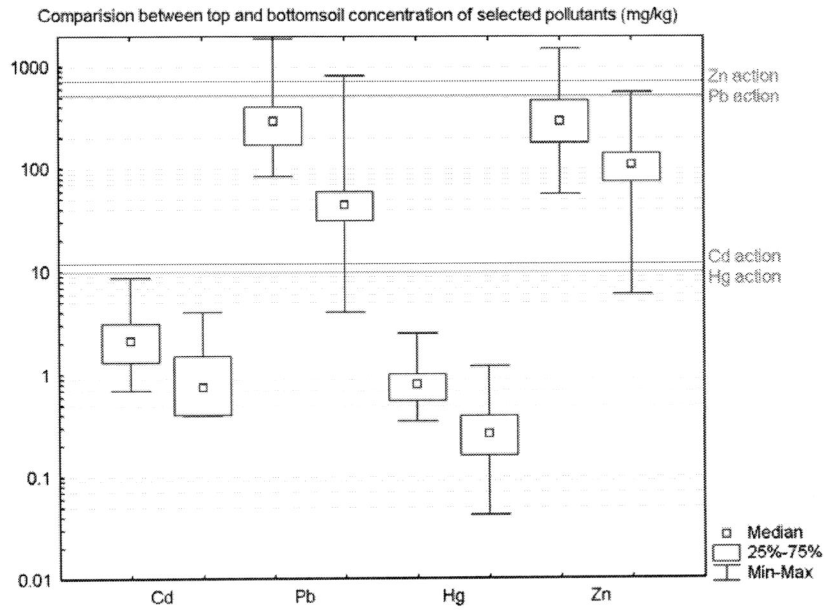

Figure 59. comparison of concentration of selected pollutants in top (0-5 cm; left box) and bottom soil (>20 cm; right box) in Jesenice area. Action values according to the New Dutchlist are marked.

5.3. Description of most intensive anomalies

5.3.1. Hg anomaly in Idrija

This anomaly has been made by more than 500-year operational period of world second largest Hg mine in Idrija where approximately 107.000 tons of Hg has been produced. Several mechanisms of Hg dispersion in the environment were present, but smelting activity was most intensive one when addressing Hg pollution in the vicinity of Idrija. Characteristic for this anomaly is that Hg is the only polluter in the area. No other elements found to be problematic or to have similar spatial distribution of concentrations in soil and attic dust. When addressing factor analysis (table 15) Hg is not included in any of the factors. It has its very own distribution.

In Idrija area it is characteristic that Hg is the only element which expresses strong correlation between Hg concentration in soil and attic dust (figure 48). Also average enrichment factors according to Slovenian background levels (average values in table 3) for Hg are greater than 100, for Hg concentration in attic dust in Idrija town are even greater than 1000 (figure 51). The extensiveness of Hg pollution is evident when comparing to other polluted and non-polluted areas in Slovenia (figure 60). Approximately 30% of the soil samples contained Hg concentration greater than the critical (OG RS, 1996) values (or action values, according to the New Dutchlist; Swartjes, 1999) and maximum values in soils exceed critical values for approx. 100-times. Action values for Hg in soils are also exceeded in Litija where Hg smeltery has also been operational for approx. one decade. But in Litija the Hg anomaly is very small comparing to the Idrija one.

Aerial distribution of Hg concentrations in soils and attic dust (figure 57) indicates that the strongest atmospheric Hg pollution is limited only inside the Idrijca River valley and on several spots where smaller furnaces have been operational in 16th and 17th century due to lack of sufficient wood supply nearby Idrija. With similar situation where Idrija would be located in flatland the anomaly would be widespread. Nevertheless Idrija contains 21 km^2 of the land which is critically polluted with Hg (Hg>10 mg/kg; Šajn & Gosar, 2005). Total investigated area has been 160 km^2.

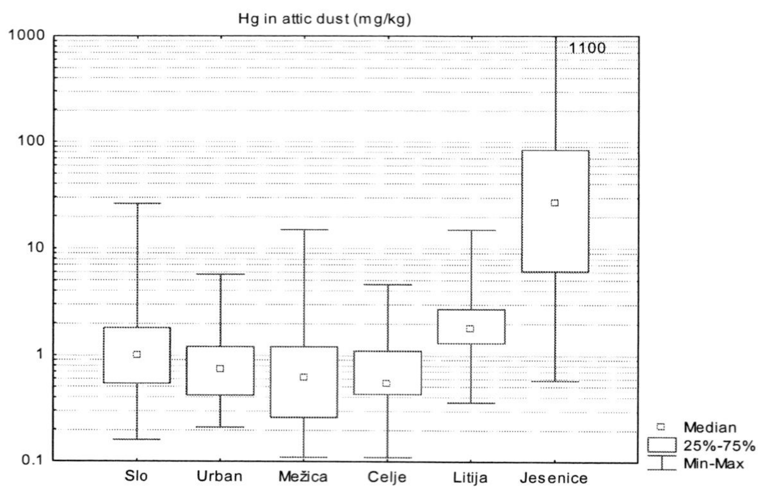

Figure 60. (Continued on next page.)

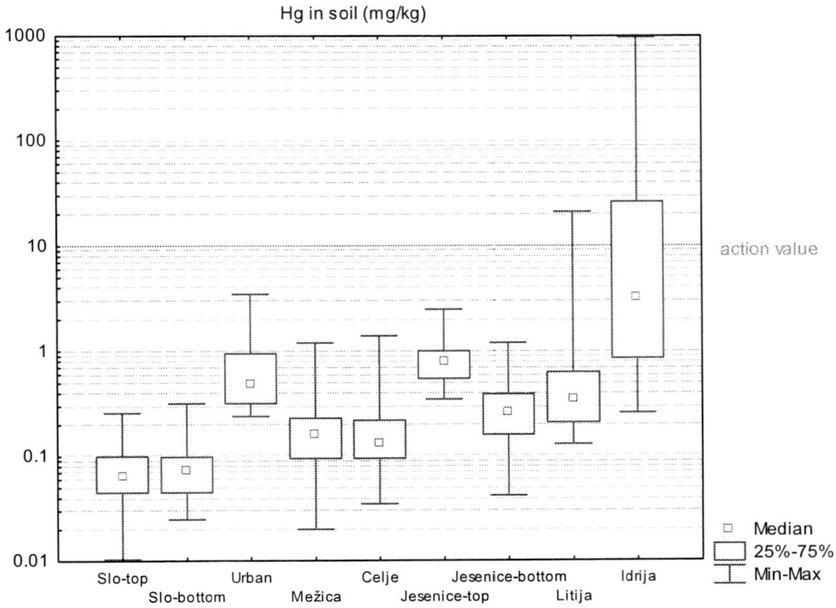

Figure 60. comparison of Hg concentrations in soils and attic dust in different polluted and unpolluted areas in Slovenia.

It would also be good to mention dumping of smelting slag into Idrijca River through centuries which has the consequences that complete Idrijca and Soča River valley and also Tržaški zaliv (Gulf of Trieste) is contaminated with Hg. Values of Hg in alluvial plains sediments of Idrijca River exceeds 2000 mg/kg (Žibret & Gosar, 2006).

5.3.2. Pb Anomaly in Mežica Caused by Pb-Zn Smelting and Cr Anomaly in Ravne Caused by Ironworks

Similar to the Idrija Hg mines Mežica Pb-Zn mines and smeltery is also situated in narrow valley of Meža River with mountains surroundings the area which are more than 2000 meters high. But this is only similarity. Because the Pb-Zn ore contains more of heavy metals the environment is contaminated not only with Pb, but also with As, Cd and Zn. The level of Pb pollution is evident from figure 61 where its concentrations in attic dust are generally the greatest comparing to other areas. Also maximum concentration of Pb in soils exceeds critical values more than 50 times and median and 75th percentile values are the biggest comparing to other polluted areas in Slovenia. Figure 61 indicates that Pb concentrations in soils in some localities in urban areas can be near or exceed 1000 mg/kg, as seen in Slovenian urban areas and Idrija case. Where anthropogenic Pb pollution is caused by smelting activities the maximum levels can be far bigger and can increase beyond 1000 mg/kg as in Celje, Jesenice and Litija.

Bypass of Mežica Pb contamination research is the discovery of new anomaly. In Ravne area there is also anomaly made by ironworks. Cr, Cu, Mo concentrations in soils, exceeding critical values, are its main characteristics. The most evident is the anomaly of Cr in attic dust (figure 62). Unfortunately the sampling pattern did not include complete anomaly as evident from figure 28. This is the reason that Md and P_{75} are not so high as expected and this is why it would be good to extend the sampling of soil and attic dust east of the Ravne in the future. Completely 101 km^2 of the Mežica-Ravne area has been surveyed and totally of 24.4 km^2 contains at least one heavy metal concentration which is found to be above critical value (Šajn & Gosar, 2005).

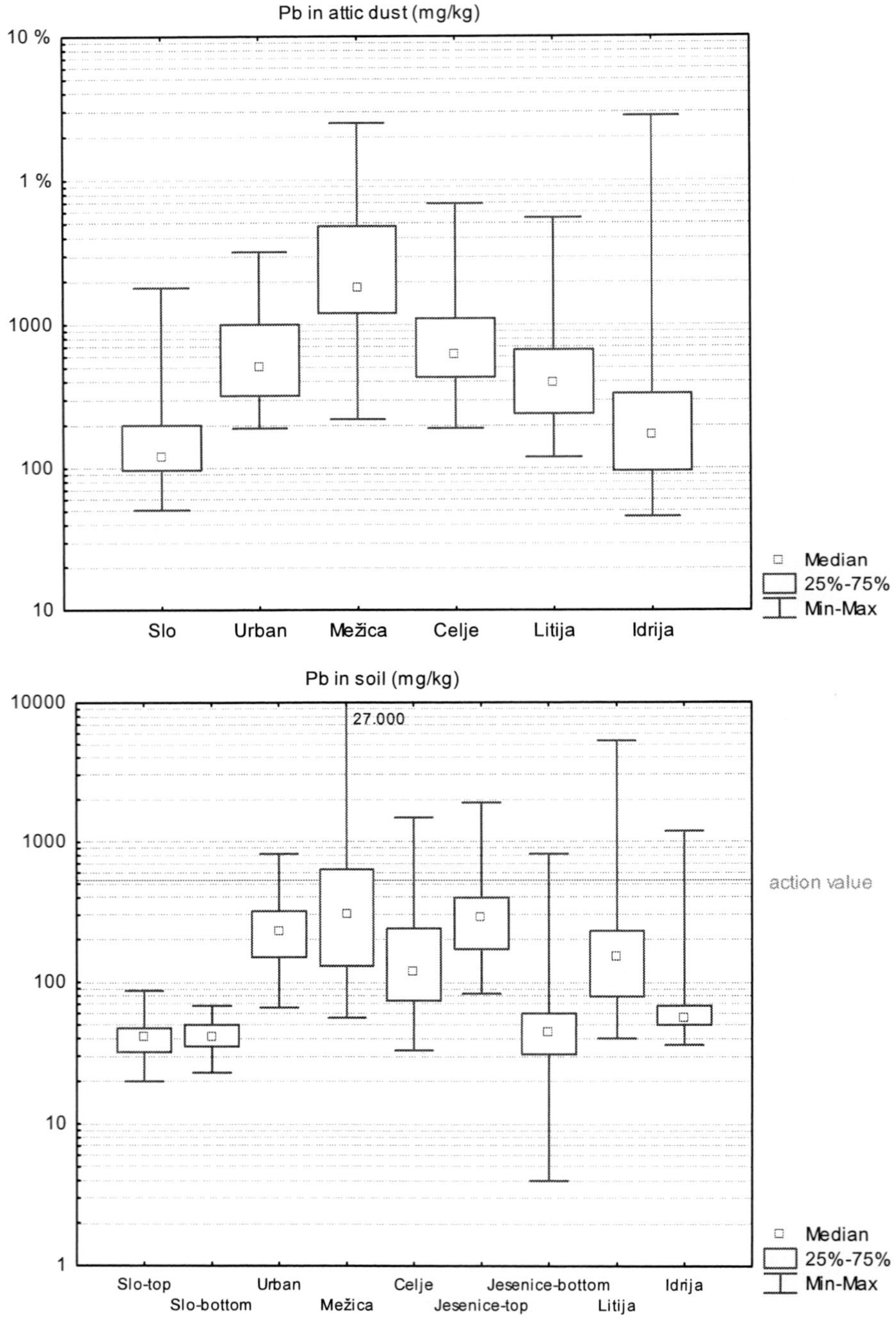

Figure 61. comparison of Pb concentrations in soils and attic dust in different polluted and unpolluted areas in Slovenia.

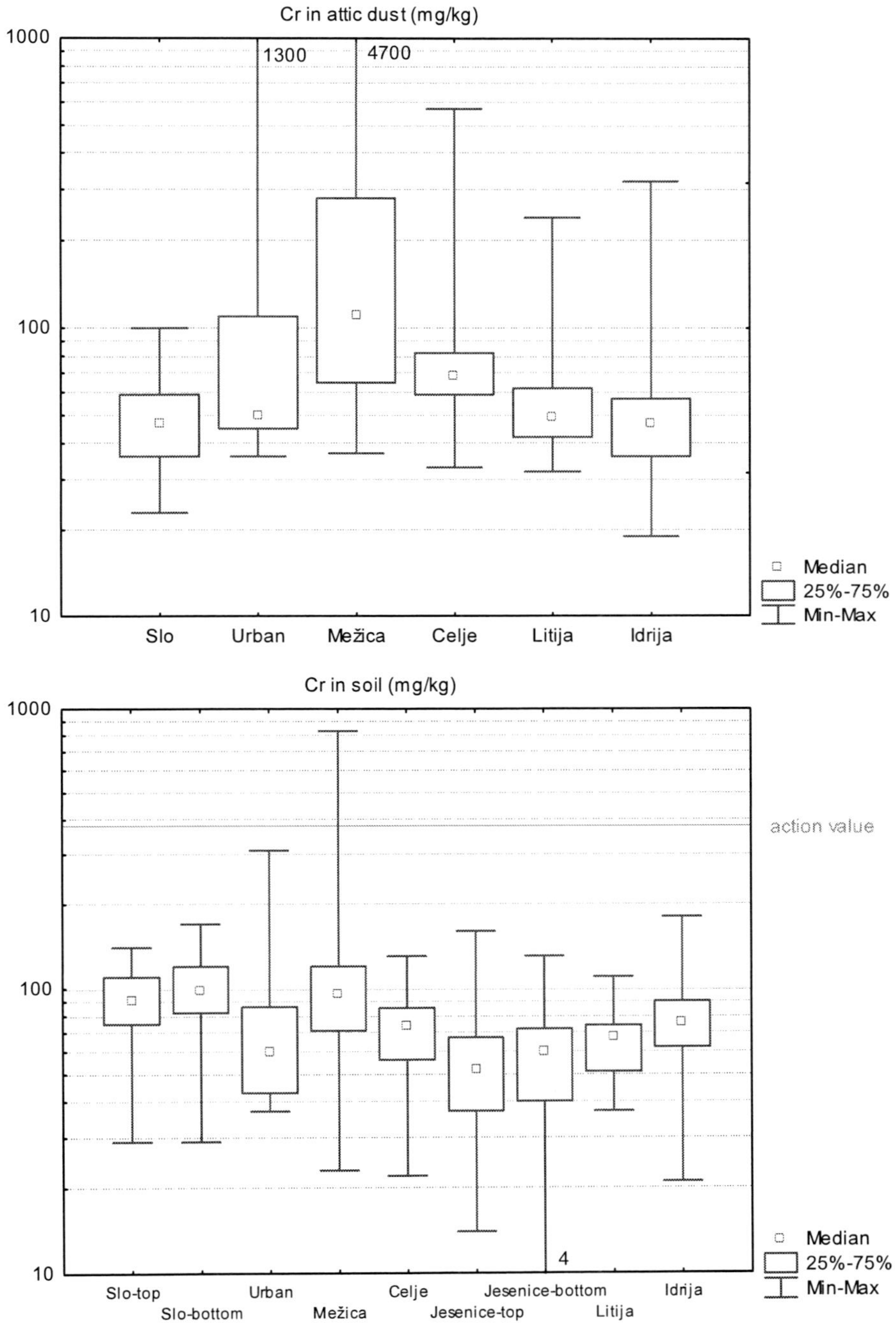

Figure 62. comparison of Cr concentrations in soils and attic dust in different polluted and unpolluted areas in Slovenia.

5.3.3. Cd and Zn anomaly in Celje

Cd and Zn anomaly in Celje is the consequence of 100 years of Zn smelting activities. Characteristic for this anomaly is that the correlation between Zn and Cd concentration in attic dust is very high (0.93; Žibret, 2002). Also great similarity can be seen from aerial distribution of Zn and Cd concentrations in soil and attic dust (figures 37 and 38). Maximum concentrations can be found around past Zn smeltery both in soil and attic dust where maximum Zn concentration in attic dust is 5.6% which characterize it as a rich Zn ore. Figures 63 and 64 compare Zn and Cd concentrations with other areas in Slovenia. Note that sampling plan in urban areas in Slovenia also contains 4 sampling points from Celje and this is the reason that maximum values for Zn and Cd in Slovenian urban areas are so high. When making comparison between Slovenian urban areas with Celje area it is better to use median values.

Zn-Cd anomaly express a lot of similarity with Pb anomaly in Mežica in the sense of geochemical association, connected with smelting. The list of heavy metals, connected with Zn and Cd is almost identical (Ag, As, Cd, Cu, Hg, Mo, Pb, S, Sb and Zn) except that geochemical association in Mežica contains also Sn. Similarity can also be found when comparing the average enrichment for these elements (figures 22 and 31). They are little higher in Mežica but nevertheless at the same scale. Differences are only that in Celje main polluters are Zn and Cd (in Mežica is Pb) and that Celje lies in more open environment (in Celje basin) than Mežica. This allows us to calculate the range of influence of Zn smelting at the basis of sampling according to the distance from the plant. Estimated range of influence is 29 and 13 km for Zn in attic dust and soil (Žibret & Šajn, 2008).

In Celje completely 92.4 km^2 of area has been geochemically surveyed. 17.7 km^2 of area has one or more heavy metals concentrations greater than the critical (action) value (Šajn & Gosar, 2005).

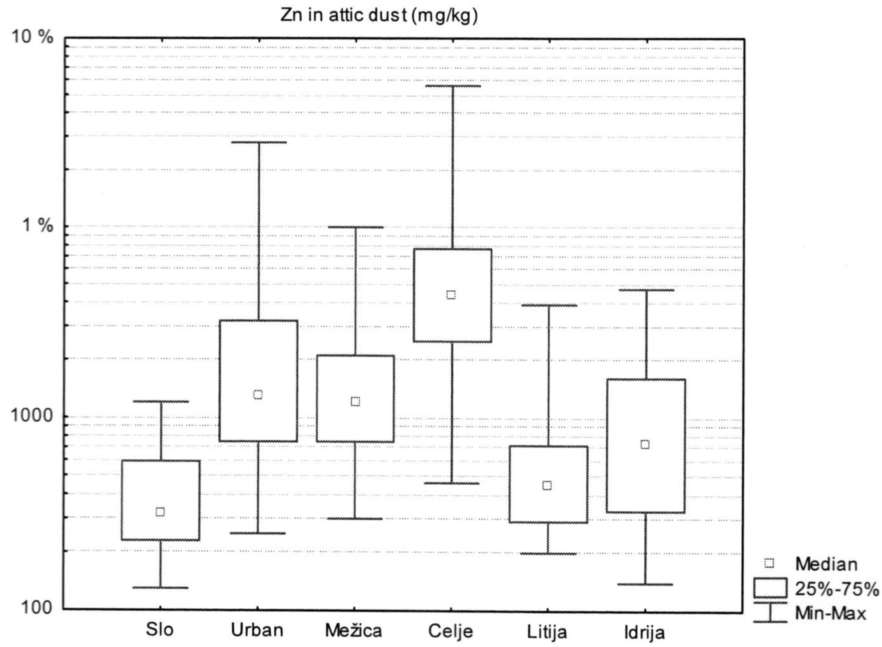

Figure 63. (Continued on next page.)

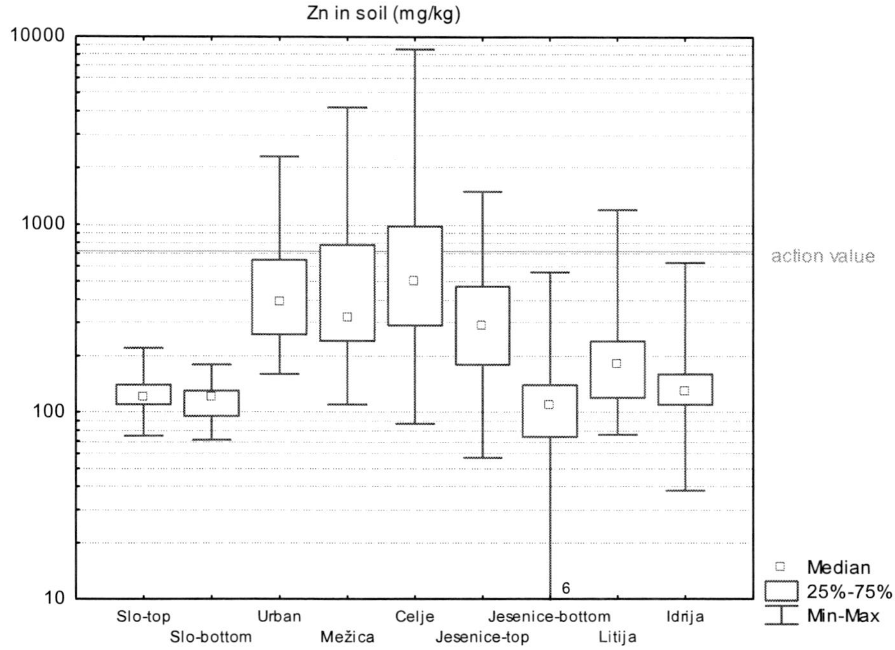

Figure 63. comparison of Zn concentrations in soils and attic dust in different polluted and unpolluted areas in Slovenia.

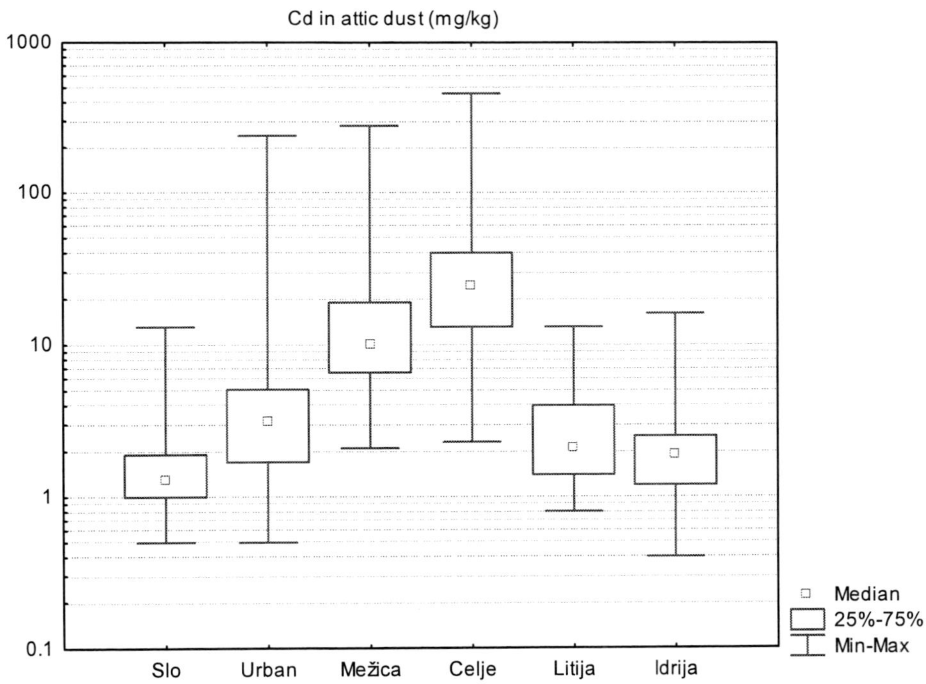

Figure 64. (Continued on next page.)

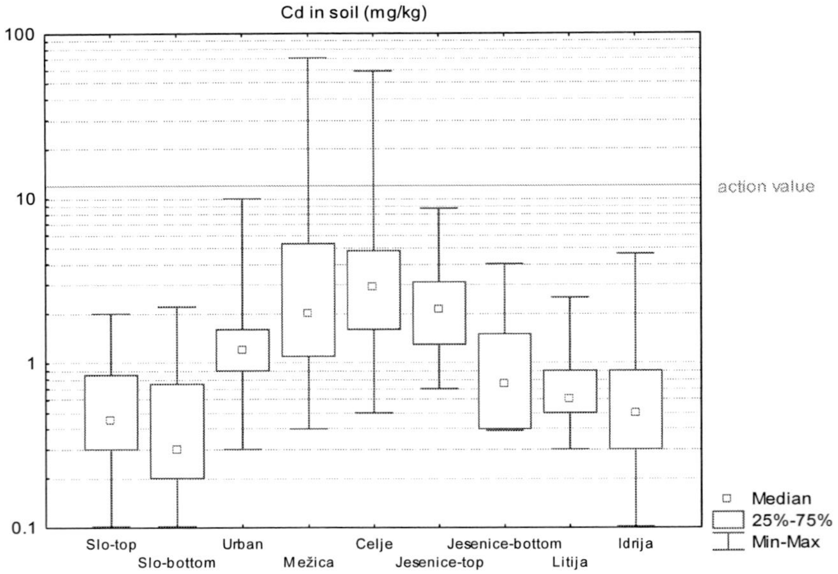

Figure 64. comparison of Cd concentrations in soils and attic dust in different polluted and unpolluted areas in Slovenia.

5.3.4. Ti anomaly in Celje

The last described anomaly is the anomaly in its formation stage in Celje. Cinkarna Celje produces titanium pigments for the last 40 years (from 1970). This production is the consequence for Ti anomaly in attic dust (figures 35 and 40) and stretches west from the production plant. Because of the fact that the emissions into the atmosphere are not very intensive the anomaly is visible only in attic dust but not in soil (figures 40 and 65). On figure 65 the P_{25}-P_{75} and also median values for Ti in attic dust in Celje are the biggest comparing to other areas. Nevertheless this is not the case for soils. This confirms that attic dust is very sensitive sampling medium to any of the present and historical atmospheric pollution.

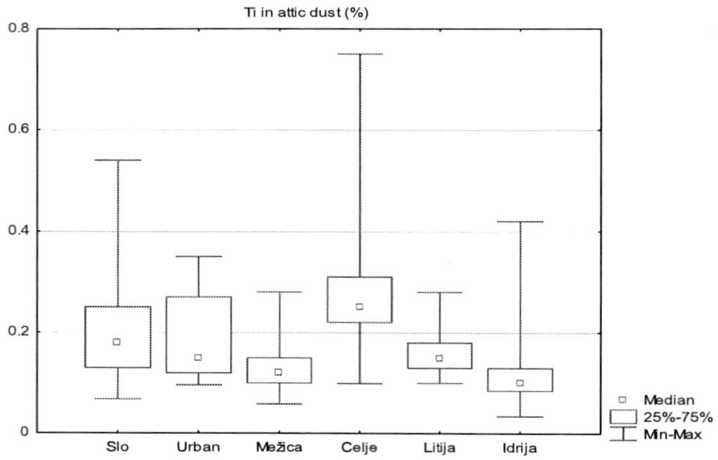

Figure 65. (Continued on next page.)

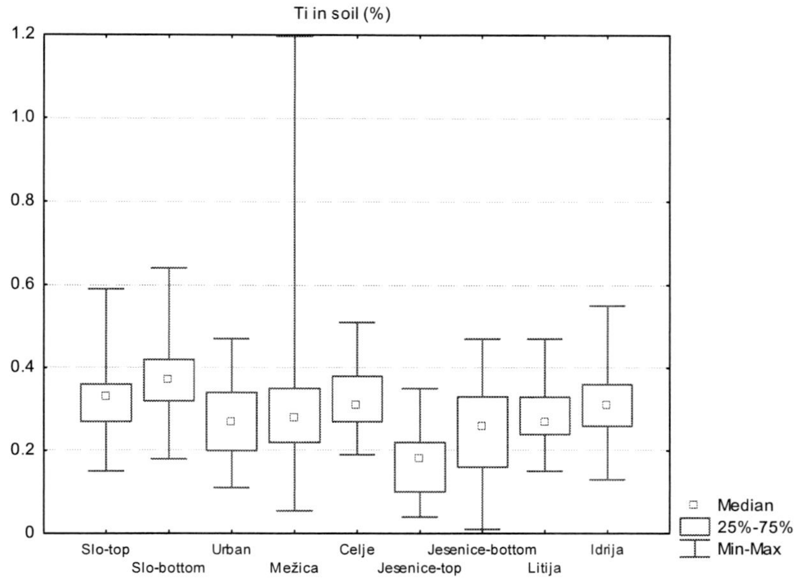

Figure 65. comparison of Ti concentrations in soils and attic dust in different polluted and unpolluted areas in Slovenia.

6. Conclusion

In this book chapter the summary of the last 10 years of geochemical research on Geological survey of Slovenia is presented. The research has been focused on the determination of contaminated areas with heavy metals and its comparison with unpolluted conditions. Several anomalies because of historical mining and smelting operations, have been extracted. Most polluted areas with heavy metals have been recognized to be Idrija where Hg concentrations in soils reach 1000 mg/kg, Mežica, where Pb concentrations in soil reach 2.7%, and Celje, where Zn concentrations in soil reach 0.86%. The article describes different geochemical associations, determined in different areas. Also, it represents good framework for future investigations and comparisons of heavy metal pollution on the basis of soil and attic dust sampling not only in Slovenia, but also in other polluted areas in the world.

7. Acknowledgements

The authors would like to thank to all of the researchers, students, and other personnel who made contributions to sampling and samples preparation. We also wish to thank all of the people, who allowed attic dust sampling inside their property and made it possible for the research to be completed. Special thanks to the Ministry of Higher education, Science and Technology (in past 10 years its name changed several times), to Slovenian Research Agency, Geological survey of Slovenia, and to the municipalities of Celje and Litija for all of their fundings. The authors would also like to thank Nova Science Publishers for their availability of resources and for their talent in making grammatical corrections and improvements.

8. References

Blagotinšek, P. Dosežek za zmanjšanje vplivov na okolje = Contribution for reduction of the environmental impacts. *Cinkarnar*. 2005, vol, 302-2.

Budkovič, T.; Šajn, R.; Gosar, M. Vpliv delujočih in opuščenih rudnikov kovin in topilniških obratov na okolje v Sloveniji = Influence of abandoned metal mines and smelters to the environment of Slovenia. *Geologija*. 2003, vol, 46-1.

Domitrovič-Uranjek, D. *Onesnaženost okolja v Celju = Environmental pollution of Celje*. Zveza društev inženirjev in tehnikov območja Celje: Celje, 1990; pp 35.

Drovenik, M.; Pleničar, M. Nastanek rudišč v SR Sloveniji = Genesis of ore deposits in Slovenia. *Geologija.* 1980, vol, 23-1.

Fabjančič, M. *Kronika litijskega rudnika = Chronology of Litija mines*. Geological survey of Slovenia: Ljubljana, unpubl. manuscript, 1972.

Godec, I. *Litija - nekoč in danes = Litija - past and present, 2nd edittion*. Publ. by the author: Litija, 1993; pp 56.

Gosar, M.; Šajn, R. Mercury in soil and attic dust as a reflection of Idrija mining and mineralization (Slovenia). *Geologija.* 2001, vol, 44-1.

Gosar, M; Šajn, R.; Biester, H. Zvrsti živega srebra v tleh in podstrešnem prahu na Idrijskem = Mercury speciation in soils and attic dust in the Idrija area. *Geologija.* 2002, vol, 45-2.

Hess, A. Verteilung, Mobilität und Verfügbarkeit von Hg in Böden und Sedimenten am Beispiel zweier hochbelasteter Industriestandorte = Distribution, mobility and availability of Hg in soils and sediments - examples of two high polluted industrial sites. *Heidelberger Geowissenschaftliche Abhandlungen*. 1993, vol, 71.

Jemec, M. *Porazdelitev kemičnih prvin v tleh in podstrešnem prahu na območju Litije = Distribution of elements in soil and attic dust on the Litija area (Diploma thesis).* NTF, dept. of geology: Ljubljana, 2006; pp 84.

Jemec, M.; Šajn, R. Geokemične raziskave tal in podstrešnega prahu na območju Litije = Geochemical research of soil and attic dust in Litija area, Slovenia. *Geologija.* 2007, vol, 50-2.

Kavčič, I. Kakšna je stopnja onečiščenosti zraka v Idriji = What is the amount of atmospheric pollution of Idrija. *Idrijski razgledi*. 1974, vol, 19-1/2.

Kosta, L.; Byrne, A. R.; Zelenko, V.; Stegnar, P.; Dermelj, V.; Ravnik, V. Studies on the uptake, distribution and transformation of mercury in living organisms in the Idrija region and comparative areas. *Acta Chimica Slovenica.* 1974, vol, 21.

Lobnik, F.; Medved, M.; Lapajne, S.; Brumen, S.; Žerjal, E.; Vončina, E.; Štajnbaher, D.; Labovič, A. *Tematska karta onesnaženosti zemljišč Celjske občine: študija = Thematic map of soil pollution of Celje community*. Biotehniška fakulteta tozd za agronomijo univerze v Ljubljani: Ljubljana, 1989; pp 159.

Mlakar, I. O problematiki Litijskega rudnega polja = On the problems of the Litija ore field. *Geologija.* 1994, vol, 36/1.

Mlakar, I. Osnovni parametri proizvodnje rudnika Idrija skozi stoletja do danes = Basic parameters of Idrija mercury mine production through centuries. *Idrijski razgledi.* 1974, vol, 19-3/4.

Mohorič, I. *Industrializacija Mežiške doline = Industrialization of Meža valley*; Založba Obzorja: Maribor, SLO, 1954; pp 315.

Mohorič, I. *Problemi in dosežki rudarjenja na Slovenskem: zgodovina rudarstva in topilništva v stoletju tehnične revolucije, knjiga 1 = Problems and achievements of mining in Slovenia: history of mining and smelting in the century of industrial revolution, book 1*. Založba Obzorja: Maribor, 1978; pp 281.

OG RS. Uredba o mejnih, opozorilnih in kritičnih imisijskih vrednosti nevarnih snovi v tleh = Decree on limit values, alert thresholds and critical levels of dangerous substances into the soil. *Official gazette of the Republic of Slovenia.* 1996, vol, 68.

Orožen, J. Oris sodobne zgodovine Celja in okolice = Modern history of Celje. *Celjski zbornik.* 1980, vol, 22.

Planinšek, F. Higienske in epidemiološke razmere v Celjski občini = Hygienic and epidemiologic conditions in Celje community. *Celjski zbornik.* 1972, vol, 14.

Reimann, C., Filzmoser, P., and Garrett, R. G. Factor analysis applied to regional geochemical data: problems and possibilities. *App Geochem.* 2002, vol, 17. DOI:10.1016/S0883-2927(01)00066-X

Šajn, R. Factor analysis of soil and attic-dust to separate mining and metallurgy influence, Meža Valley, Slovenia. *Math geol.* 2006, vol, 38-6. DOI: 10.1007/s11004-006-9039-7

Šajn, R. *Geokemične lastnosti urbanih sedimentov na ozemlju Slovenije = Geochemical properties of urban sediments on the territory of Slovenia.* Geological survey of Slovenia: Ljubljana, 1999; pp 136.

Šajn, R. Using attic dust and soil for the separation of anthropogenic and geogenic elemental distributions in an old metallurgic area (Celje, Slovenia). *Geochem explor env anal.* 2005, vol, 5-1. DOI: 10.1144/1467-7873/03-050

Šajn, R.; Gosar, M. Pollution in Slovenia owing to mining and metallurgy. In *Anthropogenic effects on the human environment in Tertiary basins in the Mediterranean / 2nd International Workshop on the UNESCO-IGCP Project.* Department of Geology, University of Ljubljana: Ljubljana, SLO, 2005; pp 21-26.

Šajn, R; Bidovec, M.; Gosar, M.; Pirc, S. Geochemical soil survey at Jesenice area, Slovenia. *Geologija.* 1999, vol, 41.

Šipec, S. *Jesenice in njihova ekološko-geografska problematika = Jesenice and it's ecological and geographical problems.* Philosophical faculty, dept. of geography, University of Ljubljana: Ljubljana, 1990; pp 232.

Souvent, P. *Pedološke, geokemične in mineraloške preiskave tal v okolici železarne Ravne = Pedological, geochemical and mineralogical soil survey in the vicinity of Ravne Ironworks (Diploma thesis).* NTF, dept. of geology: Ljubljana, 1994b; pp 115.

Souvent, P. Rudnik Mežica nekoč, danes in jutri = Mine Mežica in past, today and future. In *Okolje v Sloveniji = Environment in Slovenia;* Lah, A.; Ed.; Tehniška zalozba Slovenije: Ljubljana, SLO, 1994a; pp 533-541.

Stergar, A. V. *Sanacijski ekološki program Inexe Štore = Ecological restoration plan of Inexa Štore steel factory.* Inexa Štore: Celje, 2001; pp 4-9.

Swartjes, F.A. Risk-based Assessment of Soil and Groundwater Quality in the Netherlands: Standards and Remediation Urgency. *Risk Analysis.* 1999, vol, 19-6.

Tržan, B. Pohorje - prazgodovinski rudarski revir? = Pohorje - prehistoric mining field? *Časopis za zgodovino in narodopisje.* 1989, vol, 25-2.

Žibret, G. Determination of historical emission of heavy metals into the atmosphere: Celje case study. *Environ geol.* 2007, unpubl., online first. DOI: 10.1007/s00254-007-1151-6

Žibret, G. *Geokemične lastnosti tal in podstrešnega prahu na območju Celja = Geochemical properties of soil and attic dust in Celje area (Diploma thesis).* NTF, dept. of geology: Ljubljana, 2002; pp 78.

Žibret, G.; Gosar, M. Calculation of the mercury accumulation in the Idrijca River alluvial plain sediments. *Sci of the tot env.* 2006, vol, 368-1. DOI: 10.1016/j.scitotenv.2005.09.086

Žibret, G.; Šajn, R. Modeling of atmospheric dispersion of heavy metals in the Celje area, Slovenia. *J of geochem expl.* 2008, vol, 97. DOI: 10.1016/j.gexplo.2007.08.001.

In: Causes and Effects of Heavy Metal Pollution
Editor: Mikel L. Sanchez
ISBN 978-1-60456-900-1

Chapter 2

TREATMENT OF ACID MINE DRAINAGE BY A COMBINED CHEMICAL/BIOLOGICAL COLUMN APPARATUS: MECHANISMS OF HEAVY METAL REMOVAL

Francesca Pagnanelli [a*], ***Ida De Michelis*** [b], ***Michele Di Tommaso*** [b], ***Francesco Ferella*** [b], ***Luigi Toro*** [a] ***and Francesco Vegliò*** [b]

[a] Department of Chemistry, Sapienza University of Rome, P.le Aldo Moro 5, 00185 Rome Italy.

[b] Department of Chemistry, Chemical Engineering and Materials, University of L'Aquila, Monteluco di Roio 67040, L'Aquila Italy

ABSTRACT

Natural oxidation of sulphide minerals, exposed to the combined action of oxygen and water, results in the worst environmental problem associated with mining activities, i.e. acid mine drainage (AMD). Waters polluted by AMD are often characterised by low pH, elevated concentrations of iron, sulphates and toxic metals.

Biological remediation options in passive systems (permeable reactive barriers, PRB) usually exploit sulphur production by sulphate reducing bacteria, SRB.

In this report a combined chemical-biological treatment was tested for decontamination of synthetic AMD containing iron, arsenic, copper, manganese and zinc.

Particular attention was paid to the investigation of the mechanisms involved in pollutant removal (chemical precipitation, sorption, bioprecipitation and biosorption) as a fundamental preliminary step for permeable reactive barrier design and long term performance estimation.

Experimental tests were performed both in batch reactors and in a two-column apparatus for sequential treatment by chemical precipitation (first column filled with

* Corresponding author: e-mail: francesca.pagnanelli@uniroma1.it; fax: +39 06 490631.

natural limestone) followed by bioprecipitation/biosorption (second column filled with a natural organic mixture inoculated by sulphate reducing bacteria).

Distinct mechanisms of removal for each metal were identified by combining theoretical data of metal solution chemistry, and results obtained from independent experimental tests: batch and column tests, blank tests using natural organic mixture as biosorbing materials, acid digestions, and selective extractions of metals using solid samples of filling material after column dismantlement.

This analysis allowed isolating metal-specific mechanism of abatement and denoted the relevant contribution of biosorption phenomena in metal removal in biological column. This contribution, generally neglected in biological PRB design with respect to bioprecipitation, should be taken into account in order to avoid misleading estimation of SRB performance and also to better estimate PRB duration.

Keywords: water treatment; heavy metals; sulphate reducing bacteria; biosorption, fixed bed column reactor.

1. Introduction

1.1. Acid Mine Drainage

Natural oxidation of sulphide minerals, exposed to the combined action of oxygen and water, results in the worst environmental problem associated with mining activities, i.e. acid mine drainage (AMD) (Johnson and Hallberg, 2005). Waters polluted by AMD are often characterised by low pH, elevated concentrations of iron, sulphates and toxic metals.

AMD can be generated in underground and open pit operating mines, but also abandoned sites can remain active generation points for decades or even centuries after mine closure. The other main source of AMD is the huge amount of mill tailings (often about the 90% of treated ore): about 18 billion m^3 are produced every year, stored in impoundments or left exposed to natural weathering (Sheoran and Sheoran, 2006).

AMD is generated by biologically-accelerated oxidation of iron pyrite (FeS_2), the most abundant sulphide mineral on the planet, generally present in metal ores and coals. AMD originates from a complex series of reactions, which result in pyrite dissolution and release of sulphates and protons:

$$FeS_2 + 3.5O_2 + H_2O \rightarrow Fe^{2+} + 2SO_4^{2-} + 2H^+ \qquad (1)$$

Nevertheless, this global reaction can be misleading not showing that, in most situations, the main sulphide oxidant is Fe(III) rather than oxygen (Johnson and Hallberg, 2005; Baker and Banfield, 2003; Peppas et al., 2000):

$$FeS_2 + 14Fe^{3+} + 8H_2O \rightarrow 15Fe^{2+} + +2SO_4^{2-} + 16H^+ \qquad (2)$$

Ferrous iron regeneration by O_2 at low pH is the rate-limiting step of the whole process:

$$14Fe^{2+} + 3.5O_2 + 14H^+ \rightarrow 14Fe^{3+} + 7H_2O \qquad (3)$$

The rate of this reaction can be augmented by several orders of magnitude by the action of iron oxidising bacteria (such as *Thiobacillus ferooxidans*) resulting in the global acceleration of pyrite dissolution (reaction 1, which is the sum of reactions 2 and 3).

The same mechanisms of oxidation by O_2 and Fe(III) are responsible for dissolution of other metal sulphides releasing toxic metals in soil solution system (such as As, Zn, Cd, Pb, Cu, Ni and Mn).

The true scale of AMD pollution is of difficult assessment: an estimate of 1989 spoke about 19˙300 Km of rivers and 72˙000 ha of lake damaged by AMD (Johnson and Hallberg, 2005).

The long-term activity of pollution sources along with the predicted doubling of mine tailing production in the next 20-30 years (Sheoran and Sheoran, 2006) indicate a need for the application of an integral approach based on both prevention and treatment.

1.2. Prevention and Treatment of AMD Pollution

Operative strategies for AMD pollution involve both prevention and treatment options.

The preclusion or attenuation of AMD generation can be obtained by different ways (Peppas et al., 2000; Johnson and Hallberg, 2005; Sheoran and Sheoran, 2006; Pagnanelli et al., 2007):

- minimization of oxygen diffusion (flooding/sealing of underwater site, underwater storage of mine tailings, storage in sealed waste heap with covers of organic materials and clays);
- control of pH of mineral wastes by blending with solid material (lime, limestone, phosphates, fly ash, paper mill waste) in order to precipitate Fe(III) and reduce its oxidant action;
- total solidification of wastes;
- inhibition of iron and sulphur oxidising bacteria by such biocides as surfactants.

These prevention technologies can be integrated with the treatment of AMD polluted waters performed according to both abiotic and biological strategies.

In abiotic approaches various chemicals can be used to promote heavy metal removal by decreasing their solubility. Generally AMD polluted waters are treated by limestone in order to neutralise AMD pH and to precipitate iron:

$$2Fe(HCO_3)_2 + 1/2O_2 + H_2O \rightarrow 2Fe(OH)_3 \downarrow + 4CO_2 \tag{4}$$

$$2Fe_2(SO_4)_3 + 6CaCO_3 + 3H_2SO_4 \rightarrow 2Fe(OH)_3 \downarrow + 6CaSO_4 \downarrow + 6CO_2 \tag{5}$$

Iron removal is a fundamental step to preclude AMD mechanism going on (equation 2), but also toxic metals with low solubility in basic conditions can be removed.

Biological strategies exploit H_2S production by sulphate reducing bacteria, SRB, promoting metal precipitation as sulphides (Johnson, 2006; Neculita et al., 2007):

$$Me^{2+} + H_2S \rightarrow MeS \downarrow + 2H^+ \tag{6}$$

Precipitation of heavy metals as sulphides seems to be a better alternative with respect to metal precipitation for pH rise. In fact sulphide precipitation occurs almost regardless of wastewater characteristics, is less pH-dependent and leave lower residual metal concentrations than hydroxide precipitation (Tunay and Kabdasli, 1994). However, chemically produced H_2S is relatively expensive to handle and to use safely. The precipitation of metals with biologically produced H_2S by SRB has been then proposed as an alternative process (Foucher et al., 2001).

As for the application of abiotic and biological strategies, both active and passive systems can be adopted. Active systems require ex-situ treatment of polluted streams (by filtration, precipitation and adsorption onto active carbons), while passive systems are based on in situ technologies using permeable reactive barriers (PRB) which intercept polluted streams and determine pollutants abatement within the barrier. PRB are especially advantageous for widespread pollution sources as in the case of AMD in mine districts (Gibert et al., 2002).

Treatment in abiotic PRBs can exploit neutralizing agents, adsorbents and zero-valent iron as reactive filling materials (Table 1).

Table 1. Some examples of abiotic reactive barriers for heavy metal pollution (http://www.rtdf.org/public/permbarr/PRBSUMMS/)

Location	Installation	Scale	Reactive medium	Pollutants
Durango, CO (USA)	1995	Pilot	Fe(0)	As, Mo, Se, U, V, Zn
Elizabeth City, NC (USA)	1996	Pilot	Fe(0)	TCE, Cr(VI)
Hanford, WA (USA)	1997	Full	$Na_2S_2O_4$	Cr(VI)
Ontario, Canada	1998	Full	Zeolites	Sr-90
Ponticello, UT (USA)	1999	Full	Fe(0)	U, As, Mn, Se, V
Kolding, Denmark	1999	Full	Fe(0)	TCE, Cr(VI)
Nesquehoning, PA (USA)	1998	Full	Limestone	Pb, Cd, As, Zn, Cu

Biological PRB exploiting SRB activity are generally made up of organic mixtures as electron donor in the dissimilatory reduction of sulphate to sulphide, which generates alkalinity and promotes metal precipitation:

$$SO_4^{2-} + 2CH_2O + 2H^+ \rightarrow H_2S + 2H_2CO_3 \tag{7}$$

Organic components used in PBR are a mix of biological materials chosen on the base of the local availability: biodegradable materials (mushroom compost, manure of cow, horse and sheep, municipal compost) are generally mixed with more recalcitrant ones (sawdust, peat, straw, leaf compost) to ensure long term growth of SRB (Alvarez et al., 2007; Chang et al., 2000; Christiensen et al., 1996; Cocos et al., 2002; Gibert et al. 2003; Gibert et., 2004; Hammack et al., 1992; Peppas et al., 2000; Waybrant et al., 1998; Waybrant et al., 2002). Full scale applications of organic-carbon based sulphate reducing PRB are also characterised by the addition of gravel to improve barrier permeability and limestone to increase pH and favour SRB growth (Benner et al., 1999; Jarvis et al., 2006; Ludvig et al., 2002).

Table 2 shows some reactive mixtures reported in the literature, which have been used in lab-scale (batch and column) studies and in full-scale permeable reactive barriers for treatment of AMD and heavy metal's contaminated wastewaters.

These reducing and alkalinity-producing systems (RAPS) (Johnson and Hallberg, 2005) can be inadequate for highly iron-concentrated AMD due to limestone armouring (reduction of reactivity by deposition of iron oxides precipitates) (Simon et al., 2005) and barrier plugging for iron precipitates. In such cases, an alternative engineering configuration can be adopted by using a two-step procedure: chemical precipitation of iron by limestone (step I, pre-treatment of chemical precipitation) and then a refinement of heavy metal removal as sulphide precipitates mediated by sulphate reducing bacteria (step II, bioprecipitaton).

Chemical pretreatments of AMD aim firstly to obtain an increase of pH, which allows the precipitation of Fe(III) (involved in sulphide oxidation and then in AMD generation) and the partial removal of other toxic components. The adjustment of pH and the reduction of iron concentration are also necessary for the successive biological treatment in permeable reactive barriers to avoid the rapid plugging of the barrier (for the large amounts of iron precipitates) and to ensure neutral pH conditions for bacterial growth.

Table 2. Reactive mixtures used in lab- and in full-scale biological permeable reactive barriers for the treatment of AMD and heavy metal's contaminated wastewaters

Composition	Type	Abatement	Reference
Municipal compost Sawdust Manure Cellulose Sediments with SRB Silica sand Limestone	Batch	SO_4^{2-} ~ 100% Fe 99% Ni ~ 100% Cd 99%	Waybrant et al., (1998)
Wood chips (3%) Composted leaves (30%) Chicken manure (20%) Silica sand (5%) Sediments with SRB (37%) Limestone (2%) Urea (3%)	Batch	SO_4^{2-} 97% Ni 72 % Zn 88%	Cocos et al., (2002)
Pirite Silica sand Leaves Chips Sawdust Biological sludge Sediments with SRB	Column	SO_4^{2-} 20-60%	Waybrant et al., (2002)

Table 2. (Continued)

Composition	Type	Abatement	Reference
Limestone (50%) Compost (45%) Sediments with SRB (5%)	Column	Fe 99% Zn 55% Cd 80% Cu 97%	Gibert et al. (2003)
Compost limestone Sheep manure	Column	SO_4^{2-} 18-27%	Gilbert et al, (2004)
Gravel (50%) Municipal compost (20%) Soil (20%) Wood chips (9%) Limestone (1%)	Full-Scale PRB Ontario (Canada) 1995	SO_4^{2-} 60% Fe 85%	Benner et al. (1999)
Module 1 Limestone (50%) Municipal compost (30%) Sludge (20%) *Module 2* Limestone (50%) Municipal compost (50%) *Module 3* Zero-valent iron (0) (1%) Limestone (66%) Municipal compost (33%)	Full-Scale PRB Aznalcòllar (Spagna) 1998	Metals 90%	Carrera et al. (2001)
Composted leaves (15%) Gravel (84%) Limestone (1%)	Full-Scale PRB Vancouver (Canada) 2000	Metals 80%	Ludwig et al. (2002)
Manure and straw (25%) Municipal compost (25%) Limestone (50%)	Full-Scale PRB Northumberland (UK) 2003	SO_4^{2-} 67% Fe 95% Al 87%	Jarvis et al. (2006)

1.3. Aim of the Work

Experimental results reported in this work denote some crucial points that should be specifically addressed in order to develop new biotechnological applications for the treatment of heavy metal pollution. According to this, the research specifically focused on the investigation of the chemical and biological mechanisms responsible of heavy metal removal in biological permeable reactive barriers.

Biological permeable reactive barriers are complex continuous flow multiphase reactors in which different mechanisms can act simultaneously in heavy metal removal (Sheoran and Sheoran, 2006; Johnson and Hallberg, 2005; Johnson, 2006; Whitehead et al., 2005).

Generally bioprecipitation and chemical precipitation were addressed as the main mechanism operating. In fact different metals simply precipitate in the conditions of pH typical of SRB growth as oxyhydroxides (ferric iron, copper and aluminium) and as carbonates (copper, manganese and zinc) (Sheoran and Sheoran, 2006; Peppas et al., 2000). Mechanism investigations were generally developed according to geo-chemical studies considering both chemical equilibria in solution and metal speciation in solid phase (Herbert et al., 2000; Morrison et al., 2002; Wilkin and McNeil 2003; Johnson and Hallberg, 2005b; Swash and Monhemius 2005).

Experimental data using zero-valent iron as reactive medium denoted the importance of adsorption as initial and rapid metal uptake mechanism (Wilkin and McNeil 2003). In the same way adsorption onto organic matter (biosorption) used as reactive media in biological PRB should be accounted for in lab-studies for further development. Bioprecipitation by SRB is the long term active mechanism that should operate in metal removal. Nevertheless abatement estimates for biological permeable reactive barriers can not neglect the relevant contribution of biosorption onto organic reactive materials generally used as long term carbon source for SRB. In fact, bioprecipitation remains active in removing metals and sulphates until SRB are alive, while biosorbents tend to be saturated and undergo biodegradation processes causing the release of initially sorbed pollutants.

According to these observations the development and design of new biological PRB can not neglect the identification of the specific mechanisms involved in metal removal as key-step of lab-study research.

In this view knowledge, isolation and quantification of the different mechanisms operating in biological PRB are the primary goal of this work. To this aim experimental results of batch and column tests of SRB growth on solid media were combined with theoretical metal speciation, blank tests of biosorption and analysis of pollutants speciation in solid phase after column dismantlement.

2. Materials and Methods

2.1. Synthetic Acid Mine Drainage (AMD)

A synthetic solution was used to have a standard average composition of the influent stream to be treated. Synthetic AMD solution was obtained by dissolving weighted amounts of reagent grade chemicals ($FeSO_4$, $Fe_2(SO_4)_3$, $MnSO_4$, $CuSO_4$, $ZnSO_4$, As_2O_5) in distilled water.

Average values ± standard deviations of measured chemical composition in synthetic AMD were reported in Table 3.

Table 3. Average values of chemical composition of synthetic AMD in the feed, after chemical pre-treatment (10th output of pre-treatment for PV=70) and after biological treatment (10th output of biological treatment column for PV=60). (*Pre-treated samples of AMD after chemical treatment were upgraded to 2000±100 mg/L of sulphates; n.d.: not detectable, i.e. <0.01 ppm)

Species	Feed	After Chemical Pre-Treatment	After Biological Treatment
Ph	2.9 ± 0.2	6.1 ± 0.2	8.1 ± 0.3
So_4^{2-}	800 ± 50 Mg/L	700 ± 50 Mg/L	600 ± 50 Mg/L *
Fe	400 ± 50 Mg/L	3 ± 1 Mg/L	N.D.
Cu	50 ± 10 Mg/L	0.2 ± 0.1 Mg/L	N.D.
Zn	50 ± 10 Mg/L	46 ± 5 Mg/L	N.D.
Mn	50 ± 10 Mg/L	48 ± 5 Mg/L	N.D.
As	2.0 ± 0.5 Mg/L	N.D.	

2.2. Sulphate reducing bacteria (SRB)

Bacterial biomass was kindly furnished by the research group of Professor Groudev (Department of Engineering Geoecology, University of Mining and Geology, Sofia, Bulgaria), who collected it in the Curilo mine district located near Sophia (Groudev et al., 2001).

Bacteria used in batch and column experiments were cultivated in closed shaken flasks using standard procedures for SRB reported in the literature (Postgate, 1979). In particular medium cultivation (defined as C Medium) was used for bacterial growth and acclimatising in column tests: KH_2PO_4 0.5 g/L; NH_4Cl 1 g/L; Na_2SO_4 4.5 g/L; $CaCl_2*6H_2O$ 0.06 g/L; $MgSO_4*7H_2O$ 0.06 g/L; sodium lactate 6 g/L; yeast extract 1 g/L; $FeSO_4*7H_2O$ 0.004 g/L; sodium citrate*$2H_2O$ 0.3 g/L.

2.3. Solid Mixture

Solid mixture used both in batch and column tests was made up of 80% v/v of compost, 15% v/v of cow manure, 5% v/v of straw and traces of limestone.

2.4. Batch Tests

Preliminary batch tests were performed for both chemical precipitation and sulphate removal by SRB.

Batch tests of chemical precipitation were performed by limestone addition to synthetic AMD (250 g in 500 mL). Quarry quality limestone used in this work was mainly made up of calcium carbonates and silica (mean particle size 6 mm). Limestone-bearing suspensions for batch tests of chemical precipitation were kept under stirring and monitored during time for iron concentration and pH.

Batch tests of SRB growth in liquid medium were performed according to the cultivation procedure reported above (paragraph 2.2) and monitored during time for cell, lactate and sulphate concentration.

Batch tests of SRB growth on solid media were performed according to the following procedure: a sample (20 g) of solid mixture (see 2.3) was added in each flask, filled with 80 mL of C Medium prepared without sodium lactate and yeast extract. Therefore the flasks were sealed and 20 mL inoculum of bacteria cultivated in C Medium (in exponential growth phase) were added by a syringe through the sampling port. All experiments were conducted at 37 °C under shacking conditions. Medium characteristics (pH, E_h, SO_4^{2-} concentration and H_2S production) were monitored during the experiments, which lasted 22 days.

Measurements of pH (by CRISON GLP22), E_h (by CRISON GLP22) and H_2S (by lead acetate paper) were determined immediately after sample collection. Samples were then filtered through 0.45 μm cellulose acetate filters and used for sulphate analysis (see 2.6.1). Each test was performed twice both in presence and in absence of bacteria. Average values ± standard deviations were considered.

Sulphate removal onto single organic components (compost, olive pomace and leaves) without inoculum were carried out as those previously described, using the same amount of component as in the mixture.

2.5. Column tests

Column tests of chemical precipitation and biological precipitation were performed in two distinct fixed bed columns. Both columns were made of Plexiglas (height 1 m; diameter 0.2 m; column volume, V_b=6.65*10^{-3} m^3) with 10 equally distant outputs (0.1 m) along the axial length, numbered from the bottom to the top of the column.

Figure 1 reported a schematic representation of the 2-column system used in this work.

For chemical precipitation tests, the first column was filled with limestone and fed with synthetic AMD from the bottom (30 mL/h). The total pore volume was evaluated by pumping a known volume of water through the dry column and collecting the effluent. The difference is the total column pore water (2 L). Column effluents from the different outputs were monitored during time for pH, iron and metal's concentration. For biological precipitation tests, the second column was filled with a solid medium containing 80% v/v of compost, 15% v/v of cow manure, 5% v/v of straw and traces of limestone (total column pore water 0.9 L). SRB were inoculated along the whole length of the column through the ten outputs. After inoculation, the column was preliminary fed (2 months) with liquid C Medium with and then without lactate for bacteria acclimatisation. After acclimatisation, pre-treated sample of AMD coming from the 10th output of the first column were upgraded to 2000 mg/L of sulphates, and fed to the second column (10 mL/h) (Figure 1).

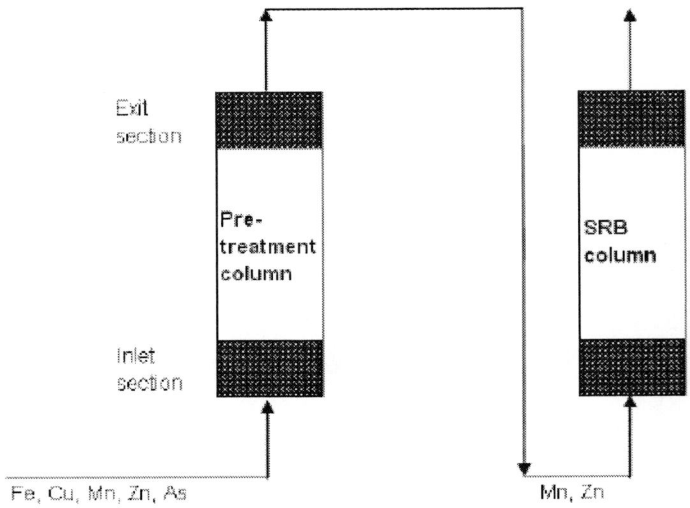

Figure 1. Schematic representation of the two-column systems for the combined chemical-biological treatment of AMD.

2.6. Analytical Methods

2.6.1. Sulphate Determination

Sulphates were determined by a turbidimetric method: 30 mL of sample to be analysed were placed in a 100 mL Erlenmeyer Maier flask and mixed under magnetic stirring with 10 mL of a solution of glycerine and sodium chloride. This solution was prepared mixing two volumes of a glycerine solution (pure glycerine diluted at 50% in distilled water) with one volume of NaCl solution (200 g of NaCl + 40 mL of concentrated HCl diluted to 1 L by distilled water). After the addition of glycerine/NaCl solution, each sample was added for 5 mL of a solution of $BaCl_2$ (prepared with 90 g of $BaCl_2*2H_2O$ in 1 litre of distilled water). After two minute stirring, samples were analysed by spectrophotometer at 420 nm. Instrument calibration was performed by standard solutions of anhydrous Na_2SO_4 (dried for 1 hour in an oven at 110 °C).

2.6.2. Metal and Metalloid Determination

Concentrations of Fe, As, Cu, Mn and Zn in feed and treated solutions were determined by an Inductively Coupled Plasma Optical Emission Spectrometer (Varian Vista-MPX CCD Simultaneous ICP-OES). Radiation wavelengths used to identify the elements were 238.204 nm for Fe, 188.89 nm for As, 327.395 nm for Cu, 257.61 nm for Mn, 213.86 nm for Zn. Solutions for calibration were prepared by diluting 1 g/L standard solutions (Fluka) with 1% nitric acid solution. Concentrations were determined by radiation intensity of three replicates at the characteristic wavelengths.

2.6.3. Cell Growth Monitoring: Turbidimetry and Lactate Determination

Qualitative information about cell growth in liquid medium during batch tests were obtained by turbidimetry measuring the absorbance change of the suspension by a spectrophotometer (440 nm).

A lactate testing kit (K-D/LATE Astori) was used for the colorimetric measurements of lactate.

2.6.4. Chemical Composition by Fluorescence Analysis

Chemical compositions of the native limestone and the precipitated solid obtained by chemical precipitation in pre-treatment column were determined by X-Ray fluorescence. After column emptying and drying of the filling material, the finest size fractionated sample (<355 micron) obtained by sieving of the filling material of the chemical treatment was taken as representative of precipitated solid. 3 g samples of grounded material (<66 micron) were pelletised using a press with an additive as a binder. The pellets were analysed in a SPECTRO XEPOS bench-top XRF spectrometer.

2.6.5. Acid Digestion of Solid Samples

Acid digestion of solid samples taken from pre-treatment and SRB columns were performed to evaluate metal concentration in solid phase after treatment of synthetic AMD. Solid samples from inlet and exit sections of both columns were analysed (inlet section: solid material in the first 20 cm of column height near the bottom inlet of feed; exit section: solid material in the last 20 cm of column height near the upper output of treated effluent). Solid samples were dried in a oven (60°C for three days) and grounded (< 74 μm). Four representative sub-samples (0.1 g) of each column section were considered. Each sub-sample was placed in a Teflon recipient and digested by using 4 mL of an oxidising mixture of concentrated acids (HNO_3:HCl=3:1) and 6 mL HF, in a microwave oven (800 W, 4 minutes; 400 W, 4 minutes; 800 W, 4 minutes; 20 minutes of ventilation) (Barbaro et al., 1995). After complete digestion of solid samples, 5.6 g HBO_3 was added to avoid silica evaporation and each liquid sample was diluted to 100 mL with deionised water. Metal and metalloid concentrations in solution were determined by an Inductively Coupled Plasma Mass Spectrometer (ICP-MS).

2.6.6. Sequential Extractions Upon Solid Samples

Solid samples extracted from the inlet and exit sections of the II column after the treatment of 70 PV were used for sequential extraction tests. Solid samples were dried in a oven (60 °C for three days) and grounded (< 74 μm). Two representative sub-samples (2.5 g) of each column section were used in the following sequential extraction procedure (Campanella et al., 1995)

I Step: 2.5 g soil sample is treated by 45 mL of ammonium acetate 1 M at pH 5 by acetic acid under stirring for 24 hours at room temperature; suspension is then centrifuged at 3000 rpm for 20 minutes, diluted at 100 mL by deionised water and analysed by ICP.

II Step: The residual solid of previous step is treated by 22.5 mL of hydroxylammonium chloride 1 M and 22.5 mL of acetic acid (25%). After 24 h stirring at room temperature a solid-liquid separation is performed by centrifugation as before and the metal-bearing solution is diluted (to 100 mL) and analysed by ICP-MS.

III Step: The residual solid of previous step is treated by 12.5 mL of HCl 0.1 M and stirred for 24 h at room temperature. As in previous steps a solid-liquid separation is performed and the solution diluted to 25 mL is analysed for metal concentrations.

IV Step: The residual solid of previous step is treated by 12.5 mL of NaOH 0.5 M under stirring for 24 h at room temperature: for soil samples with large organic content this treatment should be repeated different times until a clear solution is obtained. All the solutions separated from the solids are then dried by an IR lamp at 60 °C and then digested by using 4 mL of HNO_3 (65%) and 2 mL HF (40%) in a microwave oven (250 W, 1 min; 0 W, 2 min; 250 W, 5 min; 400 W, 5 min; 600 W, 5 min). The acid solution is then diluted to 25 mL and analysed for metals by ICP-MS.

V Step: The residual solid of previous step is added to 12.5 mL of HNO_3 8 M and digested for 3 h at 80 °C. The solution is then diluted to 25 mL and analysed by ICP-MS for metal concentrations.

The residual solid of the fifth step is finally digested as described in section 2.6.5 to determine the metals in the mineralogical matrix.

3. Results and Discussion

3.1. Preliminary batch tests of chemical precipitation

Limestone efficiency in iron removal was preliminary tested by batch tests, which also denoted the overall rate of the process. Replicated data showed a fast iron removal leading to 80% abatement within the first hour (Figure 2).

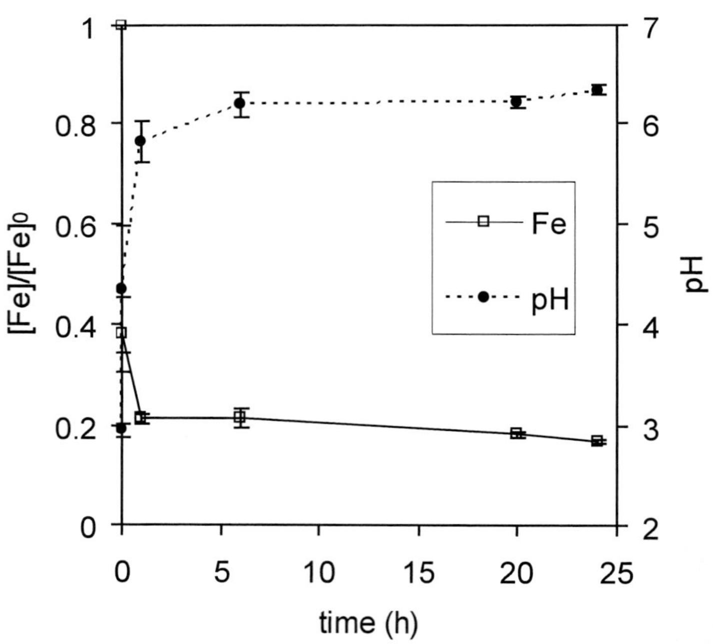

Figure 2. Preliminary batch tests of AMD treatment by chemical precipitation: iron and pH during time.

Simultaneously, a steep increase of pH was observed from 2.95 ± 0.05 to 4.4 ± 0.4 (mean values ± maximum dispersion by replicated data at the beginning of the experiment and after 1 hour). After the first hour, iron removal remained quite constant, while pH continued to raise up to 6.34 ± 0.04 (24 h) (Figure 2).

Limestone addition determined the increase of pH by carbonate dissolution and neutralisation of AMD acidity (equations 4 and 5). pH increase caused iron precipitation as a mixture of amorphous oxy and hydroxyl oxides (Kalin et al., 2006; Komnitsas et al., 2004) leading to 83% iron removal within 24 hours.

3.2. Preliminary Batch Test of Bioprecipitation

The efficacy of the SRB wild strain in sulphate removal was preliminary tested in batch tests using liquid medium. Biomass activity was evaluated by lactate and sulphate consumption (Figure 3).

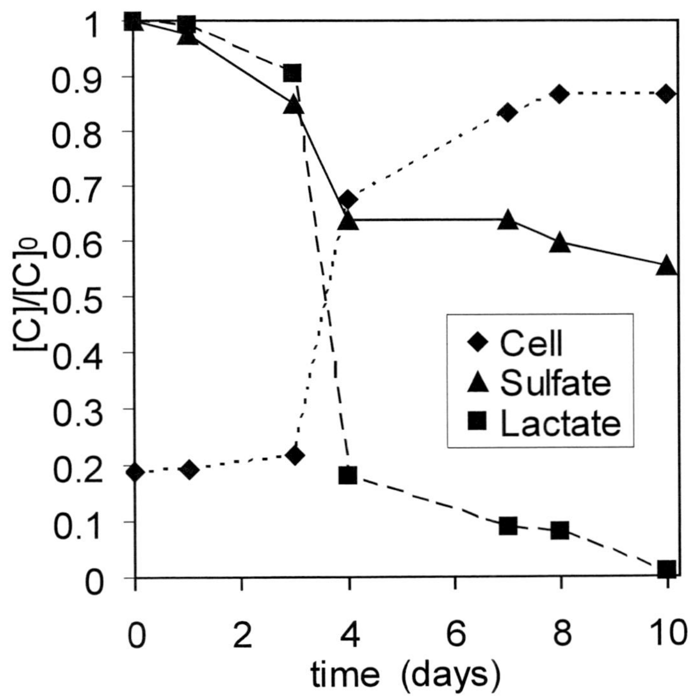

Figure 3. Preliminary batch tests of SRB growth: dimensionless concentration of biomass, lactate and sulphates during time.

Experimental data showed 55 % abatement of sulphates in 10 days with a slow decrease during the first three days, followed by a steep reduction (4th day) and then a slow further removal. The source of organic carbon (lactate) presented a similar trend during time. Optical density used to evaluate cell concentration was characterised by a typical transient of biological growth, with a latent phase (till the 3th day), an exponential phase and a final stationary phase (Figure 3).

3.3. Column Test of Chemical Precipitation: Pre-Treatment of AMD

3.3.1. Overall Performances of Pre-Treatment Column

The efficiency of AMD pre-treatment was monitored during time for pH, iron and toxic metal concentrations in the different outputs of the column. The abatement of the different species (iron and toxic metals) were reported as dimensionless concentrations versus treated volume expressed as pore volume (PV).

pH and iron concentration in the 10th output of the column for increasing values of treated volume were reported in Figure 4.

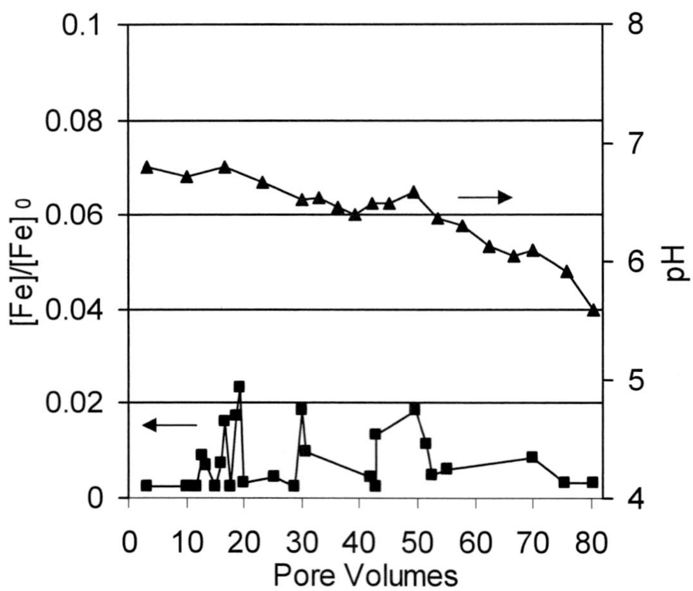

Figure 4. Column tests of chemical precipitation: iron concentration and pH in the 10th output for increasing volumes of treated AMD.

It can be seen that during time a slight decrease of pH occurred due to a reduced buffering capacity of limestone. Nevertheless, a complete removal of iron was obtained (>99%) with residual concentrations ranging from 1 to 10 ppm.

The other monitored species in treated effluents were toxic elements added in synthetic AMD: arsenic, copper, zinc and manganese. Mean values of residual metals after pre-treatment (PV=70) were reported in Table 3. Arsenic and copper were completely removed by the chemical treatment, while zinc and manganese presented residual increasing concentration during the time of this experiment. Arsenic was not detected in anyone of the output stream monitored during this column experiment (instrument detection limit 0.01 ppm). Similarly, copper presented output concentrations about 0.1-0.2 ppm.

For increasing volumes of treated AMD, the formation of dark brown-reddish precipitates of iron oxides and hydroxides was observed within the column. Semi-quantitative information about the composition of such precipitates were obtained by fluorescence analysis of the finest size fraction of the filling material separated after dismantlement of pre-treatment column. This pulverulent and reddish size fraction was chosen as representative of precipitated solid mainly made up of iron oxides and hydroxides. This assumption was confirmed by fluorescence analysis (Table 4) showing that about 22% w/w of this solid was made up of iron, while native limestone presented very low concentration of this element (0.3%).

As a general trend, all toxic elements used for synthetic AMD formulation (Cu, As, Zn and Mn) presented higher concentrations in the precipitated solid than in the native limestone. On the other side typical elements of native limestone (Al, Si and K of aluminosilicate matrix) presented comparable concentrations in precipitate and limestone.

Table 4. Element composition (%) of natural limestone and I column precipitate

Element	Limestone	I column Precipitate
Al	0.207	0.811
Si	1.459	0.403
S	0.122	0.519
K	0.229	0.223
Ca	34.470	15.090
Fe	0.310	22.520
Cu	0.080	1.075
Zn	0.308	1.535
Mn	0.220	0.348
As	0.004	0.109

Semi-quantitative nature of fluorescence analysis (restricted only to a size fractionated sample of the column filling) did not allow to verify the material mass balance of such elements in solid and liquid phase. Nevertheless, interesting information can be obtained by enrichment factors evaluated as percent concentration in the reddish precipitate divided the percent concentration in native limestone. Such estimates from fluorescence analyses denoted that iron was 73 times more concentrated in the precipitated solid extracted from the column. In the same solid, arsenic presented a concentration 28 times larger than in native limestone. Enrichment factors for copper (14), zinc (5) and manganese (1.6) were in agreement with the trend of residual concentration observed in liquid effluents.

3.3.2. Mechanisms of Abatement in Pre-Treatment Column

Abatement of toxic metals in pre-treatment column can be related to different mechanisms and, in particular, precipitation of insoluble species and sorption onto both limestone and precipitated iron oxides. In fact, iron oxides and hydroxides present high sorbent properties due to large surface area rich of hydroxyl groups able to complex species in solution (Stumm, 1987).

Further insight into mechanisms of pollutant's removal can be then obtained by:

- analysis of metal profiles along the column for increasing volumes of treated AMD,
- simulation of metal speciation
- acid digestion and selective extractions of metal concentration from solid samples taken from different sections of the columns.

As for species profiles along the column, pH showed an opposite trend with respect to iron and copper concentration (Figure 5): pH raised from the bottom to the top, while residual iron and copper concentrations decreased along the same direction.

The abatement of iron and copper only for sufficiently high pH environment denoted that chemical precipitation by pH raise is the main mechanism responsible for removal of both pollutants.

Conversely, the profiles of zinc along the column outputs denoted only a partial diminution of residual concentration from the bottom to the top of the column (Figure 6). Similar profiles were found for manganese (Figure 7), but in this case larger residual concentration were observed even during the early stage of pre-treatment.

The analysis of zinc and manganese profiles along the column for different volumes of treated AMD denoted the gradual diminution of abatement of these pollutants in the pre-treatment column (Figures 6 and 7). Such experimental finding denoted that sorption rather than precipitation is the main responsible mechanism operating in zinc and manganese removal.

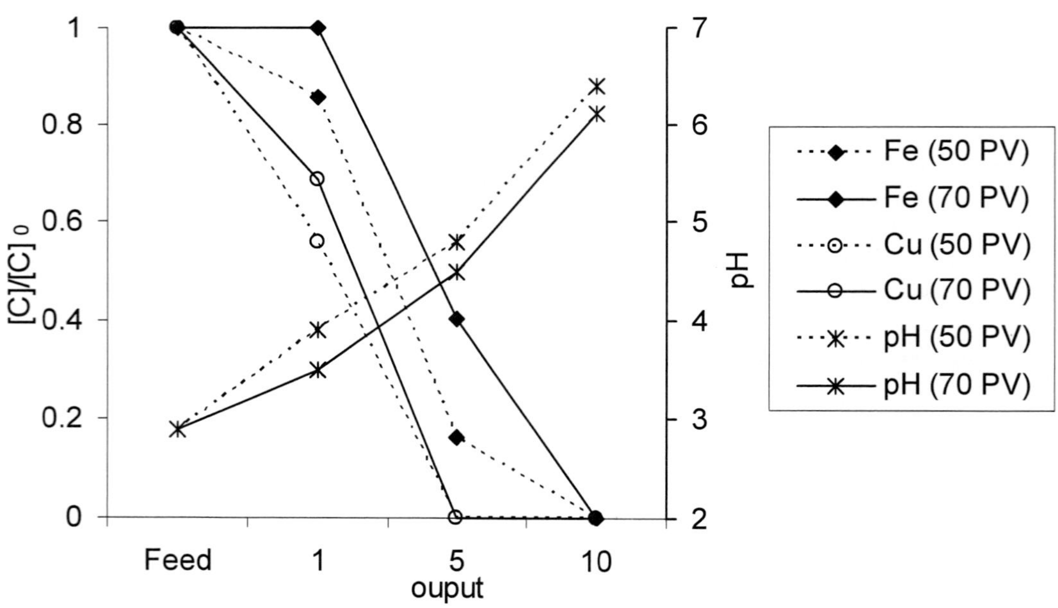

Figure 5.Column test of chemical precipitation: pH, iron, and copper profiles along the column outputs for increasing volumes of treated AMD.

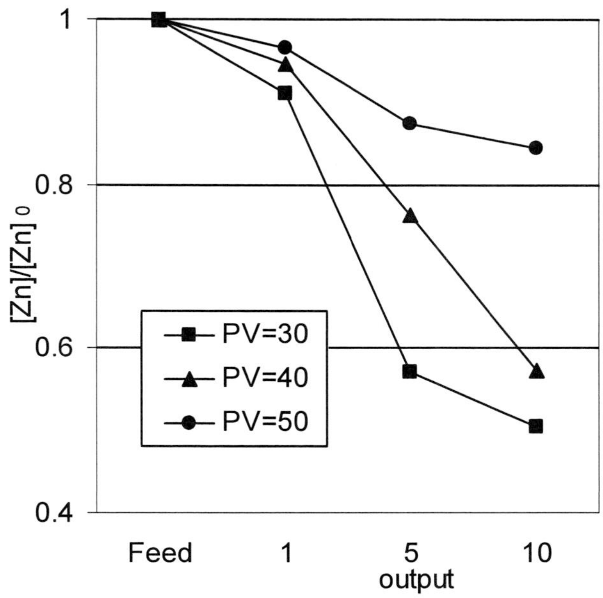

Figure 6.Column test of chemical precipitation: zinc profiles along the column outputs for increasing volumes of treated AMD.

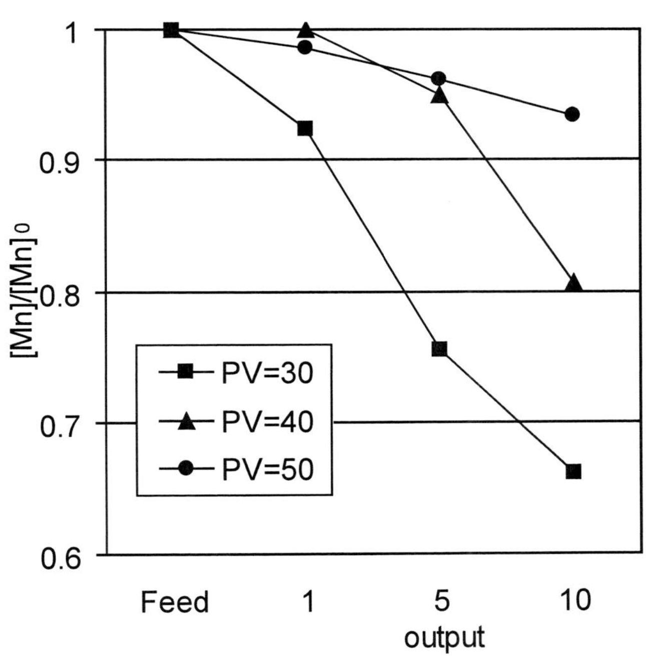

Figure 7. Column test of chemical precipitation: manganese profiles along the column outputs for increasing volumes of treated AMD.

A qualitative distinction among precipitation and sorption for the different toxic species can be obtained by considering the chemical speciation for a solution such as the synthetic AMD used here (Figures 8). Speciation diagrams (as fraction of each species versus pH) were obtained by a dedicated software for chemical equilibria (Medusa, 2001) in the investigated conditions of AMD system (mean concentrations reported in Table 3).

Fraction diagrams showed that chemical pre-treatment by limestone (pH = 6) removed completely iron and copper as Fe_2O_3 (100%) and $Cu_2CO_3(OH)_2$ (95%), respectively. On the other side, both zinc and manganese remained in solution because of larger solubility of their hydroxides, oxides and carbonate salts. Partial abatement of these metals observed along the column length (Figures 6 and 7) can be due to sorption onto limestone gravel and iron hydroxide precipitates.

Arsenic speciation (nor reported here) denoted the complete solubility of this species. Nevertheless, arsenic removal during iron precipitation was already reported in the literature (Casiot et al., 2005; Gault et al., 2005). In particular, according to speciation diagrams in the operating conditions investigated in column tests, arsenic adsorption onto iron oxides and hydroxides can be considered as the predominating mechanisms in the removal of this metalloid.

Main mechanisms occurring in metal removal were further addressed by performing the acid digestion of the filling material extracted from the pre-treatment column (PV=80). Such analyses performed on the solid filling before complete exhaustion of column, can give information about the general efficiency of the process and also the specific way each metal was accumulated in solid phase by comparing pollutant content in the solid material of the bottom (inlet section) and upper (exit section) parts of the column. Table 5 reports the average values ± standard deviations obtained from the acid digestion of four distinct solid samples of the inlet section and the exit section of the pre-treatment column.

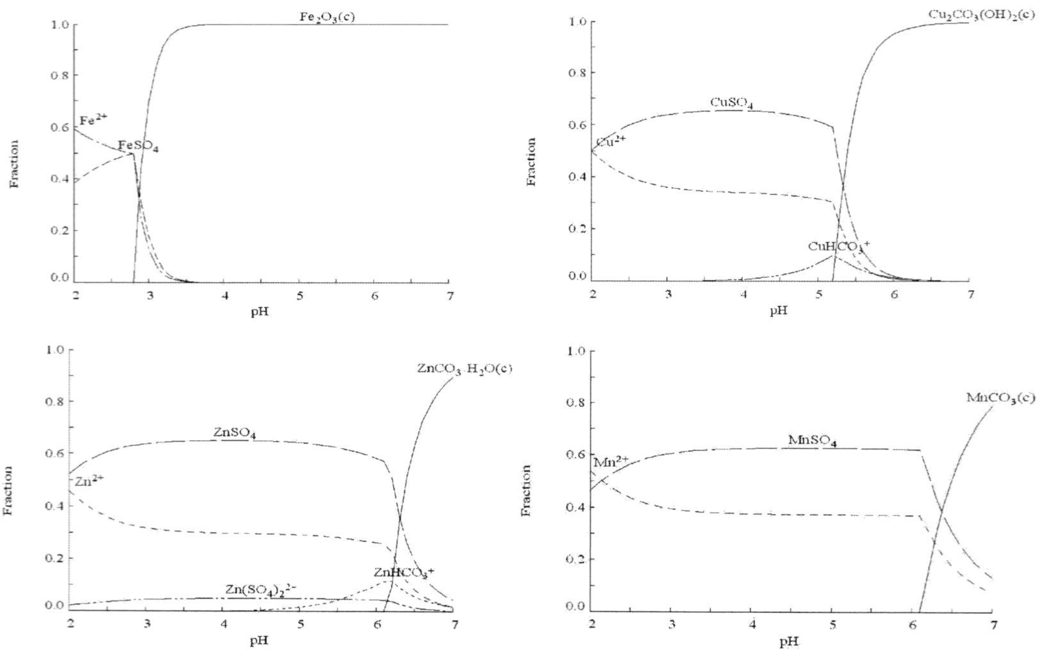

Figure 8. Speciation diagrams of iron, copper, zinc and manganese in AMD system (mean concentrations reported in Table 3; electric potential E_h=300mV).

Table 5. Acid digestion (ppm) for solid samples taken from the inlet and exit sections of the pre-treatment column (I) and the SRB column (II)

Metal	I column		II column	
	inlet section	exit section	inlet section	exit section
Fe	2500±50	5400±80	-	-
Cu	440±20	2050±50	-	-
As	28±2	53±4	-	-
Mn	325±20	290±20	870±40	4990±90
Zn	840±30	880±30	940±50	5010±60

The analysis of these data showed that iron was mainly concentrated in the exit section of the column rather than in the inlet section. The same trend was observed for copper, while manganese and zinc presented similar concentrations in solid phase both in the inlet and in the exit sections.

The different behaviour of iron and copper with respect to manganese and zinc confirmed the main hypothesis about the mechanisms of their removal deduced both by residual concentrations in the liquid phase (Figures 5, 6 and 7) and by simulation of metal speciation (Figure 8). In fact, both iron and copper were completely removed in the pre-treatment column and speciation diagrams showed that these metals become insoluble for sufficiently high pH. pH gradient along the pre-treatment column (Figure 5) denoted a gradual decrease of H^+ concentration from the inlet to the exit section of the column. As a consequence, metal precipitation due to pH rise mainly occurred the column part where sufficiently high values of pH were reached (exit section). This was the case of iron and copper. On the other hand, metals which remained soluble in the investigated range of pH conditions, manganese and zinc, presented a quite uniform concentration in solid phase along the column, because their partial removal was mainly due to adsorption phenomena not specifically related to pH gradient in the range of conditions here considered. Finally, arsenic data from acid digestion denoted a significant increase in the solid phase of the exit section: this finding confirmed this metalloid can be removed both by sorption and coprecipitation with iron (this mechanism manly occurring in the exit section when iron precipitation prevailed due to pH rise).

3.4. Column Tests of Biological Precipitation

3.4.1. Overall Performances of SRB Column

After the start-up of the SRB inoculated column, treated effluents from the column of chemical precipitation were fed to this second column. In this second step of AMD purification, SRB reduced sulphates to sulphides, which determined metal precipitation as sulphides.

Sulphate abatement in the different outputs of the column denoted the improvement of sulphate reduction for increasing volumes of treated effluents (Figure 9).

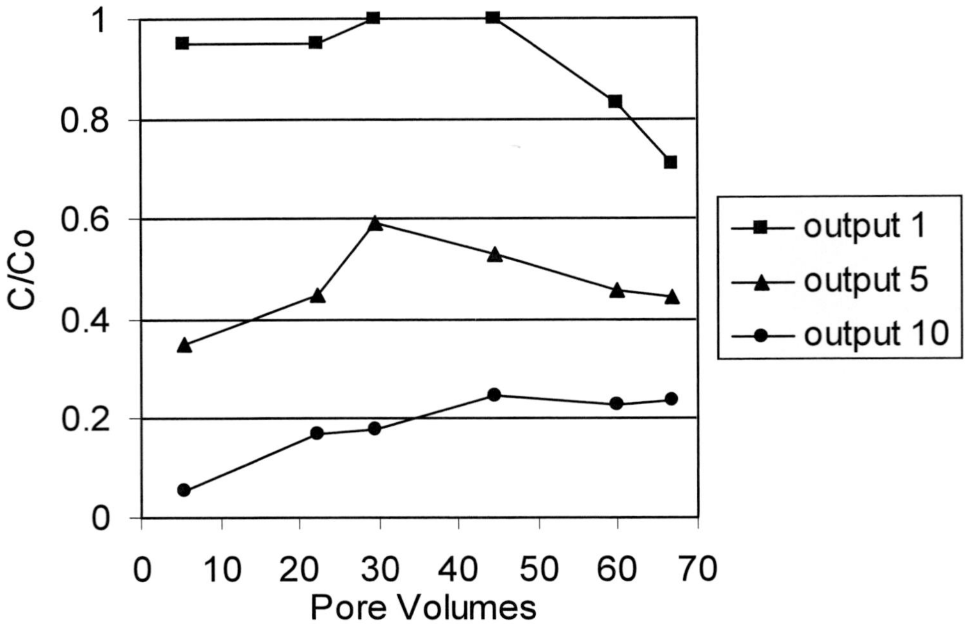

Figure 9. Column tests of biological precipitation: sulphate profiles in the different outputs for increasing volumes of treated AMD.

Such effect was especially evident for the first output where the reduction of sulphate was mainly due to SRB metabolism (PV > 60). The non monotonous trends observed for the fifth output can be related to the simultaneous occurrence of sulphate sorption and sulphate reduction. In the early stage sulphate sorption onto solid matrices prevailed over bioreduction according to the well-known biosorption properties of humic substances (Dzombak et al., 1986; Tipping, 1993; Westall et al., 1998). Gradual saturation of sorption sites determined an increase in effluent concentration. Activation of SRB activity caused the further diminution especially evident in the fifth output.

Batch growth tests using the mixture of the II column as solid media for SRB growth confirmed the relevance of biosorption in sulphate removal. In fact, blank suspensions of the solid mixture without inoculum were characterised by significant sulphate abatement, even though no sulphide release was revealed (Figure 10).

After 22 days, inoculated suspensions (releasing H_2S) showed 83% sulphate abatement, while blank samples presented 58% abatement confirming that solid components used for II column filling are characterised by significant sorbing capacities. In particular sorption properties of the main component of II column filling (compost) denoted 53% abatement after 22 days showing its relevant contribution in biosorbing properties of the filling mixtures.

Zinc was completely removed and can not be detected in any output of the SRB column (<0.01 ppm). Manganese profiles along the column denoted that sulphate reducing bacteria metabolism and simultaneous biosorption onto solid substrates completely remove this metal (Figure 11).

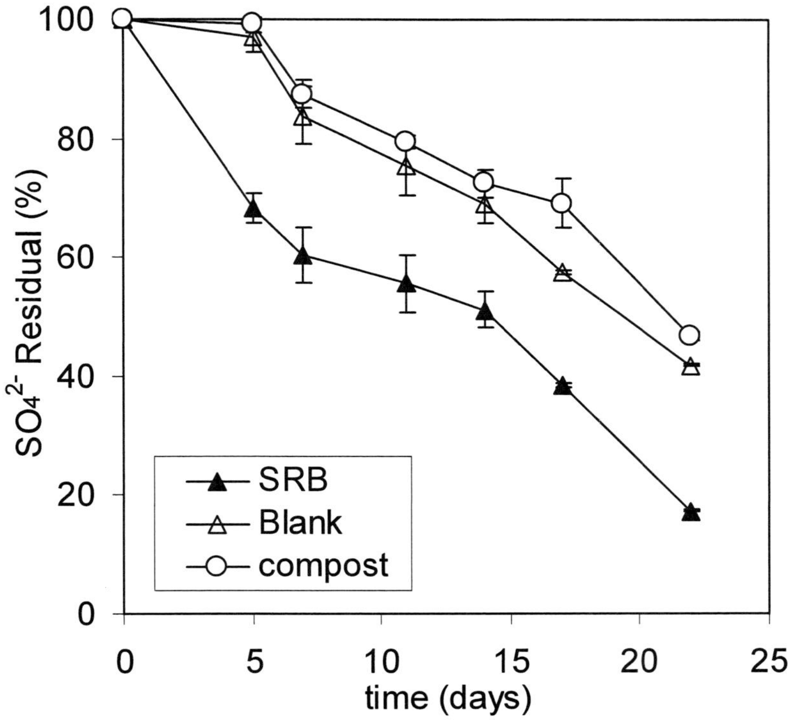

Figure 10. Sulphate abatement in batch tests using II column filling material inoculated by SRB (SRB), II column filling material without SRB inoculum (Blank), and compost used in II column filling material.

3.4.2. Mechanisms of Abatement in SRB Column

Further information about the removal of both these metals were obtained by comparing their concentrations in solid samples extracted from the inlet and exit sections of the SRB column before its exhaustion (after 60 PV).

Average values from acid digestions (Table 5) denoted that both zinc and manganese were mainly concentrated in the solid material taken from the upper part of the column (exit section).

A deeper insight into metal removal can be obtained by sequential extractions: these procedures are all based on the general principle of reacting a soil sample with chemical solutions characterised by increasing strength. In particular sequential extraction scheme used here was a modified Tessier procedure, which allows an operative speciation of metal in the solid phase according to the following scheme:

I Step: exchangeable metals;
II Step: complexed onto iron and manganese oxides;
III Step: weakly bound to organic matter;
IV Step: strongly bound to organic matter;
V Step: bound to the sulphide phase.

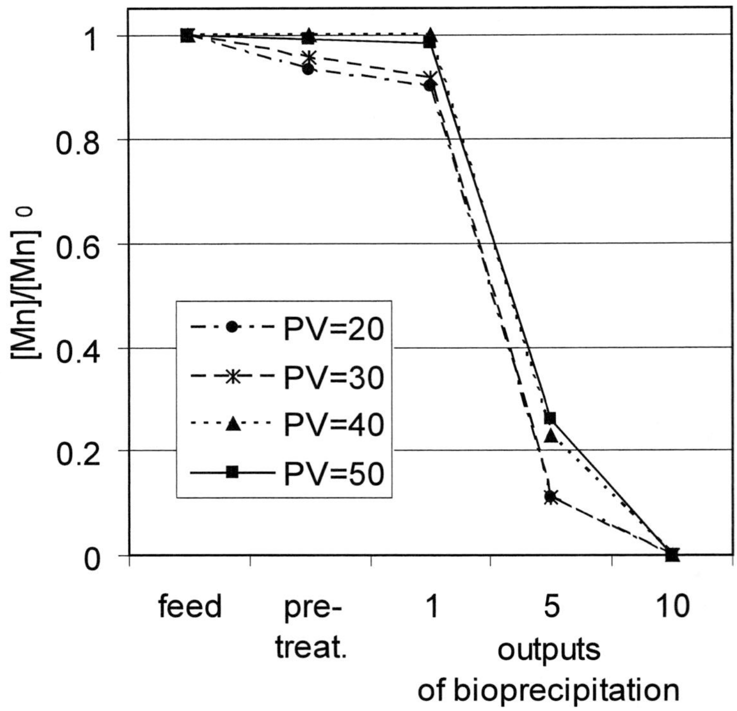

Figure 11. Column tests of biological precipitation: manganese profiles in the different outputs for increasing volumes of treated AMD.

This sequential extraction scheme was chosen to isolate the contributions due to biosorption (III+IV steps) and bioprecipitation (IV step) in Mn and Zn abatement in comparison with other extraction procedures reported in the literature (Swash and Monhemius, 2005) that are characterised by a unique extractive step for metals bound to organics and metals precipitated as sulphides.

Nevertheless, it should be pointed out that only semi-quantitative comparison can be obtained by sequential extractions due to the numerous limits of this experimental procedure (low reproducibility, error propagations, strong influence of operating conditions, effective selectivity of the extracting reagents, re-adsorption of metals during extraction) (Pagnanelli et al., 2004). Experimental results of sequential extractions were reported in Table 6 as % of extracted metal in each step. This data denoted the predominant contribution of the biosorbing properties of biological matrices both in Mn and Zn removal and the localisation of SRB activities in the upper parts of the column. This second finding can be due to the migration of gaseous H_2S in the upper part of the column and consequently the localisation in such compartment of the precipitation of metal as sulphides.

Table 6. Sequential extractions (expressed as % of extracted metals) for solid samples taken from the inlet and exit sections of the pre-treatment column (I) and the SRB column (II)

		Sequential extractions				
Metal	Section	I Step	II Step	III+IV Steps	V steps	Residue
Mn	Inlet	23	46	30	1	1
Mn	Exit	33	31	31	4	1
Zn	Inlet	22	34	43	0	1
Zn	Exit	17	23	56	3	1

4. CONCLUSION

In this study a combined chemical-biological treatment was tested for the decontamination of synthetic AMD containing Fe, Cu, As, Zn, and Mn.

Column tests denoted that this combined approach succeeded in attenuating AMD pollution. In particular, pre-treatment by natural limestone increased the pH allowing the complete removal of metals with low solubility as hydroxides and carbonates (Fe and Cu). A refining treatment in the second column (made up of natural organic matter inoculated by SRB) allowed the removal of those metals which can not be removed simply by increasing pH (Mn and Zn). Treated effluents from the second column were purified by these residual pollutants initially contained in AMD. Specific mechanisms of removal for each metal were investigated by considering profiles of pollutants along the column length, simulations of metal speciation in the investigated conditions, biosorption tests, acid digestion and sequential extractions from solid samples of column filling material.

The combination of such results led to the following main conclusions:

- iron was completely removed by precipitation in the pre-treatment column, whose active zone was only that with adequate pH conditions as evidenced by acid digestion of column sections;
- copper was completely precipitated in the first treatment column by the same chemical mechanism observed for iron;
- manganese and zinc cannot be removed only for pH rise: dynamic trends of axial profiles showed a shifting front of saturation along the column and denoted that marginal removals in the first column were due only to sorption onto limestone and iron precipitates; both metals were completely removed in the second column of biological treatment by biosorption and bioprecipitation mechanisms;
- arsenic was removed in the first column both by sorption and copreciptation with iron being concentrated in the solid from the column section active for iron precipitation.

In conclusion, this work presents a methodological approach to developing a lab-study investigation able to isolate the different mechanisms simultaneously occurring in pollutant removal in biological treatment by SRB. Such preliminary quantification should be the basis for further kinetic modelling in order to determine the half-lives of the different pollutants specifically related to SRB metabolism, which is the only long-term active mechanism in full-scale PRB. Other mechanisms such as biosorption should be isolated and not accounted for in such kinetic parameters in order to avoid overestimates of SRB performances in long term run.

Experimental results reported here showed that biosorption is not a negligible mechanism in lab-scale study generally used for half-lives estimates. Column tests can not then be used as stand alone data for kinetic characterisation of such biological systems but should be accompanied by further investigation specifically aiming to mechanism isolation and quantification.

REFERENCES

Alvarez, M. T., Crespo, C. and Mattiasson, B. (2007). Precipitation of Zn(II), Cu(II) and Pb(II) at bench-scale using hydrogen sulphide from the utilization of volatile fatty acids *Chemosphere, 66,* 1677-1683.

Barbaro, M., Passariello, B., Quaresima, S., Casciello, A. and Marabini, A. (1995). Analysis of Rare Earth Elements in Rock Samples by Inductively Coupled Plasma-Mass Spectrometry (ICP-MS). *Microchem. J., 51,* 312-318.

Baker, B. J. and Banfield, J. F. (2003). Microbial communities in acid mine drainage. *FEMS Microbiol. Ecol., 44,* 139-152.

Benner, S. G., Blowes, D. W., Gould,W. D., Herbert, R. B. and Ptacek C. J. (1999). Geochemistry of a Permeable Reactive Barrier for Metals and Acid Mine Drainage. *Environ. Sci. Technol., 33,* 2793-2799.

Campanella, L., D'Orazio, D., Petronio, B. M. and Pietrantonio, E. (1995). Proposal for a metal speciation study in sediments. *Analytica Chimica Acta, 309,* 387-393.

Carrera, J., Alcolea, A., Bolzicco, J., Knudby, C. and Ayora, C. (2001). *An experimental geochemical barrier at Aznalcollar*. Thornton and Oswald Eds Proc. 3rd Internat. Conf. Groundwater Quality Sheffield UK, 18-21 June, pp 407-409.

Casiot, C., Lebrun, S., Morin, G., Bruneel, O., Personne´, J. C. and Elbaz-Poulichet, F. (2005). Sorption and redox processes controlling arsenic fate and transport in a stream impacted by acid mine drainage *Sci. Total Environ., 347,* 122– 130.

Chang, I. S., Shin, P. K. and Kim, B. H. (2000). Biological treatment of acid mine drainage under sulphate-reducing conditions with solid waste materials as substrate. *Wat. Res., 34,* 1269-1277.

Christensen, B., Laake, M. and Lien T. (1996). Treatment of acid mine water by sulfate-reducing bacteria; results from a bench scale experiment. *Wat. Res., 30,* 1617–1624.

Cocos, I. A., Zagury, G. J., Clement, B. and Samson, R. (2002). Multiple factor design for reactive mixture selection for use in reactive walls in mine drainage treatment. *Wat. Res., 32,* 167-177.

Dzombak, D. A., Fish, W. and Morel F. M. M. (1986). Metal-humate interactions. 1. Discrete ligand and continuous distribution models. *Environ. Sci. Technol., 20,* 669-675.

Foucher, S., Battaglia-Brunet, F., Ignatiadis, I.and Morinet, D.(2001). Treatment by sulfate-reducing bacteria of chessy acid-mine drainage and metals recovery. *Chemical Engineering Science, 56,* 1639–1645.

Gault, A. G., Cooke, D. R., Townsend, A. T., Charnock, J. M. and Polya D. A. (2005). Mechanisms of arsenic attenuation in acid mine drainage from Mount Bischoff, western Tasmania *Sci. Total Environ., 345,* 219–228.

Gibert, O., de Pablo, J., Cortina, J. L. and Ayora C. (2002). Treatment of acid mine drainage by sulphate-reducing bacteria using permeable reactive barriers: A review from laboratory to full-scale Experiments. *Re/Views in Environmental Science and Bio/Technology, 1,* 327–333.

Gibert, O., de Pablo, J., Cortina, J. L. and Ayora, C. (2003). Evaluation of municipal compost/limestone/ iron mixtures as filling material for permeable reactive barriers for in-situ acid mine drainage treatment. *J. Chem. Technol. Biot., 78,* 489-496.

Gibert, O., de Pablo, J., Cortina, J. L. and Ayora, C. (2004). Chemical characterization of natural organic substrates for biological mitigation of acid mine drainage. *Water Res., 38,* 4186-4196.

Groudev, S. N., Spasova, I. I. and Georgiev, P. S. (2001). In situ bioremediation of soils contaminated with radioactive elements and toxic heavy metals. *Int. J. Miner. Process., 62,* 301-308.

Hammack, R. W. and Edenborn H.M. (1992). The removal of nickel from mine waters using bacterial sulfate reduction. *Appl. Microbiol. Biot., 37,* 674-678.

Herbert, R. B., Benner, S. G. and Blowes, D. W. (2000). Solid phase iron-sulfur geochemistry of a reactive barrier for treatment of mine drainage. *Applied Geochemistry, 15,* 1331-1343.

Jarvis, A. P., Moustafa, M., Orme, P. H. A. and Younger, P. L. (2006). Effective remediation of grossly polluted acidic, and metal-rich, spoil heap drainage using a novel, low-cost, permeable reactive barrier in Northumberland, UK. *Environ. Pollut., 143,* 261-268.

Johnson, D.B. (2006). Biohydrometallurgy and the environment: intimate and important interplay *Hydrometallurgy, 83,* 153-166.

Johnson, D. B. and Hallberg, K. B. (2005). Acid mine drainage remediation options: a review. *Sci. Total Environ., 338,* 3-14.

Johnson, D.B. and Hallberg K.B. 2005b. Biogeochemistry of the compost bioreactor components of a composite acid mine drainage passive remediation system *Science Total Environment, 338,* 81-93.

Kalin, M., Fyson, A. and Wheeler, W. N. (2006). The chemistry of conventional and alternative treatment systems for the neutralization of acid mine drainage. *Sci. Total Environ., 366,* 395-408.

Komnitsas, K., Bartzas, G. and Paspaliaris, I. (2004). Efficiency of limestone and red mud barriers: laboratory column studies. *Miner. Eng., 17,* 183-194.

Ludwig, R. D., McGregor, R. G., Blowes, D. W., Benner, S. G. and Mountjoy, K. (2002). A permeable reactive barrier for treatment of heavy metals. *Ground Water, 40,* 59-66.

Medusa Program by ©Ignasi Puigdomenech, 2001

Morrison, S. J., Matzler, D. R.and Dwyer B. P. (2002). Removal of As, Mn, Mo, Se, U, V and Zn from groundwater by zero-valent iron in a passive treatment cell: reaction progress modelling. *J. Contaminant Hydrology, 56,* 99-116.

Neculita, C. M., Zagury, G. J. and Bussiere, B. (2007). Passive treatment of acid mine drainage in bioreactors using sulphate reducing bacteria: Critical review and research needs. *J. Environ. Qual., 36,* 1-16.

Pagnanelli, F., Moscardini, E., Giuliano, V. and Toro, L. (2004). Sequential extraction of heavy metals in river sediments of an abandoned pyrite mining area: pollution detection and affinity series. *Environ. Pollut., 132,* 189-201.

Pagnanelli, F., Luigi, M., Mainelli, S. and Toro, L. (2007). Use of natural materials for the inhibition of iron oxidizing bacteria involved in the generation of acid mine drainage. *Hydrometallurgy, 87,* 27-35.

Peppas, A., Komnitsas, K. and Halikia, I. (2000). Use of organic covers for acid mine drainage control. *Miner. Eng., 13,* 563-574.

Postgate J.R. 1979. The sulphate-reducing bacteria. Cambridge University Press.

Sheoran, A. S. and Sheoran, V. (2006). Heavy metal removal mechanism of acid mine drainage in wetlands: A critical review. *Miner. Eng., 19,* 105-116.

Simon, M., Martın, F., Garcıa, I., Bouza, P., Dorronsoro, C. and Aguilar, J. (2005). Interaction of limestone grains and acidic solutions from the oxidation of pyrite tailings. *Environ. Pollut., 135,* 65–72.

Stumm W. 1987. Aquatic surface chemistry John Wiley and Sons, Inc.

Swash, P. M. and Monhemius, A. J. (2005). Characteristics and stabilities of residues from the Wheal Jane constructed wetlands. *Science Total Environment, 338,* 95-105.

Tipping, E. (1993). Modeling ion binding by humic acids. *Colloids Surf. A, 73,* 117.

Tunay, O. and Kabdasli, N. I. (1994). Hydroxide precipitation of complexed metals. *Water Research, 28,* 2117–2124.

Waybrant, K. R., Blowes D. W. and Ptacek C. J. (1998). Selection of Reactive Mixtures for Use in Permeable Reactive Walls for Treatment of Mine Drainage. *Environ. Sci. Technol., 32,* 1972-1979.

Waybrant, K. R., Ptacek, C. J. and Blowes D. W. (2002). Treatment of mine drainage using permeable reactive barrers: column experiments. *Environ. Sci. Technol., 36,* 1349-1356.

Westall, J. C., Jones, J. D., Turner, G. D. and Zachara, J. M. (1995). Models for Association of Metal Ions with Heterogeneous Environmental Sorbents. 1. Complexation of Co(II) by Leonardite Humic Acid as a Function of pH and NaClO4 Concentration. *Environ. Sci. Technol., 29,* 951-959.

Whitehead, P. G., Hall, G., Neal, C. and Prior, H. (2005). Chemical behaviour of the Wheal Jane bioremediation system. *Science Total Environ, 338,* 41-51.

Wilkin, R. T. and McNeil, M. S. (2003). Laboratory evaluation of zero-valent iron to treat water impacted by acid mine drainage. *Chemosphere, 53,* 715-725.

In: Causes and Effects of Heavy Metal Pollution
Editor: Mikel L. Sanchez
ISBN 978-1-60456-900-1

Chapter 3

Assisted Phytoextraction for Abandoned Mining Areas Remediation

Alessia Cao, Alessandra Carucci and Tiziana Lai
DIGITA, Dept. Geoengineering and Environmental Technologies, University of Cagliari, Cagliari, Italy

Abstract

The remediation of mining areas represents a relevant environmental problem in all Europe. The high concentration of heavy metals and the lack of nutrients determines the desertification of wide areas. In Sardinia (Italy) the poor management of Montevecchio – Ingurtosu mining district after mine closure caused the dispersion of high amounts of contaminants by wind and water erosion on wide areas. The wide extension of the contaminated area and the high level of contamination by heavy metals make the application of phytoextraction feasible for this area. The environmental risk related to the presence of heavy metals can be evaluated by determining the bioavailable metal fraction in soil. The Department of Geoengineering and Environmental Technologies of the University of Cagliari made experiments of phytoextraction and assisted phytoextraction both with plants having a high biomass production (*Mirabilis jalapa*) and with native species (*Cistus salvifoliius*, *Scrophularia canina* and *Teucrium flavum*). Easily biodegradable chelating agents were applied in laboratory experiences (MGDA - methylglycine diacetic acid, S,S-EDDS - [S, S]-ethylenediaminedisuccinic acid, IDSA - iminodisuccinic acid). The ability of the plant species to tolerate and accumulate heavy metals demonstrated the applicability of phytoextraction to the abandoned mining areas remediation.

Introduction

The lack in remedial measures design for eliminating or confining sources of pollution after mining activity closure may cause severe damage to the environment. Uncontrolled mine tailing dumps, exposed to water and wind erosion, generate a large amount of emissions containing heavy metals which can contaminate the surroundings environmental compartments: soil, water and air.

As a consequence the presence of mine activities altered the agronomic characteristics making these soils arid and devoid of vegetation, facilitating the spread of contaminants in large areas (Ernst, 1996; Freitas et al., 2004; Conesa et al., 2007).

Issues related to the presence of abandoned mining areas were studied by the technical scientific community with particular attention to the potential bioavailability of contaminants and the risks for the population (Ernst, 1996; Licskó et al., 1999; Cappuyns et al., 2006; Conesa et al., 2007).

Many techniques can be used for the removal of heavy metals for soil remediation. Traditional methods have in general a high environmental impact and may produce secondary pollution besides having considerable costs of maintenance (EPA, 1997; Glass, 2000). Other emerging deals involve natural remediation processes such as phytoremediation.

Phytoremediation is a technique consisting in the use of plants for the decontamination of soil and/or water using their natural capacity to assimilate, accumulate and degrade contaminants of the media considered. In the field of contaminated sites remediation, plants can be used both to mineralize and immobilize toxic compounds in the root zone, and to accumulate and concentrate metals and other inorganic compounds extracted from the ground into the air portion.

Phytoextraction, in particular, is a phytoremediation technology that uses plant species active in heavy metals absorption.

In order to improve the efficiency of conventional phytoextraction, the addition of chelating agents can be applied, with the aim to increase the bioavailable metal fraction, that is the fraction more available for plant uptake (Blaylock et al., 1997; Kambhampati and Williams, 2001; Römkens et al., 2002; Leštan et al., 2007).

This article aims to conduct a review on the phytoremediation technique and its applicability to contaminated sites remediation. The results available in literature concerning the application of this technique are presented together with those obtained by the authors in the researches carried out on the phytoremediation of some abandoned mining areas.

The following paragraphs present the characteristics of the mining sites of Sulcis-Iglesiente (south-western Sardinia, Italy) and describe the phytoremediation techniques with particular attention to the application of the assisted phytoextraction, as well as the evaluation of bioavailability as a parameter for determining the risk associated with the presence of heavy metals in soil.

Abandoned Mining Areas (The Case of Montevecchio Mining District)

Mining activity generates high amounts of wastes characterized by a low pH, high concentrations of heavy metals, lack of nutrients and low water retention capacity. This determines adverse conditions to plant establishment and soil surfaces remain bare and unvegetated. The weathering on the mining products with the combined action of wind and water erosion causes the diffusion and transport of contaminants both in sediments and soils along several kilometres in the surrounding water courses (Dessì et al., 1999; Cappuyns et al., 2006; Conesa et al., 2007).

Successful revegetation can be an attractive solution and, at the same time, be relatively inexpensive. A vegetation cover can be effective in providing the necessary surface stability to prevent wind-blow of contaminated particulates and in reducing water pollution by interception of a substantial proportion of incident precipitation (Tordoff et al., 2000). It may also allow the restoration of the biological conditions of the area before metal contamination.

The Montevecchio-Ingurtosu mining district (south-western Sardinia, Italy) has been a major source of Pb and Zn in Europe since the mid-19th century. Mining activity ceased between 1968 and 1980. The ore body consisted of galena and sphalerite veins in a gangue of quartz and carbonate minerals (siderite prevailing over ankerite, dolomite and calcite) with minor amounts of pyrite, chalcopyrite, barite, cerussite, anglesite, smithsonite and W-minerals.

In this case the main environmental concern is the presence of waste tailings generated during mineral processing. Tailings were dumped over an extensive area along the riverbeds, being now abandoned (Caredda et al., 1999). Tailings were also used for mine tunnels refilling (Concas et al., 2006). Caboi et al. (1993) and Fanfani et al. (1997) have shown that tailings are characterized by high concentrations of heavy metals such as Ni, Co, Cu, Zn, Cd, and Pb, susceptible of leaching by surface and filtrating waters. Tailings also contain residual concentrations of pyrite that enables the generation of AMD (acid mine drainage). Granulometric analysis of tailing samples mostly exhibits a grain size ranging from 500 to 120 μm and a significant percentage (~15%) of fine particles (<75 μm).

No special attempt has been made in the past to mitigate the impact of the pollution (Fanfani et al., 1997). The transport of the mine deposits along the rivers of the major water courses determined the contamination of the downstream drainage. The contaminated soils depth ranges from 0.2 up to 2.0 m. The characterization carried out during Dessì et al. (1999) project, evaluates the contaminated soil volume near Picalinna tailing dam, one of the major contamination sources of the Montevecchio area, higher than 2 Mm^3 along the riverbanks, reaching 800 m distance from the river Rio Sitzerri. The original soils of the valley, *vertisoils*, and the downstream plan were considered of good characteristics for agricultural uses. The contamination has produced a loss of fertility and desertification in a wide area, estimated in 20 km^2 (Dessì et al., 1999).

In some areas along the river basin, the level of heavy metals is higher than in the tailing dam. In general terms the survey in this wide area allowed to define two types of soil profiles: the "red soil" and the "top soil" horizon. The horizon RS (Red Soil), depth range 0.1÷1.8 m, is composed by an alternate deposit of brownish, red and grey tailing muds, the second TS (Top Soil), depth range 0.5÷1.2 m, is the original brown vertisoil, but with a high percentage of tailings. This mixing has been produced by the farmers ploughing trying to remediate the pollution of the soils, that were turned to not productive ones by the wind and water transportation of the tailing muds. Along the contaminated area both sequences described are visible. In horizon TS there is a moderate recover of crops production but with a high contamination level (Dessì et al., 1999).

The Picalinna spoil is not abiotic. Because of the gravel layer on top a pedological evolution has been observed, so some species of Mediterranean flora are present. The species collected are edible ones by the livestock or by the humans. Samples of plants taken from the horizon RS, along the drainage catchment, and on the horizon TS showed a serious pollution in oats for the livestock feeding up to 600 mg/kg in Pb in the horizon TS, while more limited values are present in the plants on the Picalinna flotation tailing dam (Montevecchio district). The heavy metals content in the edible plants sampled highlights the hazard caused by the highly bioavailable fraction of the metals in the soil that has been absorbed by the plants. This could create a major harm due to the accumulation of the toxic elements in the ecosystem and the food-chain (Dessì et al., 1999).

The ecological and human risks related to heavy metals presence in soil cannot be evaluated from the total metal concentration but from the concentration of the metal that shows some effects on organisms. A way to understand this factor is to measure the metal bioavailable fraction in soil.

METAL BIOAVAILABILITY

The bioavailable fraction of a chemical can be defined as the fraction present in a specific environmental compartment that is either available or can be made available for uptake by organisms or plants (Peijnenburg and Jager, 2003).

The concept of bioavailability, as a "fraction absorbed", was originally developed in order to quantify, in agricultural soils, the presence of trace elements (Zn, Cu, Fe, Cr, Mn, etc.) that are essential to the normal growth of plants (Leita, 2005).

At present the study of the bioavailability is not only exclusive area of agriculture, but it provides a valuable tool for assessing the risk due to the exposure of plants and organisms to high concentrations of heavy metals. A current approach of the scientific community considers the toxicity of metals depending on the bioavailable metal fraction, rather than on their total concentration in soil.

The heavy metal bioavailability depends on the different fractions of metals in soil: water soluble, exchangeable, oxide-bound, carbonate-bound, organic matter-bound, and residual that is occluded in the resistant minerals. The first two fractions, water soluble and exchangeable ones, can be considered bioavailable fractions, oxide-bound, carbonate-bound, and organic matter bound fractions could be potentially bioavailable, while the residual fraction is in general not available to organisms.

This fractionation is influenced by several soil characteristics such as pH, cation exchange capacity, organic matter content, amounts and forms of oxides and carbonates as well as mineral composition (He et al., 2005).

The mobility and availability of heavy metals in soil are controlled by many chemical and biochemical processes such as precipitation–dissolution, adsorption–desorption, complexation-dissociation, and redox reactions as well as chelation (Adriano et al., 2004; He et al., 2005).

The heavy metal bioavailable fraction assessment can be made using both direct and indirect methods. The use of direct methods enables the measurement of bioavailability *in situ*, while indirect methods such as sequential extractions allow the evaluation of natural phenomena that can occur in a contaminated area increasing the potential risks to organisms.

The indirect approach to determine bioavailability is based on the use of sequential extractions, which seem suitable to simulate the absorption of inorganic elements by the plant (Guzzi et al., 2005).

The sequential extraction procedures aim to estimate the distribution of various chemical forms of an element among the solid phases in soil. The main differences in the methods developed are the chemical extractants used, the duration of each extractive phase and the temperature at which the reactions are carried out.

One of the first sequential extraction procedures is the one proposed by Tessier et al. (1979), which parcels particulate trace metals (Cd, Co, Cu, Ni, Pb, Zn, Fe and Mn) into five distinct fractions: exchangeable, bound to carbonates, bound to iron and manganese oxides, bound to organic matter, and residual.

Some procedures, such as that proposed by Elsokkary et al. (1995) provide six stages, in order to quantify the metal fractions bound to iron and manganese oxides separately. Manganese oxides are in fact more easily reducible than iron oxides, consequently the evaluation of these metal fractions gives more elements for a correct assessment of mobility.

The sequential extraction performed by using the BCR method (standards, measurements, and testing programme of the European Commission) enables the complete fractionation of exchangeable, water- and (acetic) acid-soluble, iron/manganese oxides (hydroxylamine hydrochloride), and organic- and sulfide-bound fractions (ammonium acetate) (Rauret et al., 1998).

The procedures proposed by Barbafieri et al. (1996) and Castaldi et al. (2004) divide the metal mobile fraction into three different pools: immediately soluble, exchangeable, and complexed or adsorbed metals. The sum of these three pools represents the bioavailable metal fraction. The method of Barbafieri et al. (1996) provides for the sequential use of H_2O, KNO_3, and EDTA while that of Castaldi et al. (2004) uses H_2O, $Ca(NO_3)_2$, and EDTA.

Sequential extraction methods are useful tools for assessing the potential mobility and reactivity of heavy metals in soil. However it has to be taken into account that many sequential extraction procedures have been performed to determine the lack of essential trace elements in the soil for crop growth, not to assess metal toxicity, so the chemical extractants could be aggressive, with a consequent overestimation of heavy metal bioavailability.

This led to introduce methods to determine the concentration of metals in soil solution, such as the direct sampling of the solution present in the pores system or the use of mild extractant, such as water or alkaline salts that do not change the soil surface (Pedron and Petruzzelli, 2005).

In order to understand which metal fraction evaluated using the direct or indirect methods was better correlated to the metal bioavailability to plants, a recent study (Cao et al., 2008) compared the metal fractions with metals accumulated in leaves. The study refers to the bioavailability of lead and zinc in a mining soil collected from the Montevecchio mining district.

Direct determination of metal concentration in pore water was carried out using rhizon soil moisture samplers (Rhizosphere Research Products, Wageningen, The Netherlands). These samplers consist of a porous plastic tube capped with nylon at one end and connected, at the other end, through a polyethylene tubing, to the sampling device (a syringe).

A comparison between the lead concentration in soil solution and in leaves, after the application of chelating agents, showed that the lead concentration in soil solution is an indicator of its bioavailability. On the contrary zinc concentration in leaves was not correlated with the metal concentration in soil solution (Cao et al., 2008).

Some studies showed the different behaviour in metal absorption by plants in the presence or absence of chelants. In the absence of chelants zinc accumulation is efficient even if the concentration in soil solution is low, while only a little amount of lead is taken up by plants. On the contrary, in most cases, in a soil treated with chelants, lead uptake increases while zinc uptake decreases (Nowack et al., 2006).

The correlation between metal concentration in soil solution and the different metal fractions in soil evaluated with sequential extractions is positive only for the soluble metal fraction evaluated using the Barbafieri et al. (1996) and Castaldi et al. (2004) methods(Cao et al, 2008). The metals bioavailability to plants can therefore be more efficiently determined through the soluble metal fraction than through the total bioavailability measured with the Barbafieri et al. (1996) and Castaldi et al. (2004) methods. No correlations with the metal concentrations in soil solution were obtained with the individual metal fractions identified by using the BCR method (Rauret et al., 1998).

PHYTOREMEDIATION

Phytoremediation can be considered a plant-based remediation technology that uses plants to degrade, extract, contain, or immobilize contaminants from soil and water (EPA, 1997).

Phytoremediation applications can be classified either on the basis of process goal (degradation, extraction, containment, or a combination) or on the mechanisms involved (extraction, concentration, stabilization, degradation, volatilization) into: Phytoextraction, Phytostabilization, Rhizodegradation, Phytodegradation, Phytovolatilization.

Phytoextraction refers to the use of plants capable to accumulate metals or salts into the roots and translocate them to the aboveground shoots or leaves. The plants that can absorb unusually large amounts of metals in comparison to other plants and the ambient metal concentrations are called hyperaccumulators. A plant is a hyperaccumulator if it is able to accumulate at least 1,000 mg/kg (dry weight) of a specific metal or metalloid (for some metals or metalloids the concentration must be 10,000 mg/kg) (Baker et al., 1998).

The process that regulates the transformation of contaminants into a stable form, reducing risks related to their presence in the environment, can be applied to both organic and inorganic contaminants and is called phytostabilization. This process involves the absorption and accumulation into the roots, the adsorption onto the roots, or the precipitation and immobilization within the root zone.

Proteins and enzymes that can be produced and exuded by plants together with oxygen and water or from soil organisms, such as bacteria, yeast, and fungi, can favour degradation, metabolism, or mineralization of the contaminants in soil. When the degradation process is acted in the rhizosphere it is called rhizodegradation while it is called phytodegradation when it is acted inside the plant.

Phytovolatilization mechanism is the sequence of the uptake of a dissolved contaminant from the soil environment, the chemical speciation modification in the rhizosphere and its translocation up into the leaves where it is released to the atmosphere through the process of transpiration (EPA, 1997).

ASSISTED PHYTOEXTRACTION

The ability of plants to accumulate heavy metals has been studied since 16th century when a small perennial shrub growing in Tuscany on nickel contaminated soils was discovered (Cesalpino, 1583). This was the first study which led to the development of a whole range of new technologies and discoveries concerning mineral exploration, phytochemistry, land restoration and mining (Brooks, 1998). A series of different plants that showed the ability to accumulate metals in roots and shoots over the normal fraction necessary for plant survival and growth were discovered (Minguzzi and Vergnano, 1948; Brooks et al., 1977; Reeves and Brooks, 1983).). These plants were called hyperaccumulators. Hyperaccumulators of heavy metals are usually endemic to a given type of geological substrate and their actual presence is related to a specific type of rock or mineralization. The strict relation between the presence of hyperaccumulators and the presence of a specific type of rock led at first to the use of hyperaccumulators as mine indicators in mineral exploration.

The possibility to use plants to extract metals from soil and consequently remediate it has been studied for the first time in the early 1980s (Cunningham et al., 1995). The major advantage of phytoremediation is that the procedure can be carried out *in situ* and can be very much less expensive than physical methods such as the removal of soil (Cunningham et al., 1995).

Since hyperaccumulators are generally plants characterized by a low biomass production, the quantity of metal that plants are able to extract is low and this leads to long remediation times. Different strategies have been suggested to enhance phytoextraction.

Baker and Brooks (1989) suggested the contemporary use of the phytoextraction technique and the combustion of the dried matter to obtain energy. The first practical studies where however carried out by Nicks and Chambers (1995) in California using the nickel hyperaccumulator *Streptanthus polygaloides* that is endemic to ultramafic soils. The principle of phytomining is that metal accumulated by plants can be recovered and has consequently an economic value. The main drawback of this technique is that its applicability has been evaluated only for nickel extraction.

The main factor that limits the efficiency of phytoextraction is the availability of heavy metals to plant roots (Felix, 1997). The factors that can affect phytoavailability are: cation exchange capacity (Moore et al., 1995), pH (Hornburg et al., 1995; Reddy et al., 1995; Schmidt, 2003), organic matter content (Bliefert, 1994; Li and Shuman, 1996) and nutrients (Chen et al., 2006) as well as speciation of the metal in the different soil fractions (Reddy et al., 1995).

A way to make the phytoextraction process feasible can be also the use of high biomass yielding plants in which metal accumulation is induced by the use of chemical additives that increase the phytoavailable metal fraction contained in the soil solution (Huang and Cunningam, 1996; Blaylock and Huang, 1999). This technique is known as assisted phytoextraction.

The use of chelating agents has been efficiently applied for this objective. Both persistent (EDTA, HEDTA, DTPA, EGTA, EDDHA, HEIDA, HBED, etc.) and readily biodegradable chelating agents (EDDS, NTA etc.) are used as well as low molecular weight organic acids (Evangelou et al., 2007).

Nowack et al. (2006) evaluated 28 publications referring to soils contaminated under field conditions where the extractions were carried out in 24 hours at an approximately neutral pH. They examined the metal extracted as a function of the chelant to total metal ratio in the soil for Cu, Zn, Cd and Pb and verified that a ratio of at least 1 is needed to solubilize all the target metals and that there is a large variation among soils for a given chelant-to-metal ratio. Soil characteristics play indeed a relevant role in the metal mobilization capacity of chelants. Nowack et al. (2006) verified in particular the effect of the presence of Ca, calcite and Fe as competitors of metals for chelating agents.

The use of EDTA in phytoextraction has received considerable attention due to its low cost and high efficacy in metal solubilization. EDTA offers the best cost/performance ratio of all chelants (Nowack et al., 2006).

The studies made on the use of EDTA to enhance phytoextraction demonstrated different abilities to increase metal transport that reaches 100 fold metal accumulation and depends on metal species, metal content in the soil, the soil itself, as well as the amount of EDTA applied. Several studies have been made to understand the mechanism involved in heavy metal uptake enhancement but it has not been completely described as it is dependent on the metal and the plants used (Evangelou et al., 2007).

Although EDTA has been shown to be effective in enhancing metal accumulation by plants, EDTA and EDTA heavy metal complexes demonstrated to be toxic to soil microorganisms and to plants decreasing plant biomass production (Grcman et al., 2003; Chen et al., 2006; Ultra et al., 2005). Moreover the low biodegradability of this chelating agent determines its persistence in the environment even after soil cleaning (Wasay et al., 1998). Even if biodegradation of EDTA has been reported in pure cultures under controlled conditions, its degradation in natural conditions both in soil and in water is very low (Bucheli-Witschel and Egli, 2001). As a consequence the main drawback related to its use in assisted phytoextraction is the high value of metals (Grcman et al., 2003) and alkaline earth metals such as Ca and Mg (Barona et al., 2001) that is made bioavailable but not absorbed by the plant. These metals are released into the leachate and constitute an environmental risk for the population. EDTA persistence in environment is demonstrated by its presence in almost all natural waters (Nowack and VanBriesen, 2005).

Recent researches demonstrated that the use of easily biodegradable chelating agents allows to obtain high mobilization efficiencies both in soil washing (Tandy et al., 2004) and phytoremediation processes (Tandy et al., 2006). Examples of easily biodegradable chelants used efficiently in assisted phytoextraction are S-S EDDS and MGDA. Metal EDDS and MGDA complexes have shown to be easily biodegradable in soil (Hauser et al., 2005; Schowanek et al., 1997; Data sheet, BASF). The use of S-S EDDS with a dose of 10 mmol EDDS kg^{-1} demonstrated to increase metals in soil solution up to a factor of 840 - 4260 for Cu and 100 – 315 for Pb, and to a lesser extent for Zn (factor 23 - 50) and Cd (factor 2.5 - 38). It was found that Zn (when present as the sole metal), Cu and Pb uptake by sunflowers was increased by EDDS, but in multi-metal contaminated soil, Zn and Cd were not. EDDS was observed in the sunflower roots and shoots (Tandy et al., 2006). EDDS efficiency in metal mobilization and in inducing metal accumulation by plants seems to be comparable to EDTA one (Grcman et al., 2003). Kos and Lestan (2003) demonstrated that EDTA and EDDS addition increased Pb concentrations in the test plants by 158 and 89 times compared to the control, with a metal concentration of 817 and 464 mg/kg, respectively. The easily biodegradable chelating agent MGDA demonstrated its efficiency in the mobilization of both

Zn and Pb even if lower than those of S-S EDDS and EDTA (Tandy et al., 2004) in soil washing tests. Cao et al. (2007) verified the efficiency of MGDA in assisted phytoextraction process and its low toxicity to plants.

S-S EDDS and MGDA seem to have a positive influence on bacterial communities both in bulk soil and rhizosphere, whereas the endophytes are less affected by the treatments (Cao et al., 2007). Grcman et al. (2003) demonstrated that S-S EDDS is less toxic to fungi than EDTA and caused less stress on soil microorganisms.

Experiences on Metal Removal from Abandoned Mining Areas

Both *ex situ* and i*n situ* technologies have been applied for the decontamination of mining sites. *Ex situ* approaches imply the removal of the polluted soil from the contaminated site and its transport and cleaning in a specific plant, while *in situ* technologies allow to remediate the soil in the site itself. The *ex situ* technologies are moreover considered costly and insufficiently risk reducing (van Gestel et al., 1992). The use of biological methods like phytoremediation is acquiring increasing consideration in the scientific community (Baker et al., 1994; DOE, 1994; Ernst, 1996).

Experiences about phytoextraction and assisted phytoextraction on the contaminated areas of Sulcis Iglesiente (Sardinia, Italy) and in particular in Montevecchio and Campo Pisano have been carried out by the Department of Geoengineering and Environmental Technologies of the University of Cagliari (Italy). The areas studied in the Montevecchio site are called Pauli giuncus and Corti baccas and are located downstream the Picalinna tailing dam.

The characterization of soils used during the experiments is shown in table 1.

Easily biodegradable chelants were tested both with the use of a plant with a high biomass production (*Mirabilis jalapa*) and with native species (*Scrophularia canina* and *Cistus salviifolius*).

The experiment made using *Mirabilis jalapa* grown in the Pauli giuncus soil demonstrated that the use of chelating agents allowed to obtain a high accumulation of Pb in leaves compared to the untreated reactor in which no Pb was detected.

Table 1. Sulcis Iglesiente contaminated soils characterization

Contaminated area	C.E.C. [cmol kg^{-1}]	pH H_2O	pH KCl	N [%]	C [%]	Total [ppm] Pb	Total [ppm] Zn	H_2O [ppm] Pb	H_2O [ppm] Zn	KNO_3 [ppm] Pb	KNO_3 [ppm] Zn	EDTA [ppm] Pb	EDTA [ppm] Zn
Pauli giuncus	17.84	6.40	5.35	0.20	1.95	5526	1717	2	3	16	105	3278	443
Corti baccas	17	4.88	4.4	0.21	2	9422	3970	8	225	1276	995	4690	696
Campo Pisano	0	7.3	7.2	0.01	6.28	3258	12038	DL	776	DL	50	612	3504

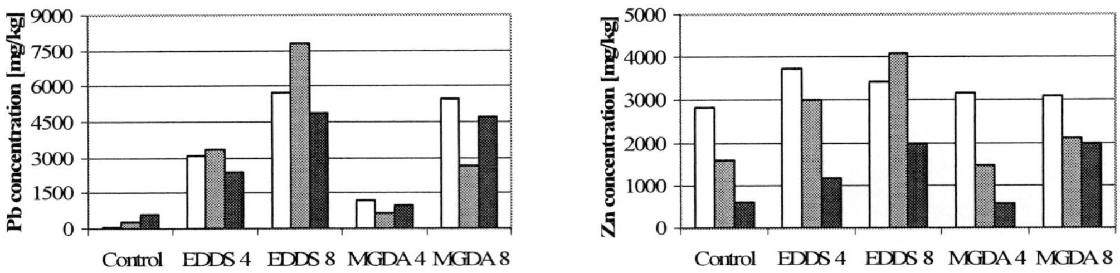

Figure 1. Distribution of Pb and Zn in *Mirabilis jalapa*: Leaves maximum metal concentration (□), Stalk (▨), and Roots (■) metal concentration at the end of the experiment, with different chelants and dosages.

8 mmol of chelant per kg of soil increased Pb concentration in leaves to 5,700 and 5,500 mg kg^{-1} with EDDS and MGDA, respectively (Cao et al., 2007). On the contrary Zn concentration in plant did not seem to be significantly related to the use of the chelating agent. Figure 1 shows the concentrations of Pb and Zn in the different plant compartments.

The high content of Pb in the plant and particularly in leaves can be due to the fact that the use of chelating agents, as stated by Vassil et al. (1998), at threshold concentrations, overcomes the physiological barrier(s) in roots that normally function to control uptake and translocation of solutes.

On the contrary Zn concentration in plant compartments is not influenced by the use of chelants. This can be probably due to a metal concentration in soil solution lower than the one necessary to overcome the physiological barrier in roots.

Experiments made in the laboratory on the Sardinian native species demonstrated that the MGDA was more efficient than EDDS both in mobilizing Pb and in increasing its accumulation by the aerial part of the plant. *Scrophularia canina* increased its Pb concentration from 100 mg/kg in the control up to 890 mg/kg for MGDA and 90 mg/kg for EDDS. *Cistus salviifolius* increased Pb concentration from 100 mg/kg in the control up to 1300 and 168 mg/kg respectively. While zinc absorption by *Scrophularia canina* was not correlated to chelants treatment, the one of *Cistus salviifolius* was increased by both MGDA (up to 5400 mg/kg) and EDDS (up to 4500 mg/kg) compared to the control reactor characterized by a concentration of 2900 mg/kg. The different efficiencies in metal phytoextraction using MGDA in the experiments with *Mirabilis jalapa* and Sardinian native species is a theme that needs to be more deeply evaluated. On the basis of the results obtained by Nowack et al. (2006) the different plant behaviour was attributed to plant-dependent leakiness of the root and the efficacy of chelate transport into the xylem.

An *in situ* application of phytoextraction was made on the site of Campo Pisano, near Iglesias. Three native species were planted on the site: *Scrophularia canina*, *Cistus salvifolius* and *Teucrium flavum*. After one year experiment the plants that showed the better capacity to accumulate Pb were *Teucrium flavum* (346 mg/kg) and *Cistus salviifolius* (185 mg/kg) while Pb accumulation in leaves of *Scrophularia canina* was about 110 mg/kg. Zinc accumulation was high in all plants: 1560 mg/kg for *Cistus salviifolius*, 1190 mg/kg for *Scrophularia canina* and 1130 mg/kg for *Teucrium flavum.*

If metal accumulation was higher in *Cistus salviifolius* and *Teucrium flavum* the survival of plants was very low while *Scrophularia canina* plants had a high percentage of survival.

This led to the selection of *Scrophularia canina* for the second phase of the experiment, that is in progress.

After the *in situ* phytoextration experiment *Scrophularia canina* plants grown in the Campo Pisano soil were transported in laboratory with the soil in which they were grown to make test of assisted phytoextraction under controlled conditions. Two easily biodegradable chelants were used: IDSA and MGDA.

Scrophularia canina accumulation of Pb after 3 days from the treatment increased from 50 mg/kg in the control to 300 mg/kg and 160 mg/kg in the plants treated with IDSA and MGDA, respectively.

CONCLUSIONS

The analysis of the environmental problem related to abandoned mining areas and of the applicability of phytoextraction as a heavy metal emerging remediation technology showed the possibility to use this technique in the tested areas of Sulcis-Iglesiente in Sardinia (Italy).

The plant species which demonstrated the best applicability to the phytoextraction technology were *Scrophularia canina* as a native species and *Mirabilis jalapa* as a high biomass production species. The plants showed a capacity to tolerate heavy metal up to a concentration of 9400 mg/kg of Pb and 3700 mg/kg of Zn. This technique is however applicable only to low depth contaminated areas, in which plant roots are able to extract contaminants, and to a limited level of heavy metal contamination determined by the toxicity of metals to plants.

The plants uptake capacity and, as a consequence, metals removal efficiency from soil depends on metal bioavailability, which is evaluated using both direct and indirect methods.

Lead concentration in soil solution shows a positive correlation with the concentration measured in leaves and can therefore be considered as an indicator of its bioavailability. The same statement cannot be made for zinc: there is not correlation between zinc concentration in soil solution and in leaves. Metal fractions in soil evaluated by using sequential extraction procedures show a correlation with the metal concentration in soil solution only for the lead and zinc soluble fractions evaluated thought Barbafieri et al. (1996) and Castaldi et al. (1994) methods (Cao et al., 2008) which can therefore be used to evaluate the bioavailability of metals to plants.

The two easily biodegradable chelating agents that demonstrated the better applicability to the assisted phytoextraction technique were MGDA and IDSA. These two chelating agents were in fact able to increase the metal bioavailable fraction and to favour metal accumulation in native plants. They also had a limited leaching effect on soil, decreasing as a consequence the possible adverse effects related to heavy metals presence in water after chelants application to soil.

The analyses made on the microbial community showed a positive effect of easily bioavailable chelating agents on bacterial communities both in bulk soil and rhizosphere (Cao et al., 2007).

REFERENCES

Adriano, D.C., Wenzel, W.W., Vangronsveld, J. and Bolan N. S. (2004). Role of assisted natural remediation in environmental cleanup. *Geoderma*, 122, 121-142.

Baker, A.J.M. and Brooks, R. R. (1989). Terrestrial higher plants which hyperaccumulate metal elements – A review of their distribution, ecology and phytochemistry. *Biorecovery*, 1, 81-126.

Baker A. J. M., McGrath S. P., Sidoli C. M. D. and Reeves R. D. (1994) The possibility of in situ heavy metal decontamination of polluted soils using crops of metal accumulating plants. *Resources, Conservation and Recycling*, 11, 41-49.

Baker A.J. M., Brooks, R.R., Reeves, R.D. (1998). Growing for Gold and Copper and Zinc. *New Science,* 177, 44–48.

Barbafieri, M., Lubrano, L. and Petruzzelli, G. (1996). Characterization of pollution in sites contaminated by heavy metals: a proposal. *Annali di Chimica*, 86, 585-594.

Barona, A., Aranguiz, I., Elias, A. (2001). Metal associations in soils before and after EDTA extractive decontamination: implications for the effectiveness of further clean-up procedures. *Environmental Pollution* 113, 79–85.

Blaylock, M. J., Salt, D. E., Dushenkov, S., Zakharova, O., Gussman, C., Kapulnik, Y., Ensley, B. D. and Raskin I. (1997). Enhanced accumulation of Pb in Indian mustard by soil – applied chelating agents. *Environmental Science and Technology,* 31, 860-865.

Blaylock, M.J., Huang, J.W. (1999). Phytoextraction of metals. In: Raskin, I., Ensley, B.D. (Eds.), *Phytoremediation of Toxic Metals: Using Plants to Clean Up the Environment* (53–70) John Wiley, New York

Bliefert, C. (1994). *Umweltchemie.* Wiley Europe-VCH, Weinheim, New York.

Brooks, R. R. (1998). *Plants that hyperaccumulate heavy metals: their role in phytoremediation, microbiology, archaeology, mineral exploration and phytomining.* New York, USA: CABI Publisher.

Bucheli-Witschel, M., Egli, T. (2001). Environmental fate and microbial degradation of aminopolycarboxylic acids. FEMS Microbiology Review 25, 69–106.

Caboi, R., Cidu, R., Cristini, A., Fanfani, L., Massoli-Novelli, R., Zuddas, P. (1993). The abandoned Pb–Zn mine of Ingurtosu, Sardinia (Italy). *Environmental Geology.*, 34, 211–218.

Cao, A., Carucci, A., Lai, T., La Colla, P., Tamburini, E. (2007). Effect of biodegradable chelating agents on heavy metals phytoextraction with Mirabilis jalapa and on its associated bacteria *European Journal of Soil Biology*, 43, 200- 206.

Cao, A., Carucci, A., Cappai, G., Lai, T. (2008). Heavy metals bioavailability and chelate mobilization efficiency in an assisted phytoextraction process. Special issue in *Environmental Geochemistry and Health*. 30, 2, 115-119.

Cappuyns, V., Swennen, R., Vandamme, A. and Niclaes, M. (2006). Environmental impact of the former Pb–Zn mining and smelting in East Belgium. *Journal of Geochemical Exploration*, 88, 6-9.

Caredda, A.M., Cristini, A., Ferrara, C., Lobina, M.F., Baroli, M., 1999. Distribution of heavy metals in the Piscinas beach sediments (SW Sardinia, Italy). *Environmental Geology*, 38, 91–100.

Castaldi, P., Santona, L., Cozza, C., Giuliano, V., Abruzzese, C. and Melis, P. (2004). Recovering of soils contaminated with heavy metals: preliminary results. *Fresenius Environmental Bulletin*, 13, 1232-1236.

Cesalpino, A. (1583). De plantis libri n. 16. Florentiae, 369.

Chen, Y., Wang, Y., Wu, W., Lin, Q., Xue S. (2006). Impacts of chelate-assisted phytoremediation on microbial community composition in the rhizosphere of a copper accumulator and non-accumulator. *Science of the Total Environment*, 356, 247– 255.

Concas, A., Ardau, C, Cristini, A., Zuddas, P., Cao, G. (2006). Mobility of heavy metals from tailings to stream waters in a mining activity contaminated site. *Chemosphere*, 63, 244–253.

Conesa, H. M., Garcıa, G., Faz, A. and Arnaldos, R. (2007). Dynamics of metal tolerant plant communities' development in mine tailings from the Cartagena-La Union Mining District (SE Spain) and their interest for further revegetation purposes. *Chemosphere*, 68 (6), 1180-1185.

Cunningham, S. D., Berti, W. R. and Huang, J. E. (1995). Phytoremediation of contaminated soils. *Trends in biotechnology*, 13, 393-397.

Dessì, R., Boi, M., Persod, P. and Ullu F. (1999) Environmental characterisation and rehabilitation proposals of Montevecchio mine district. Proceedings of REWAS 99, San Sebastian, Vol. III, 2541-2550.

DOE (1994) Phytoremediation Research Needs. Workshop report. Department of Energy, USA, Santa Rosa.

EPA. "Recent developments for in situ treatment of metals contaminated soils". (1997). Available from: http://www.epa.gov/tio/download/remed/metals2.pdf.rom.

Elsokkary, I. H., Amer, M. A. and Shalaby, E. A. (1995). Assessment of inorganic lead species and total organo-alkyllead in some Egyptian agricultural soils. *Environmental Pollution,* 87, 225-233.

Ernst, W. H. O. (1996). Bioavailability of heavy metals and decontamination of soils by plants. *Applied Geochemistry*, 11, 163-167.

Evangelou, M.W.H., Ebel M., Schaeffer A. (2007). Chelate assisted phytoextraction of heavy metals from soil. Effect, mechanism, toxicity, and fate of chelating agents. *Chemosphere*, 68, 989–1003.

Fanfani, L., Zuddas, P., Chessa, A. (1997). Heavy metals speciation analysis as a tool for studying mine tailings weathering. *Journal of Geochemical Exploration*, 58, 241–248.

Felix, H.R.Z. (1997). Field trials for in situ decontamination of heavy metal polluted soils using crops of metal-accumulating plants. *Z. Pflanzenernahr. Bodenk*. 160, 525–529.

Freitas, H., Prasad, M.N.V. and Pratas, J. (2004). Plant community tolerant to trace elements growing on the degraded soils of São Domingos mine in the south east of Portugal: environmental implications. *Environment International,* 30, 65-72.

Glass, D. J. (2000). Economic potential of phytoremediation. In I. Raskin and B.D. Ensley (Eds), *Phytoremediation of toxic metals: using plants to clean up the environment* (15-31). New York, John Wiley and Sons Inc. Publisher.

Grcman, H., Vodnik, D., Velikonja-Bolta, S., Lestan, D. (2003). Ethylenediaminedissuccinate as a new chelate for environmentally safe enhanced lead phytoextraction. *Journal of Environmental Quality*, 32, 500–506.

Guzzi, L., Martignon, G., Martinetti, W., Perotti, M. and Petruzzelli, G. (2005). La biodisponibilità dei microinquinanti inorganici nella VRE (valutazione rischio ecologico): verifica sperimentale di approcci metodologici di misura e stima. Proceedings of III Conferenza Inquinamento da metalli pesanti: la biodisponibilità, Sassari, 5-6 May 2005.

Hauser, L., Tandy, S., Schulin, R. and Nowack, B. (2005) Column extraction of heavy metals from soils using the biodegradable chelating agent EDDS. *Environ. Sci. Technol.*, 39, 6819–6824.

He, Z. L., Yang, X. E. and Stoffella, P. J. (2005). Trace elements in agroecosystems and impacts on the environment. *Journal of trace elements in medicine and biology,* 19, 125-140.

Hornburg, V., Welp, G., Bru¨mmer, G., 1995. Verhalten von Schwermetallen in Bo¨den (2): Extraktion mobiler Schwermetalle mittels $CaCl_2$ und NH_4NO_3. *Z. Pflanzenernaehr. Bodenkd.* 158, 137–145.

Huang, J.W., Cunningham, S.D., 1996. Lead phytoextraction: species variation in lead uptake and translocation. *New Phytologist* 134, 75–84.

Kambhampati, M. S. and Williams, L. (2001). Phytoremediation of lead-contaminated soils using *Mirabilis jalapa*. . , A. Leeson, E. A. Foote, M. K. Banks, and V. S. Magar (Eds.), *Phytoremediation, wetlands and sediments*. San Diego, California, Battelle Press.

Kos, B., Lestan, D. (2003). Induced phytoextraction/ soil washing of lead using biodegradable chelate and permeable barriers. *Environ. Sci. Technol.*, 37, 624-629.

Leita, L. (2005). La biodisponibilità nel suolo. Proceedings of III Conferenza Inquinamento da metalli pesanti: la biodisponibilità, Sassari, 5-6 May 2005.

Leštan, D., Luo, C.L. and Li, X. D. (2007). The use of chelating agents in the remediation of metal-contaminated soils: A review. Environ. Pollution, DOI:10.1016/j.envpol. 2007.11.015

Li, Z., Shuman, L.M. (1996). Heavy metal movement in metal contaminated soil profiles. *Soil Science* 161, 656–666.

Licskò, I., Lois, L. and Szebenyi, G. (1999). Tailings as a source of environmental pollution. *Water Science and Technology*, 39 (10-11), 333-336.

Minguzzi, C. and Vergnano, O. (1948). Il contenuto di nichel nelle ceneri di Alyssum bertolonii. *Atti della Società Toscana di Scienze Naturale*, 55, 49-74.

Moore, R., Clark, W.D., Stern, K.R. (1995). Botany. WCB Publishers, Dubuque, Iowa.

Nicks, L. and Chambers, M. F. (1995). Farming for metals. Mining Engineering Management, September, 15-18.

Nowack, B., Schulin, R. and Robinson, B. H. (2006). Critical assessment of chelant-enhanced metal phytoextraction. *Environmental Science and Technology*, 40 (17), 5225-5232.

Nowack, B., VanBriesen, J. M. (2005). Chelating agents in the environment. In: Nowack, B., VanBriesen, J. M., (Eds), *Biogeochemistry of Chelating Agents* (1-18) ACS Symposium Series; American Chemical Society, Washington, DC,; Vol. 910.

Pedron, F. and Petruzzelli, G. (2005). Potenzialità dei metodi di passive approach per la valutazione della biodisponibilità dei metalli pesanti nei siti contaminati. Proceedings of III Conferenza Inquinamento da metalli pesanti: la biodisponibilità, Sassari, 5-6 May 2005.

Peijnenburg, W.J.G.M. and Jager, T. (2003). Monitoring approaches to assess bioaccessibility and bioavailability of metals: matrix issues. *Ecotoxicology and Environmental Safety*, 56, 63–77.

Rauret, G., Lòpez-Sànchez, J. F., Sahuquillo, A., Rubio, R., Davidson, C., Ure, A. and Quevauviller Ph. (1998). Improvement of the BCR three step sequential extraction procedure prior to the certification of new sediment and soil reference materials. *Journal of Environmental Monitoring,* 1, 57-61.

Reddy, K.J., Wang, L., Gloss, S.P. (1995). Solubility and mobility of copper, zinc and lead in acidic environments. *Plant Soil*, 171, 53–58.

Reeves R. D and Brooks, R. R. (1983). Hyperaccumulation of lead and zinc by two metallophytes from a mining area of Central Europe. *Environmental Pollution Series* A31, 277-287.

Römkens, P., Bouwman, L., Japenga, J. and Draaisma, C. (2002). Potentials and drawbacks of chelate-enhanced phytoremediation of soils. *Environmental Pollution*, 116, 109-121.

Schowanek, D., Feijtel, T.C.J., Perkins, C.M., Hartman, F.A., Federle, T.W., Larson, R.J. (1997). Biodegradation of [S,S], [R,R] and mixed stereoisomers of ethylene diamine disuccinic acid (EDDS), a transition metal chelator. *Chemosphere* 34, 2375–2391.

Schmidt, U. (2003). Enhancing phytoextraction: the effect of chemical soil manipulation on mobility, plant accumulation, and leaching of heavy metals. *Journal of Environmental Quality*, 32, 1939–1954.

Tandy, S., Bossart, K., Mueller, R., Ritschel, J., Hauser, L., Schulin, R. and Nowack, B. (2004). Extraction of Heavy Metals from Soils using Biodegradable Chelating Agents. *Environmental Science and Technologies*, 38,937-944.

Tandy, S., Schulin, R., and Nowack B. (2006). Uptake of Metals during Chelant-Assisted Phytoextraction with EDDS Related to the Solubilized Metal Concentration. *Environmental Science and Technologies*, 40,2753-2758

Tessier, A., Campbell, P. G. and Bisson, M. (1979). Sequential extraction procedure for the speciation of particulate trace metals. *Analytical Chemistry*, 51 (7), 844-851.

Tordoff, G.M., Baker, A.J.M., and Willis, A.J. (2000). Current approaches to the revegetation and reclamation of metalliferous mine wastes. *Chemosphere*, 41, 219–228.

Ultra, V. U., Yano, A., Iwasaki, K., Tanaka, S., Kang, Y., Sakurai, K. (2005). Influence of chelating agent addition on copper distribution and microbial activity in soil and copper uptake by Brown Mustard (Brassica juncea). *Soil Science and Plant Nutrition* 51, 193-202.

van Gestel C. A. M., Dirven-Van Breemen E. M. and Kamerman J. W. (1992) Evaluation of Decontaminated Soils. National Institute Public Health and Environmental Protection, No. 216402005. Bilthoven.

Vassil, D. A., Kapulnik, Y., Raskin, I., Salt, E.D. (1998). The role of EDTA in lead transport and accumulation by Indian Mustard, *Plant Physiology*, 117, 447-453.

Wasay, S.A., Barrington, S.F., Tokunaga, S. (1998). Remediation of soils polluted by heavy metals using salts of organic acids and chelating agents. *Environmental Science and Technologies,* 19, 369–379.

In: Causes and Effects of Heavy Metal Pollution
Editor: Mikel L. Sanchez
ISBN 978-1-60456-900-1

Chapter 4

PHYTOCHELATINS IN WILD PLANTS FROM GUANAJUATO CITY – AN IMPORTANT SILVER AND GOLD MINING CENTER IN MEXICO

Kazimierz Wrobel, Julio Alberto Landero Figueroa and Katarzyna Wrobel*

Instituto de Investigaciones Científicas, Universidad de Guanajuato,
L. de Retana N° 5, 36000 Mexico

ABSTRACT

Phytochelatins (PCs) are a group of small, metal-binding peptides that are biosynthesized by higher plants, some fungi and algae in the response to heavy metal exposure. One actual research topic focuses on better understanding the global effect that all elements present in natural environments exert on the PCs production by plants.

In this work, PCs levels were evaluated in the wild plants, chronically exposed to low or moderate levels of heavy metals. The quantification of total PCs in plant extracts was carried out by HPLC with fluorimetric detection, after derivatization of free –SH groups with monobromobimane. Additionally, the distribution of metals in molecular mass (MM) fractions of these same extracts was studied by size exclusion chromatography with on-line UV and ICP-MS detection. All samples were collected in Guanajuato city (Mexico), which has long been an important silver and gold mining area. Among different metals reported in Guanajuato soils, lead, cadmium, copper and silver were selected in this study, because of their capability to induce phytochelatins in plants. The common plants from this region were analyzed, namely: *Ricinus communis* (castor bean), *Tithonia diversifolia* (Mexican sunflower) and *Opuntia ficus* (nopal). The analytical approach involved the ICP-MS analysis of total elements in soil, soil fractions and wild plants and also the evaluation of relationships between PCs, metal levels found in plants/soil and different soil parameters.

---

* Corresponding autor: katarzyn@quijote.ugto.mx.

In the analysis of plants, PC-2, PC-3 and PC-4 were detected in nopal, PC-2 in castor bean, while in Mexican flower no phytochelatins were found. In further development, the extracts of soil humic substances were obtained and the distribution of metals in molecular mass (MM) fractions was studied by size exclusion chromatography with on-line UV and ICP-MS detection. The soil humic substances (HS) were also assessed.

In search of possible relationship between the parameters measured, the statistical analysis of correlation was performed. The results obtained indicate that the binding of metals to soil HS contributes in lowering their uptake by castor bean plant. On the other hand, the soils collected at nopal roots presented low HS levels and no correlation with metals in plant was found. The results obtained in the sequential extraction of soils and the abundance of sulfide minerals in Guanajuato indicate that the sulfide bound metals were the primary forms of Pb, Cu and Cd in soil adjacent to nopal roots. Owing to their generally poor solubility, rizosphere processes should be important in mobilizing metals and their uptake by nopal.

Our results provide further evidence on the role of environmental conditions in the accumulation of heavy metals in relation to PCs production in different plant genotypes. In particular, multi-elemental approach is necessary in studies on PCs induction in actual field situations, where plants (or other organisms) are exposed to a variety of metals and metalloids.

INTRODUCTION

Phytochelatins (PCs) are a group of small, metal-binding peptides of the general structure $(Glu\text{-}Cys)_n\text{-}Gly$, where n varies from 2 to 11 (Figure 1a) [1,2]. Other families of PCs, differing in the type of C-terminal amino acid have also been characterized [3]. PCs are biosynthesized by higher plants, some fungi and algae in the response to heavy metal exposure, which is considered a major detoxifying pathway for heavy metals in these organisms [4]. The PCs role in maintaining the homeostasis of intracellular levels of essential metal ions has also been highlighted [5]. Depending on the coordination capacity of metal ion and on the specific ability of plant for PCs synthesis, up to four sulfur atoms can be coordinated from either single or multiple PC molecules, resulting in PC-1, PC-2, PC-3 and PC-4 complexes (Figure 1b). The formation of PC-5 and PC-6 has also been reported, but only in transgenic plants [6], in cell cultures [7] and freshwater green alga [8].

The term phytochelatins was introduced by Zenk et al. in 1985, after isolation of novel metal-bining peptides from cell suspension cultures of a higher plant after exposure to Cd [9]. Since then, the extensive studies have been carried out aiming structural characterization of PCs and their metal complexes [10,11] as well as the elucidation of induction mechanisms [12,13]. In the first approach, the isolation and characterization of PC synthases from different organisms have been carried out. Most commonly, the experimental approach has involved plant growth (or other types of culture) under controlled exposure conditions, typically using the salt of one element at relatively high concentrations.

The expression of PCs not only enhances the tolerance to heavy metals but also causes their accumulation in plant tissues, so the feasibility of PCs for phytoremediation of toxic heavy metals from contaminated soil and water has been explored [14,15]. In particular, plant engineering to assure higher rates of PCs induction has been studied [16-18].

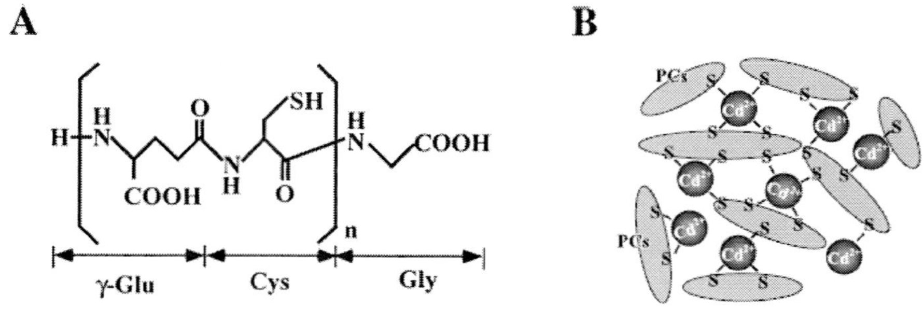

Figure 1. Primary structures of PCs (A) and PC-Cd complexes (B) [2].

However, before the routine use of plants in phytoremediation becomes possible, several questions regarding genetic regulation of PC synthase, its localization in plants and interaction of PCs and PCs synthase with different metal/metalloid forms, need to be answered. Furthermore, the mechanisms responsible for metal uptake and accumulation in plants in relation to PCs production are not fully understood. It should be realized that, in natural environments, plants are exposed to many elements and their forms at different concentration levels. In general, processes occurring in the soil-root system are considered important [19,20], yet the role of environmental parameters, possibly affecting metal bioavailability to plants, remains unclear. The increased production of phytochelatins has been observed in the presence of various organic chelators (ethylene diamine tetraacetate, ethylene diamine disuccinate, citric acid, etc.) as added to nutrient solution and/or soil [21-24]. On the other hand, metal binding by humic matter in soils has been shown to affect the mobility of metals in the environment. Depending on the experimental design and the type of metal, the increased or decreased uptake of Cd, Cu, Pb have been observed [25-30]. Importantly, the data relating soil humic substances with phytochelatin production in plants exposed to heavy metals are lacking.

In order to gain further insight on the induction of PCs and element complexation by these compounds in real-field situation, the global effect of different elements present at low concentrations should be studied together with possible effects of relevant environmental parameters.

Within this context, the present work has been focused on the wild plants growing in Guanajuato city, located in central Mexico. Guanajuato city has long been an important silver and gold mining center and the leaching of several metals and metalloids from the mine tailings to different environmental compartments has been reported [31,32]. Consequently, the plants growing in this area are chronically exposed to various metals, that are present in soils in relatively wide range of concentrations. Among those elements, we have focused on lead, cadmium, copper and silver, because of their capability to induce phytochelatins in plants [3,10,33-37] and also because of their abundance in the region of Guanajuato [31,32]. For the purposes of this work, three wild plants have been selected, namely *Ricinus communis* (castor bean), *Tithonia diversifolia* (Mexican sunflower) and *Opuntia ficus* (nopal). These plants are common in Guanajuato and they can grow within small distance one from another, which is important to assure similar exposure conditions of different plants. The accumulation of metals in castor bean has already been reported [38,39]. Its possible application in phytoremediation has been explored [38,40], yet data available for Mexican sunflower and nopal are scarce [41]. Quite importantly, no information on possible induction of PCs in any of the three plant types has been found.

In a course of this study, we have focused on answering the following questions:

1. Do the wild plants chronically exposed to low/moderate concentrations of Ag, Cd, Cu and Pb induce PCs?
2. How to distinguish among elements that are important for PCs induction and those that are simply bound to PCs already existing?
3. Are the soil parameters important in PCs induction?

In the following sections, the experimental design and the results obtained in this study are systematically presented.

Experimental Design

The sampling sites were selected in order to cover a wide range of element concentrations in soil. Thus, the site 1 corresponded to the place where the silver ores had been processed in the past; the site 2 was located at the Guanajuato river close to the mine tailing; the site 3 was at a hill, relatively distant from the mines; the site 4 was in the urban area developed on main tailing; the site 5 was in the part of city far from the mines and the site 6 was at the riverside, but far from the mining zone. Additionally, nopal samples were also collected in the "clean area", which corresponded to the countryside, about 40 km from Merida city, peninsula of Yucatán.

The plants were collected in November 2005 (just after the wet season), taking at each site three samples of roots and three samples of leaves from plants of similar size and color of leaves. To avoid possible variations of soil parameters, the area of any sampling site did not exceed 9 m^2 At each sampling site, the pH of topsoil (0 - 30 cm) was measured. The relative variability of pH values for several measurements at one site did not exceed 2 % and the values found at different sites were in the range pH 7.7 to pH 8.0 (pH 8.0 – 8.1 in the soil from Merida).

For each plant species, the roots and leaves collected at one site were pooled. Roots were carefully separated from the adjacent soil and pooled. Three washing agents were tested, namely deionized water, 0.05 M $CaCl_2$ and 0.05 M EDTA (with and without ultrasonication). Since the washing procedure had no significant effect on the results, the roots were washed with deionized water in ultrasonic bath (10 min). Both, roots and leaves were cut into small pieces, freeze-dried and homogenized by mortar grinding in liquid nitrogen.

The soil samples were collected in these same sites, from which the plants were taken. (six samples of about 250 g per 9 m^2 site). For the determination of total Ag, Cd, Cu and Pb, the samples from each site were pooled and dried. Soil fractionation was carried out according to the simplified Tessier method [42]. Fresh soils, without pooling, were used for the assay of humic substances.

In plant material, total element concentrations were determined by inductively coupled plasma mass spectrometry (ICP-MS) after wet acid digestion. The extraction of PCs from freeze-dried root and leave homogenates was carried out with ammonium acetate (pH 7.4) [43]. Metal binding to PCs was studied by size exclusion liquid chromatography (SEC) with on-line UV and ICP-MS detection. Total PCs were assayed by reversed phase high performance liquid chromatography (RPHPLC) with fluorimetric detection [44,45].

The analytical approach in soil involved: (i) the determination of total element concentrations by ICP-MS after acid digestion; (ii) ICP-MS determination of elements in soil fractions after Tessier sequential extraction, (iii) assessment of soil humic substances (HS) by UV/Vis spectrophotometry [46] and (iv) SEC – UV/Vis – ICP-MS analysis of HS extracts for examination of metal binding to HS.

In search of possible relationship among parameters measured, the statistical analysis of correlation was performed, based on the results obtained in six sampling sites (software package Statistica for Windows, StatSoft Inc. 2000, Tulsa OK).

Capability of the Wild Plants Chronically Exposed to Low or Moderate Concentrations of Metals in Soil for the Induction of Phytochelatins

In the first approach, cadmium, copper, silver and lead were determined in the pooled soils and plant roots. As an example, the results obtained in sites 1-3 are presented in Table 1. The metal levels found in soil were in agreement with the ranges reported previously [31,32], yet substantial differences can be observed among different sampling sites. Among four metals, lead was found to be the most abundant in soil (maximum concentration 362 ± 8 $\mu g\ g^{-1}$, site 1), copper levels were generally lower with the highest value also at the site 1 (221 ± 7 $\mu g\ g^{-1}$), Ag levels were about 4 to 5 times lower that those for Cu (highest value 42.2 ± 0.7 $\mu g\ g^{-1}$, site 1) and cadmium presented concentrations below 1.0 $\mu g\ g^{-1}$ at any site. Even though the three plant species were collected simultaneously from these same sites, the highest concentrations of metals were found in *Opuntia f.*, followed by *R. communis* and *T. diversifolia*, (Table 1). As could be expected, the concentrations of metals in roots were higher in plants collected from more contaminated sites (site 1 and 2 versus site 3).

In order to examine if any of three wild plants can produce PCs, the root and stem extracts were obtained and analyzed for phytochelatins by reversed phase chromatography. It should be stressed that, at this point, we were interested in the total content of PCs and not in specific metal – PCs binding. Thus, in the analytical procedure applied, metals bound to PCs were eliminated by DPTA complexation and, after reduction, the fluorescent derivatization of the free thiol groups was achieved [44,45,47]. In the analysis of plant extracts, phytochelatins were found in *Opuntia f.* and in *R. communis*, but not in *T. diversifolia* [36]. Furthermore, nopal extracts contained PC-2, PC-3 and PC-4 and castor bean extract only PC-2. Total PCs levels in the plant samples collected at sites 1-4 are presented in Table 2. It can be observed that nopal contained relatively high concentrations of PCs as compared to castor bean. Also, the PCs levels in roots were always higher than in leaves.

In the context of the first question put in this work, the results obtained indicate that the wild plants, chronically exposed to low or moderate metal levels in soil do biosynthesize phytochelatins. However, our results seem to confirm that this process is genotype - dependent [18,48].

Table 1. Total cadmium, copper, lead and silver levels in soils and plant roots collected from three different sites in Guanajuato (1 - Hacienda when the silver ores had been processed in the past, 2 - Riverside close to the mine tailing, 3 – Site at a hill, relatively distant from the mines, each analysis was performed in triplicate)

Sample	Site	Mean concentration of heavy metal ± SD, μg g^{-1} (dried weight)			
		Ag	Cd	Cu	Pb
Soil	1	42.20 ± 0.65	0.373 ± 0.018	221 ± 7	362 ± 8
	2	3.40 ± 0.07	0.483 ± 0.010	20.6 ± 1.0	194 ± 7
	3	3.20 ± 0.32	0.296 ± 0.011	17.0 ± 0.5	19.4 ± 0.5
Mean concentration of heavy metal ± SD, ng g^{-1} (freeze-dried roots)					
R. communis	1	131 ± 6	16 ± 3	2580 ± 80	2740 ± 60
	2	141 ± 9	123 ± 8	2600 ± 70	1720 ± 50
	3	85 ± 6	12 ± 2	992 ± 17	198 ± 6
T. diversifolia	1	23 ± 2	9 ± 1	530 ± 16	805 ± 18
	2	35 ± 4	15 ± 2	592 ± 20	445 ± 10
	3	< 5 (DL)	< 0.5 (DL)	187 ± 7	20.5 ± 0.6
Opuntia f.	1	27 ± 2	224 ± 9	5030 ± 110	3250 ± 87
	2	8.8 ± 1.2	173 ± 7	4050 ± 98	2190 ± 84
	3	5.3 ± 1.5	124 ± 7	5750 ± 103	854 ± 46

nf – not found.

Table 2. Total PCs in the extracts from plants collected at three sampling sites characterized by different soil metal contents. (1 - Hacienda when the silver ores had been processed in the past, 2 - Riverside close to the mine tailing, 3 – Site at a hill, relatively distant from the mines, 4 – urban area developed on the mine tailing, each analysis was performed in triplicate)

Sampling site	μg of total PCs per gram of the freeze-dried plant homogenates		
	Nopal roots	Nopal leaves	Castor bean roots
1	12.2 ± 0.4	8.4 ± 0.6	2.3 ± 0.2
2	16.8 ± 0.4	10.0 ± 0.3	5.1 ± 0.3
3	27.7 ± 0.3	10.4 ± 0.2	2.2 ± 0.4
4	74.3 ± 0.3	20.3 ±0.2	28.3 ± 0.8

Specific Role of Metals in the Production of Phytochelatins in Plants, Naturally Exposed to a Wide Range of Elements

In this work, four metals (Ag, Cd, Cu, Pb) were selected because of their abundance in the Guanajuato region and because of their known capability to induce phytochelatins in plants [10,33-35]. Since the highest levels of total phytochelatins were found in nopal (Table 2), the extracts of this plant were used to study specific metal-PCs binding. As an example, the chromatograms obtained for the plant material from site 3 in Guanajuato are presented in

Figure 2 (cadmium could not be detected). For lead, lower element levels were observed in stems than roots. Furthermore, it eluted mainly as bound to high molecular mass compounds (between 75 – 95 % of total area of chromatogram), but also in the low molecular mass fraction. For copper, only one elution peak was observed in the region of low molecular mass (Figure 2b). Contrary to lead, in many samples the copper abundance in stems was higher than in roots. The elution profile of silver (Figure 2a) indicates its association with high molecular weight compounds; however, the chromatographic peaks were relatively broad suggesting non-specific binding. The results obtained indicate different profiles of metal association with PCs. Based on peak area measurements, the relative distribution of three metals between high and low molecular mass fractions was estimated for each plant sample.

It should be stressed that the procedure by SEC – ICP-MS enabled to get data on specific binding of each metal to PCs, while RPHPLC with fluorimetric detection provided complementary quantitative information on total pool of PCs in plant material (Table 2). In the context of the second question to be answered in this work, which regards the role of individual metals in PCs production, it was considered that the correlation between total PCs and a metal bound to PCs would indicate that this metal is important in the induction of PCs. On the contrary, when no relationship between these two parameters exists, a metal would be just bound to PCs already existing.

The correlation analysis was performed taking the data obtained in the analysis of PCs-rich plant extracts by SEC – ICP-MS and RPHPLC with fluorimetric detection. For lead, significant correlation was found between total PCs and SEC-ICP-MS areas ($r = 0.7026$, $p = 0.088$ for roots and $r = 0.8581$, $p = 0.071$ for stems). Furthermore, total PCs correlated with lead eluted in SEC high molecular mass fraction ($r = 0.6835$, $p = 0.023$ for roots and $r = 0.8818$, $p = 0.048$ for stems). These results confirm the association of Pb with nopal phytochelatins and suggest its possible role in PCs induction. For copper, correlation was observed between total PCs and entire SEC–ICP-MS area in roots ($r=0.7568$, $p = 0.129$) and between total PCs and high molecular mass fraction of SEC in roots ($r= 0.7455$, $p=0.138$). Apparently, copper could be involved in the induction of PCs in roots, however the translocation of lead and copper through the plant is apparently different. No statistically important correlation was found for silver indicating that it is not associated with phytochelatins production.

Soil Parameters in Relation to the Induction of PCs in Wild Plants

As discussed in the Introduction, various processes occurring in the soil-root system are considered important for PCs induction, however the role of environmental parameters, possibly affecting metal bioavailability to plants, remains unclear [19,20]. Binding of polyvalent cations to humic substances derived from different environmental compartments has often been reported and the following order of decreasing affinity has been proposed: Cu(II) > Ni(II) > Co(II) > Pb(II) > Cd(II) > Cr(III) >> (Mn(II), Mo(VI), Zn(II)) [49-51].

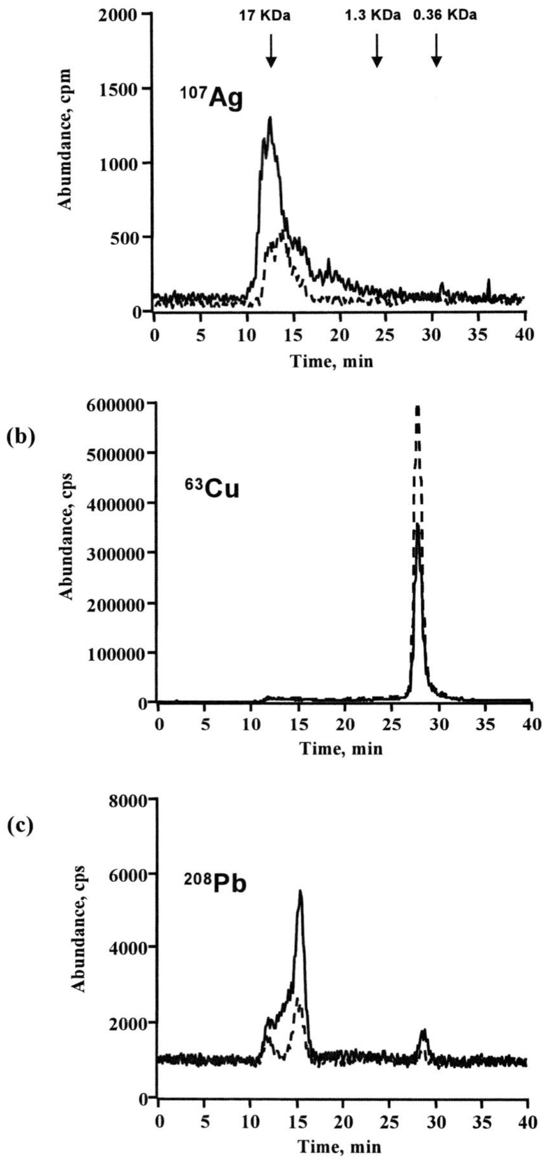

Figure 2. SEC – ICP-MS chromatograms of the (----) nopal stem extract; (____) nopal root extract: (a) ^{107}Ag; (b) ^{63}Cu; (c) ^{208}Pb .

Several authors observed that humus matter is capable to reduce the bioavailability of certain metals in soils [25,30,52,53]. Based on the above reports, the focus of this work was to evaluate possible relationship between soil humic substances (HS) and PCs.

The HS were assayed in fresh soil samples taken at the roots of nopal and castor bean (two plant synthesizing PCs) and the results obtained are summarized in Table 3. As can be observed, the HS levels were significantly lower in the samples collected at cactus roots, which is in agreement with literature data [54,55]. Consequently, possible relationship between soil HS and PCs production was studied in castor bean plant.

Table 3. The concentrations of soil humic substances found at the roots of two plants in different sampling sites

Sampling site	Soil humic substances, mg g^{-1} (fresh soil)	
	Nopal	Castor bean
1	75 ± 13	481 ± 38
2	453 ± 19	415 ± 42
3	175 ± 23	1345 ±100
4	43 ± 8	553 ± 47
Merida	10 ± 6	-

The statistical analysis of correlation was performed, using all data obtained for Ag, Cd, Cu, Pb in soils and in plant roots, PC-2 in plant extracts and HS in soils. For Cd and Ag, the statistically important positive correlation between metal concentration in soils and in plants was observed ($r = 0.8092$, $p = 0.051$; $r = 0.9261$, $p = 0.009$), while for lead and copper this relation presented low statistical significance ($r = 0.5232$, $p = 0.287$ and $r = 0.5849$, $p = 0.223$ respectively). Strong positive correlation can also been observed between cadmium and lead levels in plant ($r = 0.9664$, $p = 0.002$). On the other hand, the inverse correlation of metal in plants and soil humic substances was found. The statistical significance of such relation decreased in the following order: Cu > Pb > Cd > Ag ($r = -0.7457$, $p = 0.089$; $r = -0.6558$, $p = 0.157$; $r = -0.5280$, $p = 0.282$; $r = -0.2084$, $p = 0.692$, respectively). It should be stressed that the decreasing order of the statistical significance observed in this work is in agreement with the order of heavy metals affinity to humic substances, cited above [49-51]. In the view of our results, the binding of metals to soil humic substances contributes in lowering their uptake by *R. communis*. Apparently, the metals presenting strong affinity to HS have lower bioavailability to this plant. Indeed, there was no important relationship between soil and plant levels for Cu and Pb, but statistically important correlation between these parameters was found for other two elements, presenting lower affinity to HS (Ag and Cd).

On the other hand, the inverse correlation between soil HS and plant PC-2 was found ($r = -0.7825$, $p = 0.066$). Among four metals studied, cadmium levels in soil and in plant presented strong positive correlation with PC-2 ($r = 0.7857$, $p = 0.064$; $r = 0.9395$, $p = 0.005$, respectively). For lead, the correlation was significant only between PC-2 and metal in plants ($r = 0.9573$, $p = 0.003$) and other two metals (Cu, Ag) did not correlate with PC-2 in plants. These results suggest that both, cadmium and lead promote phytochelatin induction in *R. communis*. The lack of correlation between soil Pb and plant PC-2 ($r = 0.4261$, $p = 0.400$) suggest that soil humic substances contribute in lowering the metal uptake by *R. communis*.

To examine the association of metals with soil humic substances, these compounds were extracted from soil with 0.1 M sodium pyrophosphate and analyzed by SEC – ICP-MS as described elsewhere [46]. As an example, two chromatograms are presented in Figure 3 that were obtained in the analysis of soil from site 5 and 6 (HS concentrations 553 mg g^{-1} and 143 mg g^{-1} respectively). The relative distribution of metals in molecular mass fraction corresponding to HS was estimated at each sampling site, by calculating the fraction of metal co-eluted with HS in size exclusion chromatography (peak area between 10 and 18 min of chromatogram with respect to total area of ICP-MS chromatogram) and the results obtained are presented in Table 4.

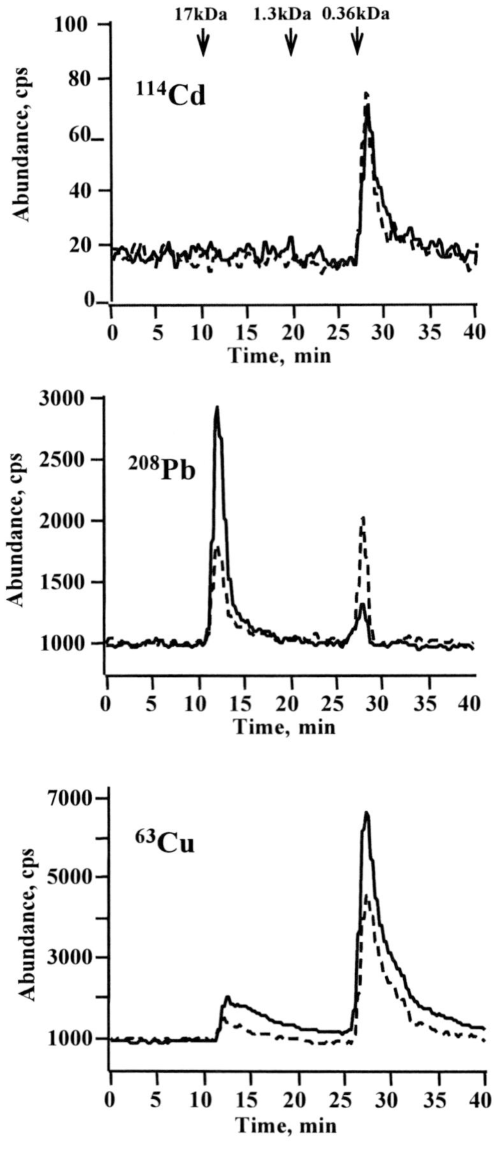

Figure 3. SEC – ICP-MS chromatograms of the soil extracts from site 5 (—) and site 6 (----): (a) ^{114}Cd; (b) ^{208}Pb; (c) ^{63}Cu.

It can be observed that, independently of the HS level, cadmium co-eluted with low molecular mass fraction, so this metal apparently is not bound to HS (Figure 3a, Table 4). For copper and lead, their elution occurred in different molecular mass fractions of the soil extracts (Figure 3b, 3c). As shown in Table 4, higher relative contributions of lead with respect to copper were always observed in the elution region of high molecular mass. Furthermore, the relative distribution of lead was clearly affected by the concentration of HS (Figure 3b). In particular, for higher HS concentrations in soil extract, relatively lower contributions of metal in low molecular mass fractions was observed. These results provide further evidence that the soil humic substances contribute in lower bioavailability of Pb and Cu, that present high affinity to HS.

Table 4. Humic substances in soils and a fraction of metal co-eluted with HS in size exclusion chromatography (based on the peak area measurement vs total area of chromatogram).

Sampling site	HS, mg g^{-1}	Metal association with soil HS[(a)] Cd	Cu	Pb
1	481	0	1.4	14
2	415	0	15	100
3	1345	0	24	72
4	372	0	7.4	59
5	553	0	17	94
6	143	0	32	27

[(a)]- % of metal in soil extract, which co-eluted with HS in size exclusion chromatography.

Since soils adjacent to the nopal roots contained low concentrations of humic substances, it cannot be expected that this parameter would affect metal uptake by plant. To get more information on the bioavailable metal forms, sequential extraction of soils from sites 1 – 6 was performed and the four metals of interest were determined in each fraction. The simplified Tessier method enables to extract the following fractions: F1 – exchangeable; F2 - fraction bound to carbonates; F3 - associated with oxides of iron and manganese; F4 - organic matter and sulfides and F5 – residual. The results obtained showed that the major fractions of Pb and Cu were bound to F4 at any sampling site (62.9 % - 86.3 % of total soil Pb and 60.4 % - 65.1 % of total soil Cu respectively) Significantly lower levels of these two metals were found in carbonate fractions (F2: less than 10 % of total Pb and less than 10 % of Cu in soil) and only minute amounts corresponded to "free" Pb or Cu (F1). For Ag, practically all the metal was bound to the organic matter and sulfides (F4, more than 95 % of total soil Ag). As already mentioned, the Cd soil levels were lower as compared to other metals, however, relatively uniform metal distribution among F1 – F5 fractions was observed. At sites 1 and 4, the relative distribution of "free" Cd was 36 % and 32 % respectively.

In search of any relationship between Pb, Cu and Ag levels in soil fractions F1 – F5 and root/leave concentrations of these metals, the statistical correlation analysis was carried out taking analytical data obtained at six sampling sites (Cd was excluded, because of its low concentrations in soil and in plant tissues). No statistically important correlation was found between metal levels in soil versus leaves. The correlation parameters obtained for roots and soil data are presented in Table 5. At first, for Pb and Cu, total (acid-digested) metal in soil did not correlate strongly with its concentration in plant ($r = 0.7321$, $p = 0.176$ and $r = 0.8169$, $p = 0.091$, respectively). Furthermore, no statistically important correlation was found between any metal in the fractions F1 (exchangeable), F2 (carbonates), F3 (manganese and iron oxides) or F5 (residual) and root metal, indicating that these fractions were not important for the uptake of Cu, Pb and Ag by nopal. On the other hand, statistically significant correlations found between metal in F4 and root metals ($r = 0.8436$, $p = 0.072$ for Pb; $r = 0.8648$, $p = 0.058$ for Cu; $r = 0.8842$, $p = 0.046$ for Ag, Table 5) suggest that organic matter and sulfide bound metals (fraction F4) should be bioavailable to nopal. The results obtained in spectrophotometric assay revealed low levels of humic substances (primary soil organic matter) in soils adjacent to the roots of nopal (Table 3). On the other hand, sulfide minerals are abundant in Guanajuato region [56].

Table 5. Statistical analysis of correlation between total metal concentrations in soil and soil fractions F1 – F5 and their levels found in nopal roots

metal in soil vs metal in roots	Pb	Cu	Ag
soil *vs.* root	0.7321	r = 0.8169	r = 0.9368
	0.176	p = 0.091	p = 0.019
F1 *vs.* root	nf	nf	nf
F2 *vs.* root	nf	nf	nf
F3 *vs.* root	nf	nf	nf
F4 *vs.* root	r = 0.8436	r = 0.8648	r = 0.8842
	p = 0.072	p = 0.058	p = 0.046
F4 *vs.* root	nf	nf	nf

nf – no statistically important correlation was found.

Our results suggest that sulfides are primary Pb, Cu and Cd forms in soil close to the cactus roots (at least 20 cm from the roots). The solubility of these compounds is generally poor, so the rhizosphere processes and, in particular root exudation of low-molecular weight carboxylic acids should be important in mobilizing sulphide bound metals [57,58]. However, the literature data on these processes in nopal or other cactaceous is scarce. Puente et al.[59] proposed the contribution of bacteria and fungi in the rhizoplane of nopal roots (the area at the plant and soil interface) as a factor increasing metal bioavailability.

CONCLUSION

The results obtained in this study enable to answer the interrogations set up at the beginning. In particular it can be concluded that the wild plants, chronically exposed to low/moderate metal levels in soil biosynthesize phytochelatins. In the analysis of three different plants collected simultaneously at these same sites, PCs were found in nopal and in castor bean, but not in Mexican sunflower, which seems to confirm that phytochelatin induction is genotype dependent. Furthermore, lead and cadmium were found to be important for PCs induction in nopal. It is proposed that the uptake of these metals could be controlled by rizosphere processes. In the case of castor bean, the metals promoting PCs induction in plant were also cadmium and lead, however, their uptake is apparently controlled by the association of metals with soil HS.

In general, the results obtained provide evidence that PCs induction in plants growing in natural environments depends on: (i) total element concentrations in soil; (ii) actual physicochemical forms of elements in soil; (iii) organic matter content in soil and (iv) different processes occurring in rizosphere.

REFERENCES

[1] Zenk, M.H. *Gene* 1996, 179, 21-30.

[2] Hirata, K.; Tsuji, N.; Miyamoto, K. *J Biosci Bioeng* 2005, 100, 593–599.

[3] Persson, D.P.; Hansen, T.H.; Holm, P.E.; Schjoerring, J.K.; Hansen, H.C.B.; Nielsen, J.; Cakmak, I.; Husted, S. *J. Anal. At Spectrom.* 2006, 21, 996-1005.
[4] Clemens, S. *Planta* 2001, 212, 475-486.
[5] Thumann, J.; Grill, E.; Winnacker, E.L.; Zenk, M.H. *FEBS Lett* 1991, 284, 66-91.
[6] Takagi, M.; Satofuka, H.; Amano, S.; Mizuno, H.; Eguchi, Y.; Hirata, K.; Miyamoto, K.; Fukui, K.; Imanaka, T. *J Biochem* (Tokyo) 2002, 131, 233-239.
[7] Yen, T.Y.; Villa, J.A; DeWitt, J.G. *J. Mass Spectrom.* 1999, 34, 930-941.
[8] Le Faucheur, S.; Behra, R.; Sigg, L. *Environ.Toxicol. Chem.* 2005, 24, 1731-1737.
[9] Grill, E.; Winnacker, E.L.; Zenk, M.H. *Science* 1985, 230, 674–676.
[10] Scarano, G.; Morelli, E. *Biometals* 2002, 15, 145-151.
[11] Spain, S.M.; Rabenstein, D.L. *Anal. Chem.* 2003, 75, 3712-3719.
[12] Cobbett, C.S. *Curr. Opin. Plant. Biol.* 2000, 3, 211-216.
[13] Clemens, S. *J. Plant. Physiol.* 2006, 163, 319-332.
[14] Mejare, M.; Bülow, L. *Trends Biotechnol* 2001, 19, 67-73.
[15] Meagher, R.B. *Curr. Opin. Plant Biol.* 2000, 3, 153-162.
[16] Bovet, L.; Feller, U.; Martinoia, E. *Environ. Int.* 2005, 31, 263-267.
[17] Yang, X.; Feng, Y.; He, Z.; Stoffella, P.J. *J. Trace Elem. Med. Biol.* 2005, 18, 339-353.
[18] Clemens, S. *Int. J. Occup. Med. Environ. Health* 2001, 14, 235-239.
[19] Koster, M.; Reijnders, L.; van Oost, N.R.; Peijnenburg, W.J. *Environ. Pollut.* 2005, 133, 103-116.
[20] Lucho Constantino, C.A.; Prieto García, F.; del Razo, L.M.; Rodriguez Vazquez, R.; Poggi Varaldo, H.M. *Agric. Ecosys. Environ.* 2005, 108, 57-71.
[21] Piechalak, A.; Tomaszewska, B.; Baralkiewicz, D. *Phytochemistry* 2003, 64, 1239-1251.
[22] Sun, Q.; Wang, X.R.; Ding, S.M.; Yuan, X.F. *Chemosphere* 2005, 60, 22-31.
[23] Meers, E.; Ruttens, A.; Hopgood, M.J.; Samson, D.; Tack, F.M. *Chemosphere* 2005, 58, 1011-1022.
[24] Turgut, C.; Katie Pepe, M.; Cutright, T.J. *Environ. Pollut.* 2004, 131, 147-154.
[25] Lorenzo, J.I.; Beiras, R.; Mubiana, V.K.; Blust, R. *Environ. Toxicol. Chem.* 2005, 24, 973-980.
[26] Voets, J.; Bervoets, L.; Blust, R. *Environ. Sci. Technol.* 2004, 38, 1003-1008.
[27] Garcia-Mina, J.M.; Antolin, M.C.; Sanchez-Diaz, M. *Plant. and Soil.*2004, 258, 57-68.
[28] Misra, V.; Pandey, S.D. *Bull. Environ. Contam. Toxicol.* 2005, 74, 725-731.
[29] Inaba, S.; Takenaka, C. *Environ. Int.* 2005, 31, 603-608.
[30] Kungolos, A.; Samaras, P.; Tsiridis, V.; Petala, M.; Sakellaropoulos, G. *J. Environ. Sci.* Health A Tox Hazard Subst *Environ. Eng.* 2006, 41, 1509-1517.
[31] Garcia-Meza, V.; Ramos, E.; Carrillo-Chavez, A.; Duran-de-Bazua, C. *Bull. Environ. Contam. Toxicol.* 2004, 72, 170-177.
[32] Morton-Bermea, O.; Carrillo-Chavez, A.; Hernandez, E.; Gonzalez-Partida, E. B*ull. Environ. Contam. Toxicol.* 2004, 73, 770-776.
[33] Morelli, E.; Scarano, G. *Mar. Environ. Res.* 2001, 52, 383-395.
[34] Keltjens, W.G.; Van Beusichem, M.L. *Plant and Soil* 1998, 203, 119-126.
[35] Maitani, T.; Kubota, H.; Sato, K.; Yamada, T. *Plant Physiol.* 1996, 110, 1145-1150.
[36] Landero Figueroa, J.A.; Wrobel, K.; Afton, S.; Caruso, J. A.; Gutierrez Corona, J.F.; Wrobel, K. *Chemosphere* 2008, 70, 2084-2091.

[37] Mehra, R.K.; Tran, K.; Scott, G.W.; Mulchandani, P.; Saini, S.S. *J Inorg Biochem* 1996, 61, 125-142.
[38] Giordani, C.; Cecchi, S.; Zanchi, C. *Environ Manage* 2005, 36, 675-681.
[39] Khan, A.G.; Chaudhry, W.J.; Hayes, W.J.; Khoo, C.S.; Hill, L.; Fernández, R.; Gallardo, P. *Water Air Soil Poll* 1998, 104, 389-402.
[40] Gupta, A.K.; Sinha, S. *Bioresour. Technol.* 2006, 98, 1788-1794.
[41] Olivares, E.; Peña, E.; Aguiar, G. *J. Plant Physiol.* 2002, 59, 743–749.
[42] Tessier, A.; Campbell, P.G.C.; Bisson, *N.J. Anal. Chem.* 1979, 51, 844.
[43] Pereira Navaza, A.; Montes-Bayon, M.; LeDuc, D.L.; Terry, N.; Sanz-Medel, A. *J. Mass Spectrom.* 2006, 41, 323–331.
[44] Doring, S.; Korhammer, S.; Oetken, M.; Markert, B. Fresenius *J. Anal. Chem.* 2000, 366, 316-318.
[45] Sneller, F.E.; van Heerwaarden, L.M.; Koevoets, P.L.; Vooijs, R.; Schat, H.; Verkleij, J.A. *J. Agric. Food Chem.* 2000, 48, 4014-4019.
[46] Wrobel, K.; Sadi, B.B.M.; Wrobel, K.; Castillo, J.R.; Caruso, *J.A. Anal. Chem.* 2003, 75, 761-767.
[47] Tang, D.; Shafer, M.M.; Vang, K.; Karner, D.A.; Armstrong, D.E. *J. Chromatogr.* A 2003, 998, 31-40.
[48] Hall, J.L. *J. Exp. Bot.* 2002, 53, 1-11.
[49] Evangelou, V.P.; Marsi, M. *Plant and Soil* 2001, 229, 13-24.
[50] Pandey, A.K.; Pandey, S.D.; Misra, V. *Ecotoxicol. Environ. Safe* 2003, 47, 195.
[51] Sadi, B.B.M.; Wrobel, K.; Wrobel, Z.; Kannamkumarath, S.S.; Castillo, J.R.; Caruso, J.A. *J. Environ. Monit.* 2002, 4, 1010-1016.
[52] Lamelas, C.; Wilkinson, K.J.; Slaveykova, V.I. *Environ. Sci. Technol.* 2005, 39, 6109-6116.
[53] Remon, E.; Bouchardon, J.L.; Cornier, B.; Guy, B.; Leclerc, J.C.; Faure, O. *Environ. Pollut.* 2005, 137, 316-323.
[54] Galizzi, F.A.; Felker, P.; Gonzalez, C.; Gardiner, D. *J Arid Environ* 2004, 59, 115-132.
[55] Burke, I.C.; Lauenroth, W.K.; Riggle, R.; Brannen, P.; Madigan, B.; Beard, S. *Ecosystems* 1999, 2, 422-438.
[56] Carrillo-Chavez, A., Morton-Bermea, O.; Gonzalez-Partida, E.; Rivas-Solorzano, H.; Oesler, G.; Garcia-Meza, V.; Hernandez, E.; Morales, P.; Cienfuegos, E. Ore *Geol. Rev.* 2003, 23, 277–297.
[57] Wenzel, W.W.; Lombi, E.; Adriano, D.C. in: *Heavy Metal Stress in Plants*, 2004, pp. 313-344 (Prasad, M.N.V., Ed.) Springer-Verlag, Berlin, Germany.
[58] Jones, D.L.; Hodge, A.; Kuzyakov, Y. *New Phytologist* 2004, 163, 459-480.
[59] Puente, M.E.; Bashan, Y.; Li, C.Y.; Lebsky, V.K. *Plant Biol.* (Stuttg) 2004, 6, 629-642.

In: Causes and Effects of Heavy Metal Pollution
Editor: Mikel L. Sanchez

ISBN 978-1-60456-900-1

Chapter 5

Fate of Trace Elements in the Venice Lagoon Watershed and Conterminous Areas (Italy)

C. Bini*

Dept. of Environmental Sciences, University of Venice - Dorsoduro, 2137.30123-Venezia, Italy

Abstract

Element mobility is of major importance with regard to bioavailability and the potential risk for contamination. Different factors control the ultimate fate of a toxic element in the environment, that is, if it will precipitate or will be adsorbed , or released, transported and taken up by organisms.

The objectives of this work are:

- To evaluate background levels of heavy metals in soils of highly vulnerable area in northern Italy.
- To ascertain metal mobility and possible contamination of some sites, and the related environmental hazard, with special reference to the pollution of the Venice lagoon, which is a unique and delicate ecosystem.
- The Venetian territory is characterized by different pedolandscapes:
- A wide plain formed by alluvial deposits. Most soils here (Entisols, Inceptisols, Alfisols) are cultivated with extensive agriculture;
- Gently ondulating conglomerates, marls and limestones with shallow soils (Entisols and Inceptisols) frequently cultivated with vineyards, or forested;
- Mountain ranges with steep morphology. Forestry and grassland are the main land utilization types on these soils (mostly Inceptisols and Mollisols).

Approximately 900 soil samples from 300 representative soil profiles were analyzed for As, Cd, Co, Cr, Cu, Hg, Mn, Ni, Pb, Zn. Data were statistically processed to find close relationships among elements.

---

* E-mail: bini@unive.it.

Agriculture soils. The soils examined (more than 200 sites) contain generally high levels of anthropogenic Cu, Zn, Pb, and As.

Forest soils. The heavy metal contents in the soils examined (more than 100 sites) are generally below the target values and depend mainly upon local physico-chemical and geological conditions. Anthropogenic Pb and Cd are concentrated especially in organic layers.

Increasing acidic conditions, redox status, organic matter content and pore solution are the factors responsible for trace elements mobilization within the soil. From the soil, trace elements move to groundwater and to lagoon, where they are concentrated in sediments or transferred to organisms.

The elements with the lowest relative mobility (bioavailability) are Co, Cr, Hg, Pb, As; intermediate elements are Cu, Ni and the most bioavailable are Zn and Cd.

The soils investigated have heavy metal concentrations that are generally within the regulatory guidelines. Exceptions are anthropogenic Cu and Zn in agricultural soils, Cd and Pb at forest sites.

The ecological risk posed by single elements is limited for As and very high for Cd. The cumulative toxic risk indicates a relevant bioaccumulation of trace elements in the lagoon ecosystem.

Keywords: soil contamination, Venice Lagoon watershed, environmental pollution, environmental hazard, pedometrics.

1. Introduction

Research on distribution and circulation of trace elements in different environmental contexts, since the last decades, has received a remarkable attention from various scientific domains (e.g. geology, pedology, chemistry, agronomy, botany, pharmacy). The increasing interest concerned particularly the following topics:

- background knowledge and soil genesis;
- environmental contamination;
- groundwater pollution;
- soil remediation;
- toxicology and human health.

The results achieved in these different items allow a general insight on the distribution and circulation of trace elements in natural environments, though quantitative estimation of elemental fluxes and resident time in the different geochemical spheres, especially the pedosphere, is still deficient. As a matter of fact, present knowledge not only does not permit to have reliable data on the levels of some soil microelements (e.g. As, Cd, Cr, Hg, Tl), but also it does not highlight the main processes which control kinetics, equilibria, circulation and fluxes of elements in the environment.

Element mobility in the exogenic environment is of major importance with regard to their bioavailability and the potential risk for contamination. Indeed, supergenic alteration processes may lead to the release of potentially toxic elements, particularly heavy metals, in the environment. The factors that control the ultimate fate of a toxic element, that is, if it will precipitate as an insoluble phase or will be adsorbed on the surface of some other phase, or will be released, transported and eventually taken up by plants, depend mainly upon local physico-chemical, climatic, biologic and geologic conditions (Brummer, 1986; Langmuir, 1997).

In order to examine the behaviour of such elements, it is necessary to study the interface between rocks, the biosphere and the hydrosphere, that is, the *pedosphere*. This is a multicomponent complex system whose chemical equilibria are frequently in a thermodynamic steady state (Sposito, 1983). The rate of redox reactions, for example, may be influenced by mic roorganisms. Reduced or oxidized elements may, in turn, influence the geochemical behaviour of other elements (Cambier and Charlatchka, 1999).

Concerning heavy metals, these are "dispersed" trace elements with generally limited concentrations in the earth's crust and the related soils (Kabata Pendias and Pendias, 1992). However, metal accumulation in the environment may occur at some locations, owing to different sources (Angelone and Bini, 1992). Possible "natural' accumulation may be related to heavy metal-bearing rocks (e.g. Ni and Cr in serpentine: Angelone et al., 1993) or to mineralized areas (e.g. Pb and Zn from mixed sulfide mines: Benvenuti et al., 1997), while anthropogenic accumulation is related to industrial activities (e.g. Cd in metallurgy, Cr in varnish and leather factories: Bini et al., 2008), agriculture and urban sewage sludge (e.g. Zn and Cu from fertilizers: Deluisa et al., 1996; Cd, Pb, Cr from sludge: Petruzzelli, 1989).

Especially the last item is paying great attention at present, since increasing quantities of urban sludge are produced and extensively introduced in the environment. Moreover, atmospheric input from industrial emissions, heat power plants, heavy traffic and acid rains may account for increasing heavy metal concentration in soils (Norra et al., 2006). Therefore, identification of the sources responsible for soil contamination is an important issue, since high loads of heavy metals applied to soils, or stored in soils, may determine soil quality degradation, surface and groundwater pollution, accumulation in plants, phytotoxicity and successive transfer to the food chain.

All trace elements are toxic if their intake through ingestion or inhalation is excessive. In particular Ag, As, Be, Cd, Ce, Ge, Hg, Pb, Tl are good examples of potentially harmful elements (PHEs) that have no proven essential functions, and are known to have adverse physiological effects at relatively low concentrations (Abrahams, 2002).

Examples of toxicity by heavy metals are known since the Antiquity (Nriagu, 1983). For instance, one of the supposed causes for the Roman Empire drop is the increasing lead toxicity from Pb-bearing potteries and wine containers, as it was found in Roman findings and bones. Lead (saturnism) and Hg (hydrargillism) poisoning cases were frequently recorded in workers employed in mining industry and even in hat factories in Tuscany (Dall'Aglio et al., 1966). At present, diseases and toxicity related to microelement contamination (Cr, Cu, Ni, Pb, Tl, Zn,) of air, water and soil from industrial activities are well established (Thornton, 1993; Abrahams, 2002). For example, the most notable cause of Tl poisoning occurred adjacent to a cement works in Germany (Abrahams, 2002).

The risk arising from metals depends on their bioavailability, which in turn depends in the form in which they occur (Adriano et al., 1995): this is the reason why the risk to human health cannot be assessed on the basis of the total concentration of the toxic metal. Background values correspond to the total content of metals in soils not affected by human activities, i.e. they are the *reference values* for most countries. Soil guide values have been introduced in the late1950s in Japan, in 1980 in The Netherlands, in 1986 in Switzerland, in 1987 in Great Britain, in 1994 in Germany. Since that time, many countries, notably the

U.S.A., Canada, Great Britain and the Netherlands, have progressed further in setting standards for hazardous constituents in soil, health-risk based soil screening levels and soil remediation. However, legislation on maximum admissible levels of heavy metals in the environment in the EU is rather confusing. Indeed, a general regulatory guideline on the maximum trace element concentration in soils has not yet been established, the current references being related to the total metal content in waste and sewage sludge to be spread on soil (Adriano et al., 1995). Moreover, there is little agreement among the members in their implementation of the EC Directive of 1986. Several attempts have been made to adopt background values in order to obtain more viable reference values for regulatory decision. Although there are some similarities, standard criteria on the background level of metals in soils, however, are not yet established. In many EU countries, the target values proposed by the Dutch Ministry of Housing, Spatial Planning and Environment (Table 1) have been accepted as the reference national values, although the strict application of the Dutch target values is difficult to achieve for several reasons (parameters influencing the bioavailability of metals, site-specific conditions, the use of land after restoration, etc).

The present values, although related to the total metal concentration, are based not only on considerations of the natural contaminant concentration, but also take into account the local circumstances. Moreover, they are regarded as having been exceeded, and the soil seriously contaminated. Furthermore, they take into account both human toxicological and ecotoxicological considerations (Adriano et al., 1995).

Recent legislative regulations in EU were devoted to polluted sites reclamation, with particular reference to different land use.

Table 1. Dutch target values (also referred to as A-value or reference value of the prior regulatory guidelines) and intervention values (also referred to as C-value) for selected metals in soil (mg/kg dry matter). (Source: Dutch Ministry of Housing, Spatial Planning and Environment. The Hague, The Netherlands)

Metal	target value	intervention value
Arsenic	29	55
Barium	200	625
Cadmium	0.8	12
Chromium	100	380
Cobalt	20	240
Copper	36	190
Mercury	0.3	10
Lead	85	530
Molybdenum	10	200
Nickel	35	210
Zinc	140	720

Table 2. Maximum metal concentration values recordable in soil and subsoil of contaminated sites, with reference to specific land utilization (L. A. n° 471/99, annexe 1)

	Green and residential areas mg/kg d.m.	Commercial and industrial areas mg/kg d.m.
Antimony	10	30
Arsenic	20	50
Berillium	2	10
Cadmium	2	15
Cobalt	20	250
Chromium (total)	150	800
Chromium VI	2	15
Mercury	1	5
Nickel	120	500
Lead	100	1000
Copper	120	600
Selenium	3	15
Tin	1	350
Thallium	1	10
Vanadium	90	250
Zinc	150	1500
Cianides	1	100
Fluorides	100	2000

The Legislation Act 471/99, promoted by the Italian Ministry of Environment, deals with contaminated site restoration and soil protection, and introduces regulatory thresholds (Table 2), which, when exceeded, would require mandatory clean-up operations.

Soil contaminants constitute a known global problem, and more knowledge is required of them, their behaviour, and their pathways to humans. Nevertheless, information on the chemical status of the elements is difficult to achieve.

Though in most cases reference is made to their total level in the different compartments, the fate of heavy metals in the environment depends upon their available fraction. Whilst soils may contain high total concentrations of elements, many factors, including soil pH and redox potential, clay and organic matter content, influence the speciation, mobility and bioavailability of elements to plants. The implication is that there are often a number of effective barriers operative in the transfer of potentially harmful elements from soils to food.

Metal concentration generally is not directly related to the potential hazard: indeed, not all the metal present in the soil is available to plants, part of it being trapped in the crystal lattice of minerals (e.g. Ni in serpentine soils may achieve 800 mg/kg, but most of it is unavailable and therefore does not contribute to Ni toxicity). The bioavailability, however, is related to the chemical bonding of the metal with the soil mineral, the inorganic and organic colloids, the kind of humic substances, the physico-chemical conditions, etc. Sequential extraction, therefore, is needed to separate different fractions of heavy metals, in order to evaluate the actual environmental hazard of the different elements. Analytical protocols have already been established for this purposes, though doubts remain on the real efficiency and significance of different extractions.

Several extractants have been indicated in the literature in order to obtain fractionation and speciation of the different forms of metal (see f.i. Tessier et al., 1979). Among them, EDTA or DTPA has been suggested by several authors (e.g. Goupta et al., 1994) as he most significant to evaluate the metal fraction potentially available to plants over a long period of time, while electrolytes like $CaCl_2$ or NH_4COOCH_3 are considered to extract soluble and extractable fractions over a short time.

A profound insight of the above mentioned items, therefore, is needed, in order to achieve the following finalisations:

a) knowledge on the distribution and circulation of trace elements in the different geochemical domains (lithosphere, pedosphere, hydrosphere, biosphere, atmosphere) contributes to better understand the natural processes responsible for the soil genesis and evolution, the relations with landscape and vegetation, and the ecosystems equilibria;
b) human activities, technological processes and modern industrial products pose major environmental pollution concerns. Water, soil, vegetation may be appreciably affected by toxic or critical substances emitted in the atmosphere, or introduced in surface and groundwater, by these activities. Successively, after a variable time interval, they may be deposited at the earth surface, posing environmental hazard. Potentially harmful elements (As, Cd, Cr, Cu, Hg, Ni, Pb, Zn, etc.) may have toxic effects on living organisms, including humans;
c) Research on living organisms occurring at sites naturally and/or anthropically contaminated may allow identification of individuals (plants and animals), that are indicators of degraded environmental systems, or that may be utilized for restoration of contaminated sites, e.g. with phytoremediation techniques.

Based on the above mentioned assumptions, in the frame of soil surveys carried out in recent years (1997-2003) by the EPA of the Veneto Region and the University of Venice (see f.i. Giandon et al., 2000; Bini and Zilocchi, 2001; Giandon et al., 2001) as a part of the intervention programs against pollution of the lagoon of Venice, a survey of agricultural, industrial and undisturbed (forest, grassland and wetland) areas, including the lagoon, was carried out in the Venice drainage basin and the conterminous territory.

The Lagoon of Venice is a shallow transitional environment located in a densely populated industrial/agricultural area (population 1,500,000), in the northern part of the Adriatic sea (Figure 1), from which it is separated by some flat and narrow islands. Three inlets allow water exchange between the lagoon and the Adriatic sea. The whole area of the lagoon and the conterminous land has historically undergone severe anthropogenic pressure, with direct inputs of pollutants from both industrial and urban discharges. Other sources of contamination are river inputs, which collect industrial, domestic and agricultural polluting substances coming from the drainage basin of the lagoon.

Figure 1. Location of the investigated region. Inset: the Venice Lagoon watershed and conterminous areas.

The objectives of this work were:

- To evaluate the background level of heavy metals in soils of the Venice drainage basin and conterminous areas;
- To estimate the environmental impact of agricultural and industrial activity;
- To ascertain metal mobility and possible contamination of some sites, and the related environmental hazard, with special reference to the pollution of the Venice lagoon, which is a unique and delicate ecosystem.

2. Materials and Methods

2.1. Study Area

The investigated territory extends for approximately 2000 km^2, from the pre-Alpine fringe to the lagoon of Venice (Figure 1, inset), and is characterized by different pedolandscapes (Giandon et al., 2001):

- mountain ranges with steep morphology. The geology of this area is different from place to place, but limestone and dolostone dominate over crystalline or silicate rocks. Forestry and grassland are the main land utilisation types on these mountain soils (mostly inceptisols and mollisols);

- gently ondulating conglomerates, marls and volcanic rocks of tertiary age, with shallow hills frequently cultivated with vineyards, or forested when slope gradient is high;
- a wide alluvial plain originated by the sedimentary activity of alpine rivers during the Quaternary. It consists of alluvial deposits of mixed lithology (depending on erosional contribution from uplands), size and age. Most soils are cultivated with extensive agriculture.
- The south-eastern margins of the mainland are characterized by sand dunes, urban soils and reclaimed land with variable texture (from coarse-grained to fine-grained) and composition, which form the borderline of the lagoon.

The soil reference system is based on the typological approach adopted by EPA-Veneto (ARPAV, 2005), and focuses on both land characteristics, soil morphology and land use. Urban and undisturbed soils under annual crops, permanent grassland and forests were selected.

In the alluvial plain, which is part of the Soil Region 18.8 (Cambisols-Luvisols Region), the surveyed soils are characterized by a high degree of heterogeneity, ranging from reddish Alfisols (Hapludalfs in the USDA Soil Taxonomy, 1999; Chromic Luvisols in the FAO-ISRIC WRB, 1998) in the Upper Pleistocene plain to fine-textured Inceptisols with vertic or gleyic characters (Typic, Vertic and Aquic/Oxyaquic Eutrudepts in S.T.; Cambisols, Calcisols and Gleysols in the WRB) in the late Wurmian-Holocene plain, and to fine, gleyed Entisols (Fluvents, Aquents in S.T.; Fluvisols, Gleysols in the WRB) and coarse-textured Entisols (Psamments in S.T.; Arenosols in WRB) in recent sandy deposits which border the lagoon. In lowland and wetland areas, where organic matter accumulates, Mollisols (Phaeozems in WRB) and Histosols occur.

In the gently ondulating landscape of Tertiary age (Soil Region 16.5, Cambisols-Leptosols Region), soils are mostly Inceptisols (Eutrudepts or Dystrudepts in S.T.; Eutric or Dystric Cambisols with some andic characters in the WRB). Entisols are mostly Typic and Lithic Udorthents in S.T., and Eutric or Rendzic Leptosols in the WRB.

In mountain areas (Soil Region 34.3, Leptosols Region with Cambisols), Inceptisols and Mollisols with weakly differentiated profile (Udepts, Udolls, Rendolls in S.T.; Calcari-Mollic Cambisols or Phaeozems in the WRB) are the prevailing soils, with subordinate podzolic soils (Haplorthods in S.T., Haplic Podzols in the WRB) and eroded, shallow Entisols (Udorthents in S.T., Rendzic or Dystric Leptosols or Regosols in the WRB).

2.2. Laboratory Methods

900 soil samples (both topsoil and subsoil) from more than 300 representative soil profiles were selected for standard soil characterization and trace elements analysis. The sampling sites (446 from agricultural areas, 454 from rangeland and forest areas, 10 from urban areas and the lagoon border) were selected following the soil typological approach (ARPAV, 205) within the four pedolandscapes previously described, comprehensive of parent material, relief and land use.

Samples were air-dried and passed through a 2 mm sieve in order to separate coarse fragments from the fine earth. Standard laboratory methods established by the Italian Ministry of Agriculture (MIRAAF, 1994) were applied on the fraction <2 mm, and included: particle-size distribution (pipette method following organic matter removal with H_2O_2 and dispersion with sodium metaphosphate); pH (1:2.5 soil-water suspension); total carbonates (gas-volumetric method); readily oxidizable organic carbon (Walkley-Black method); cation exchange capacity (titration with EDTA after extraction with triethanolamine and $BaCl_2$ buffered at pH 8.2). Exchangeable bases were determined by AA spectrometry, after $BaCl_2$ and triethanolamine extraction.

A fraction of the fine earth (<2mm) was finely ground in a agate mortar to obtain a homogeneous powder (<100μm). Heavy metal concentrations in soils were quantified, after digestion in acidic mixture for total ("aqua regia": ISO 11466) and DTPA-extractable (ISO 14870) fractions. Instrumental determination of metals (As, Cd, Co, Cr, Cu, Fe, Hg, Mn, Ni, Pb, Zn) was performed by atomic absorption spectroscopy on the fine fraction (<100μm) of the soils previously selected.

2.3. Statistical Analyses

Univariate and multivariate statistical analyses were applied to the data set. The recorded data were statistically processed with the software STATISTICS for Windows, version 2002.

Frequency distribution and Pearson correlation matrices were calculated in order to evidence simple and multiple correlations between variable pairs. Principal Component Analysis (PCA) was applied for reducing the number of variables in the data set to a few components or factors, and therefore identifying the relationships between the variables, to enhance data variability interpretation. Varimax rotation of the factors was used to maximize the variance explained by each principal component (Hair et al., 1998). Hierarchical cluster analysis was also used to highlight similarities between sampling sites, and to verify the consistent associations between sampling sites and groups of elements (pedogeochemical fingerprints).

3. Results and Discussion

3.1. Agriculture Soils

The territory investigated covers approx 160,000 ha and extends from the lagoon borderline up to 50 kilometers over about 1/4 of the whole agriculture areas in North Eastern Italy, mostly in the alluvial plain.

The soils examined (200 sites, totaling 446 soil samples) were grouped in 5 orders and 11 subgroups of the Soil Taxonomy (USDA, 1999); Inceptisols (77%) are the most represented, followed by Entisols (15%) and Alfisols (5%); Mollisols (2%) and Vertisols (1%) complete the whole set of soils in the agricultural area. (Figure 2).

The mean levels, variance and ranges of soil properties (pH, organic matter, total carbonates, CEC and particle size distribution) are reported in Table 3.

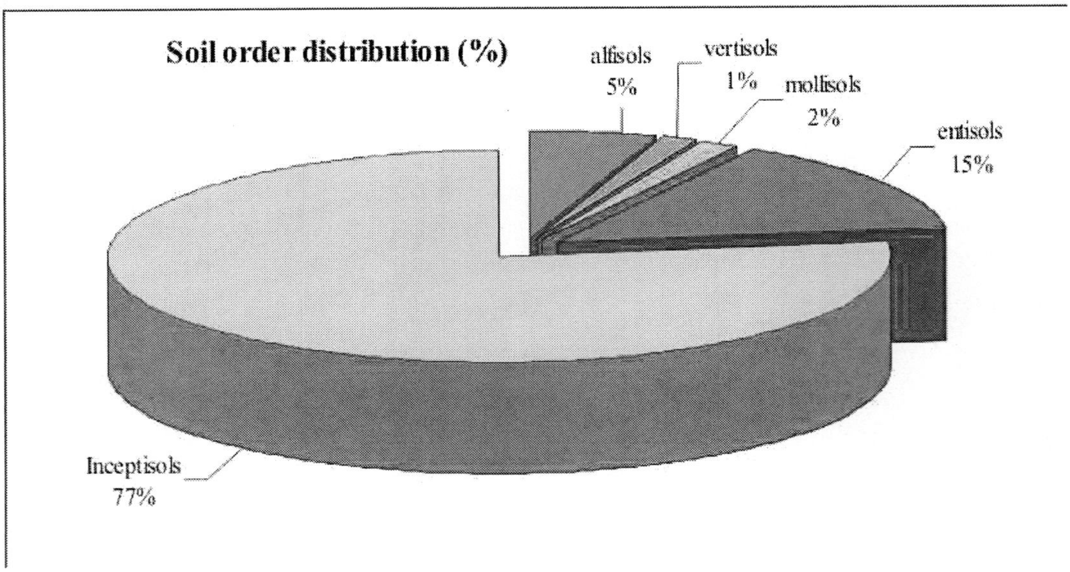

Figure 2. Sketch diagram of the distribution of agricultural soils as taxonomic USDA orders.

Table 3. Descriptive statistics for agricultural soil properties (446 soil samples, 223 topsoil, 226 subsoil)

	pH	Organic Carbon %	Total Carbonates %	CEC cmol(+)/kg	Clay %	Silt %	Sand %
Mean	8.11	2.15	17,25	19.54	27.51	45.14	27.35
Median	8.23	1.06	10	17.71	25.54	44.26	24.88
Standard deviation	0.45	3.82	17.56	9.28	12.74	11.67	18.99
minimum	4.83	0.2	0	2.30	1.92	1.17	0.4
maximum	8.9	4.3	82	64.71	64.30	70.05	96.35

A wide range of values was found for all the determined parameters. The pH value ranges between 4.8 and 8.9, accordingly to the nature of parent material and pedogenic processes. Mean Organic Carbon level is 2.1%, with the maximum 4.3% in surface horizons of Mollisols; total Carbonates also have a wide range (0-82%), with very high variance, in relation to the parent material characteristics and to leaching processes within the soil; CEC ranges from 2.30 to 65 cmol(+)/kg, depending on the particle size composition. The soils have also a great variability in grain size distribution (2 - 64 % clay, 1-70% silt, 0.4-96% sand), in relation to the parent material composition and the site morphology.

The mean values, variance and ranges of total heavy metal concentrations (mg/kg d.m.) in the soils examined are summarized in Table 4.

Table 4. Descriptive statistics for total heavy metal concentrations (mg/kg d.m.)

	Cd	Co	Cr	Cu	Mn	Ni	Pb	Zn	As	Hg
Mean	0.52	9.95	32.55	52.97	581.98	32.55	24.14	90.27	13.72	0.22
Median	0.51	9.77	28.55	36.2	529	22.5	22.6	88.35	10.75	0.15
Standard deviation	0.248	4.12	19.00	49.35	257.24	40.64	15.64	32.55	11.05	0.23
minimum	0	1.57	4.67	3.05	147	3.1	0.72	12.6	0	0
maximum	1.65	31.5	149	362	1573	413	203	201	65.1	1.24

The soils present generally high levels of heavy metals, with mean values of 53 mg/kg total Cu (range 3 - 362), 90 mg/kg total Zn (range 13 -201), 32 mg/kg total Ni (range 3 - 413), 32 mg/kg total Cr (range 5 - 149); Pb (mean 24 mg/kg), Co (mean 10mg/kg) and Cd (mean 0.5 mg/kg) present higher concentrations in organic horizons than in mineral horizons (data not shown). Of the volatiles, mean total As is 14 mg/kg (range 0 - 65*)*, while Hg (mean 0.2 mg/kg) presents low and uniform concentration along the soil profile. The highest metal concentrations, far above the regulatory threshold for contamination (see Table 1), were recorded especially in surface horizons (data not shown), as it was reported also in a recent work on soils of a portion of the alluvial plain (Ungaro et al., 2008).

Arsenic is an environmental issue in the Veneto region, since its concentration is currently (in 30% of the samples examined) above the regulatory threshold for contamination (20 mg/kg). The origin of As high concentration in soils is related to two main factors: rock composition and human activities. Volcanic and mine areas contribute to naturally high As concentrations in soils, whilst combustion of fossil fuels, agrochemical application, smelting etc. are the main anthropic As sources (Ungaro et al., 2008). The spatial distribution of As concentration in topsoil in the Venice watershed is reported in Figure 3. A clear anthropogenic input is localized in the central area, not far from the Venice industrial area, while some hotspots of lithogenic origin are localized in the volcanic area of the Euganean hills (southwestern zone).

Nickel also presents, in the same volcanic area, a few hotspots (3.4% frequency within the whole data set) with concentration levels higher than 75 mg/kg, although in the whole region most soils have low Ni concentrations (<25 mg/kg in 65% of the examined samples).

Zinc and copper are known as critical micronutrients for plants, and are largely utilized in agriculture to combat vineyard diseases. Their concentration in surface soil horizons, therefore, is frequently above the regulatory threshold for contamination. In the whole set of samples, Zn contents (Figure 4) account for 72% in the > 75 mg/kg concentration class, and for 5% in the >150 mg/kg class; Cu contents account for 30% in the >50 mg/kg, and for 15 % in the >100 mg/kg concentration class (Giandon et al., 2000). It is very likely that zinc and copper pollution of surface horizons could be a consequence of agrochemical compounds containing these metals and even As (Deluisa et al*.,* 1996).

Some Pb hotspots (3% of the whole set of samples with concentration >50 mg/kg, with a maximum value up to 203 mg/kg; Figure 5) and Cd accumulation (17% of the whole set of samples with concentration >1 mg/kg) were recorded in areas close to the Venice industrial zone, and in the northern part, close to the forested area, suggesting an anthropogenic input from both industrial emissions and atmospheric pre-alpine deposition.

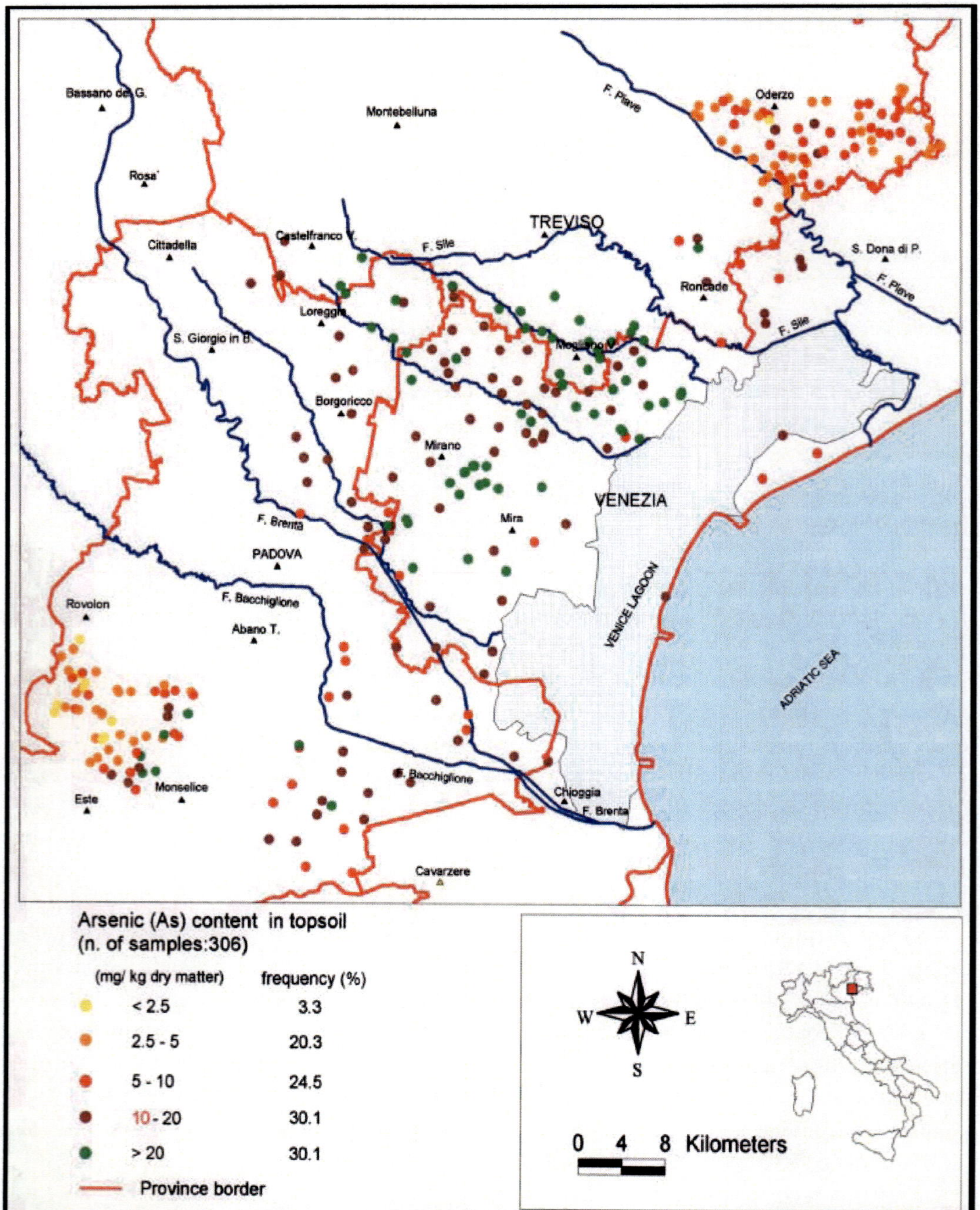

Figure 3. Spatial distribution of As concentration in agricultural soils of the Venice lagoon watershed, with indication of the concentration classes.

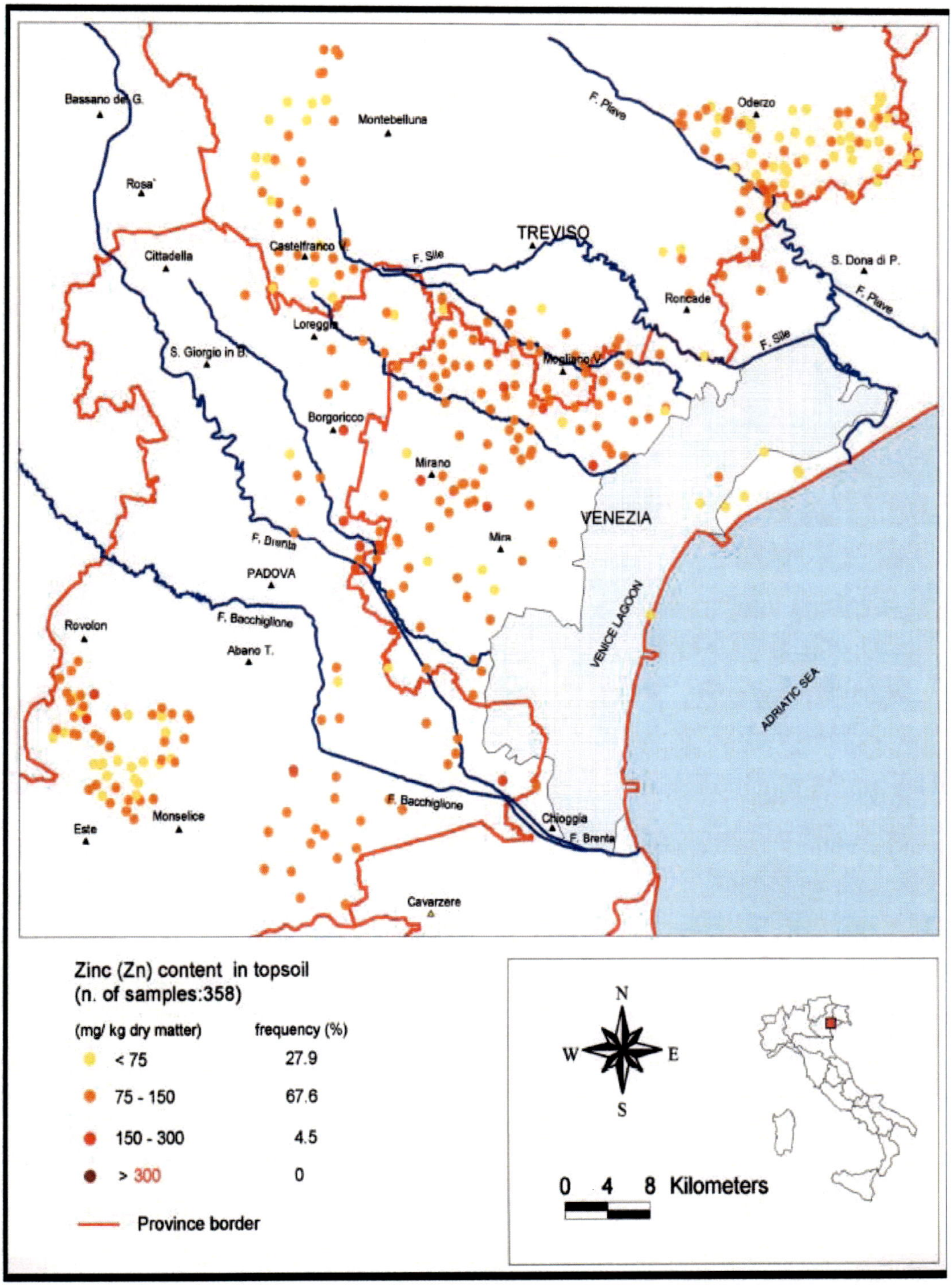

Figure 4. Spatial distribution of Zn concentration in agricultural soils of the Venice lagoon watershed, with indication of the concentration classes.

Metal speciation is an useful tool to identify different chemical bonding of the metal with the soil inorganic and organic colloids (Goupta et al., 1994). The DTPA- extractable metal concentrations were determined in the same set of samples as above, in order to assess possible soil contamination and to distinguish lithogenic and anthropogenic metal inputs. The results are summarized in Table 5.

The DTPA-extractable metal concentration presents a significant reduction with respect to the total amount, as expected. Mean extractable Cu was 12mg/kg, Zn 2.5mg/kg (very low concentration, possible deficiency in plants), Pb 3.7 mg/kg, Cd 0.08 mg/kg, Ni 0.4 mg/kg, Cr 0.06 mg/kg. However, important concentrations were recorded especially for Cu (112 mg/kg), Pb (29 mg/kg) and Cd (0.8 mg/kg).

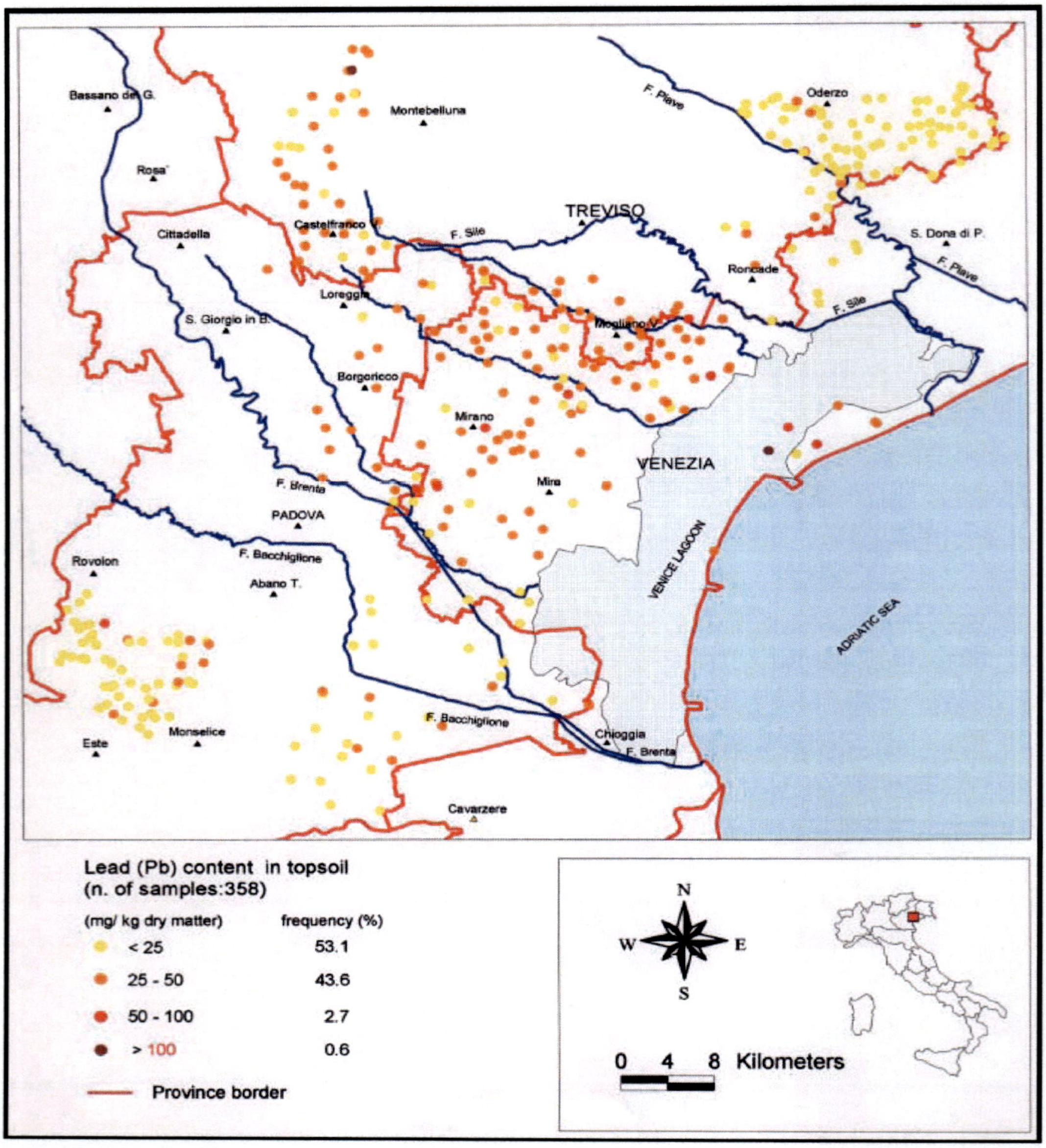

Figure 5. Spatial distribution of Pb concentration in agricultural soils of the Venice lagoon watershed, with indication of the concentration classes.

Table 5. Descriptive statistics for DTPA-extractable heavy metal concentration (mg/kg d. m.) in the soils examined

	Fe	Mn	Zn	Cu	Ni	Pb	Cd	Cr
Mean	28.53	15.69	2.51	12.09	0.43	3.66	0.077	0.058
Median	18.2	12.10	1.51	6.31	0.315	3.20	0.080	0
Standard deviation	65.99	13.24	3.07	16.70	0.61	2.84	0.057	0.080
range	1307.31	112.39	24	112	7.04	28.6	0.82	0.31
minimum	1.69	0.61	0	0	0	0	0	0
maximum	1309	113	24	112	7.04	28.6	0.82	0.31

This suggests that these elements could be, almost in part, linked to organic colloids of anthropic origin (Cu from agrochemicals, Pb and Cd from atmospheric and industrial emissions). Instead, Ni and Cr concentrations could be in relation to the lithological matrix. Similarly, the low iron and manganese extractable fraction (mean 28 mg/kg, and 16 mg/kg, respectively) could be related to the limited mobility of these two elements, as determined by the oxidation-reduction conditions.

The whole data set (soil characteristics, total and DTPA-extractable metal concentrations) was statistically processed in order to find relationships between the variables, which could better explain environmental processes and possible soil contamination. The correlation matrix is shown in Table 6.

Statistically significant correlations ($p<0.05$) are observed between soil properties and metal concentration. Clay content exhibits positive correlation with Cr ($r=0.648$), and Co ($r=0.542$), sand content relates negatively to Cr ($r=-0.43$) and Co ($r= -0.38$), suggesting these two metals to be mostly of lithogenic origin; CEC relates to Cr ($r=0.69$) and Co ($r=0.53$) again, confirming their lithogenic origin; total carbonates relate to Zn ($r=-0.52$) and Cd ($r=-0.52$), suggesting antagonism between Ca and the two microelements.

A statistically ($p<0.05$) positive correlation between some element couples (e.g. Ni/Cr 0.802; Zn/Cd 0.601; Mn/Co 0.656; Cr/Co 0.699; Zn/Co 0.534) and some soil parameters (e.g. Clay/CEC 0.735) was found.

Negative correlations are exhibited with sand and clay ($r= -0.727$), sand and CEC ($r=-0.516$), sand and silt ($r=- 0.847$), as expected. Bioavailable element fraction too exhibits positive correlations with total concentration (Cu tot/Cu av 0.927; Pb tot/Pb av 0.587; Cu av/Zn av 0.595). The correlations exhibited with pH and O.M. are insignificant. These results suggest the bioavailable fraction to be most likely related to the anthropic origin of the metals examined.

The principal component analysis (PCA) was applied to the whole data set. The components were rotated using a varimax rotation (Hair et al., 1998), which maximizes the variance of the squared normalized factor loadings across variables for each factor. The multivariate statistics allowed separation of the 23 variables considered into seven principal components, which account globally for 73.88 % of the variance. Only the first sixth principal components with eigenvalues greater than one were selected for factor analysis.

Table 6. Correlation coefficients matrix. Values >0.6 are in bold

	pH	O.C.	CaCO3	clay	silt	sand	CEC	Fe_av	Mn_tot	Mn_av	Zn_tot	Zn_av	Cu_tot	Cu_av	Ni_tot	Ni_av	Pb_tot	Pb_av	Cd_tot	Cd_av	Cr_tot	Cr_av	Co_tot
pH	1																						
O.C.	-0,1350	1																					
$CaCO_3$	0,4434	-0,0606	1																				
clay	-0,0183	-0,0099	-0,1741	1																			
silt	0,2444	0,0294	0,3179	0,3935	1																		
sand	-0,1520	0,0043	-0,0439	-0,7271	**-0,8479**	1																	
CEC	-0,1155	0,1608	-0,2962	0,7358	0,2460	-0,5168	1																
Fe av	-0,2048	0,0660	-0,1117	0,0403	-0,0323	-0,0060	0,0608	1															
Mn tot	-0,1738	-0,0584	-0,4292	0,2600	0,0316	-0,1890	0,2533	0,0572	1														
Mn av	-0,4348	-0,1092	-0,3404	0,0539	-0,1820	0,0931	0,0358	0,0791	0,4842	1													
Zn tot	-0,1424	0,2930	-0,5168	0,2525	0,0198	-0,1889	0,3173	0,0638	0,3074	-0,0054	1												
Zn av	-0,0891	0,0329	0,1474	-0,1947	-0,1069	0,1673	-0,1011	0,0500	-0,0191	0,1037	0,1878	1											
Cu tot	0,0092	0,1104	0,0124	0,0347	0,0554	-0,0852	0,1056	-0,0224	0,1577	-0,0284	0,1995	0,4878	1										
Cu av	-0,0032	0,0709	0,1037	-0,1101	0,0096	0,0159	-0,0478	-0,0155	0,0548	-0,0134	0,1024	0,5951	**0,9269**	1									
Ni tot	-0,0658	-0,0217	-0,1168	0,3542	0,0985	-0,2429	0,4867	0,0041	0,3075	-0,0707	0,1593	-0,1182	0,0592	-0,0755	1								
Ni av	-0,2669	0,1377	-0,0696	0,2359	0,0342	-0,1358	0,4604	0,1999	-0,1277	0,0652	-0,0318	0,1331	-0,0087	-0,0413	0,1979	1							
Pb tot	-0,2260	0,0654	-0,4121	0,0758	-0,1280	0,0296	0,0765	0,0955	0,2956	0,2648	0,4570	0,1344	0,0287	0,0178	-0,1563	-0,0333	1						
Pb av	-0,0988	0,1983	-0,1705	0,0179	-0,079803	0,0354	0,0397	0,0771	0,0070	0,0544	0,3895	0,2524	0,2020	0,1873	-0,1660	-0,0058	0,5877	1					
Cd tot	-0,1408	0,3052	-0,5184	0,0542	-0,0912	-0,0209	0,1188	0,0662	0,3498	0,2009	**0,6010**	-0,0322	0,0821	0,0117	-0,1598	-0,1677	0,4789	0,3452	1				
Cd av	-0,1919	0,3083	-0,2083	0,0045	-0,1199	0,0387	0,0878	0,1127	0,1827	0,2969	0,2379	0,3189	0,3649	0,3479	-0,1212	0,1713	0,2542	0,3104	0,3641	1			
Cr tot	-0,0790	0,0221	-0,3094	**0,6486**	0,1575	-0,4309	**0,6923**	0,0034	0,3563	-0,0101	0,2920	-0,1676	0,0721	-0,1062	**0,8027**	0,2735	-0,0639	-0,1148	-0,0499	-0,074173	1		
Cr av	0,0144	-0,1224	0,3251	0,2935	0,3499	-0,3740	0,2673	0,0130	0,0037	-0,1702	-0,2275	0,3125	0,4017	0,3711	0,3807	0,3440	-0,4186	-0,2029	-0,5078	-0,037687	,362733	1	
Co tot	-0,1371	0,1338	-0,5450	0,5429	0,1159	-0,3825	0,5937	0,0404	**0,6567**	0,1650	0,5344	-0,1673	0,1647	-0,0335	0,4715	0,1235	0,2392	0,0464	0,4310	0,127998	**,699727**	,042608	1

Due to the large heterogeneity of the data set, and the high number of variables (23), the loading factors show (Table7) values generally lower than 0.5. Nevertheless, significant information could be achieved from the PCA.

The largest loadings for the first component, which accounts for 20 % of the observed variance, were recorded for total Co, Cr, Cu, Mn and Zn, an for CEC, clay and $CaCO_3$. The second component, which accounts for 17.7 of the observed variance, showed the largest loadings for Cd, Pb, sand and negative loadings for silt. The third component was mostly represented by bioavailable metal concentrations, with negative loadings for Cd, Cr, Zn, Cu and total Cu. The fourth component included pH, and available Ni with negative loading. The fifth component had negative loadings for Organic Carbon and available Pb. The sixth component accounted for only 5.5 of the variance, and exhibited negative loadings for available Mn and Fe, and positive loading for total Ni.

The separation of most of the total and available heavy metal concentrations in different groups of components (PC1 and PC3, respectively) indicates that the prevailing origins of the total elements should be different from those of the available ones, as previously discussed, and found also by Ungaro et al. (2008) in soils of the Brenta river alluvial plain..

Table 7. Factor loading matrix from Principal Component Analysis of 23 soil variables. Values with loading factors >0.25 are in bold (tot=total; av=available)

Variables	PC1	PC2	PC3	PC4	PC5	PC6
Co tot	0.389	0.003	0.062	0.029	0.194	0.086
CEC	0.350	-0.152	0.004	-0.048	-0.184	0.072
Cr tot	0.339	-0.196	0.065	-0.161	0.107	0.220
Clay	0.324	-0.210	0.044	-0.129	-0.129	-0156
Zn tot	0.278	0.179	-0.059	0.238	-0.017	-0.248
Mn tot	0.273	0.088	0.024	-0.040	0.268	-0.268
$CaCO_3$	-0.266	-0.259	-0.160	0.118	-0.104	-0.082
Cd tot	0.189	0.295	0.044	0.024	0.024	0.036
Pb tot	0.156	0.282	0.006	0.115	-0.015	-0.188
Sand	-0.231	0.267	0.061	0.027	0.027	-0.238
Silt	0.089	-0.286	-0.086	0.059	0.059	-0.272
Cd av	0.115	0.219	-0.252	-0.148	-0.148	-0.109
Cr av	0.058	-0.307	-0.311	0.032	0.032	-0.053
Zn av	-0.028	0.098	-0.442	-0.046	-0.046	-0.031
Cu tot	0.090	0.030	-0.506	0.199	0.085	0.085
Cu av	0.006	0.054	-0.532	0.174	0.028	0.028
pH	-0.128	-0.172	-0.020	0.358	0.177	0.159
Ni av	0.128	-0.089	-0.073	-0.323	-0.459	0.017
O. C.	0.088	0.093	-0.088	0.137	-0.347	0.351
Pb av	0.087	0.222	-0.159	0.227	-0.258	-0.004
Mn av	0.112	0.187	0.046	-0.248	0.089	-0.558
Fe av	0.058	0.057	-0.014	-0.133	-0.275	-0.338
Ni tot	0.236	-0.202	0.039	-0.223	0.181	0.280
eigenvalue	5.0	4.4	2.9	2.0	1.6	1.4
Variance%	20.0	17.7	11.6	8.4	6.4	5.5
Cum. Var.%	20.0	37.7	49.3	57.4	63.8	69.3

The distribution of the scores of the loading factors in the scatterplot of the principal components 1 and 3 shows (Figure 6) that total heavy metals (with the exception of Cu), CEC and clay present positive scores for both PC1 and PC3, while available heavy metals present very high negative scores for the PC3. Soil total carbonates present negative scores for both PC1 and PC3.

Soil horizon differentiation in the PC1-PC3 scatterplot (Figure 7) shows a concentration of values related to the A horizon around the zero, with a dispersion towards negative PC3 scores (anthropic origin). On the contrary, samples representing the B horizons have positive scores for PC3, and samples from the C horizons have PC1 scores approximately = zero.

The spatial distribution of the soil samples in the PC2/PC3 scatterplot according to their specific properties (corresponding to the taxonomic level of subgroup: *vertic, aquic, oxyaquic, fluventic, alfic, typic*) allows identification of individual domains where every soil property (i. e: pedogenic character) is prominent (Figure 8).

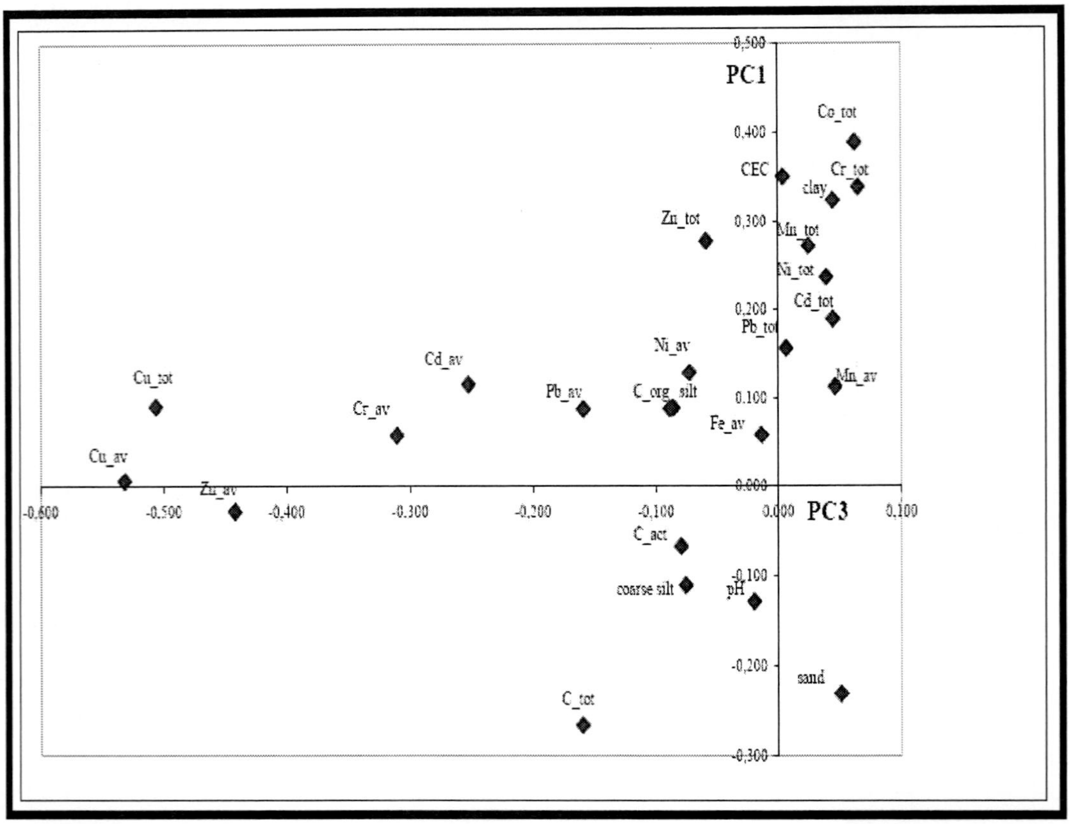

Figure 6. Distribution of variable loadings of agricultural soils in the scatterplot of principal components 1 and 3.

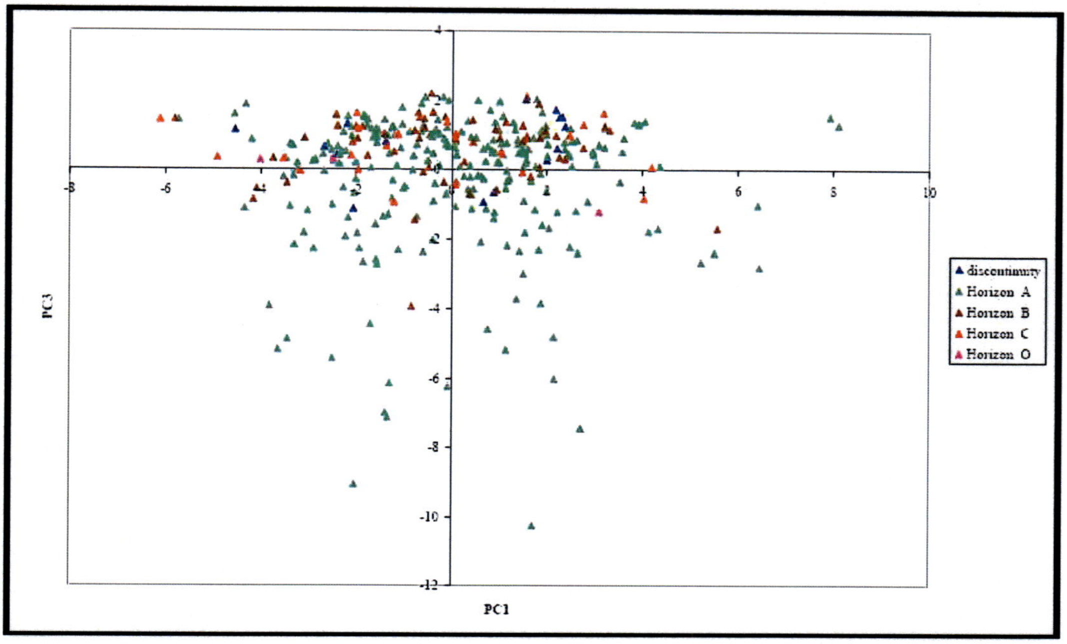

Figure 7. Soil horizon differentiation and spatial distribution in the scatterplot of principal components 1 and 3 of agricultural soils. O, A, B, C, are symbols for the regular succession of pedogenic soil horizons; discontinuity marks a difference between surface and subsurface horizons within the soil profile.

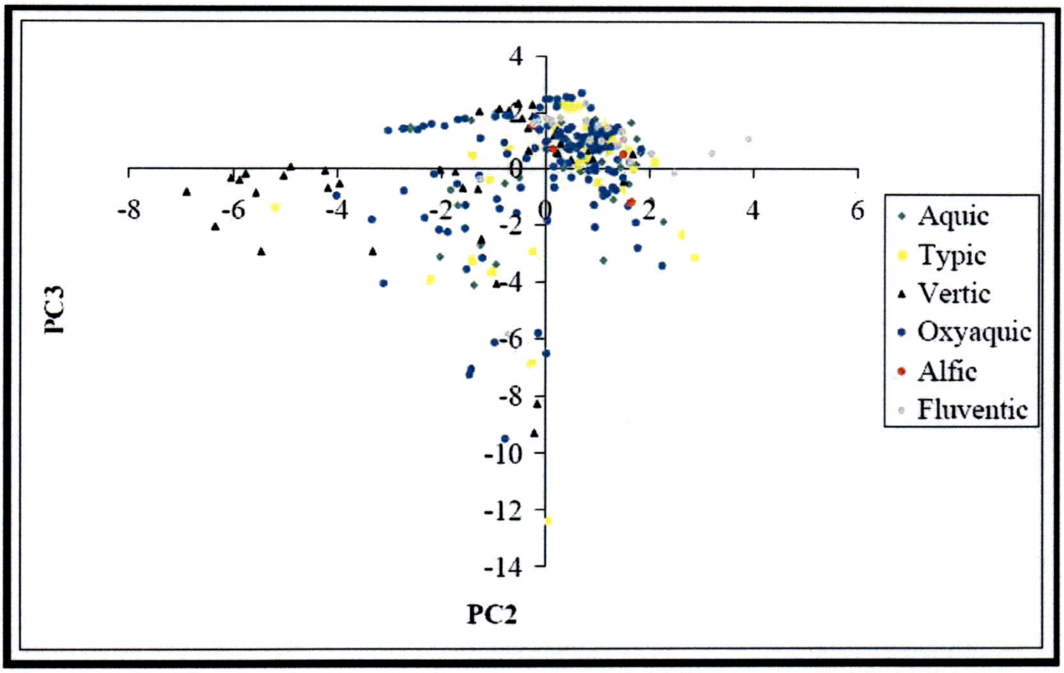

Figure 8. Spatial distribution of soil samples in the scatterplot of principal components 2 and 3 of agricultural soils, according to the different USDA Soil Taxonomy subgroup level, as revealed by specific soil features.

Soils having *vertic* properties (i. e. the geological factor) are dispersed from the zero towards negative PC3 values, while soils of the subgroups *aquic* and *oxyaquic* (i. e. hydrological factor) are located in the vicinity of the zero, the subgroup *fluventic* (i. e. organic matter factor) presents positive scores for PC3, and the subgroups *alfic* and *typic* (i. e. leaching factor) are scattered, being independent from both PC2 and PC3. These results suggest that every pedogenic factor prevails over the others in determining element mobility and therefore in controlling the soil evolution (Bini and Zilocchi, 2001).

Interpreting the PC analysis in terms of processes occurring in the soil, and fate of trace elements in the soil environment, the PC1 accounts for elements of lithogenic origin, as indicated by their association with CEC and clay. On the contrary, PC2 is expression of the anthropogenic origin for total Cd and Pb, which are positively associated with sand, and negatively with silt: This suggests that waste disposal and atmospheric deposition are the main sources of these metals in the area, as suggested also by Scazzola et al. (2004) . Moreover, because of its low mobility, Pb is still enriched in soils polluted by traffic emissions (Norra et al., 2006).

The PC3 accounts for bioavailable microelements of likely anthropogenic origin. Agrochemicals from agricultural areas and metallurgy from local plants are the main trace element sources. Most of the sampled sites are located at distance less than 30 km from the industrial area of Porto Marghera, and therefore contamination may explain the enrichment in available metals, which are likely contributed by surface reactions, adsorption onto other mineral phases (Norra, 2006) or by biogeochemical cycles (Sterckeman et al., 2004). The same is true of Factor 5, where a strong association of available Pb with organic carbon is observed, in accordance with the results obtained by Sterckeman et al. (2004) in soils developed on recent marine deposits of Northern France.

Factor 4, which relates pH to available Ni, does not give more information with respect to the correlation coefficients (Table 6), confirming the pH to be ineffective with respect to element mobility in the present soil environment.

Oxidation-reduction conditions seem to have a selective effect on the association of available Mn and Fe in Factor 6.

3.2. Rangeland and Forest Soils

The investigated rangeland and forested areas extend over approximately 40,000 ha (20% of the whole area), mostly in mountain areas. Mesophile mixed woods (oak, elm, holmbean, ash, maple), with subordinate beech and spruce stands, are the most important forest components, while limestone is the dominant parent material.

The soils examined (120 sites were established, totaling 454 samples (188 from topsoil and 266 from subsoil, of which 142 were from the C horizon) were grouped in seven Great Groups of the Soil Taxonomy (1999), as a function of parent material and pedogenetic features. Their partitioning is shown in Figure 9. Inceptisols (50%) and Mollisols (42%) are the dominant soil orders. Entisols (4%), Alfisols (2%) and Spodosols (2%) complete the soil geography of the area.

The mean levels, variance and ranges of pH, organic matter, total carbonates, CEC, particle size distribution are reported in Table 8.

Table 8. Descriptive statistics for rangeland and forest soil properties

	pH	O.C. %	Total Carbonates %	CEC cmol(+)/kg	Clay %	Silt %	Sand %
Number of samples	454	188	119	312	454	454	454
Mean	5.0	25.5	32	22.45	14.5	21.20	20.88
Variance	19	31	14	116	129	55	87
Min	3.3	10.1	0	5.60	1.6	1.35	0.60
Max	7.4	44.6	64	52.34	56.2	47.92	91.80

Table 9. Descriptive statistics for total heavy metal concentrations (mg/kg d.m.) in rangeland and forest soils examined

	Al	Fe	Mn	Zn	Cu	Ni	Pb	Cd	Cr
Mean	13793	17020	1044	166	23	22	120	2.4	27
Variance	57	48	81	42	38	72	61	61	64
Min	2320	3490	94	73	12	3	33	0.7	4
Max	36500	36680	4715	588	77	70	549	6.7	87

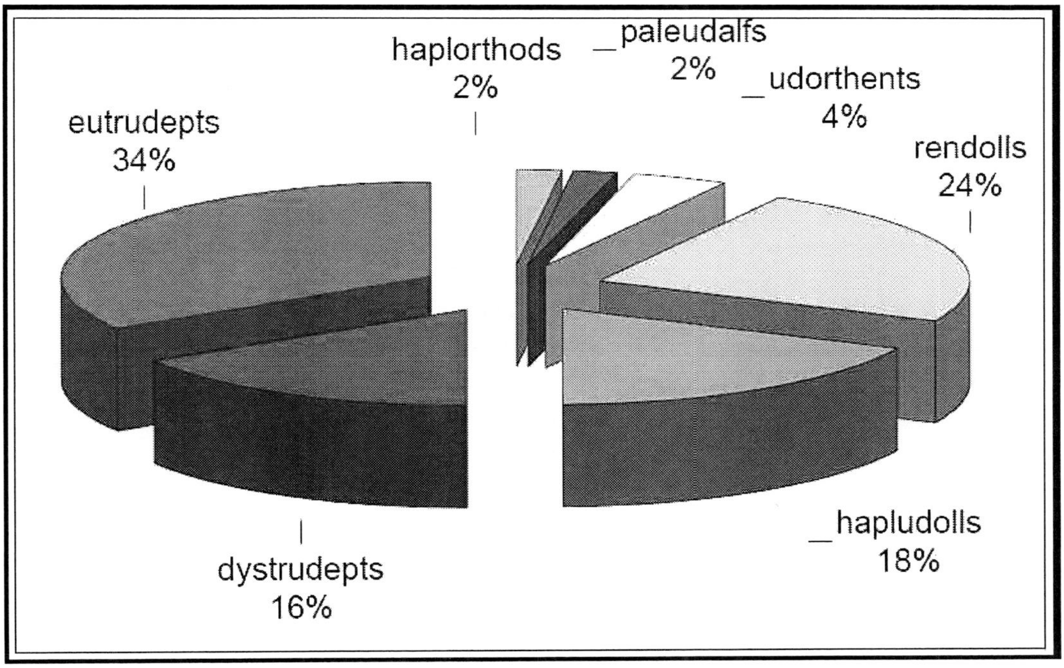

Figure 9. Sketch diagram of the distribution of rangeland and forest soils as taxonomic USDA great groups.

Most soil horizons (61%) have acidic reaction, with mean pH = 5.0 (range 7.4 - 3.3), though many are developed from calcareous rocks; this suggests that decarbonation and leaching occurred in the investigated area as a consequence of high annual precipitation (map =1200 mm).

The organic matter content is generally high (mean 255 g/kg, range 446-101) at surface, and decreases with depth, the richest horizons being those from calcareous parent material (mollic epipedon).

Total carbonates range from 0 to 64% (in very calcareous soils), but $CaCO_3$ levels higher than 20% occur only in 33% of the samples. The CEC presents a mean value of 22 cmol(+)/kg and a very high variance (range 5-52 cmol(+)/kg), with 71% of the examined samples having more than 18 cmol(+)/kg CEC. The particle size distribution points to large differences in the percentage of single components, with high variance for clay and also for sand, depending on the parent material characteristics, the site morphology and the soil evolution.

The results obtained with the total content of the elements investigated (Table 9) show that heavy metals present contents generally within the reference values reported in Table 1, with the exception of Zn (mean 166mg/kg), and Pb (mean 120 mg/kg). However, total Cr (mean 27 mg/kg, range 4-87 mg/kg) exceeds the threshold in 12% of samples, Ni (mean 22 mg/kg, range 3-70 mg/kg) in 8%, while Mn presents some high contents (up to 4700 mg/kg) at some sites.

The heavy metal concentration in the "C" horizon may be considered the pedogeochemical background in the pedosphere (Sterckeman et al., 2004). The mean metal concentration in the C horizon of the profiles investigated (Zn 77mg/kg; Cu 12 mg/kg; Cr 27 mg/kg; Ni 15 mg/kg; Pb 40 mg/kg; Cd 0.5 mg/kg; Mn 400mg/kg; Fe 3490 mg/kg) is within the "normal" range for soils of western Europe (Angelone and Bini, 1992). Surface horizons, however, present elemental concentrations higher than C horizons (Zn up to 132 mg/kg; Cu 21 mg/kg; Cr 55 mg/kg; Ni 37 mg/kg; Pb 69 mg/kg; Cd 1.4 mg/kg), while mean levels of trace elements in the B horizon are lower than those in the A and C horizons. Moreover, anthropogenic Pb (mean 120 mg/kg, range 33-549 mg/kg) and Cd (mean 2.4 mg/kg, range 0.7-6.7 mg/kg) are strongly accumulated (50% and 70% of the examined samples, respectively) especially at surface, due to their known affinity with organic matter, as reported also by Sterckeman et al. (2004). Such accumulation is very likely related to atmospheric inputs from acidic precipitation which has affected Northern Italy for many years (Bini and Michelutti, 1997; Bini et al, 2003). A confirmation of this statement is that heavy metal contribution by acid rains to the forest soil in the Cansiglio Integral Biological Reserve (80km North-East from Venice) was recorded in two experimental plots under beech and spruce, during 1997 (Bini and Bresolin, 1998).

As regards Pb, however, it is important to point out that highest concentrations in the soils investigated were recorded in a mine area at a distance of approximately 120km from Venice. Similarly, it was demonstrated (Dubois et al., 1998) that Cd in soils may derive partly from the calcareous parent material. Hence, in both cases the geological matrix effect must be summed to that of anthropogenic contribution.

The univariate statistical analysis of the whole data set accounted for significant ($p<0.05$) positive correlation of some element couples (Table 10). Chromium correlates well with Ni ($r= 0.896$) and with Fe ($r=520$), and moderately with Zn and Cd; Zn is well correlated with Mn ($r=0.694$) and Pb($r= 0.697$), and moderately correlated with Cd and Cr; Pb and Zn are well correlated with Mn ($r=0.618$; $r=0.694$, respectively). Moreover, Cr and Ni are significantly correlated with particle size (negative correlation with sand, and positive correlation with silt and clay). Silt and clay are negatively correlated with sand, as expected.

The organic carbon and the pH do not show significant correlations, as with the agricultural soils.

Table 10. Correlation coefficient matrix (values at $p<0.05$ are in bold)

	O. C.	Fe	Mn	Cd	Cr	Cu	Ni	Pb	Zn	pH	Sand	Silt	Clay
O. C.	1												
Fe	-0,11	1											
Mn	-0,17	0,37	1										
Cd	0,197	0	0,265	1									
Cr	-0,10	**0,520**	0,336	0,493	1								
Cu	0,001	0,13	0,112	0,035	0,095	1							
Ni	-0,18	0,460	0,403	0,36	**0,896**	0,149	1						
Pb	-0,04	0,01	**0,618**	0,139	-0,015	0,014	-0,019	1					
Zn	0,002	0,35	**0,694**	0,414	0,407	0,157	0,3805	**0,697**	1				
pH	-0,25	-0,46	-0,08	0,35	0,17	-0,028	0,1842	-0,071	-0,08	1			
Sand	-0,26	-0,2	-0,22	-0,407	**-0,636**	0,03	**-0,526**	-0,058	-0,24	-0,14	1		
Silt	0,286	0,16	0,234	0,46	**0,539**	-0,018	**0,496**	0,0221	0,19	0,232	**-0,630**	1	
Clay	0,105	0,27	0,302	0,242	**0,647**	0,05	**0,646**	0,1098	0,29	0,004	**-0,576**	0,309	1

The multivariate statistics allowed subdivision of the variables (totally 18 variables) into five principal components (Table 11), which cover globally more than 90% of the cumulative variance. Only the first three principal components have eigenvalues >1, and will be discussed here. PC1 accounts for 50% variance, PC2 19%, PC3 12%.

PC1 accounts for positive scores of trace elements (Cr, Ni, Zn, Cd, Mn), silt and clay, and a strong negative score for sand; therefore, it may be considered an expression of mineral weathering and mobilization of elements of lithogenic origin. PC2 accounts for negative scores for carbonates and pH, and positive scores for Fe, suggesting to be expression of specific pedogenetic processes (i. e. carbonation, brunification) and redox conditions within the soil. Therefore, PC2 distinguishes well calcareous soils from acid ones. PC3 accounts for positive scores of the soil organic components (organic carbon and nitrogen), and therefore is an expression of humification processes, and Pb, confirming the strong affinity of this metal with organic matter, as previously stated. PC4 accounts for highly negative scores for exchange acidity and amorphous iron. This component, therefore, illustrates the highest weathering processes in acidic environment, like leaching and podzolisation.

Interpreting the PCA in terms of pedogenesis, the distribution of variables in the PC1/PC2 scatterplot (Figure 10) allows distinction of strongly weathered horizons from silicate rocks (podzolic soils, group 5, and acidic soils, group 4) with positive loadings for PC1 and PC2, from slightly altered horizons from calcareous parent material (Mollisols, group 1, and Inceptisols, group 2), brunified and leached subsurface horizons (groups 2, 3) and from organic-matter rich horizons of less developed soils (A horizons of Mollisols, Entisols and Inceptisols, Groups 1, 2, 3).

Table 11. Principal Component Analysis of 18 soil variables ($p<0.05$). Values with loading factors >0.6 are in bold

	PC1	PC2	PC3	PC4	PC5
Sand	-0.9439	0.0552	-0.2328	0.0789	-0.1465
Cr	0.9419	0.2133	0.1533	0.1504	0.1186
Ni	0.9227	0.1810	-0.1397	0.1603	0.1944
Clay	0.9029	0.2028	0.2004	-0.0768	-0.0858
Silt	0.8702	-0.1171	0.4401	0.0760	0.0769
Zn	0.7200	0.5130	0.3608	0.1197	0.1101
Mn	0.6987	0.6194	-0.1084	0.1619	0.0584
Cd	0.6305	-0.3112	0.4431	0.2431	0.1032
$CaCO_3$	-0.1218	-0.9337	-0.1138	0.1515	-0.2079
Fe	0.3620	0.8910	0.0497	-0.1003	0.1550
pH	0.1743	-0.8620	-0.1677	0.4018	0.0804
O.C.	0.1444	0.1473	0.9107	-0.2955	0.1053
N	0.1072	-0.0487	0.8911	0.0296	0.1200
Pb	0.5067	0.3420	0.6733	0.0049	0.0813
Acidity	-0.2175	0.2202	-0.0296	-0.8696	-0.0561
Fe ox	0.0618	0.4985	0.3224	-0.6990	0.1004
CEC	0.1806	0.3371	0.5364	0.0586	0.6629
Cu	0.5575	0.3618	-0.0163	-0.0311	0.6046
eigenvalue	9,9	3.8	2.4	0.9	0.8
Variance %	49.9	19.3	12.2	4.9	4.1
Cum. Var.%	49.9	69.2	81.4	86.3	90.4

In the PC1/PC3 scatterplot (Figure 11), the component 3 identifies soil samples involved in humification processes: the A horizons, indeed, account for high PC3 loadings for soils of groups 1, 2, 3, and slight loadings for soils of groups 4 and 5. The variable "sand" in the component 1 shows a strong loading in the C horizons, which are obviously influenced by the parent material composition. Conversely it shows little loading with horizons having marked pedogenic evolution, as in soil groups 4 and 5.

In the PC2/PC3 scatterplot (Figure 12), the component 2 separates thoroughly subalkaline soils (groups 1-2) from acid soils (groups 4 – 5). The component 3, instead, allows separation of Acidic soils (groups 4, 5) and leached soils (group 3) from calcareous ones (groups 1, 2) and, within the latter, the surface horizons originated by humification processes, from the subsurface ones, which are contributed by parent material.

The cluster analysis carried out on the same set of samples (Figure 13) gives adjunctive information on the similarity among soil properties and soil horizons. Soil properties are clearly grouped according to their affinity (e. g. organic carbon with CEC and Pb, pH with $CaCO_3$, exchange acidity with extractable Fe, total Fe, Ni, Cr, with Mn, Cu with Zn), as expected. Some pedogeochemical fingerprints, therefore, are highlighted with clustering. In terms of pedogenic horizons genesis, subalkaline surface horizons (A horizons, groups 1, 2, 3) are clearly separated from the acid ones (A4-A5 horizons). The parent material too is clearly distinguished in calcareous (C1-C2-C3) and acidic (C4-C5) components. Moreover, the pedogenetic continuity within the different soil profiles is evident in leached and podzolic soils (A3-B3, A5-B5 horizons),as well as in subsoil horizons of acidic soils (B4-C4 horizons) and of calcareous ones (B1-C1 horizons). The only B2 horizons do not show any similarity. This may be explained by the intermediate character of these horizons, originated by slight weathering during the brunification processes. However, at higher level of similarity, they are clearly related to subalkaline parent materials of soil groups 1, 2 and 3.

3.3. Inland Coastal Areas and Lagoon Sediments

The inland coastal areas and the lagoon of Venice constitute a complex system which has been, and is currently, affected by human activities (industry, agriculture, settlements, tourism) that impact severely the ecosystem.

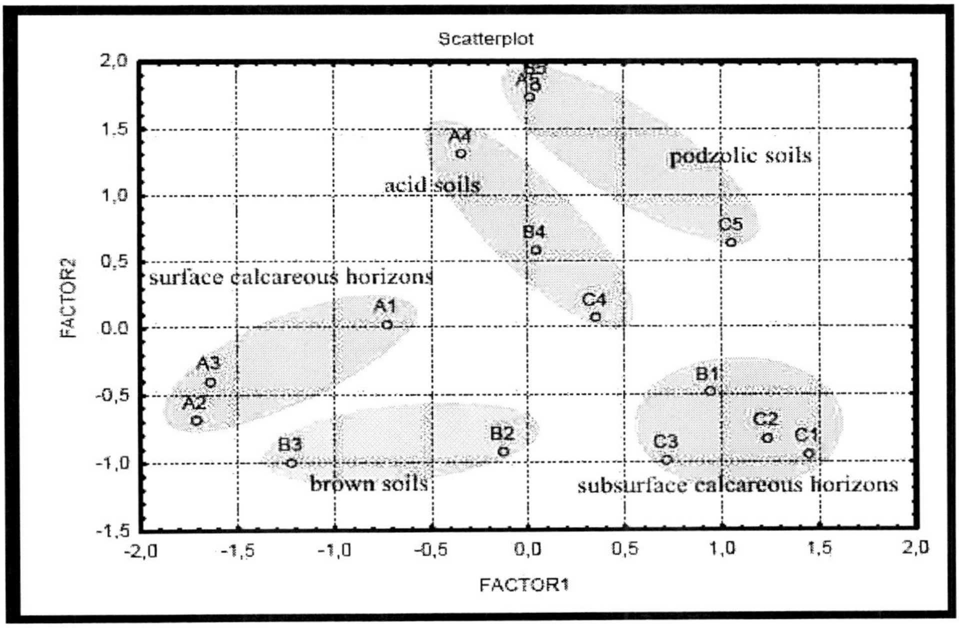

Figure 10. Spatial distribution of soil samples in the scatterplot of principal components 1 and 2 of rangeland and forest soils (rotated varimax raw), with indication of different soil groups. A, B, C are symbols for soil horizons. Numbers represent different soil groups, as follows: 1: calcareous soils (Rendolls, Hapludolls); 2: brown soils (Eutrudepts, Udorthents); 3: leached soils (Paleudalfs); 4: Brown acid soils (Dystrudepts); 5: Podzolic soils (Spodosols).

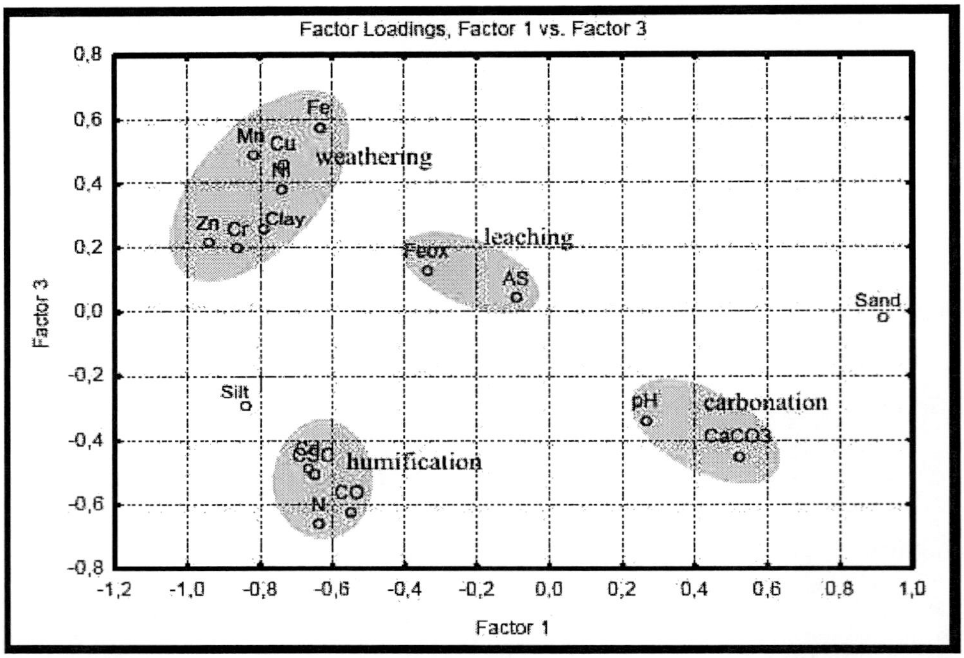

Figure 11a. Distribution of variable loadings of rangeland and forest soils in the scatterplot of principal components 1 and 3 (rotated varimax raw), with indication of the main pedogenic processes as revealed by different variables aggregation.

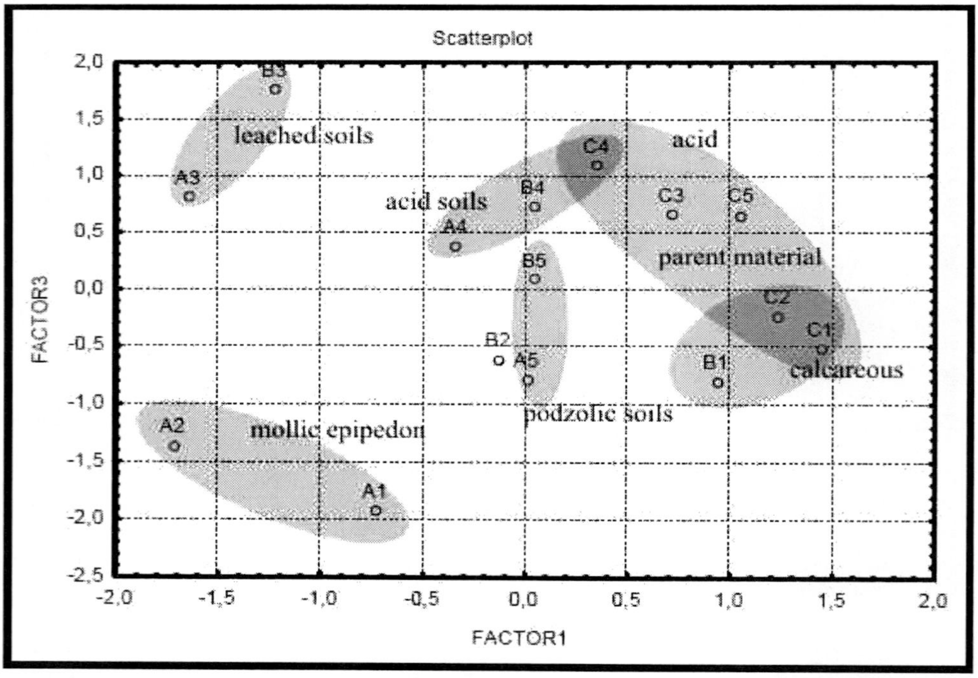

Figure 11b. Spatial distribution of soil samples in the scatterplot of principal components 1 and 3 of rangeland and forest soils (rotated varimax raw), with indication of different soil groups. Symbols are as in Figure 10.

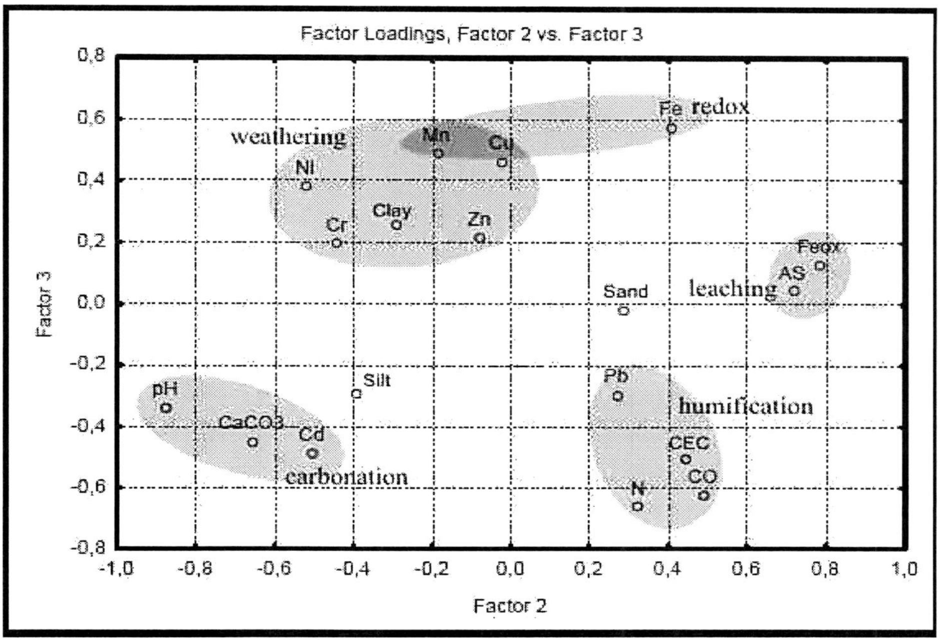

Figure 12a. Distribution of variable loadings of rangeland and forest soils in the scatterplot of principal components 2 and 3 (rotated varimax raw), with indication of the main pedogenic processes as revealed by different variables aggregation.

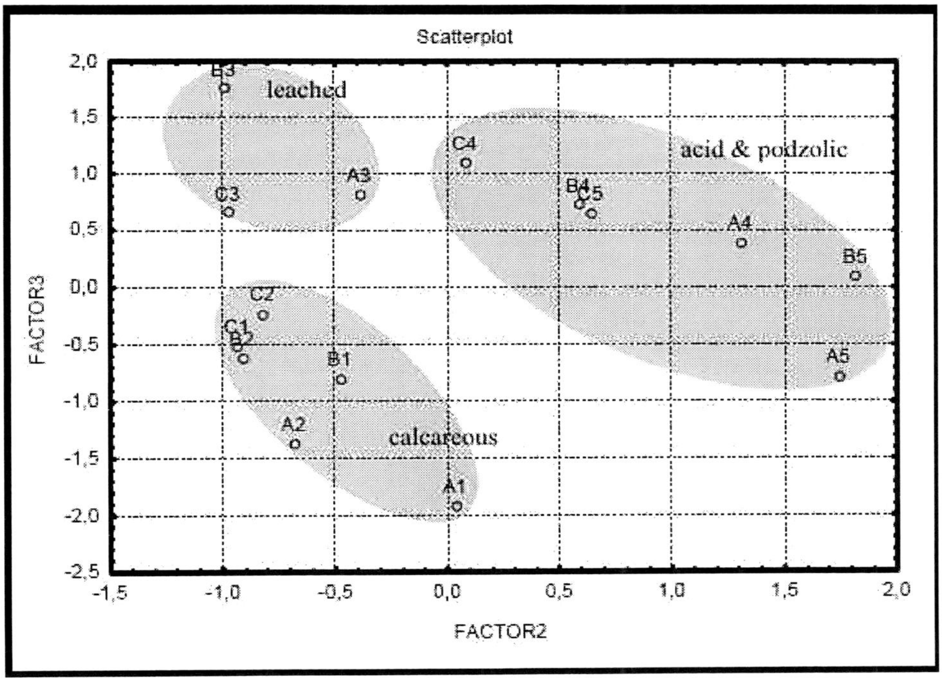

Figure 12b. Spatial distribution of soil samples in the scatterplot of principal components 2 and 3 of rangeland and forest soils (rotated varimax raw), with indication of different soil groups. Symbols are as in Figure 10.

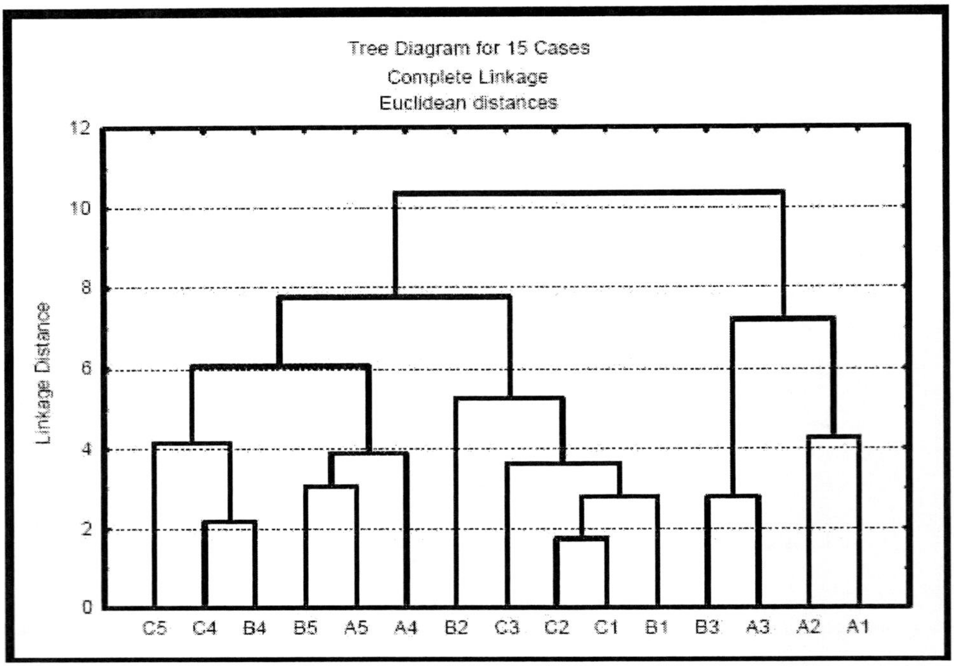

Figure 13. Tree diagram illustrating the clustering of different soil horizons according to their similarity. Symbols are as in Figure 10.

Discharge from the agricultural drainage basin affect particularly the area North of the city of Venice, where the Dese river is one of the main inputs of agrochemical contaminants (Argese and Bettiol, 2001). The central and southern areas, instead, receive important pollutant inputs from the industrial zone of Porto Marghera, where a number of chemical and petrochemical plants operate since the '50s (Bernardello et al., 2006). Additional sources of pollution are domestic sewage and waste disposal from the urban area. These inputs have progressively deteriorated the land quality, and are chronologically related to the development of the industrial and urban areas (Pavoni et al, 1987; Bernardello et al., 2006). As a consequence of the increasing land contamination, significant amounts of contaminants are accumulated in soils of the borderline and in the lagoon sediments, which may constitute a potential source of secondary pollution (Turetta et al., 2005).

Results of surveys carried out in recent years on the mainland (Scazzola et al., 2003; 2004) and on the lagoon sediments (Bernardello et al, 2006; Zonta et al., 2007) show that heavy metal concentration increased from the beginning of the industrial activities in the area until the '90s, when Porto Marghera declined. From that period, following new regulatory guidelines, a survey of the whole watershed and systematic monitoring of water, air and soil pollution have been conducted. Increasing heavy metal concentration was detected along a transect from the mainland (undisturbed soils of agricultural land, more than 40 km far from the city of Venice) to urban soils close (less than 5 km) to Porto Marghera industrial area, and to the inner border of the lagoon (Scazzola et al., 2004). A set of 10 soil profiles was selected from areas with different land use, in order to assess previous metal concentrations and possible contamination levels.

Most of the metals have concentrations above background values (Table12), with Cd, Cr, Hg, Zn in lagoon sediments showing the highest values. The highest metal concentrations were found in an area between Porto Marghera and the city of Venice, where both industrial and urban sewage are discharged. The concentration gradient for most metals (Cd, Cr, Cu, Hg, Mn, Ni, Zn) decreases from the inner margin of the lagoon to both the outer border and the mainland, suggesting that the inputs from the industrial area represent the main factor influencing the metal spatial variability. The spatial gradient of concentrations, therefore, is an index of overall pollution. However, sediment re-suspension and dispersion of fine particles by tidal currents may contribute to the recorded distribution pattern.

Most of the trace elements in samples examined were positively correlated with fine particles (<53 μm), supporting the importance of particle size in influencing metal accumulation, as it was found in previous works by Sterckeman et al. (2004) and Bernardello et al. (2006). Adsorption, surface diffusion, formation of precipitates onto other mineral phases like Fe oxides could be the form the heavy metals accumulate.

Two groups of metals (Fe, Mn, Cr, Ni, and Cd, Cu, Pb, Zn, respectively) have been found by Zonta et al. (2007) showing high correlation coefficients among them (Fe-Mn =0.94; Fe-Ni =0.92; Fe-Cr= 0.93; Pb-Cu=0.97; Cd-Zn=0.80; Cu-Zn=0.89) and with clay (clay-Ni= 0.87; clay-Cr=0.79; clay-Cu=0.83; clay-Pb=0.87). Similar results were obtained by Turetta et al. (2005), who performed a PCA on samples from the lagoon sediments and found a clear separation between two groups of metals: high positive scores for As, Mn, (factor 1) and Zn and Cd (factor 2), low score for Cu.

The impact of anthropogenic activities on the composition of surface soils can be demonstrated by metal abundances in sediments of the inner lagoon margin, as well as in urban soils (Table 12), as demonstrated also by Norra et al. (2006) and Fitzpatrick et al. (2007) in urban soils from Germany and U.S.A. In a PCA carried out on soil samples from different sites, Norra et al. (2006) found that the spatial distribution of metal concentration increases from forested areas to agricultural soils and from low density settlements to high density settlements and industrial areas. Similar results have been recorded in samples from the surroundings of the Venice lagoon area, as reported in this paper. The studied area is intensively cultivated and situated in one of the most populated and industrialised regions. Most of the sampling sites are located less than 40 km from the industrial and urban area. Several industries still emit considerable amounts of various elements into the atmosphere. Therefore, contamination could explain part of the enrichment of the surface horizons, al least for some elements. The concentration values for most of the trace elements (e. g. As, Cr, Cu. Hg, Zn) are higher in urban soils than in undisturbed and reference sites. Iron and Mn are higher at the lagoon inner margin than at the outer margin, , suggesting that oxidation-reduction status plays a significant role in their mobility.

Comparing the relative contribution of the three main sources of contaminants (industry, agriculture, atmosphere), the industrial area of Porto Marghera is still the primary source for Cd (68%) and Hg (39%), and the second source for Ni (31%) and Zn (32%), as stated also by Bellucci et al. (2000) and Critto et al. (2005). The origin of Cd, as well as Zn, Co and Cu, is related to the production of metallic zinc from the mineral sphalerite processing (1936-1978) in industrial plants of Porto Marghera (Pavoni et al., 1987).

Mercury is considered one of the most potentially harmful elements in environmental (particularly aquatic) ecosystems, due to its high toxicity, the potential to accumulate to high levels in aquatic organisms, and therefore the impact on the human health. The average concentration found in the sediments of the Venice Lagoon is 0.6 mg/kg (Zonta et al., 2007). Its concentration at contaminated sites in the lagoon ranges between 40 and 500 mg/kg total

Hg. However, no correlation was found between high levels of historical Hg contamination and organism MetylHg concentrations (Bloom et al., 2003).

Table 12. Mean values (mg/kg) of microelements in soils and lagoon sediments at selected sites (data processed from Argese and Bettiol, 2001; Bernardello et al., 2006; Pavoni et al., 1992; Scazzola et al., 2004; Ungaro et al., 2008; Zonta et al., 2007).

	Outer border	Central lagoon	Inner margin	Urban soils	Undisturbed soils	Background (pre-industrial)
As	7.1	n.d.	13	19	12	10
Cd	0.59	1.8	2.0	0.7	0.9	1.0
Co	5.1	n.d.	3	10	9	15
Cr	61	66	107	23	17	20
Cu	14	37	32	41	21	20
Fe	12421	18375	20385	n.d.	n.d.	20.000
Hg	0.7	1.7	1.1	0.1	0.04	0.1
Mn	237	341	361	n.d.	n.d.	250
Ni	31	32	50	21	20	20
Pb	19	78	40	45	53	25
Zn	71	167	255	95	73	70

Instead, the highest relative inputs for As (91%), Cr (77%), Cu (65%), Ni (56%) and Pb (61%) come from the drainage basin, and are related to their geochemical affinity and to the mineralogical composition of soils and sediments. Arsenic has been already discussed as accumulating by primary geogenic origin, followed by agrochemical input (Ungaro et al., 2008). Arsenic, as well as Hg, is an element of major environmental concern, due to its toxicity and potential to accumulate in living organisms. High As concentrations were found in algae and mussels living in the Venice lagoon (Argese et al., 2005). Total As concentration in *Ulva lactuca* and *Mytilus galloprovincialis* ranged from 10 mg/kg to almost 30 mg/kg (mean 30mg/kg and 22 mg/kg, respectively), with a significantly greater accumulation in the mussels digestive gland, where As content achieved 75 mg/kg. Moreover, As is present in mussels predominantly in the organic and less toxic forms (Argese et al., 2005).

Chromium and Ni accumulation is likely related mostly to their geogenic origin (Cr 77%, Ni 56%). These metals are probably conveyed to the lagoon by runoff water from the drainage basin, or are released from fine particle lagoon muds during erosion and re-suspension consequent to clam harvesting (Sfriso et al., 2005). A second source for Cr (13%) and Ni (31%) is that these metals are typically related to emissions of metal processing manufacturers. Metallurgy contributes Cu (16%) and Zn (32%), which are components of alloys. Traffic is a further important source of Cu, Zn, Pb. Lead was an indicator of traffic emissions in former times since Pb was added to fuel as an antiknock (Norra et al., 2006), and it is likely the present high Pb concentrations to be due to its low mobility.

The high relative input of Cu (65%) comes from runoff of agricultural areas, where it is widely used, together with Zn, against vineyard diseases.

4. Conclusion

Four different factors are responsible for the fate of trace elements in soils: acidic conditions, redox status, organic matter content and soil solution chemistry.

Acidification: increasing acidic conditions, determined by heavy precipitation occurring in the area, cause the progressive replacement of cations on mineral surfaces by protons; decrease in pH results in increased Cu, Zn, Cd and Pb mobilization and accumulation at rangeland and forest sites.

Redox status: oxidation-reduction seems to have a noticeable effect on the distribution of some elements, like Fe, Mn, As. Oxidizing conditions determine precipitation of hydrous oxides, and increased metal sorption capacity (e.g. Cd, Co, As); conversely, reducing conditions cause dissolution of hydrous oxides, liberation of adsorbed metals and release of elements (e.g. Fe, Mn, Co, Cr, Ni) from the soil.

Organic Matter: increased concentration of organic complexing ligants from litter decomposition, plant root exudates and sewage sludge application is the major cause of metal accumulation in surface horizons, especially Cu and Zn at agricultural sites, Pb and Cd at forest sites.

Pore solution chemistry: changes in the ionic composition of the soil solution accounts for competition at sorption sites (e.g. Zn inhibits Cd adsorption), and for the formation of soluble or insoluble compounds. Moreover, a geological matrix effect must be accounted for metal release by hydrolysis of parent material and accumulation in soils (e.g. Ni, Cr, As, Fe, Mn from silicate minerals, Cd from limestone).

Several of the above mentioned mobilization mechanisms are interlinked. For example, oxidation of insoluble sulfide minerals leads both to an increase in acidification by sulfuric acid formation and to the liberation of cations, particularly As (Mascaro et al., 2000). Likewise, the decomposition of freshly incorporated litter and soil organic matter, accompanied by the formation of low molecular weight organic acids, contributes to leaching of metals (e.g. Fe, Mn) down the soil profile in a chelate form, and influences the soil evolution, through the podzolisatios process.

From the soil, trace elements move to surface and groundwater, and through water to the lagoon, where they are concentrated in sediments, or transferred to organisms.

The group of elements with the lowest relative mobility (bioavailability) includes Co, Cr, Hg, Pb, As; intermediate elements are Cu, Ni, and the most bioavailable are Zn and Cd.

The level of contamination of the soils investigated is low to moderate, with only some hotspots. Soil heavy metal concentrations are generally within the regulatory guidelines. Exception to this trend is given by *anthropogenic* Cu and Zn and some As at agricultural sites, and by *natural* enrichment of As, Ni, Cr. The concentration values for most of the trace elements are higher in both the agriculture and urban sites than in forest sites. The recorded accumulation of Pb and Cd at forest sites is probably related to acidic precipitation which affected the investigated areas. However, the geological matrix effect must be summed to that of anthropogenic contribution.

Pollution sources could not be detected with certainty at present. Metal concentration decreases with increasing distance from the lagoon, and there are differences between undisturbed, agricultural and urban soils. Surface horizons are enriched with Pb, Cu, Cd, Hg, and this enrichment coincides with an enrichment in soil organic matter, suggesting Pb and

Cu to enter, at least in part, the biogeochemical cycle; conversely, Cd and Hg should be introduced in the environment by human activities. This is in agreement with the scheme proposed by Sterckeman et al. (2004) for heavy metals in undisturbed soils: the anthropogenic elements at surface are more likely to be bound to organic matter, while the geogenic elements would mainly be associated with silicates and Fe oxy-hydroxides, with a sensitivity to redox conditions.

Comparison of the correlations enables inferences concerning the interactions of trace elements with soil properties. This allows a first explanation of the distribution of heavy metal contents in soils. Metals are essentially associated with the soil organo-mineral fraction in the surface horizon, and have positive correlation with soil properties like clay and CEC, and negative correlation with sand, suggesting a lithogenic origin (i. e. they are pedogeochemical fingerprints).

Principal component analysis supports the differentiation between lithogenic and anthropogenic origin of trace elements, and the association between metals and fine-size grains, and that of Pb with organic matter, is evident in the principal components of PCA of both agricultural and forest soils. The positive loadings for most metals on PC1 indicate that this component is associated with a lithogenic origin, and the negative loadings of bioavailable metals at agricultural sites may be considered as an index of overall contamination of anthropic origin.

The lagoon acts as a sink where heavy metals contributed by agricultural land or originated by waste disposal or by industrial emissions and atmospheric deposition, are conveyed. The ultimate fate of heavy metals, therefore, is to contribute substantially to the lagoon and sediment pollution, and pose environmental hazard for living organisms, including humans.

The ecological risk as expressed by the hazard quotient (HQ = metal in organisms/metal in soils or sediments) posed by single elements is little for As (HQ <2 at Porto Marghera, <1.5 in central lagoon; <1 on the mainland, on the basis of 10 points), although at some places its concentration is higher than the regulatory threshold, and very high for Hg and Cd (HQ <2.5 in central and outer border, <2 in the inner margin).

The cumulative toxic risk for As+Cd+Hg (+Zn) ranges between 6 and 9 toxic units at some hot spots in the central and southern lagoon (between Porto Marghera and Chioggia harbors), indicating a relevant bioaccumulation of trace elements in the lagoon ecosystem, and ranks <3 at other sites on the mainland.

In conclusion, the investigation of soils of the Venice lagoon watershed and the conterminous areas provides valuable information for a comprehensive environmental assessment. Contamination is a major concern to the lagoon, and is probably a residual effect of severe historical inputs since the '50s. Instead, on the mainland slight diffuse contamination occur, ant is likely due to anthropic as well as lithogenic origin.

ACKNOWLEDGEMENTS

The Environmental Protection Agency of the Veneto Region (ARPAV) is kindly acknowledged for providing agricultural soils sampling and laboratory analyses. Data processing was carried out by G. Camici at the Dept of Physics, University of Florence. S.

Gemignani helped in laboratory analyses and computer processing of forest soils data, at the Dept. of Environmental Sciences, University of Venice.

REFERENCES

Abrahams P. W. (2002) - Soils: their implications to human health. *Sci Tot. Envir.*, 291, 1-32.

Adriano, D. C., Chlopecka, A., Kapland, D. I., Clijsters, H., Vangrosvelt, J. (1995) – Soil contamination and remediation philosophy, science and technology. In: R. Prost, editor, *Contaminated Soils*, 466-504, Paris, INRA.

Angelone, M. and Bini, C. (1992) – Trace elements concentrationin soils and plants of Western Europe. In: D. C. Adriano, editor, *Advances in trace substances research,* 19-60, Boca Raton, FL, Lewis Publisher

Angelone, M., Vaselli, O., Bini, C., Coradossi, N. (1993) – Pedogeochemical evolution and trace element availability to plants in ophiolitic soils from Tuscany (Italy). *Z. Pflanzen. Bodenk.*, 154, 217-223.

Argese, E. and Bettiol, C. (2001) – Heavy metal partitioning in sediments from the lagoon of Venice (Italy). *Toxicol. Environ. Chemistry*, 79, 157-170.

Argese, E., Bettiol, C., Bertini, S., Gobbo, L., Rigo, C. (2002) – Optimization of metal speciation techniques in various environmental matrices and application to the study of pollution in the lagoon of Venice. In: P. Campostrini, editor, *Scientific research and safeguarding of Venice*. Venezia, 387-391, Istituto Veneto di Scienze, lettere ed Arti.

ARPAV (2005) – *Carta dei suoli del Veneto*. Firenze, S.E.L.C.A..

Bellucci, L., Frignani, M., Paolucci, D., Ravanelli, M. (2002) - Distribution of heavy metals in sediments of the Venice lagoon: the role of the industrial area. *Sci. Tot. Envir.*, 295, 35- 49.

Benvenuti, M., Mascaro, I., Corsini, F., Lattanzi, P., Tanelli, G. (1997) – Mine waste dumps and heavy metal pollution in abandoned mining district of Boccheggiano (southern Tuscany, Italy). *Envir. Geol.*, 30, 238-243.

Bernardello, M., Secco, T., Pellizzato, F., Chinellato, M., Sfriso, A., Pavoni, B. (2006) – The changing state of contamination in the Lagoon of Venice. Part 2: heavy metals. *Chemosphere*, 64, 1334-1345.

Bini, C. and Bresolin, F.(1998). Soil acidification by acid rain in forest ecosystems. *Sci. Tot. Envir.*, 222, 1-15.

Bini, C. and Michelutti, G., (1997) - Heavy metal bioaccumulation in forest soils of alpine environment (eastern Alps, Italy). In: A. Chang editor, *Biogeochemistry of trace elements*, IV, 365-366, Berkeley, Cal. Berkeley Univ. Press.

Bini, C. and Zilocchi, L. (2001) - Fate of trace elements in the pedosphere: Venetian territory, Italy. In: R. Cidu editor, *Water-Rock Interaction*. Lisse, 1047-1050, Swets and Zeitlinger.

Bini, C., Giandon, P., Vinci, I. (1999) – Pedology and heavy metals. A regional application in Italy. In: W. Wenzel, D.C. Adriano, B., Alloway, H.E., Doner, C., Keller, N., Lepp, M., Mench, R., Naidu, G., Pierzynski editors, *Proc.* 5th *ICOBTE*, 1, 102-103, Wien.

Bini, C., Maleci, L., Romanin, A. (2008) – The chromium issue in soils of the leather tannery district in Italy. *Jour. Geoche. Expl.*, 96, 2-3: 194-202.

Bini C., Gemignani S., Zilocchi L. Corradini F., Sartori G. (2003) – Fate of trace elements in the pedosphere of forest soils in alpine environment, Italy. In: G., Gobran and N., Lepp, editors, *Proc. 7th ICOBTE,* 1, 26-28. Uppsala.

Bloom, N.S., Moretto, L.M., Ugo, P (2003) – Speciation and biogeochemistry of Hg in two shallow estuaries: Lavaca bay (Texas) and Venice Lagoon (Italy). In: G., Gobran and N., Lepp, editors, *Proc. 7th ICOBTE,* 1, 20-21. Uppsala.

Brummer, G.W. (1986). Heavy metal species, mobility and availability. In: M., Bernhard, R., Brinkman and P. J., Sadler, editors, *The importance of chemical speciation in environmental processes.* Berlin, Springer Verlag.

Cambier, P. and Charlatchka, R. (1999). Influence of reducing conditions on the mobility of divalent trace metals in soils. In: H. M., Selim and I. K., Iskandar, editors, *Fate and transport of heavy metals in the vadose zone.* Chelsea, MI, Ann Arbor Press.

Critto, A., Carlon, C., Marcomini, A. (2005) – Screening ecological risk assessment for the benthic community in the Venice Lagoon (Italy). *Environ. Int.*, 31, 1094-1100.

Dall'Aglio, M., Da Roit, R., Orlandi, C., Tonani, F. (1966) – Prospezione geochimica del mercurio. Distribuzione del mercurio nelle alluvioni della Toscana. *L'industria Mineraria*, XVII, 391-398.

Deluisa, A., Giandon, P., Aichner, M., Nardelli, F., Stringari, G. (1996) - Copper pollution in Italian vineyard soils. *Comm. Soil Sci. Plant Anal.*, 27, 5/8: 1537-1548.

Dubois, J. P., Okopnik, F., Benitez, N., Vedy J. C. (1998) – Origin and spatial variability of Cd in some soils of the Swiss Jura. *Proc. 16th ISSS Congress*, II, 25, 476, Montpellier.

FAO-ISRIC (1998) - *World Reference Base for Soil Resource*. Rome, pp98,. FAO.

Fitzpatrick, M. L., Long, D. T., Pijanowski, B. C. (2007) – Exploring the effects of urban and agricultural land use on surface water chemistry, across a regional watershed, using multivariate statistics. *Appl. Goechem.*, 22, 1825-1840.

Giandon, P., Vinci, I., Fantinato, L. (2000) – Heavy metal concentration in soils of the basin draining in the Venice lagoon. *Boll. Soc. It. Sci. Suolo*, 49, 1-2, 359-366.

Giandon, P., Ragazzi, F., Vinci, I., Fantinato, L., Garlato, A., Mozzi, P., Bozzo, G. P. (2001) – La carta dei suoli del bacino scolante in laguna di Venezia. *Boll. Soc. It. Sci. Suolo,* 50 (suppl), 273-280.

Goupta, S. K., Vollmer, M. K., Krebs, R. (1994) - The importance of mobile, mobilizable and pseudo total heavy metal fractions in soil for three level risk assessment and risk management. *Sci. Tot. Envir.*, 130.

Hair, J. F., Anderson, R. E., Tatham, R. L., Black, W. (1998) – *Multivariate Data Analysis* (fifth edition), New York, 87-138, Prentice-Hall, U.S.A.

Kabata-Pendias, A. and Pendias, H. 1992 – *Trace elements in soils and plants*, 2nd ed., Boca Raton, FL, CRC Press.

Langmuir, D. (1997) - *Aqueous Environmental Geochemistry.* New York, pp600. Prentice Hall, U.S.A.

Mascaro, I., Benvenuti, M., Bini, C., Corsini, F., Costagliela, P., Ferrari, M., Manieri, C., Parrini, P., Tanelli, G., Vitiello, G. (2000) – Studio ambientale dell'area mineraria dimessa del Bottino (Alpi Apuane – Toscana Settentrionale). *Geologia Tecnica ed Ambientale*, 2, 3-12.

MIRAAF (1994) – *Metodi ufficiali di analisi chimica del suolo.* Ministero delle Risorse Agricole, Alimentari e Forestali, pp276, Roma.

Norra, S., Lanka-Panditha, M., Kramar, U., Stuben, D. (2006) – Mineralogical and geochemical patterns of urban surface soils, the example of Pforzheim, Germany. *Appl. Geochem.*, 21, 2064-2080.

Nriagu, J.O. (1983) – *Lead and lead poisoning in antiquity*. Wiley-Interscience, New York.

Pavoni, B., Donazzolo, R., Marcomini, A., Degobbis, D., Orio, A. A. (1987) – Historical development of the Venice lagoon contamination as recorded in radiodated sediment cores. *Marine Pollution Bulletin*, 18, 18-24.

Pavoni, B., Marcomini, A., Sfriso, A., Donazzolo, R., Orio, A. A. (1992) - Changes in an estuarine ecosystem – the lagoon of Venice – as a case study. In: D. Dunnette, R. J. O'Brien editors, *The Science of Global Change*. Amer. Chem. Soc. Symp. Series, n°483, pp 287-305.

Petruzzelli, G., (1989) – Recycling wastes in agriculture: heavy metal bioavailability. *Agr., Ecosys. Envir.*, 27, 493-503.

Scazzola, R., Avezzù, S., Biancotto, R., Chiamenti, E., Chiozzotto, E., Gerotto, M., Palonta, M., Roiter, S. (2003) – Assessment of heavy metal background values in the soils of inland coastal areas of Venice, Italy. *Ann. Chim.*, 93, 465-470.

Scazzola, R., Matteucci, G., Guerzoni, S., Chiamenti, E., Possini, P., Molinaroli, E. (2004) – Evaluation of trace metal fluxes to soils in hinterland of Porto Marghera industrial zone: comparison with direct measurements in the lagoon of Venice. *Water, Air and Soil Pollution*, 153, 195-203.

Sfriso, A., Facca, C., Marcomini, A. (2005) – Sedimentation and erosion processes in the lagoon of Venice (Italy). *Environ. Int.*, 31, 983-992.

Sposito G. (1983). The chemical forms of trace metals in soils. In: *Applied Environmental Geochemistry*. Academic Press, UK, 504 p.

Sterckeman, T., Douay, F., Baize, D., Fourrier, H., Proix, N., Schvartz, C. (2004) – Factors affecting trace element concentrations in soils developed on recent marine deposits from northern France. *Appl. Geochem.*, 19, 89-103.

Tessier, A., Campbell, P.G. C., Bisson, M. (1979) – Sequential extraction procedure for speciation of particulate trace minerals. *Anal. Chem.*, 7: 844-851.

Thornton, I. (1993) – Environmental geochemistry and health in the 1990s: a global perspective. *Appl. Geochem.*, suppl., Issue n°2, 203-210.

Turetta, C., Capodaglio, G., Cairns, W., Rabar, S., Cescon, P. (2005) - Benthic fluxes of trace metals in the lagoon of Venice. *Microchemical Journal*, 79, 149-158.

Ungaro, F., Ragazzi, F., Cappellin, R., Giandon, P. (2008) – Arsenic concentration in the soils of the Brenta Plain (Northern Italy): mapping the probability of exceeding contamination thresholds. *Jour. Geochem. Explor.*, 96, 117-131.

USDA – Soil Conservation Service (1999) – Soil Taxonomy (2th edit.) U. S. Printing Office, Washington D. C.

Zonta, R., Botter, M., Cassin, D., Pini, R., Scattolin, M., Zaggia, L. (2007) – Sediment chemical contamination of a shallow water area close to the industrial zone of Porto Marghera (Venice Lagoon, Italy). *Marine Pollution Bulletin*, 55, 529-542.

Reviewed by J. Bech Borras, Chair of Soil Science, University of Barcelona, Spain.

In: Causes and Effects of Heavy Metal Pollution
Editor: Mikel L. Sanchez
ISBN 978-1-60456-900-1

Chapter 6

Anthropogenic Mercury Pollution in Aquatic Systems: A Review of Enviromental Fate and Human Health Risks

S. Michele Harmon
University of South Carolina Aiken, 471 University Parkway, Aiken, SC 29801, USA

Abstract

The environmental chemistry of mercury is complex and difficult to predict because it is controlled by a multitude of environmental processes which includes photochemical reactions, chemical oxidation and reduction, microbial transformation, and physiological fractionation. Mercury vapor from natural and anthropogenic sources is released to the atmosphere and distributed globally before being oxidized to a water-soluble form and returned to terrestrial and aquatic systems via deposition. This results in the distribution of mercury contamination to all parts of the earth. Mercury's natural cycle has been significantly disrupted by anthropogenic activities, and atmospheric mercury concentrations have steadily increased since the Industrial Revolution. In aquatic environments, mercury is methylated through a micobially-mediated process primarily involving sulfate-reducing bacteria. Therefore, methylation is strongly influenced by factors that affect these bacterial consortia. Of equal importance are variables that affect the availability of inorganic mercury for uptake by these bacteria. Conditions that favor mercury methylation include low pH, high DOC concentrations, and low redox conditions. However, the factor that asserts the most control is sulfur chemistry and its link to sulfate-reducing bacteria and inorganic mercury availability. Methylmercury is of great concern in aquatic environments because of its ability to bioconcentrate and biomagnify through trophic webs. Human exposure to this neurotoxin ultimately results from consumption of contaminated fish.

Introduction

Mercury exists in a variety of inorganic and organic forms. Mercury transitions are controlled by a multitude of environmental process, including photochemical reactions, chemical oxidation and reduction, microbial transformations, and physiological fractionation [1] leading to complex and highly mobile cycles of transitions between molecular forms, each with varying degrees of toxicity.

Global atmospheric mercury cycling begins with mercury vapor released from natural sources, such as volcanoes [2] and from anthropogenic activities such as fossil fuel combustion (coal combustion, in particular) and municipal waste incineration [3]. In the elemental form (Hg^0), mercury vapor is chemically stable. Estimates of residence times in the general atmosphere range from 100 days [4] to more than one year [5-6], allowing for the distribution of mercury over wide areas. After oxidation in the upper atmosphere, water-soluble inorganic mercury (Hg^{2+}) returns to the earth via atmospheric deposition, resulting in global dissemination to the surface of both aquatic and terrestrial systems. Evidence of mercury contamination from atmospheric deposition can be found in even the most remote polar regions of the world, where there are few people and no local industrial sources.

Other mercury contributions to the aquatic environment include the natural weathering of igneous rocks and cinnabar (HgS) deposits, as well as the erosion of surface soil that has been contaminated with mercury through atmospheric deposition, historic agricultural activities [7-8], and surface applications of sewage sludge [9]. The natural erosion processes of these contaminated soils are further enhanced through human activities such as deforestation and poor agricultural practices. Surface runoff laden with mercury-contaminated soil eventually makes its way into aquatic systems.

More direct anthropogenic mercury contributions come from municipal wastewater treatment plants [10-11] and a number of industrial processes. These include chloralkali production, mining operations and ore processing, metallurgy and electroplating, chemical manufacturing, ink manufacturing, leather tanning, pharmaceutical production, pulp and paper mills, and textile manufacture [12]. Prior to the discovery of the dangers from mercury biomethylation and bioaccumulation, these industries discharged large quantities of inorganic mercury waste into the aquatic environment, leaving a legacy of mercury accumulation that can be observed in the sediment record [13-18]. While many industrial discharges have been either eliminated or are highly regulated, there is ongoing concern regarding less-regulated activities such as extensive small-scale goldmining operations in the Amazon basin [19-20].

Globally, concentrations and fluxes of mercury are considered to be elevated and unstable due to the significant disruption of mercury's natural cycle by anthropogenic activities [1]. It is believed that anthropogenic mercury inputs may comprise as much as one-half of the mercury entering the world ecosystem [21], and there are indications that atmospheric mercury concentrations have steadily increased since the beginning of the industrial age. Of particular importance in this chapter are the environmental factors that influence the microbial transformation of inorganic mercury species into the more toxic methylated form.

MERCURY IN THE AQUATIC ENVIRONMENT

In the natural aquatic environment, dissolved mercury occurs in four important forms: elemental mercury (Hg^0), divalent inorganic mercury (Hg^{2+}), mercurous mercury (Hg^{1+}), and methylmercury (CH_3-Hg^+) [22-23]. Of these species, methylmercury is of primary environmental concern because it is the most toxic in aquatic systems; it is more readily accumulated by aquatic organisms than inorganic mercury [24-26]; and it is the fraction of the total mercury pool that is most efficiently transferred up the food chain to higher trophic levels [27-30]. Although methylmercury may occur in natural waters and sediments in extremely low concentrations, it is very efficiently enriched by aquatic animals. Therefore, the aqueous concentration of mercury or methylmercury in a particular aquatic system is not a true indicator of the body burdens found in aquatic organisms. Although synthetic organomercurials have been banned for some decades, more than 90% of total mercury in muscle tissue of top aquatic predators is methylmercury [31-33]. Methylmercury is of great concern from a human health standpoint because of its ability to pass through important biological membranes such as the blood-brain and placental barriers.

THE FORMATION OF METHYLMERCURY IN THE AQUATIC ENVIRONMENT

The behavior of mercuric pollutants in the natural environment is difficult to predict because the chemistry of mercury is so complex. As described previously, mercury can occur in the aquatic environment in three inorganic forms: elemental (Hg^0), divalent (Hg^{2+}), and mercurous (Hg^{1+}). The nature and reactions of these species determine their solubility, mobility, and toxicity in aquatic ecosystems. Depending on the prevailing ambient conditions, mercury compounds in aquatic systems can be interconverted between sediment and water phases, taken up by aquatic biota, lost to the atmosphere, or transported with sediment particulate matter to new locations (Figure 1) [34].

The proportion of the total mercury pool that is methylated in surface water depends on the rates of methylation and demethylation. These processes vary with water quality parameters which include salinity, redox potential, pH, temperature, and complexing agents such as humic and fulvic acids [35-36]. Although mercury methylation can occur through abiotic processes [37-40], biological methylation has proven to be the major contributor of methylmercury in natural aquatic environments [41]. Mercury methylation has been linked to microbial activity for some time [42-43], and numerous studies have indicated that sulfate-reducing bacteria are primarily responsible for mercury methylation [44-47]. Therefore, methylation is controlled by factors that influence not only the bacterial consortia, but the availability of inorganic mercury for uptake by these bacteria as well. With that in mind, it is not surprising that the methylation of mercury in most aquatic systems is largely controlled by a combination of pH, organic matter cycling, and sulfur chemistry.

Many studies have shown that methylmercury concentrations in water and sediment increase along with a decrease in pH [48-51]. Consequently, low pH has been correlated with high mercury levels in fish tissue in a number of studies [32, 52-59]. In a 1990 review article, Winfrey and Rudd [57] noted that a decreased pH generally stimulates methylmercury production at the sediment-water interface. They hypothesized that lower pH enhances the bioavailability of mercury for methylation by decreasing Hg^{+2} binding with particulates, while at the same time decreasing the loss of volatile elemental mercury (Hg^0) from surface water. Laboratory studies later confirmed that bacterial uptake of Hg^{2+} increases as the pH of water containing dissolved organic carbon is decreased [60]. An additional trend correlated with low pH was the release of methylmercury from the sediment surface and into the water column [48, 61].

Organic carbon affects mercury methylation and demethylation in two ways. First, some forms of dissolved organic carbon (DOC) are known sources of decomposable carbon substrates for microbial populations. Inputs of such carbon may stimulate microbial activity and subsequent mercury methylation if an available mercury pool is present. Secondly, inorganic mercury is strongly bound by DOC, notably humic substances [62-64]. This binding reduces the bioavailability of inorganic mercury to the microbial methylators, thus decreasing methylmercury production. However, as conditions become more acidic, this mercury-DOC binding is reduced [61], increasing mercury bioavailability to bacterial methylators and enhancing methylation. It is therefore assumed that low pH and high DOC conditions are ideal for mercury methylation.

This assumption has been explored in several laboratory studies. Miskimmin et al. [65] used radioisotopic techniques to measure rates of mercury methylation and demethylation in lake water and determine how these rates were affected by DOC concentration, pH, and microbial respiration. Increases in DOC resulted in decreased methylation; however, reducing pH from 7.0 to 5.0 resulted in moderate increases in net methylation rate at both low and high DOC concentrations (500-2600 uM). These results were in agreement with Barkay et al. [66] who used a Hg^{2+} bioindicator to demonstrate that DOC affects the rate of methylmercury synthesis by reducing the availability of Hg^{2+} to methylating bacteria. The reduction in bioavailable mercury was more pronounced at pH 7 than at pH 5. *In situ* studies of the Elbe River ecosystem in Europe demonstrated an "intimate interaction between regional geochemical cycles of organic carbon and mercury" [67]. When fate and behavior of mercury in the U.S. Patuxent River estuary was studied [68], it was concluded that organic matter content appeared to control the total mercury concentration in sediments and that methylmercury in sediments was positively correlated with both total mercury concentration and organic matter.

Because sulfate-reducing bacteria thrive under anaerobic conditions, environments with low oxidation-reduction potential (redox) are associated with increased mercury methylation. Several laboratory studies [69-71] demonstrated that mercury methylation was favored at lower levels of redox. Numerous *in situ* studies have indicated that mercury methylation is most intense at the sediment /surface water interface where steep redox gradients generally combine with high levels of microbial activity [57, 71-78].

The combination of low redox and high organic matter make wetlands an ideal environment for methylmercury production. Hurley et al. [79] found that methylmercury yields were highest from catchments containing wetlands, and that percent wetland surface area was positively correlated with methylmercury yield. A study conducted on the Sudbury River in the northeastern U.S. indicated that riparian wetlands contaminated with mercury from an industrial point source were important sites of production and release of methylmercury and that total mercury concentrations increased significantly as the river passed through the riparian wetland reach [80]. A bioaccumulation study in this same river system [81] concluded that the potential entry of methylmercury into the benthic food chain was greater in contaminated palustrine wetlands than in the reservoirs, despite the fact that reservoir sediments contained higher mercury concentrations. Other studies that have observed the importance of wetlands as sources of methylmercury to downstream lakes and streams include St Louis et al. [82-83], Rudd [84], Branfireun et al. [85-87], and Galloway and Branfireun [88]. The relationship between mercury methylation and wetlands has been an issue of some concern due to the fact that constructed wetlands are becoming a popular option for the treatment of industrial effluents, stormwater, and municipal wastewater. Several researchers have noted the potential for mercury methylation in constructed wetlands [89-94] and have noted that the benefits of this treatment option should be carefully weighed against the risk of providing an additional methylmercury source.

Research conducted in newly flooded areas have shown increases in methylmercury concentrations in both fish [95-96] and aquatic insects [97]. It is likely that these increases are the consequence of increased mercury methylation due to the formation of ideal microbial conditions (low redox, high DOC) combined with the decomposition of biomass, which enhances the release of labile mercury from inundated soils. This has raised some apprehension regarding the buildup of mercury contamination in arctic regions. It has been hypothesized that as global climate change melts polar snows and transforms large regions of permafrost into wetland, a significant accumulation of atmospherically-deposited mercury will enter the methylation cycle and be transported into arctic food webs [98].

Other factors that influence mercury methylation include temperature and salinity. Bisogni and Lawrence [99] concluded that temperature affects methylation through its effect on overall microbial activity. Callister and Winfrey [73] studied methylmercury biosynthesis in sediment and water of the upper Wisconsin river in the U.S. and determined that mercury methylation was higher at an optimum temperature of 35 C°. Other studies have shown that mercury methylation is favored as salinity decreases [69-70, 100].

With ties to both temperature and the presence of DOC for bacterial stimulation, there are strong seasonal patterns that accompany mercury methylation. Hurley et al. [101] found that methylmercury production in the Florida Everglades was highest in July, while total mercury concentrations remained relatively constant through the year. In littoral zone sediments of Pallette Lake in northern Wisconsin, the highest methylmercury concentrations in porewater were observed in the spring, while highest bulk phase concentrations occur later in the summer [102]. Temporal studies of Texas' mercury-contaminated Lavaca Bay sediments showed a short, active period of methylation in early spring, followed by a slow year-long decrease as methylmercury was degraded back to Hg^{2+} [103]. Balogh et al. [104] noted elevated methylmercury concentrations in two south-central Minnesota prairie streams following early May algal blooms and then again in early October after autumnal leaf fall. Eckley and Hintelmann [105] observed seasonal changes in mercury methylation rates in five lakes across Canada, with the greatest methylation occurring after autumn lake turnover. Seasonal changes in stream flow also affect mercury methylation rates [51] in certain systems where flow rate has the potential to change pH as well as change the concentration of other elements (e.g., Se, Mo, W) that have been shown to interfere with the activity of sulfate reducing bacteria and the associated mercury methylation [106].

While, pH, DOC, redox, temperature, and salinity influence environmental mercury methylation, the factor that asserts the most control is sulfur chemistry and its link to sulfate-reducing bacteria and inorganic mercury availability.

SULFUR CHEMISTRY, SULFATE-REDUCING BACTERIA, AND MERCURY METHYLATION

The primary experiment that demonstrated the efficiency of sulfate reducers in mercury methylation was performed by Compeau and Bartha [44] who used mercury-spiked slurries of saltmarsh sediment incubated in an anaerobic chamber to identify the microorganisms responsible for mercury methylation. The addition of 2-bromomethane sulfonate (a methanogen inhibitor) increased methylmercury synthesis, and sodium molybdate (an inhibitor of sulfate reducers) decreased methylation by 95%. Enrichment and isolation procedures yielded a *Desulfovibrio desulfuricans* culture that vigorously methylated mercury. This research also determined that the methylation activity of sulfate-reducing bacteria is fully expressed only when sulfate is limiting and when fermentable organic substrates (pyruvate, lactate, and acetate) are available.

Sulfate reducing bacteria (SRB) are obligate anaerobes that obtain energy for growth by oxidation of organic substrates. They remove hydrogen atoms from these organic molecules and use sulfate as the terminal electron acceptor during respiration, thus reducing sulfate to sulfide:

$$SO_4^{2-} + \text{organic matter} \xrightarrow{\text{SRB}} HS^{2-} + H_2O + HCO_3^-$$

Once in solution, sulfide forms strong bonds with cationic metals, such as Hg^{2+}, making them unavailable for biological uptake [107], thus explaining why Compeau and Bartha [44] noted greater methylation efficiency in treatments with low (limiting) sulfate concentrations. This agreed with an earlier series of experiments in which Compeau and Bartha [108] established that the lack of mercury methylation in the presence of sulfide concentrations as low as 10 mg/L was entirely due to the binding of Hg^{2+} ions as HgS. Gilmour and Henry [45] summarized this relationship as follows: "It is an ecological paradox that sulfate-reducing bacteria mediate mercury methylation while the sulfide they produce inhibits the process."

Sulfate-reducing bacteria generally fall into two phylogenetic families based upon metabolism. The first group is the Desolfovibrionaceae family. Genera within this family are incomplete oxidizers of organic substrates. Members of this family utilize a number of organic acids as carbon sources (e.g., lactate, pyruvate, fumarate, proprionate, ethanol) but produce acetate as a final metabolic product. A second group, the Desulfobacteriaceae family is comprised of complete substrate oxidizers which utilize the above mentioned organic acids in addition to acetate and they produce CO_2 as a final metabolic product [109]. Many of the earlier mercury methylation studies focused upon *Desulfovibrio*, an incomplete oxidizer. A more recent laboratory study of pure cultures and marine sediment [110] used molecular probes to identify bacterial consortia and noted that members of the Desulfobacteriaceae family (acetate utilizers) methylated mercury more efficiently than members of the *Desulfovibrio* group. Macalady et al. [111] used polar lipid fatty acid analysis in lake sediments and observed that *Desulfobacter*, another member of the acetate-utilizing Desulfobacteriaceae family, were important methylators and were more abundant than *Desulfovibrio*. While all sulfate reducers might not methylate mercury, it is probably safe to assume that methylation is common across a broad range within this bacterial consortium. It is also likely that the carbon substrate influences mercury methylation by controlling the predominant genera present as well as the efficiency of respiration and, thus, methylation.

Recent studies have suggested that other groups of bacteria, including the iron-reducing *Geobacter sp.*, methylate mercury in natural systems [112-113]. However, the total contributions of these other methylating microbes is still unknown, as sulfate reducers have been the most heavily studied to date.

The Balance of Sulfate, Sulfide, and Mercury Methylation

In estuarine and marine sediments, where sulfate is plentiful, bacterial sulfate reduction may be limited by the availability of organic matter. This is not the case in freshwater sediments, where organic substrates may be plentiful, but sulfate reduction is limited by sulfate concentration [114-116]. This is important to mercury cycling, because mercury methylation in sediment has been significantly correlated with sulfate reduction rate [110, 117-119]. Freshwater sulfate concentrations typically range from 50-450 μM [120] but will increase with eutrophication or acid deposition.

While reviewing the literature, Gilmour and Henry [44] speculated that the optimal sulfate concentration for mercury methylation by sulfate-reducing bacteria in sediments was in the range of 200-500 μM. Above this concentration, sulfide would inhibit methylation, while at lower sulfate levels, sulfate reduction and hence methylation would be limited by available sulfate. This speculation led to concern for freshwater sediments impacted by acid deposition, which have a low pH and are sulfate rich. While this is a vast issue complicated by the concept that the optimal sulfate level will vary from system to system as a function of other factors affecting sediment sulfate reduction rates, a number of *in situ* studies have attempted to address the sulfate/methylmercury dynamic in particular systems.

Branfirun et al. [121] used a mesocosm experimental design to investigate the effects of long-term sulfate deposition in peatlands of southern Sweden. Investigators added sulfur in amounts equivalent to the 1980s dry and wet deposition for the area. Methylmercury concentrations in peat porewater were up to six times above background after three years. In an investigation of the Quabbin Reservoir in the northeastern U.S., Gilmour et al. [46] found that additions of sulfate to either anoxic sediment slurries or lake water above intact sediment cores resulted in increased microbial production of methylmercury from added inorganic mercury. King et al. [89] reported statistically significant methylmercury increases in the effluent of a pilot-scale constructed wetland treatment system after a surface application of gypsum ($CaSO_4$).

The testing and development of radiotracer techniques using $^{203}HgCl_2$ by Gilmour and Riedel [122] allowed much more detailed observations of *in situ* mercury methylation rates. These methods were later used in an extensive study of the Florida Everglades to explore the factors that control the methylation process in that system [78]. These researchers determined that maximum methylation rates were generally within centimeters of the sediment surface, and methylation was never observed in water overlying sediment cores. They also observed that methylmercury concentration and production were inversely related to sulfate reduction rate and pore water sulfide.

Sulfide and Mercury Bioavailability

The final component of the sulfate-mercury relationship that must be considered is how sulfide affects the bioavailability of mercury. Traditional thinking was that dissolved mercury concentration is controlled strictly by direct precipitation as solid HgS as well as adsorption to solid phases (organic particulates, clays, etc.) in the water column. While the reaction of sulfide with Hg^{2+} to produce insoluble cinnabar (HgS) certainly decreases mercury availability for methylation [78], there is possibly a range of low sulfide concentrations that may actually enhance mercury bioavailability. Because there is believed to be no mechanism for the active uptake of mercury by bacteria, it must be assumed that mercury enters the cell through diffusive mechanisms. Several studies have noted that the concentration of dissolved (bioavailable) inorganic mercury in sulfidic porewaters increases to a certain degree along with sulfide concentration [68, 103]. Several research groups [68, 123-125] have shown that under low sulfidic conditions, the availability of mercury for methylation is increased by the formation of neutral dissolved mercury-sulfide complexes such as HgS0 and Hg(SH)20. The concept behind the neutral mercury sulfide hypothesis is that these uncharged chemical species readily enter bacteria through passive diffusion. As the concentration of sulfide increases, an equilibrium shift will replace the neutral species with charged disulfide complexes [123-124]. Mercury methylation in a pure culture experiment has been shown to decrease about fourfold as sulfide concentration increased from 10-6 to 10^{-3} M [126].

Demethylation

The decomposition of methylated mercury is also an important process regulating the methylmercury concentration in sediment and water. The measured concentrations of methylmercury in environmental media represent the balance of the rates of mercury methylation and demethylation [33-34]. The combined effect of methylmercury production and demethylation leads to a state of equilibrium with a near constant level of methylmercury in sediments that rarely exceeds 1 to 1.5% of the total mercury concentration [127]. But because methylmercury biomagnifies in the food chain, the proportion of methylmercury in fish and other aquatic biota is often much higher.

Although abiotic demethylation may occur through photodegradation [128], demethylation is predominantly microbially mediated [36, 43, 129]. Bacterial demethylation has been documented in most organic environments such as sediments [42, 130], soil [131], surface water [49, 57, 132-133], and wetlands [82]. Numerous bacterial strains capable of demethylating methylmercury are known, and include both aerobic and anaerobic species [36, 129-130, 134-135]. However, demethylation appears to be predominantly accomplished by aerobic organisms, and demethylation seems to be highest in oxic waters [1].

The microbial degradation of methylmercury is known to proceed by two pathways, methyl cleavage or oxidative demethylation. Methyl cleavage is the most commonly accepted mechanism of microbial methylmercury decomposition and involves cleavage of the carbon-mercury bond by the organomercurial lyase enzyme. This produces Hg^{+2} and CH_4, followed by the reduction of Hg^{2+} to Hg^0 by the mercuric reductase enzyme, which is encoded by the *merB* and *merA* genes in bacteria possessing mercury resistance [129, 136-137]. Oremland et al. [130] observed that anoxic sediments and bacterial cultures could also form $^{14}CO_2$ as well as $^{14}CH_4$ when incubated with ^{14}C-labeled methylmercury. This revealed the existence of another bacterial mechanism for demethylation, termed oxidative demethylation, that produces Hg^{+2}, CH_4, and CO_2 as end products. Inhibitor experiments suggested the involvement of both sulfate-reducing and methanogenic bacteria in the oxidative demethylation of sediments [138]. When applying these ^{14}C labeling methods along an eutrophication gradient in the Florida Everglades, Marvin-Dipasquale and Oremland [139] noted that $^{14}CO_2$ production in all samples indicated that oxidative demethylation was an important degradation mechanism performed by both sulfate-reducing bacteria and methanogens, although the data suggested that methanogens were dominant at *in situ* methylmercury concentrations. It is still unclear as to which of the two demethylation mechanism dominates under specific environmental conditions. It has also not been determined whether the Hg^{2+} produced via oxidative demethylation is further reduced to Hg^0, remethylated, or bound to particulates [34].

Methylmercury Bioaccumulation by Aquatic Organisms

Bioaccumulation of methylmercury by aquatic organisms is affected by the availability of both inorganic mercury and methylmercury and by the kinds and biomass of organisms that accumulate methylmercury [17]. Although pathways include direct uptake from water, most accumulation is through the food web [140-142]. The relative importance of either pathway depends on trophic level, duration and intensity of exposure, and environmental factors. Once accumulated in biota, methylmercury biomagnifies with each trophic level.

The majority of published results relate to mercury bioaccumulation in fish [143]. According to Spry and Wiener [59], the trophic route appears to represent more than 90% of the mercury accumulated in carnivorous fish and, based on the work of Bloom [144], it is generally accepted that virtually all (>95%) of the mercury burden measured in edible fish tissue is in the form of methylmercury. Extensive studies have been carried out to identify the factors that influence mercury accumulation in fish [95, 145-146]. Mercury concentrations in fish are typically inversely related to ambient pH, alkalinity, and primary production [54, 56, 59, 147-149]. For example, Haines et al. [150] studied perch (*Perca fluviatilis*) in 26 Russian lakes and concluded that lake acidity and DOC concentration were highly related to fish mercury burdens, with the highest concentrations being found in fish from acidic, high-DOC waters. Ponce and Bloom [151] exposed trout (*Oncorhynchus mykiss*) to low concentrations of methylmercury (1.38 ± 0.49 ng/L) at four pH levels (8.2, 7.0, 6.3, 5.8) for 8 weeks to determine if pH has a direct effect on the rate of bioaccumulation. Results indicated that pH had a significant inverse effect on methylmercury bioaccumulation at only the lowest exposure level. This suggested that pH affects bioaccumulation to a certain degree, but a threshold may exist above which pH does not play a significant role.

For some time, food web transfer has been acknowledged as important for the accumulation of mercury in fish, and mercury bioconcentration and biomagnification have been clearly assessed in fish tissue. This was largely due to the ease with which methylmercury concentrations could be measured in large organisms such as fish. There remained considerable uncertainty, however, about the biological behavior of mercury at lower trophic levels until Watras and Bloom [30] developed ultrasensitive methods for measuring methylmercury concentrations in individual zooplankton. This allowed trophic level comparisons among these small organisms. Zooplankton can accumulate methylmercury through direct uptake from the water column and via ingestion of phytoplankton [152-153]. Several researchers have used experimentally-manipulated lakes to determine that both acidification and trophic position enhanced the bioaccumulation of methylmercury in the plankton [30] and confirmed that there were large increases in bioconcentration factors between trophic levels, clearly indicating biomagnification. It has also been shown that some planktonic species, such as those of the genus *Daphnia*, are more important in this process than others [154]. As with fish, environmental factors play a large part in mercury bioaccumulation in these lower trophic levels. Westcott and Kalff [155] collected zooplankton from 24 lakes in south-central Ontario and noted that methylmercury concentrations were highest in zooplankton from acidic brown-water (high DOC) lakes.

Phytoplankton, algae, and diatoms inhabit the base of aquatic food webs, and it is here that methylmercury enters the trophic system. The concentration of methylmercury in phytoplankton is related to the concentration of methylmercury in the water, but this relationship is strongly influenced by dissolved organic carbon. Tracer studies have indicated that methylmercury not only binds to the outside of cell membranes of these organisms (both living and dead), but it is also actively taken into the cytoplasm of living cells [156]. It is thought that DOC may compete with algal cells for methylmercury binding; therefore, higher DOC concentrations limit methylmercury's bioavailability to phytoplankton. Mason et al. [157] reported that uptake of mercury and methylmercury by phytoplankton was a function of pH, salinity, and ligand concentration, all of which influenced mercury speciation and complexation. Additional studies [158-159] determined that an increase in algal biomass dilutes the concentration of mercury consumed by zooplankton, thus resulting in a lower dietary input to grazers and reduced bioaccumulation in algal-rich eutrophic systems.

Sediments often act as a sink for contaminants such as mercury, and resuspension into the water column can occur through a variety of processes including diffusion, advection, or biotransfer through organisms feeding at the sediment-water interface. Benthic invertebrates will readily and quickly bioaccumulation methylmercury from aquatic sediment and pass it into food webs [160-162]. As with the planktonic organisms discussed previously, the bioavailability and subsequent bioaccumulation of methylmercury by sediment-dwelling organisms is strongly affected by sediment organic carbon content [163-164]. Isotope tracer studies have confirmed that bioaccumulation of methylmercury by sediment dwelling organisms is also an important step for introducing atmospherically-deposited mercury into aquatic food webs [160].

Toxicity of Methylmercury in Humans

The initial indications of the powerful biomagnifying properties of methylmercury appeared in the form of neurologic disorders in predatory birds in Sweden. These birds were the top predators in a trophic web that included small mammals and seed-eating birds that consumed freshly planted grain from agricultural fields [165]. It was common at this time for agricultural seed to be treated with fungicides made of alkyl mercury compounds. Further study of birds soon demonstrated that fish-eating birds with no direct connection to mercury-treated seeds, also demonstrated elevated levels of mercury in feathers. This finding led to the landmark discovery that mercury could be converted by bacteria in the aquatic environment to methylmercury, which would bioconcentrate and biomagnify up trophic levels [42, 166]. Since the 1970s, there have been countless studies to quantify the bioaccumulation and effects of methylmercury in wildlife. However, it is the effects of this compound on humans that brings about the greatest attention and concern.

Mercury's threat to humans from aquatic food webs was first noted in Japan during the 1950s when a cluster of neurologic disorders appeared amongst the families of fishermen of Minamata Bay. Fish in the bay were being exposed to methylmercury in industrial effluent from the manufacture of acetaldehyde. Once in the bay, methylmercury biomagnified in fish to levels that would prove lethal to human consumers [167]. In North America, methylmercury concerns were first noted in 1969 when the Ontario Water Resources Commission noted elevated levels of mercury in the sediments of the St. Clair River. Subsequent studies identified mercury bioaccumulation in fish and ultimately resulted in the loss of the commercial fisheries of the Great Lakes [168].

Treated seed grain also triggered several large-scale acute human poisonings in Pakistan, Guatemala, and Iraq during the 1960s and 1970s [169]. In these cases, people mistakenly used mercury-treated grains for direct consumption rather than for planting. While these incidents were not related to fish consumption, they are important to note because much of what is known about the toxicology of high-level exposure to mercury and methylmercury in humans has been learned from these poisonings. More direct data from aquatic systems have come from extensive methylmercury exposures in Minamata, Japan. In addition, three large epidemiological studies have been conducted in the Seychelle Islands [170-171], the Faroe Islands [172-173], and in New Zealand [174-175] to establish dose-response relationships and evaluate the more subtle endpoints of methylmercury toxicity. All three of these populations depend upon fish and marine mammals as a major food source, thus providing an ongoing source of methylmercury exposure.

Chronic exposure to mercury results in toxicity to the central nervous system, and the extensive use of mercury in the fur, felt and hat industry in the eighteenth century was the cause of "mad hatters disease," a condition characterized by delerium and hallucinations. As with elemental mercury, the major human health risks from methylmercury exposure are neurotoxic effects in adults and toxicity to the fetus when mothers are exposed during pregnancy [176]. The target organ of methylmercury toxicity is the brain in both adults and fetuses.

Methylmercury crosses the blood-brain barrier by attaching to the thiol ligand in the amino acid cysteine, creating a molecular complex sufficiently similar in structure to methionine that it is actively transported through neutral amino acid carriers [177-178]. Methylmercury also crosses the placental barrier to accumulate in the fetal brain. By replacing the hydrogen ions in sulfhydryl groups of proteins, methylmercury inhibits protein synthesis, thus disturbing enzyme systems and affecting structural and transport proteins. This results in the loss of neural cells in specific areas of the adult brain, commonly referred to as focal lesions. In the developing human brain, mercury disrupts the formation of the microtubules necessary for both cell division and the migration of developed cells into their proper orientation in the fetal brain [179].

Methylmercury exposure has also been associated with cardiovascular problems in children [180] and adults [181-183], as methylmercury exposure seems to increase atherosclerosis by promoting the formation of free radicals and by compromising cellular mechanisms for coping with oxidative stress [184]. At least two epidemiological studies have linked methylmercury exposure to deaths from leukemia [185-186]. Recent studies in Hong Kong, where fish consumption is high, have linked high mercury body burdens in children with skin disorders and autism [187]. There are also indications that the developing immune system may be sensitive to methylmercury exposure [188-189] leading to allergic disorders [187].

Chronic low-dose methylmercury exposure to the developing brain produces no apparent morphological alterations, but has been linked to more subtle biochemical changes. One of these changes involves variations in the normal production and secretion of neurotransmitters such as serotonin, dopamine, gamma-amniobutyric acid (GABA), norepinephrine, acetylcholine, and glutamate [190]. These neurotransmitters play an important role in the regulation of brain development; therefore, these alterations may be related to observed memory and learning deficits in children exposed *in utero* [191].

Recommended Safe Mercury Exposure Levels and Fish Consumption

There is much debate over the safety of fish in the diet, as fish consumption offers a number of health benefits as a source of protein, vitamin E, selenium, and omega-3 fatty acids. The conflict between these health benefits versus methylmercury risks is confusing to consumers, and finding the appropriate balance has been difficult for regulatory and public health agencies. The United States Government-recommended safe intake level is referred to as the reference dose (RfD) and is generally understood to be the dose that can be absorbed daily for a lifetime without a significant risk of adverse effects. In its 1997 report to the U.S. Congress, the U.S. EPA [192] estimated a RfD of 0.1 μg methylmercury/kg body weight per day. Based on this guideline, there are currently 3,080 fish consumption advisories in effect across the United States due to mercury contamination [193]. Advisories may be specific for a particular lake or stream and typically recommend limiting intake or avoiding certain fish, depending upon trophic level and local mercury contamination. These advisories are often specific for sensitive subgroups such as children, pregnant women, and nursing mothers. In the United States, fish consumption advisories cover more than 14 million acres of lakes and over 800,000 miles of river throughout the country [193]. Allowable dose levels set by public health agencies in selected countries are summarized in Table 1. Adding to the confusion are issues of fish size and trophic level. Because of methylmercury's ability to biomagnify within food webs, certain fish species will naturally contain more mercury than others. Predatory fish such as tuna, shark, and swordfish are at the top of marine food webs; while bass, walleye, and pike dominate freshwater systems. Fish size is also an issue, as larger fish are typically older and have had more time to accumulate a greater body burden of methylmercury. Therefore, local fish consumption advisories often refer specifically to fish species and size.

Table 1. Recommended safe mercury consumption levels as set by public health agencies. This information was adapted from the United Nation's Global Mercury Assessment [195] and from the United Kingdom's Scientific Advisory Committee on Nutrition [196]

Country or Organization	Recommended Safe Mercury Consumption Level
Australia	Tolerable Weekly Intake: 2.8 μg mercury/kg body weight per week for pregnant women.
Canada	Provisional Tolerable Daily Intake: 0.47 μg mercury/kg body weight per day for most of the population and 0.2 μg mercury/kg body weight per day for women of child-bearing age and young children
European Union and United Kingdom	Provisional Tolerable Weekly Intake: 1.6 μg mercury/kg body weight per week
United States	US EPA reference dose: 0.1 μg methylmercury/kg body weight per day
United Nations' World Health Organization and Food and Agriculture Organization	ECFA Provisional Tolerable Weekly Intake: 3.3 μg methylmercury/kg body weight per week

While this section focuses upon methylmercury exposure via dietary consumption (presumably from contaminated fish), it is important to note that mercury is toxic in every chemical form and that there are variations in the degree of toxicity as well as the critical target. Much depends upon the route of exposure. Significant exposures may also come from inhalation, dermal contact, or direct mercury consumption (as with mercury-treated grain consumption). Less-defined mercury toxicity might come from exposure to mercury through dental amalgams and exposure through thimerosol preservatives in vaccines. Complete reviews of mercury toxicity have been published by a number of authors and government agencies including ATSDR [12]; US EPA [192]; Clarkson [194] and the United Nations Environment Programme [195].

CONCLUSION

The major source of atmospheric mercury contamination is combustion of fossil fuels, coal in particular. As people and their governments have become more aware of the risks presented by mercury contamination, initiatives and legislation at local and national levels have been put in place to limit certain uses and releases. While these are somewhat effective, the environmental persistence and global spread of mercury contamination indicates that national measures may not be sufficient in the long term.

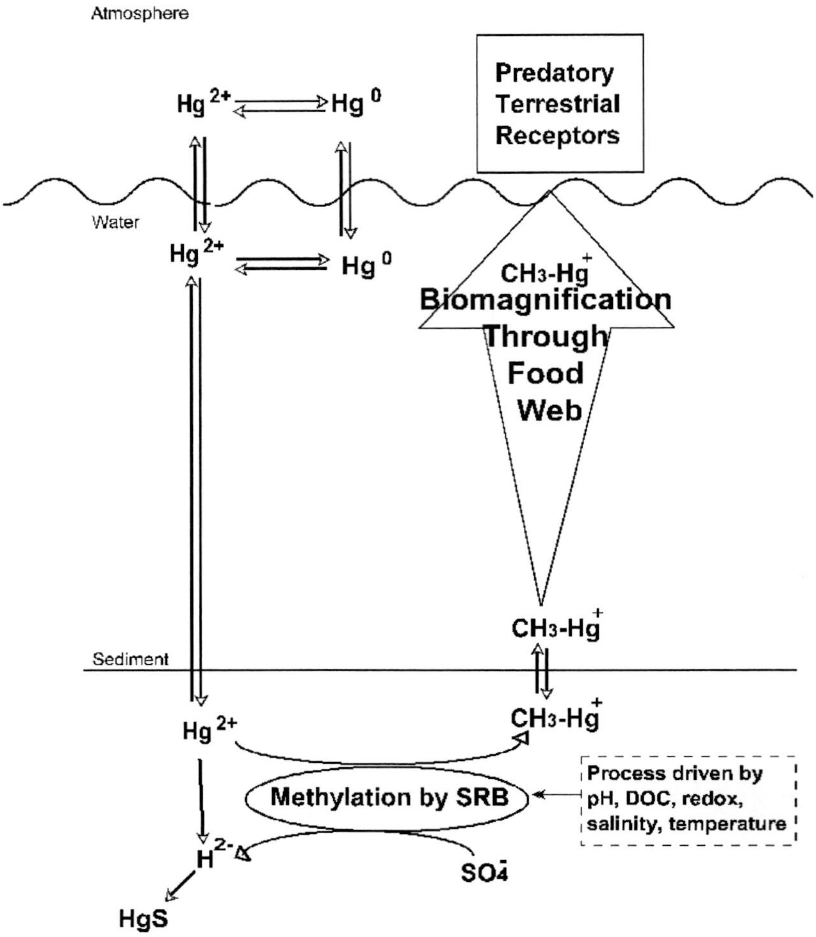

Figure 1. Mercury cycle in aquatic systems. Methylation of inorganic mercury (Hg^{2+}) is driven by sulfate-reducing bacteria (SRB) and is dependent upon environmental conditions such as pH, dissolved organic carbon (DOC) concentration, sediment reduction-oxidation potential (redox), salinity, and temperature. Methylmercury (CH_3-Hg^+) then biomagnifies through food webs, where it becomes dangerous to humans via fish consumption.

Several countries and international organizations have initiated regional measures to reduce mercury contamination, but because mercury can cross borders to accumulate in food webs far distant from the source, all nations will have to reduce fossil fuel combustion and consider cleaner, safer technologies for industry and commercial energy production before this issue can be fully resolved.

Acknowledgments

The author would like to thank William Jackson for his careful review of the manuscript. Sincere appreciation also goes to Jeff King for initiating the author's interest in mercury methylation and providing much instruction on this topic over several years.

REFERENCES

[1] Meili, M. Mercury in lakes and rivers. In: Sigel, A; Sigel, H, editors. *Metal Ions in Biological Systems*, Volume 34: Mercury and Its Effects on Environment and Biology, Marcel Dekker Inc, New York, NY. pp 29-51. 1997.

[2] Nriagu, JO. A Global Assessment of Natural Sources of Atmospheric Trace-Metals. *Nature*, 1989, 338, 47-49.

[3] Wang, QR; Kim, D; Dionysiou, DD; Sorial, GA; Timberlake, D. Sources and remediation for mercury contamination in aquatic systems - a literature review. *Environmental Pollution*, 2004, 131, 323-336.

[4] Radke, LF; Friedli, HR; Heikes, BG. Atmospheric mercury over the NE Pacific during spring 2002: Gradients, residence time, upper troposphere lower stratosphere loss, and long-range transport. *Journal of Geophysical Research-Atmospheres*, 2007, 112 (D19).

[5] Lindqvist, O; Rodhe, H. Atmospheric Mercury - A Review. *Tellus Series B-Chemical and Physical Meteorology*, 1985, 37, 136-159.

[6] Slemr, F; Langer, E. Increase in global atmospheric concentrations of mercury inferred from measurements over the Atlantic Ocean. *Nature*, 1992, 355, 434-436.

[7] Smart, NA. Use and residues of mercury compounds in agriculture. *Residue Reviews*, 1968, 23, 1.

[8] Cooper, CM; Gillespie, WB. Arsenic and mercury concentrations in major landscape components of an intensively cultivated watershed. *Environmental Pollution*, 2001, 111, 67-74.

[9] Carpi, A; Lindberg, SE; Prestbo, EM; Bloom, NS. Methyl mercury contamination and emission to the atmosphere from soil amended with municipal sewage sludge. *Journal of Environmental Quality*, 1997, 26, 1650-1655.

[10] Glass, GE; Sorensen, JA; Schmidt, KW; Rapp, GR; Yap, D; Fraser, D. Mercury deposition and sources for the upper Great Lakes region. *Water Air and Soil Pollution*, 1991, 56, 235-249.

[11] Hermanson, MH. Anthropogenic mercury deposition to arctic lake sediments. *Water Air and Soil Pollution*, 1998, 101, 309-321.

[12] ATSDR. *Toxicological Profile for Mercury.* Agency for Toxic Substances and Disease Registry. Atlanta, GA. 1999.

[13] Meger, SA. Pollution precipitation and the geochronology of mercury deposition in lake sediment of northern Minnesota. *Water Air and Soil Pollution*, 1986, 30, 411-419.

[14] Norton, SA; Dillon, PJ; Evans, RD; Mierle, G; Kahl, JD. The history of atmospheric deposition of Cd, Hg, and Pb in North America: evidence from lake and peat bog sediments. In: Lindberg, S.; Page, AL; Norton, SA, editors. *Sources, Deposition, and Canopy Interactions Vol 3, Acidic Precipitation*, Springer-Verlag, New York, NY. 1990.

[15] DeLacerda, LD; Salomons, W; Pfeiffer, WC; Bastos, WR. Mercury distribution in sediment profiles from lakes of the high Pantanal, Mato Grosso State, Brazil. *Biogeochemistry*, 1991, 14, 91-97.

[16] Swain, EB; Engstrom, DR; Brigham, ME; Hening, TA; Brezonik, PL. Increasing rates of atmospheric mercury deposition in midcontinental North America. *Science*, 1992, 257, 784-787.

[17] Zillioux, EJ; Porcella, DB; Benoit; JM. Mercury cycling and effects in freshwater wetland ecosystems. *Environmental Toxicology and Chemistry*, 1993, 12, 2245-2264.

[18] Rood, BE; Gottgens, JF; Delfino, JJ; Earle, CD. Increasing mercury accumulation in Florida Everglades and Savannahs Marsh flooded soils. *Water Air and Soil Pollution*, 1995, 80, 1-4.

[19] Pfeiffer, WC; Lacerda, LD. Mercury Inputs in the Amazon Region. *Environmental Technology Letters*, 1988, 9, 325-330.

[20] Branches, FJP; Erickson, TB; Aks, SE; Hryhorczuk, DO. The price of gold - mercury exposure in the Amazonian rain-forest. *Journal of Toxicology-Clinical Toxicology*, 1993, 31, 295-306.

[21] Fitzgerald, WF; Clarkson, TW. Mercury and monomethyl mercury: present and future concerns. *Environmental Health Perspectives*, 1991, 96, 159-166.

[22] EPRI (Electric Power Research Institute). *Measurement of Bioavailable Mercury Species in Freshwater and Sediments*. EPRI EA-5197. Prepared by Battelle, Pacific Northwest Laboratories. Richland, Washington. 1987.

[23] Rood, BE. Wetland mercury research: a review with case studies. *Current Topics in Wetland Biogeochemistry*, 1996, 2, 73-108.

[24] Olson, KR; Bergman, HL; Fromm, PO. Uptake of methylmercury chloride and mercuric chloride by trout: a study of uptake pathways into the whole animal and by erythrocytes in vitro. *Canadian Journal of the Fisheries Research Board*, 1973, 30, 1293-1299.

[25] DeFreitas, ASW; Lloyd, KM; Quadri, SU. Mercury bioaccumulation in the detritus-feeding benthic invertebrate *Hyalella azteca. Proceedings of the N.S. Institute of Science*, 1981, 31, 217-236.

[26] Paulose, PV. Bioaccumulation of inorganic and organic mercury in a freshwater mollusc. *Lymnaea acumintata. Journal of Environmental Biology*, 1987, 8, 185-189.

[27] Boudou, A; Ribeyre, F. Comparative study of the trophic transfer of two mercury compounds, $HgCl_2$ and CH_3HgCl, between *Chlorella vulgaris* and *Daphnia magna*. Influence of temperature. *Bulletin of Environmental Contamination and Toxicology*, 1981, 27, 624-629.

[28] Boudou, A; Ribeyre, F. Experimental study of trophic contamination of *Salmo gairdneri* by two mercury compounds, $HgCl_2$ and CH_3HgCl, analysis at the organism and organ level. *Water Air and Soil Pollution*, 1985, 36, 137-148.

[29] Saouter, E; Ribeyre, F; Boudou, A. Bioaccumulation of mercury compounds ($HgCl_2$ and CH_3HgCl) by *Hexagenia rigid*a (Ephemeroptera). In: Vernet, JP, editor. *Heavy Metals in the Environment* (VII) Vol 1. CEP Consultants, Edinburgh, UK. pp 378-381. 1989.

[30] Watras, CJ; Bloom, NS. Mercury and methylmercury in individual zooplankton: implications for bioaccumulation. *Limnology and Oceanography*, 1992, 37, 1313-1318.

[31] Bloom, NS. Determination of picogram levels of methylmercury by aqueous phase ethylation, followed by cryogenic gas chromatography with cold vapor atomic fluorescence detection. *Canadian Journal of Fisheries and Aquatic Sciences*, 1989, 46, 1131-1140.

[32] Grieb, TM; Driscoll, CT; Gloss, SP; Schofield, CL; Bowie, GI; Porcella, DB. Factors affecting mercury accumulation in fish in the upper Michigan peninsula. *Environmental Toxicology and Chemistry*, 1990, 9, 919-930.

[33] Baldi, F. Microbial transformation of mercury species and their importance in the biogeochemical cycle of mercury. In: Sigel, A; Sigel, H, editors. *Metal Ions in Biological Systems*, Volume 34: Mercury and Its Effects on Environment and Biology, Marcel Dekker Inc, New York, NY. pp 213-257. 1997.

[34] Ullrich, SM; Tanton, TW; Abdrashitova, SA. Mercury in the aquatic environment: A review of factors affecting methylation. *Critical Reviews in Environmental Science and Technology*, 2001, 31, 241-293.

[35] Watras, CJ; Huckabee, JW, editors. *Mercury as a Global Pollutant: Towards Integration and Synthesis*. CRC Press, Lewis Publishers, Boca Raton, FL. 1994.

[36] Matilainen, T; Verta, M. Mercury methylation and demethylation in aerobic surface waters. *Canadian Journal of Fisheries and Aquatic Sciences*, 1995, 52, 1597-1608.

[37] Weber, JH. Review of possible paths for abiotic methylation of mercury (II) in the aquatic environment. *Chemosphere*, 1993, 26, 2063-2077.

[38] Hamasaki, T; Nagase, H; Yoshioka, Y; Sato, T. Formation, distribution, and ecotoxicity of methylmetals of tin, mercury, and arsenic in the environment. *Critical Reviews in Environmental Science and Technology*, 1995, 25, 45-91.

[39] Celo,V; Lean, DRS; Scott, SL. Abiotic methylation of mercury in the aquatic environment. *Science of the Total Environment*, 2006, 368, 126-137.

[40] Chen, BW; Wang, T; Yin, YG; He, B; Jiang, GB. Methylation of inorganic mercury by methylcobalamin in aquatic systems. *Applied Organometallic Chemistry*, 2007, 21, 462-467.

[41] Berman, M; Bartha, R. Levels of chemical vs biological methylation of Hg in sediments. *Bulletin of Environmental Contamination and Toxicology*, 1986, 36, 401-404.

[42] Jensen, S; Jernelov, A. Biological methylation of mercury in aquatic organisms. *Nature*, 1969, 223, 753-754.

[43] Olson, BH; Cooper, RC. In situ methylation of mercury by estuarine sediment. *Nature*, 1974, 252, 682-683.

[44] Compeau, G; Bartha, R. Sulfate reducing bacteria: Principal methylators of mercury in anoxic estuarine sediments. *Applied and Environmental Microbiology*, 1985, 50, 498-502.

[45] Gilmour, CC; Henry, EA. Mercury methylation in aquatic systems affected by acid deposition. *Environmental Pollution*, 1991, 71, 131-169.

[46] Gilmour, CC; Henry, EA; Mitchell, R. Sulfate stimulation of mercury methylation in freshwater sediments. *Environmental Science and Technology*, 1992, 26, 2281-2287.

[47] Devereux, R; Winfrey, MR; Winfrey, J; Stahl, DA. Depth profile of sulfate-reducing bacterial ribosomal RNA and Hg methylation in an estuarine sediment. *FEMS Microbiological Ecology*, 1996, 20, 23-31.

[48] Miller, DR; Akagi, H. pH affects mercury distribution not methylation. *Ecotoxicology and Environmental Safety*, 1979, 3, 36-38.

[49] Xun, L; Cambell, NER; Rudd, JWM. Measurement of specific rates of net methyl mercury production in the water column and surface sediments of acidified and circumneutral lakes. *Canadian Journal of Fisheries and Aquatic Sciences*, 1987, 44, 750-757.

[50] Bloom, NS; Watras, CJ; Hurley, JP. Impact of acidification on the methylmercury cycle of remote seepage lakes. *Water Air and Soil Pollution*, 1991, 56, 477-491.

[51] Bonzongo, JCJ; Nemer, BW; Lyons, WB. Hydrologic controls on water chemistry and mercury biotransformation in a closed river system: The Carson River, Nevada. *Applied Geochemistry*, 2006, 21, 1999-2009.

[52] Scheider, WA; Jeffries, DS; Dillon, PJ. Effects of acidic precipitation on Precambrian freshwaters in southern Ontario. *Journal of Great Lakes Research*, 1979, 5, 45-51.

[53] Akielaszek, JJ; Haines, TA. Mercury in the muscle tissue of fish from 3 Northern Maine lakes. *Bulletin of Environmental Toxicology*, 1981, 27, 201-208.

[54] Wren, CD; MacCrimmon, HR. Mercury levels in the sunfish, *Lepomis gibbosus*, relative to pH and other environmental variables of Precambrian Shield Lakes. *Canadian Journal of Fisheries and Aquatic Sciences*, 1983, 40, 1737-1744.

[55] Hakanson, L; Nilsson, A; Andersson, T. Mercury in fish in Swedish lakes. *Environmental Pollution*, 1988, 49, 145-162.

[56] Cope, WG; Wiener, JG; Rada, RG. Mercury accumulation in yellow perch in Wisconsin seepage lakes; relation to lake characteristics. *Environmental Toxicology and Chemistry*, 1990, 9, 931-940.

[57] Winfrey, MR; Rudd, JWM. Environmental factors affecting the formation of methylmercury in low-pH lakes: Review. *Environmental Toxicology and Chemistry*, 1990, 9, 853-870.

[58] Lindqvist, O. Mercury in the Swedish environment. Recent research on causes, consequences, and corrective methods. *Water Air and Soil Pollution*, 1991, 55, 1-261.

[59] Spry, DJ; Wiener, JG. Metal bioavailability and toxicity to fish in low-alkalinity lakes: A critical review. *Environmental Pollution*, 1991, 71, 243-304.

[60] Kelly, CA; Rudd, JWM; Holoka, MH. Effect of pH on mercury uptake by an aquatic bacterium: Implications for Hg cycling. *Environmental Science and Technology*, 2003, 37, 2941-2946.

[61] Hintelmann, H; Welbourne, PM; Evans, RD. Binding of methylmercury compounds by humic and fulvic acids. *Water Air and Soil Pollution*, 1995, 80, 1031-1034.

[62] Kerndorff, H; Schnitzer, M. Sorption of metals on humic acid. *Geochim Cosmocim Acta*, 1980, 44, 1701-1708.

[63] Lodeniius, M; Seppanen, A; Autio, S. Sorption of mercury in soils with different humus content. *Bulletin of Environmental Contamination and Toxicology*, 1987, 39, 593-600.

[64] Jackson, TA. The influence of clay minerals, oxides, and humic matter on the methylation and demethylation of mercury by microorganisms in freshwater sediments. *Applied Organomettalic Chemistry*, 1989, 3, 1.

[65] Miskimmin, BM; Rudd, JWM; Kelly, CA. Influence of dissolved organic carbon, pH, and microbial respiration rates on mercury methylation and demethylation in lake water. *Canadian Journal of Fisheries and Aquatic Sciences*, 1992, 49, 17-22.

[66] Barkay, T; Gillman, M; Turner, RR. Effects of dissolved organic carbon and salinity on bioavailability of mercury. *Applied and Environmental Microbiology*, 1997, 63, 4267-4271.

[67] Wallschlaeger, D; Desai, MVM; Spengler, M; Wilken, RD. Mercury speciation in floodplain soils and sediments along a contaminated river transect. *Journal of Environmental Quality*, 1998, 27, 1034-1044.

[68] Benoit, JM; Gilmour, CC; Mason, RP; Riedel, GS; Riedel, GF. Behavior of mercury in the Patuxent River Estuary. *Biogeochemistry*, 1998, 40, 249-265.

[69] Blum, JE; Bartha, R. Effects of salinity on methylation of mercury. *Bulletin of Environmental Contamination and Toxicology*, 1980, 25, 404-408.

[70] Compeau, G; Bartha, R. Methylation and demethylation of mercury under controlled redox, pH, and salinity conditions. *Applied and Environmental Microbiology*, 1984, 48, 1203-1207.

[71] DeLaune, RD; Jugsujinda, A; Devai, I; Patrick, WH. Relationship of sediment redox conditions to methyl mercury in surface sediment of Louisiana Lakes. *Journal of Environmental Science and Health Part a-Toxic/Hazardous Substances and Environmental Engineering*, 2004, 39, 1925-1933.

[72] Wright, DR; Hamilton, RD. Release of methylmercury from sediments: Effects of mercury concentration, low temperature and nutrient addition. *Canadian Journal of Fisheries and Aquatic Sciences*, 1982, 39, 1459-1466.

[73] Callister, SM; Winfrey, MR. Microbial methylation of mercury in upper Wisconsin river sediments. *Water Air and Soil Pollution*, 1986, 29, 453-465.

[74] Korthals, ET; Winfrey, MR. Seasonal and spatial variations in mercury methylation and demethylation in an oligotrophic lake. *Applied and Environmental Microbiology*, 1987, 53, 2397-2404.

[75] Steffan, RJ; Korthals, ET; Winfrey, MR. Effect of acidification on mercury methylation, demethylation, and volatilization in sediments from an acid-susceptible lake. *Applied and Environmental Microbiology*, 1988, 54, 2003-2009.

[76] Watras, CJ; Bloom, NS; Claas, SA; Morrison, KA; Gilmour, CC; Craig, SR. Methylmercury production in the anoxic hypolimnion of a dimictic seepage lake. *Water Air and Soil Pollution*, 1995, 80, 735-745.

[77] Gagnon, C; Pelletier, E; Mucci, A; Fitzgerald, WF. Behavior of methylmercury in organic-rich sediments. *Canadian Journal of Limnology and Oceanography*, 1996, 41, 428-434.

[78] Gilmour, CC; Riedel, GS; Ederington, MC; Bell, JT; Benoit, JM; Gill, GA; Stordal, MC. Methylmercury concentration and production rates across a trophic gradient in the northern Everglades. *Biogeochemistry*, 1998, 40, 327-345.

[79] Hurley, JP; Benoit, JM; Babiarz, CL; Shafer, MM; Andren, AW; Sullivan, JR; Hammond, R; Webb, DA. Influences of watershed characteristics on mercury levels in Wisconsin rivers. *Environmental Science and Technology*, 1995, 29, 1867-1875.

[80] Waldron, MC; Colman, JA; Breault, RF. Distribution, hydrologic transport, and cycling of total mercury and methylmercury in a contaminated river-reservoir-wetland system. *Canadian Journal of Fisheries and Aquatic Sciences*, 2000, 57, 1081-1091.

[81] Naimo, TJ; Wiener, RG; Cope, WG; Bloom, NS. Bioavailability of sediment-associated mercury to Hexagenia mayflies in a contaminated floodplain river. *Canadian Journal of Fisheries and Aquatic Sciences*, 2000, 57, 1092-1102.

[82] St. Louis, VL; Rudd, JW; Kelly, CA; Beaty, KG; Bloom, NS; Flett, RJ. Importance of wetlands as sources of methylmercury to boreal forest ecosystems. *Canadian Journal of Fisheries and Aquatic Sciences*, 1994, 51, 1065-1076.

[83] St. Louis, VL; Rudd, JW; Kelly, CA; Beaty, KG; Flett, RJ; Roulet, NT. Production and loss of methylmercury and loss of total mercury from boreal forest catchments containing different types of wetlands. *Environmental Science and Technology*, 1996, 30, 2719-2729.

[84] Rudd, JWM. Sources of methylmercury to freshwater ecosystems: A review. *Water Air and Soil Pollution*, 1995, 90, 697-713.

[85] Branfireun, BA; Heyes, A; Roulet, NT. The hydrology and methylmercury dynamics of a Precambrian Shield headwater peatland. *Water Resources Research*, 1996, 32, 1785-1974.

[86] Branfireun, BA; Hilbert, D; Roulet, NT. Sources and sinks of methylmercury in a boreal catchment. *Biogeochemistry*, 1998, 41, 277-291.

[87] Branfireun, BA; Roulet, NT; Kelly, CA; Rudd, JWM. In situ sulphate stimulation of mercury methylation in a boreal peatland: Toward a link between acid rain and methylmercury contamination in remote environments. *Global Biogeochemical Cycles*, 1999, 13, 743-750.

[88] Galloway, ME; Branfireun, BA. Mercury dynamics of a temperate forested wetland. *Science of the Total Environment*, 2004, 325, 239-254.

[89] King, JK; Harmon, SM; Fu, TT; Gladden, JB. Mercury removal, methylmercury formation, and sulfate reducing bacteria profiles in wetland mesocosms. *Chemosphere*, 2002, 46, 859-870.

[90] Harmon, SM; King, JK; Gladden, JB; Chandler, GT; Newman, LA. Methylmercury formation in a wetland mesocosm amended with sulfate. *Environmental Science and Technology*, 2004, 38, 650-656.

[91] Harmon, SM; King, JK; Gladden, JB; Chandler, GT; Newman, LA. Mercury body burdens in *Gambusia holbrooki* and *Erimyzon sucetta* in a wetland mesocosm amended with sulfate. *Chemosphere*, 2005, 59, 227-233.

[92] Stamenkovic, J; Gustin, MS; Dennett, KE. Net methyl mercury production versus water quality improvement in constructed wetlands: Trade-offs in pollution control. *Wetlands*, 2005, 25, 748-757.

[93] Gustin, MS; Chavan, PV; Dennett, KE; Donaldson, S; Marchand, E; Fernanadez, G. Use of constructed wetlands with four different experimental designs to assess the potential for methyl and total Hg outputs. *Applied Geochemistry*, 2006, 21, 2023-2035.

[94] Chavan, PV; Dennett, KE; Marchand, EA; Gustin, MS. Evaluation of small-scale constructed wetland for water quality and Hg transformation. *Journal of Hazardous Materials*, 2007, 149, 543-547.

[95] Bodaly, RA; Hecky, EE; Fudge, RJP. Increases in fish mercury levels in lakes flooded by the Churchill River Diversion, Northern Manitoba. *Canadian Journal of Fisheries and Aquatic Sciences*, 1984, 41, 682-686.

[96] Johnston, TA; Bodaly, RA; Mathias, JA Predicting fish mercury levels from physical characteristics of boreal reservoirs. *Canadian Journal of Fisheries and Aquatic Sciences*, 1991, 48, 1468-1475.

[97] Hall, BD; Rosenberg, DM; Wiens, AP. Methyl mercury in aquatic insects from an experimental reservoir. *Canadian Journal of Fisheries and Aquatic Sciences*, 1998, 55, 2036-2047.

[98] Macdonald, RW; Harner, T; Fyfe, J. Recent climate change in the Arctic and its impact on contaminant pathways and interpretation of temporal trend data. *Science of the Total Environment*, 2005, 342, 5-86.

[99] Bisogni, JJ; Lawrence, AW. Kinetics of mercury methylation in aerobic and anaerobic aquatic environments. *Journal of the Water Pollution Control Federation*, 1975, 47, 135-152.

[100] Compeau, G; Bartha, R. Effect of salinity on mercury methylating activity of sulfate reducing bacteria in estuarine sediments. *Applied and Environmental Microbiology*, 1987, 53, 261-265.

[101] Hurley, JP; Krabbenhoft, DP; Cleckner, LB; Olson, ML; Aiken, GR; Rawlik Jr., PS. System controls on the aqueous distribution of mercury in the northern Florida Everglades. *Biogeochemistry*, 1998, 40, 293-311.

[102] Krabbenhoft, DP; Gilmour, CC; Benoit, JM; Babiarz, CL; Andren, SW; Hurley, JP. Methylmercury dynamics in littoral sediments of a temperate seepage lake. *Canadian Journal of Fisheries and Aquatic Sciences*, 1998, 55, 835-844.

[103] Bloom, NS; Gill, GA; Cappellino, CS; Dobbs, C; McShea, L; Driscoll, C; Mason, R; Rudd, J. Speciation and cycling of mercury in Lavaca Bay, Texas sediments. *Environmental Science and Technology*, 1999, 33, 7-13.

[104] Balogh, SJ; Huang, Y; Offerman, HJ; Meyer, ML; Johnson, DK. Episodes of elevated methylmercury concentrations in prairie streams. *Environmental Science and Technology*, 2002, 36, 1665-1670.

[105] Eckley, CS; Hintelmann, H. Determination of mercury methylation potentials in the water column of lakes across Canada. *Science of the Total Environment*, 2006, 368, 111-125.

[106] Chen, Y; Bonzongo, JCJ; Lyons, WB; Miller, GC. Inhibition of mercury methylation in anoxic freshwater sediment by group VI anions. *Environmental Toxicology and Chemistry*, 1997, 16, 1568-1574.

[107] Perry, KA. Sulfate reducing bacteria and immobilization of metals. *Marine Georesources and Geotechnology*, 1995, 13, 33-39.

[108] Compeau, G; Bartha, R. Effects of salt sea anions on the formation and stability of methylmercury. *Bulletin of Environmental Contamination and Toxicology*, 1983, 31, 486-491.

[109] Rooney-Varga, JN; Sharak Genthner, BR; Devereux, R; Willis, SG; Friedman, SD; Hines, ME. Phylogenetic and physiological diversity of sulphate-reducing bacteria isolated from a salt marsh sediment. *Systematic and Applied Microbiology*, 1998, 21, 557-568.

[110] King, JK; Kostka, JE; Frischer, ME; Saunders, FM. Sulfate reducing bacteria methylate mercury at variable rates in pure culture and in marine sediments. *Applied and Environmental Microbiology*, 2000, 66, 2460-2437.

[111] Macalady, JL; Mack, EE; Nelson, DC; Scow, KM. Sediment microbial community structure and mercury methylation in mercury-polluted Clear Lake, California. *Applied and Environmental Microbiology*, 2000, 66, 1479-1488.

[112] Fleming, EJ; Mack, EE; Green, PG; Nelson, DC. Mercury methylation from unexpected sources: Molybdate-inhibited freshwater sediments and an iron-reducing bacterium. *Applied and Environmental Microbiology*, 2006, 72, 457-464.

[113] Kerin, EJ; Gilmour, CC; Roden, E; Suzuki, MT; Coates, JD; Mason, RP. Mercury methylation by dissimilatory iron-reducing bacteria. *Applied and Environmental Microbiology*, 2006, 72, 7919-7921.

[114] Smith, RL; Klug, MJ. Reduction of sulfur compounds in the sediments of a eutrophic lake basin. *Applied and Environmental Microbiology*, 1981, 41, 1230-1237.

[115] Kuivila, KM; Murray, JW. Organic matter diagenesis in freshwater sediments: the alkalinity and total CO_2 balance and methane production in the sediments of Lake Washington. *Limnology and Oceanography*, 1984, 29, 1218-1230.

[116] Carignan, R. Quantitative importance of alkalinity flux from the sediments of acid lakes. *Nature*, 1985, 317, 158-160.

[117] Choi, S-C; Bartha R. Environmental factors affecting mercury methylation in estuarine sediments. *Bulletin of Environmental Contamination and Toxicology*, 1994, 53, 805-812.

[118] King, JK; Saunders, FM; Lee, RF; Jahnke, RA. Coupling mercury methylation rates to sulfate reduction rates in marine sediments. *Environmental Toxicology and Chemistry*, 1999, 18, 1362-1369.

[119] King, JK; Kostka, JE; Frischer, ME; Saunders, FM; Jahnke, RA. A quantitative relationship that demonstrates mercury methylation rates in marine sediments are based on the community composition and activity of sulfate-reducing bacteria. *Environmental Science and Technology*, 2001, 35, 2491-2496.

[120] Feng, J; Hsieh, YP. Sulfate reduction in freshwater wetland soils and the effects of sulfate and substrate loading. *Journal of Environmental Quality*, 1998, 27, 968-972.

[121] Branfireun, BA; Bishop, K; Roulet, NT; Granberg, G; Nilsson, M. Mercury cycling in boreal ecosystems: The long-term effect of acid rain constituents on peatland pore water methylmercury concentrations. *Geophysical Research Letters*, 2001, 28, 1227-1230.

[122] Gilmour, CC; Riedel, GS. Measurement of Hg methylation in sediments using high specific activity ^{203}Hg and ambient incubation. *Water Air and Soil Pollution*, 1995, 80, 747-756.

[123] Benoit, JM; Gilmour, CC; Mason, RP; Heyes, A. Sulfide controls on mercury speciation and bioavailability to methylating bacteria in sediment pore waters. *Environmental Science and Technology*, 1999, 33, 951-957.

[124] Benoit, JM; Mason, RP; Gilmour, CC. Estimation of mercury-sulfide speciation in sediment pore waters using octanol water partitioning and implications for availability to methylating bacteria. *Environmental Toxicology and Chemistry*, 1999, 18, 2138-2141.

[125] Drott, A; Lambertsson, L; Bjorn, E; Skyllberg, U. Importance of dissolved neutral mercury sulfides for methyl mercury production in contaminated sediments. *Environmental Science and Technology*, 2007, 41, 2270-2276.

[126] Benoit, JM; Gilmour, CC; Mason, RP. The influence of sulfide on solid-phase mercury bioavailability for methylation by pure cultures of *Desulfobulbus propionicus* (1pr3). *Environmental Science and Technology*, 2001, 35, 127-132.

[127] Pak, KR; Bartha, R. Mercury methylation and demethylation in anoxic lake sediments and by strictly anaerobic bacteria. *Applied and Environmental Microbiology*, 1998, 64, 1013-1017.

[128] Sellers, P; Kelly, CA; Rudd, JMW; MacHutchon, AR. Photodegradation of methylmercury in lakes. *Nature*, 1996, 380, 694-697.

[129] Robinson, JB; Tuovinen, OH. Mechanisms of microbial resistance and detoxification of mercury and organomercury compounds: Physiological, biochemical and genetic analyses. *Microbiological Reviews*, 1984, 48, 95-124.

[130] Oremland, RS; Culbertson, CW; Winfrey, MR. Methylmercury decomposition in sediments and bacterial cultures: involvement of methanogens and sulfate reducers in oxidative demethylation. *Applied and Environmental Microbiology*, 1991, 57, 130–137.

[131] Beckert, WF; Moghissi, AA; Au, FHF; Bretthauer, EW; McFarlane, JC. Formation of methylmercury in a terrestrial environment. *Nature*, 1974, 249, 647-675.

[132] Furutani, A; Rudd, JWM. Measurement of mercury methylation in lake water and sediment samples. *Applied and Environmental Microbiology*, 1980, 40, 770-776.

[133] Matilainen, T. Involvement of bacteria in methylmercury formation in anaerobic lake waters. *Water Air and Soil Pollution*, 1995, 80, 757-764.

[134] Spangler, WE; Spigarelli, JM; Rose, JM; Flippin, RS; Miller, HH. Degradation of methylmercury by bacteria isolated from environmental samples. *Applied and Environmental Microbiology*, 1973, 25, 488-493.

[135] Spangler, WE; Spigarelli, JM; Rose, JM; Miller, HH. Methylmercury: Bacterial degradation in lake sediments. *Science*, 1973, 80, 192-193.

[136] Summers, A. Organization, expression and evolution of genes for mercury resistance. *Annual Review of Microbiology*, 1986, 40, 607-634.

[137] Walsh, CT; Distefano, MD; Moore, MJ; Shewchuk, LM; Verdine, GL. Molecular basis of bacterial resistance to organomercurial and inorganic mercuric salts. *Federation of American Societies for Experimental Biology Journal*, 1988, 2, 124-130.

[138] Oremland, RS; Miller, LG; Dowdle, P; Connell, T; Barkay, T. Methylmercury oxidative degradation potentials in contaminated and pristine sediments of the Carson river, Nevada. *Applied and Environmental Microbiology*, 1995, 61, 2745-2753.

[139] Marvin-DiPasquale, MC; Oremland, RS. Bacterial methylmercury degradation in Florida Everglades peat sediment. *Environmental Science and Technology*, 1998, 32, 2556-2563.

[140] Watras, CJ; Bloom, NS; Fitzgerald, WF; Wiener, JG; Rada, R; Hudson, RJM; Gherini, SA; Porcella, DB. Sources and fates of mercury and methylmercury in remote temperate lakes. In: Watras, CJ; Huckabee, JW, editors. *Mercury as a Global Pollutant: Towards Integration and Synthesis*. CRC Press, Lewis Publishers, Boca Raton, FL. 1994.

[141] Hall, BD; Bodaly, RA; Fudge, RJP; Rudd, JWM; Rosenberg, DM. Food as the dominant pathway of methylmercury uptake by fish. *Water Air and Soil Pollution*, 1997, 100, 13-24.

[142] Hudson, RJM; Gherini, SA; Watras, CJ; Porcella, DB. A mechanistic model of the biogeochemical cycle of mercury in lakes. In: Watras, CJ; Huckabee, JW, editors. *Mercury as a Global Pollutant: Towards Integration and Synthesis*. CRC Press, Lewis Publishers, Boca Raton, FL. 1994.

[143] Boudou, A; Ribeyre, F Mercury in the Food Web: Accumulation and Transfer Mechanisms. In: Sigel, A; Sigel, H, editors. *Metal Ions in Biological Systems*, Volume 34: Mercury and Its Effects on Environment and Biology, Marcel Dekker Inc, New York, NY. pp 289-319. 1997.

[144] Bloom, NS. On the chemical form of mercury in edible fish and marine invertebrate tissue. *Canadian Journal of Fisheries and Aquatic Sciences*, 1992, 49, 1010-1017.

[145] Verta, M. Changes in fish mercury concentrations in an intensively fished lake. *Canadian Journal of Fisheries and Aquatic Sciences*, 1990, 47, 1888-1897.

[146] Nilsson, A; Hakanson, L. Relationships between mercury in lake water, water color, and mercury in fish. *Hydrobiologia*,1992, 235/236, 675-683.
[147] McFarlane, GA; Franzin, WG. An examination of Cd, Cr and mercury concentrations in livers of northern pike and white sucker from five lakes near a base metal smelter in Manitoba. *Canadian Journal of Fisheries and Aquatic Sciences*, 1980, 37, 1573-1578.
[148] Haines, TA; Komov, VT; Jagoe, CH. Lake acidity and mercury content of fish in Darwin National Reserve, Russia. *Environmental Pollution*, 1992, 85, 823-828.
[149] Lange, T; Royals, H; Connor, L. Influence of water chemistry on mercury concentration in largemouth bass from Florida lakes. *Transactions of the American Fisheries Society*, 1993, 122, 74-84.
[150] Haines, TA; Komov, VT; Matey, VE; Jagoe, CH. Perch mercury content is related to acidity and color of 26 Russian lakes. *Water Air and Soil Pollution*, 1995, 85, 823-828.
[151] Ponce, RA; Bloom, NS. Effect of pH on the bioaccumulation of low-level, dissolved methylmercury by rainbow trout (*Oncorhynchus mykiss*). *Water Air and Soil Pollution*, 1991, 56, 631-640.
[152] Monson, BA; Brezonik, PL. Influence of food, aquatic humus, and alkalinity on methylmercury uptake by *Daphnia magna*. *Environmental Toxicology and Chemistry*, 1999, 8, 560-566.
[153] Tsui, MTK; Wang, WX. Uptake and elimination routes of inorganic mercury and methylmercury in *Daphnia magna*. *Environmental Science and Technology*, 2004, 38, 808-816.
[154] Pickhardt, PC; Folt, CL; Chen, CY; Klaue, B; Blum, JD. Impacts of zooplankton composition and algal enrichment on the accumulation of mercury in an experimental freshwater food web. *Science of the Total Environment*, 2005, 339, 89-101.
[155] Westcott, K; Kalff, J. Environmental factors affecting methyl mercury accumulation in zooplankton. *Canadian Journal of Fisheries and Aquatic Sciences*, 1996, 53, 2221-2228.
[156] Pickhardt, PC; Fisher, NS. Accumulation of inorganic and methylmercury by freshwater phytoplankton in two contrasting water bodies. *Environmental Science and Technology*, 2007, 1, 125-131.
[157] Mason, R; Reinfelder, JR; Morel, FMM. Uptake, toxicity, and trophic transfer of mercury in a coastal diatom. *Environmental Science and Technology*, 1996, 30, 1335-1345.
[158] Pickhardt, PC; Folt, CL; Chen, CY; Klaue, B; Blum, JD. Algal blooms reduce the uptake of toxic methylmercury in freshwater food webs. *Proceedings of the National Academy of Sciences of the United States of America*. 2002.
[159] Chen, CY; Folt, CL. High plankton densities reduce mercury biomagnification. *Environmental Science and Technology*, 2005, 39, 115-121.
[160] Orihel, DM; Paterson, MJ; Blanchfield, PJ; Bodaly, RA; Hintelmann, H. Experimental evidence of a linear relationship between inorganic mercury loading and methylmercury accumulation by aquatic biota. *Environmental Science and Technology*, 2007, 41, 4952-4958.
[161] Paterson, MJ; Blanchfield, PJ; Podemski, C; Hintelmann, HH; Gilmour, CC; Harris, R; Ogrinc N; Rudd, JWM; Sandilands, KA. Bioaccumulation of newly deposited mercury by fish and invertebrates: an enclosure study using stable mercury isotopes. *Canadian Journal of Fisheries and Aquatic Sciences*, 2006, 63, 2213-2224.

[162] Zizek, S; Horvat, M; Gibicar, D; Fajon, V; Toman, MJ. Bioaccumulation of mercury in benthic communities of a river ecosystem affected by mercury mining. *Science of the Total Environment*, 2007, 377, 407-415.

[163] Nuutinen, S; Kukkonen, JVK. The effect of selenium and organic material in lake sediments on the bioaccumulation of methylmercury by *Lumbriculus variegatus* (oligochaeta). *Biogeochemistry*, 1998, 40, 267-278.

[164] Mason, RP; Lawrence, AL. Concentration, distribution and bioavailability of mercury and methylmercury in sediments of Baltimore Harbor and Chesapeake Bay. *Environmental Toxicology and Chemistry*, 1999, 18, 2438-2447.

[165] Borg, K; Wanntorp, H; Erne, K; Hanko, E. Mercury poisoning in Swedish wildlife. *The Journal of Applied Ecology*, 1966, 3, 171-172.

[166] Wood, JM; Kennedy, FS; Rose, CG. Synthesis of methyl mercury compounds by extracts of a methonogenic bacterium. *Nature*, 1968, 220, 173-174.

[167] Niigata Report. Report on the cases of mercury poisoning in Niigata, Tokyo. Ministry of Health and Welfare. 1967.

[168] Hartig, JH; Zarull, MA; Ciborowski, JJH; Gannon, JE; Wilke, E; Norwood, G; Vincent, A, editors. *State of the Strait: Status and Trends of Key Indicators*. Great Lakes Institute for Environmental Research, Occasional Publication No. 5, University of Windsor, Ontario, Canada. 2007.

[169] Bakir, F; Damluji, SF; Amin-Zaki, L; Murtadha, M; Khalidi, A; al Rawi, NY; Takriti, S; Dhahir, HI; Clarkson, TW; Smith, JC; Doherty, RA. Methylmercury poisoning in Iraq. *Science*, 1973, 181, 230-241.

[170] Davidson, PW; Meyers, GJ; Cox, C; Shamlaye, CF; Marsh, DO; Tanner, MA; Berlin, M; Sloane-Reeves, J; Cernichiari, E; Choisy, O; Choi, A; Clarkson, TW. Longitudinal neurodevelopmental study of Seychellois children following in utero exposure to methylmercury from maternal fish ingestion: outcomes at 19 and 29 months. *Neurotoxicology*, 1995, 16, 677-688.

[171] Davidson, PW; Meyers, GJ; Cox, C; Axtell, C; Shamlaye, CF; Sloane-Reeves, J; Cernichiari, E; Needham, L; Choi, A; Wang, Y; Berlin, M; Clarkson, TW. Effects of prenatal and postnatal methylmercury exposure from fish consumption on neurodevelopment: outcomes at 66 months of age in the Seychelles Child Development Study. *Journal of the American Medical Association*, 1998, 280, 701-707.

[172] Grandjean, P; Budt-Jorgensen, E; White, RF; Weihe, P; Debes, F; Keiding, N. Methylmercury exposure biomarkers as indicators of neurotoxicity in children aged 7 years. *American Journal of Epidemiology*, 1999, 150, 301-305.

[173] Grandjean, P; Weihe, P; White, RF; Debes, F; Araki, S; Yokoyama, K; Murata, K; Sorensen, N; Dahl, R; Jorgensen, PJ. Cognitive deficit in 7-year-old children with prenatal exposure to methylmercury. *Neurotoxicological Teratology*, 1997, 19, 417-428.

[174] Kjellstom, T; Kennedy, P; Wallis, S; Mantell, C. Physical and Mental Development of Children with Prenatal Exposure to Mercury from Fish. Stage I: Preliminary Tests at Age 4. National Swedish Environmental Protection Board Report 3080. Solna, Sweden. 1986.

[175] Kjellstom, T; Kennedy, P; Wallis, S; Stewart, A; Friberg, L; Lind, B; Wutherspoon, T; Mantell, C. Physical and Mental Development of Children with Prenatal Exposure to

Mercury from Fish. Stage II: Interviews and Psychological Tests at Age 6. National Swedish Environmental Protection Board Report 3642. Solna, Sweden. 1989.

[176] Goyer, RA. Toxic effects of metals. In: Amdur, MO; Doull, J; Klaassen, CD, editors. *Casarett and Doull's Toxicology: The Basic Science of Poisons.* Fourth Edition. McGraw-Hill Inc. New York, NY. pp. 623-680. 1993.

[177] Kerper, LE; Ballatori, N; Clarkson, TW. Methylmercury transport across the blood-brain barrier by an amino acid carrier. *American Journal of Physiology*, 1992, 262, R761-R765.

[178] Clarkson, TW; Vyas, JB; Ballatorl, N. Mechanisms of mercury disposition in the body. *American Journal of Industrial Medicine*, 2007, 50, 757-764.

[179] Castoldi, AF; Coccini, T; Ceccatelli, S; Manzo, L. Neurotoxicity and molecular effects of methylmercury. *Brain Research Bulletin*, 2001, 55, 197-203.

[180] Sorensen, N; Murata, K; Budtz-Jorgensen, E; Weihe, P; Grandjean, P. Prenatal methylmercury exposure as a cardiovascular risk factor at seven years of age. Epidemiology, 1999, 10, 370-375.

[181] Salonen, JT; Seppanen, K; Lakka, TA; Salonen, R; Kaplan, GA. Mercury accumulation and accelerated progression of cartoid atherosclerosis: a population-based prospective 4-year follow-up study in men in Eastern Finland. *Atherosclerosis*, 2000, 148, 265-273.

[182] Guallar, E; Sanz-Gallardo, MI; van't Veer, P; Bode, P; Aro, A; Gomez-Aracena, J; Kark, JD; Riemersma, RA; Martin-Moreno, JM; Kok, FJ; et al. Mercury, fish oils, and the risk of myocardial infarction. *New England Journal of Medicine*, 2002, 347, 1747-1754.

[183] Virtanen, JK; Voutilainen, S; Rissanen, TH; Mursu, J; Tuomainen, TP; Korhonen, MJ; Valkonen, VP; Seppanen, K; Laukkanen, JA; Salonen, JT. Mercury, fish oils, and risk of acute coronary events and cardiovascular disease, coronary heart disease, and all-cause mortality in men in eastern Finland. *Arteriosclerosis Thrombosis and Vascular Biology*, 2005, 25, 228-233.

[184] Virtanen, JK; Rissanen, TH; Voutilainen, S; Tuomainen, TP. Mercury as a risk factor for cardiovascular diseases. *Journal of Nutritional Biochemistry*, 2007, 18, 75-85.

[185] Janicki, K; Dobrowolski, J; Krasnicki, K. Correlation between contamination of the rural environment with mercury and occurrence of leukemia in men and cattle. *Chemosphere*, 1987, 16, 253-257.

[186] Kinjo, Y; Akiba, S; Yamaguchi, N; Mizuno, S; Watanabe, S; Wakamiya, J; Futatsuka, M; Kato, H. Cancer mortality in Minamata disease patients exposed to methylmercury through fish diet. *Journal of Epidemiology*, 1996, 6, 134-138.

[187] Zhang, L; Wong, MH. Environmental mercury contamination in China: Sources and impacts. *Environment International*, 2007, 33, 108-121.

[188] NRC. *Toxicological Effects of Methylmercury.* National Research Council. National Academy Press. Washington, D.C. 2000.

[189] Sweet, LI; Zelikoff, JT. Toxicology and immunotoxicology of mercury: A comparative review in fish and humans. *Journal of Toxicology and Environmental Health-Part B-Critical Reviews*, 2001, 4, 161-205.

[190] Mottet, NK; Vahter, ME; Charleston, JS; Friberg, LT. Metabolism of methylmercury in the brain and its toxicological significance. In: Sigel, A; Sigel, H, editors. *Metal Ions in Biological Systems*, Volume 34: Mercury and Its Effects on Environment and Biology, Marcel Dekker Inc, New York, NY. pp 370-403. 1997.

[191] Cagiano, R; De-Salvia, MA; Renna, G; Tortella, E; Braghiroli, D; Parenti, C; Zanoli, P; Baraldi, M; Annau, Z; Cuomo, V. Evidence that exposure to methyl mercury during gestation induces behavioral and neurochemical changes in offspring of rats. *Neurotoxicology and Teratology*, 1990, 12, 23-28.

[192] US EPA. Mercury Report to Congress. Office of Air Quality and Standards. Washington, DC. 1997.

[193] US EPA. National Listing of Fish Advisories. US EPA Fact Sheet. EPA-823-F-07-003. Office of Water, Washington, DC. August, 2007.

[194] Clarkson, TW. The three modern faces of mercury. *Environmental Health Perspectives*, 2002, 110, 11-23.

[195] UNEP. Global Mercury Assessment. United Nations Environment Programme. UNEP Chemicals. Geneva, Switzerland. 2002.

[196] SACN. 2004. Advice on Fish Consumption. Scientific Advisory Committee on Nutrition. UK Food Standards Agency and the Department of Health. London, England.

In: Causes and Effects of Heavy Metal Pollution
Editor: Mikel L. Sanchez

ISBN 978-1-60456-900-1

Chapter 7

IN SITU MEASUREMENT OF METAL CONCENTRATION IN RIVER WATER USING PORTABLE EDXRF SYSTEM

Fábio L. Melquiades[1,2*], Carlos R. Appoloni[2], Paulo S. Parreira[2] and Wislley D. Silva[2]

[1] State University of Center-West /Department of Physics, P.O.Box 3010, Zip Code 85015-430 - Guarapuava – PR – Brazil

[2] State University of Londrina /Department of Physics, P.O.Box 6001, Zip Code 86051-990 Londrina – PR - Brazil

ABSTRACT

Development of new analytical techniques and methodologies capable to identify and quantify the composition of complex samples, as the ones related to environmental problems, is an actual tendency. The objectives of this work were: to use X-ray fluorescence technique, with portable system, to identify and quantify the chemical elements present in the water and its concentration in the sampling place, to optimize the preconcentration methodology and to adapt it for field use. The analysis were realized at laboratory and in situ, measuring water samples in natura and pre-concentrated in membranes. It was employed a portable X-ray tube (Ag target, 50 μm Ag filter, 4 W) to excite the samples and a Si-PIN detector (221 eV resolution for 5.9 KeV energy and 25 μm Be window) with standard electronics for acquisition and evaluation of the spectra. The samples were filtered for suspended particulate matter retention. After this, the optimized preconcentration procedure, with APDC precipitation, was applied. The standard reference materials SRM1640 and SRM1643e, prepared in the same conditions of the samples, were analyzed for methodology validation. Samples from several points at Londrina city, Paraná State, Brazil, were analyzed. It was possible to identify and quantify Ca, Ti, Mn, Fe, Cu, Zn and Pb. The equipment performance and robustness were very good and the results satisfactory for in situ analysis employing a portable system. Considering membrane measurements, the system detection limits are below the

* fmelquiades@unicentro.br.

maximum values established by national and international legislation for drinking water. Besides this, the quantification limit, that is around 0.01 mg L-1 for the majority of the elements, makes viable the methodology application for water analysis. Portable EDXRF has shown to be an useful tool for environmental analysis, as it is a fast, efficient and convenient technique, with potential to substitute high cost and time consuming laboratory methods.

Keywords: Water, X-ray fluorescence, Portable EDXRF, Metal.

1. Introduction

One of the most dangerous kinds of pollution in aquatic systems is due to the dumping of heavy metals. Their increasing use in industries and other activities considered essential to modern human life has resulted in modifications in the natural geochemical cycle of these elements, generating several environmental problems[1].

The determination of heavy and toxic elements permits the study of their distributions, the pollution level as well as the risk assessment in the investigated ecosystems. Various techniques have been applied for the analysis of trace heavy metals in aqueous samples. Among them the most common are atomic absorption spectrometry (AAS), inductively coupled plasma mass spectrometry (ICP-MS), liquid chromatography (LC) and instrumental neutron activation analysis (INAA)[2,3]. The necessity of a nuclear reactor for INAA and the exhaustively sample preparation for AAS, ICP and LC, stimulate the interest towards X-ray fluorescence (XRF) in environmental investigations[4,5].

A current tendency is the development of new analytical techniques and methods with the ability to identify and quantify complex sample constituents as those related environmental problems, and also been able to provide high analytical velocity answer and/or possible to be used on field for *in situ* analysis[6]. XRF is one of the atomic spectrometric techniques that can be adapted for true field portable use [7].

The classical high resolution cryogenic detectors, like Si(Li) and high purity germanium (HPGe) detectors are not suitable for portable instrumentation. The technology of room temperature detectors, cooled by Peltier effect, like CdTe PIN or Silicon PIN detectors, which was developed and advanced in the last years, has led to spectrometers with performances close to those for cryogenic detectors. They are been employed in portable systems for XRF analysis, with improvements in terms of size and weight of the instrumentation [8,9]. These works suggest that portable XRF seems ideal for non-destructive analysis on field.

The basic principle of XRF is to excite a sample with an electromagnetic beam (X-rays or gamma rays) or charged particles and to measure directly the characteristic X-ray energies emitted by the sample in a detector [10]. It is indicate for inorganic chemical elements identification and quantification in different kinds of samples and for a wide range of atomic numbers and concentrations.

X-ray spectrometry, in its various forms, is now a powerful, well established and mature technique for environmental analysis. It offers multi-element capability, economy, high speed and easy operation, and its physical principles, advantages and limitations are now well understood [11,12,13]. More specifically EDXRF has increasingly been applied, in the last 20

Table 1. Summary of papers reporting results from *in situ* XRF analysis in environmental samples

Author/year	XRF technique	Excitation Source	Matrix	Sample preparation/ Comments	Determined elements (LLD#)
Cesareo et al[12] 1998	EDXRF and in situ EDXRF	Miniaturized X-ray tube	Aerosol, algae, microalgae, marine sediments and paint	Thin and intermediate samples were prepared mixing powder with epoxy resins	Pb, K, Ca, Mn, Fe, Ni, Cu, Zn, Sr, Ti, Rb, Zr, Ba LLD value is 0.2%of the concentration found.
Kalnicky and Singhvi[14] 2001	FPXRF	^{55}Fe, ^{109}Cd, ^{241}Am	Soil, sediment, thin films paints, reference materials	Analysis of NIST standards	K(362), Ca(211), Ti(120), Mn(255), Co(274), Ni(49), Cu(17), Zn(32), As(17), Se(6), Sr(15), Zr(7), Mo(2), Hg(17), Pb(11), Rb(6), Cd(46), Sn(36), Sb(17), Ba(16), Fe(459) values in mg/kg
Khusainov et al[8] 1999	FPXRF	^{241}Am, ^{238}Pu ^{109}Cd, ^{57}Co	Metal Alloy, Sand and Environmental control	Portable γ and X-ray analyzers were tested.	Ti, V, Cr, Mn, Fe, Co, Ni, Cu, W, Nb, Mo,Zr, Nb, Hf, Ta, Hg, Tl, Bi, Pb(1 μg/cm^2), As(1 μg/cm^3)
Longoni et al[9] 1998	FPXRF	X-ray tube	Metal alloy and paints	Direct measurement	Cu, Zn, Pb, Fe, Sn, Cd, Ba, Hg, Se, Sr, Zn
Kirtay et al [22] 1998	FPXRF	^{55}Fe, ^{109}Cd, ^{241}Am	Marine sediment	The sample is digested and the extract is cleaned up and analyzed	Cu(50-100 ppm), Zn(50-100 ppm), Pb(25-50 ppm)
McDonald et al [23] 1999	Cone penetrometer - CPXRF	Miniaturized X-ray tube	Soil	The system was introduced directly in the soil and tested for lead contamination. Comparison of average analysis of AAS and CPXRF have slopes ranging from 0.84 – 1.02	Pb (not informed)
Bernick et al [24] 1995	FPXRF	^{244}Cm, ^{241}Am	Soil and sediment hazardous waste	Thin films gravimetrically prepared	Cd(35.5), Cr(12.9), Cu(16.1), Ni(16.1), Pb(16.1), Zn(16.1) values in μg/10.75cm^2
Economou [25] 2001	APXS, FPXRF	^{244}Cm	Martian soil and Martian rocks	Direct measurement	Na_2O, MgO, Al_2O_3, SiO_2, P_2O_5, SO_3, Cl, K_2O, CaO, TiO_2, Cr_2O_3, MnO, FeO (not informed)
Rieder et al [26] 1997	APXS, FPXRF	^{244}Cm	Martian soil and Martian rocks	Direct measurement and comparison of results with terrestrial soil and rocks	Na_2O, MgO, Al_2O_3, SiO_2, SO_3, Cl, K_2O, CaO, TiO_2, FeO (not informed)
Fiorini and Longoni [27] 1999	EDXRF *in situ*	Miniaturized X-ray tube	Glass, paints	Direct measurement	Cu, Zn, Pb, K, Ba, Cr, Fe, Mn, Ca, Ni (not informed)
McComb andGesser [28] 1999	EDXRF, WDXRF	^{241}Am	Water	Chelating group bound to a cloth in a passive monitor. Values of correlation between WDXRF and ICP-OES ranging from 0.84 – 1.02	Pb, Cu, Cd, Fe, Mn, Zn, Mg (not informed)
Clark et al. [29] 2006	EDXRF	^{109}Cd	Urban soil	Direct analysis after drying in specific recipient	Pb (400 mg Kg^{-1})
Kilbridge et al. [30] 2006	EDXRF	^{109}Cd and ^{241}Am	Soil	Direct analysis after drying in specific recipient	Cd, Pb, As, Ni, Zn, Mn, Fe and Cu (not informed).

#In parenthesis are the Low Level Detection (LLD), when present at the referred work, with independent units.

years, to the analysis of aerosols, waters, sediments, soils, solid waste and other environmental samples. Field portable XRF (FPXRF) technology has gained widespread acceptance in the environmental community. It provides a viable, cost- and time-effective approach for *in situ* analysis of a variety of environmental samples and offers some advantages compared with conventional methods that are been applied in the analysis of environmental samples [14].

There are various review works that deals with environmental analysis like air, water, soil, biological samples and others [15,16,17] using different techniques. Other reviews pay attention specifically in XRF in a general way [18,19,20], covering instrumentation, detectors, spectral analysis, pre-concentration techniques, etc.

Due to the cited characteristics, portable XRF systems has been applied *in situ* for trace metal analysis in different kinds of samples [21]. Table 1 presents some additional information about portable XRF papers and show that the methodology presented in those references are used, in its majority, for solid samples. Only one work deals with water analysis, even so with *in situ* sample collection.

The objectives of the present work is to use X-ray fluorescence technique, with portable system, to identify and quantify the chemical elements present in the water and its concentration in the sampling place, to optimize the pre-concentration methodology and to adapt it for field use. In this work are presented the detection limits, the quantification limits and metal concentration results for river and lake waters. The analysis were in situ, measuring in natura water and samples pre-concentrated in membranes.

2. Material And Methods

2. 1. Sampling

The samples were collected from Igapó lakes and its affluent at Londrina city, Paraná, Brazil. Igapó lakes cross the central region of the city and are part of the micro basin of Cambé river, which supplies the lakes. They are very used for recreation, aquatic sports and fishing. Figure 1 show the region of sampling.

Laboratory tests and *in situ* measurements were performed. For laboratory analysis, polyethylene bottles were used for superficial water collection and the samples prepared in the same day. Water samples were collected in plastic beakers for *in situ* measurements.

For *in natura* analysis, 10 mL of sample was separated. For membrane analysis, 1 L of sample was filtered in 0.45 μm ester cellulose membranes for suspended particulate matter retention (non dissolved metals determination). After this APDC precipitation methodology was applied for sample pre-concentration (dissolved metals determination). A diagram of sample preparation is in Figure 2.

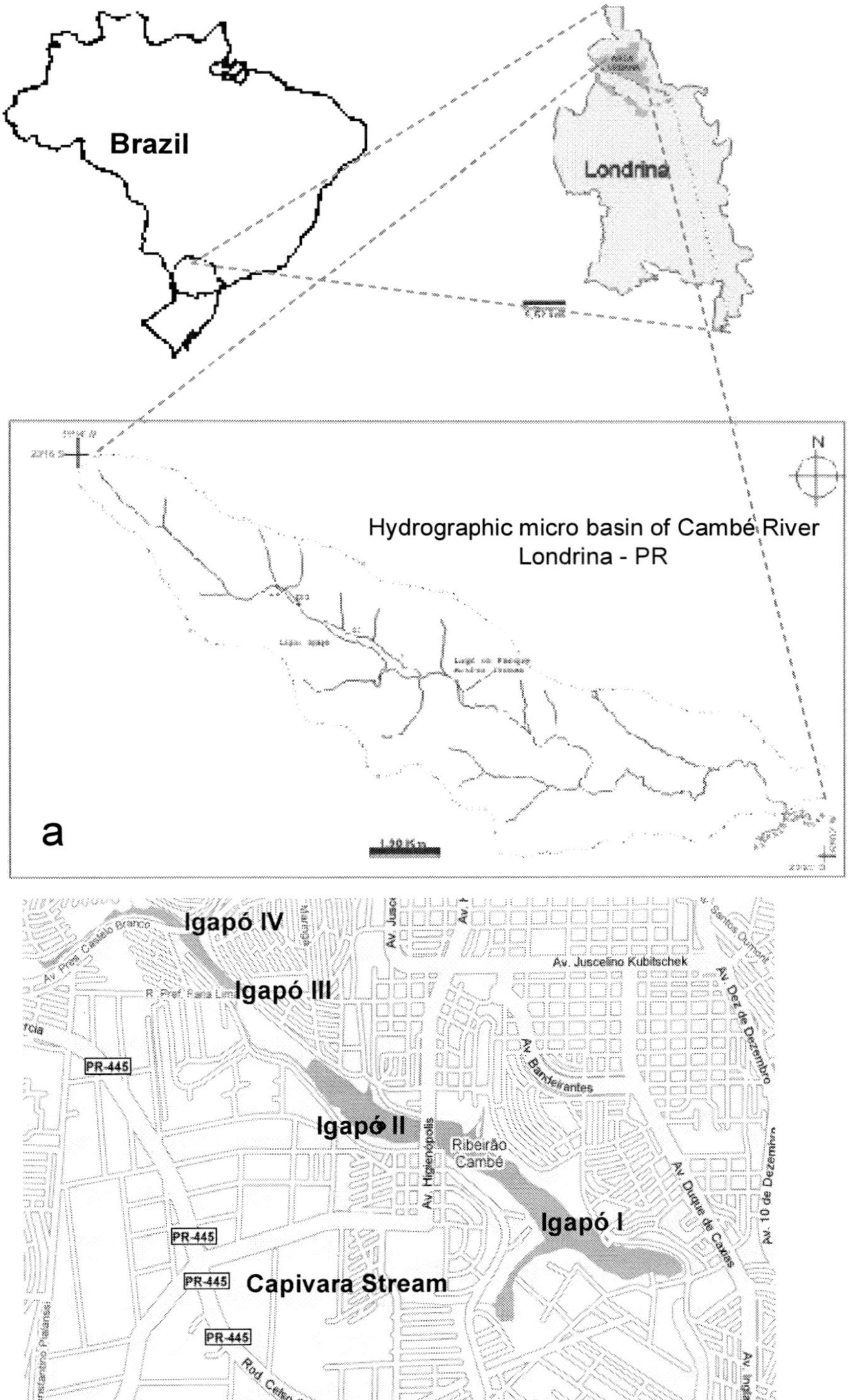

Figure 1. (a) Region of analysis identification. (b)Delimitation of the sampling area at Londrina, PR, Brazil.

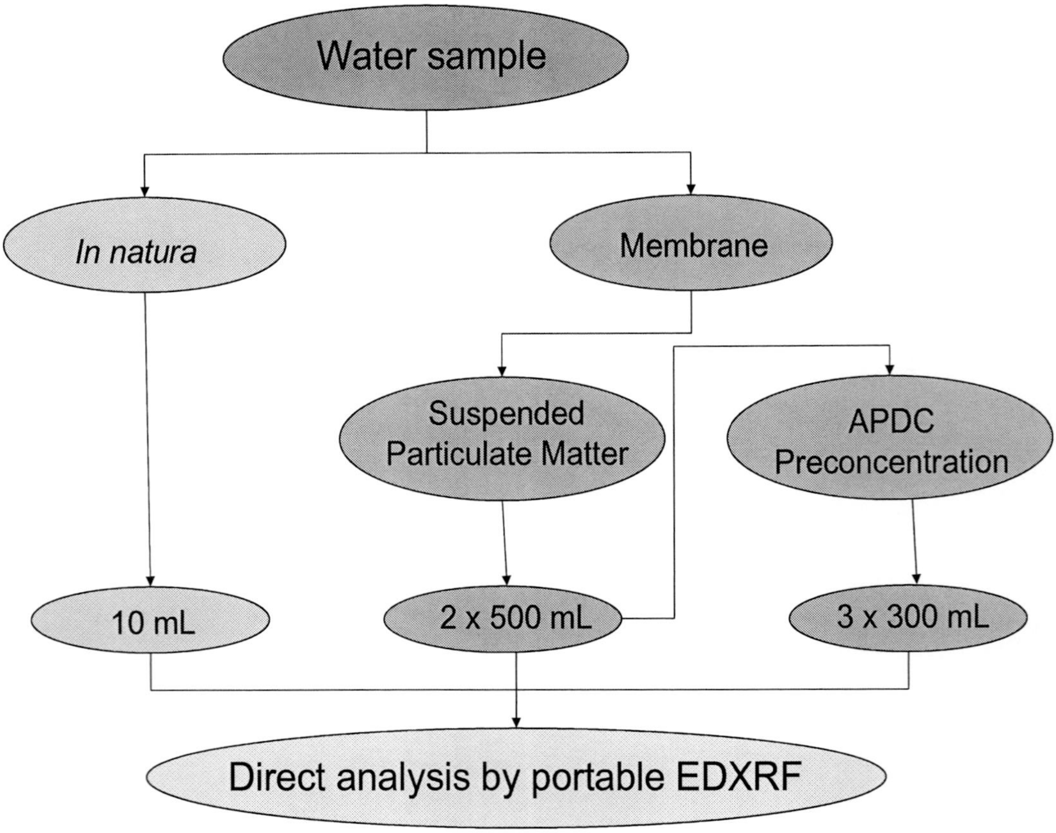

Figure 2. Squematic diagram of sample preparation procedure.

2. 2. Instrumentation

The measurement system comprises a Si-PIN detector (221 eV FWHM at 5.9 keV and 25 μm Be window), coupled to a pre-amplifier, both thermoelectrically cooled, a high voltage source with amplifier, multichannel analyzer [31]. A notebook was used for data acquisition and evaluation of the spectra. The sample excitation was done with a mini X ray tube (Ag target, 4 W) [32]. The whole system is portable, and can be used in the sampling place, as show Figure 3.

The measurement conditions were 28 kV, 10 μA and 50 μm Ag filter, on the tube. Ag collimator with 3 mm diameter aperture on detector and 500 s of irradiation time.

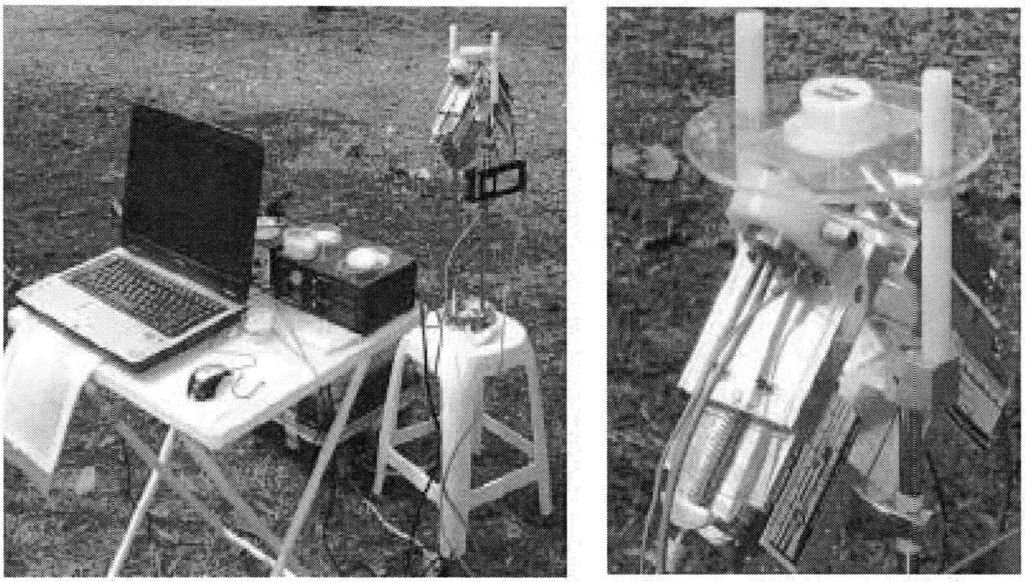

Figure 3. Measurement system and sample geometry.

2. 3. Pre-Concentration Methodology

To reach the national and international maximum permissive level of concentration the majority of the water analysis methods require a pre-concentration step [33]. Theses methods can reduce the matrix effect, improving detection limits and results accuracy in EDXRF methodology. Several methods were considered[6],[34,35,36,37], but as the pre-concentration needs to be accomplished at field it was opted by the precipitation with Ammonium Pyrrolydine Dithiocarbamate (APDC), which is a non specific chelating agent, that reacts with metallic ions forming a very stable complex with the majority of the transition metals [38].

Preconcentration with APDC consists in adding a quantity of APDC in a specific concentration to a liquid sample volume with corrected pH; after that, the solution is stirred for some time and, in the sequence, filtered in membrane filter with a vacuum pump.

For pH adjustment, HNO_3 and NaOH were used. The APDC solution (Sigma Aldrich Inc.) was freshly prepared. After the pH adjustment, the APDC solution was added to the sample, stirred for chelating and filtered with a vacuum pump in cellulose ester membranes with 0.45μm pore size. According to Alvarez et al. (2000) [39] the pore diameter does not influence the procedure and likewise Narin and Soylak (2003) [40], quantitative results are better with cellulose nitrate or acetate membranes. The membranes were dried at room temperature for 24h or for 30 min. at Sun (See Apendix II for details).

Figure 4. present the equipments adapted for in situ use.

Figure 4. Pre-concentration equipments and reagents for in situ application.

2.3.1. Optimization of the APDC Pre-Concentration Methodology

Literature shows several works that deal with APDC as a precipitation agent, but with differences in the parameter values used in the pre-concentration methodology, as shown Table 2.

Generally, studies that deal with the optimization of experimental variables are realized through procedures that evaluate the variable effect individually, which in general, impedes optimum true values establishment. The unvariate optimization does not consider the interaction of variables [41]. In the last years, systems of multivariate optimization have demonstrated their use in several knowledge fields. The observation of the variables effects and the interactions between them are of great importance to understand the process that is been tested or monitored in an experiment [42].

So, it was accomplished a factorial design for the optimization of heavy metal pre-concentration methodology with APDC chelating agent, aiming to obtain conclusive information about the factor conditions for APDC usage. At the same time, it also aims at evaluating the performance of a portable EDXRF system to analyze the samples [43].

Table 2. Abstract of APDC pre-concentration methodologies in the literature

Reference	Sample Volume (mL)	pH	APDC Concentration	APDC Volume (mL)	Stirring time (min)
[35]	100	4	1%	1	15
[38]	200	3	1%	4	20
[39]	1000	3-7	1%	4-25	30
[40]	250	2	0-10 mg		10
[44]	100	3	2%	1	20
[45]	200	4	1%	2	15
[46]	50	4	0.5%	2	30

Table 3. Factorial design sequence for optimization procedure determination.

Design number	Factor number	Factor	Level (-)	Level (+)
Design 1	1	pH	3	5
	2	Sample volume	100 mL	200 mL
	3	APDC concentration	1%	2%
	4	APDC volume	1 mL	4 mL
	5	Stirring time	5 min	20 min
Design 2	1	pH	5	6
	2	Sample volume	200 mL	250 mL
	5	Stirring time	5 min	10 min
Design 3	1	pH	4	5
	2	Sample volume	300 mL	250 mL

2.3.1.1. Factorial Design

It was performed a fractional factorial design $2^{(5-1)}$ and two complete factorial designs: 2^3 and 2^2, as shown in Table 3. An unvariate design was performed using the established conditions for a pH curve construction of the pre-concentration procedure to finalize the optimization process, seeking to establish the best pH for each element individually. This factor is highly important in the process.

2.3.1.2. Sample Preparation

Multi-element solutions from stock mono-element solutions (Sigma Aldrich Inc.) were prepared. The multi-element solutions consisted of the following elements, with concentrations in mg L^{-1}: Mn (0.5), Fe (0.3), Cu (0.5), Zn (0.3), Se (0.3), Pb (0.5). The dilution was in distilled water.

2.3.1.3. Factorial Design Results

Figure 5 presents the normal graphics for $2^{(5-1)}$ factorial design, in which it can be noticed that the effects of factors 1 and 2 move away from the tendency. That indicates that the values should be higher than the ones tested. This is confirmed by analyzing the values starting from the ones in which the effects have a real meaning. The values for significant effect are in the lower right corner of each graphic of Figure 4. This behavior was verified for all elements except Se, which pH value must be reduced. From this analysis, the values of 1 mL APDC solution with 2% concentration were fixed, since these factors do not have a significant influence on the final results.

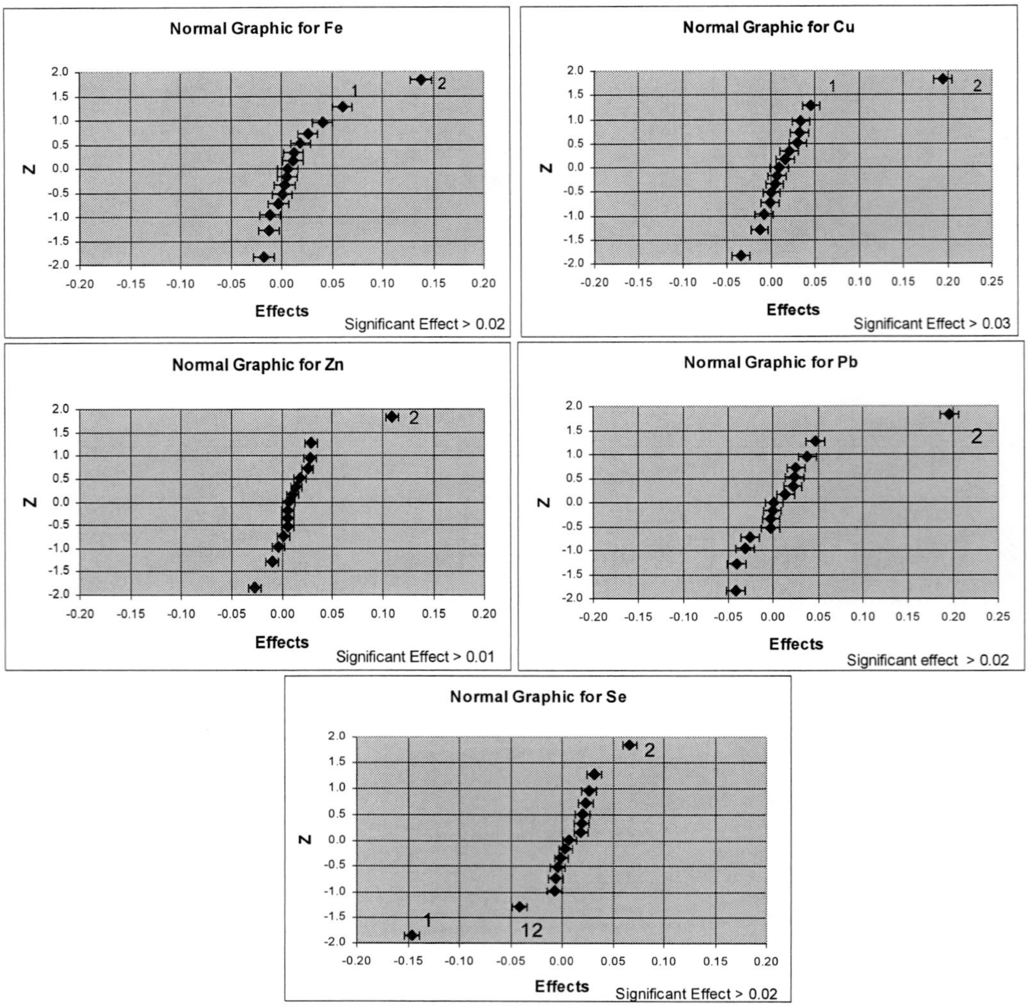

Figure 5. Normal graphic with $2^{(5-1)}$ factorial design result.

For the 2^3 factorial design, factors 1 and 2 values were higher, as shown in Table 3. The results are in Figure 6, where it can be seen that the same factors still influence the result. Here, factor 1 must be reduced and factor 2 increased. Again, the behavior for Se is different, hindering its determination in these conditions.

Table 4 shows the results for the 2^2 factorial design. It was chosen the L e assay, in which the five elements were determined with 10% deviation. This assay corresponds to 300 mL sample volume at pH 4.

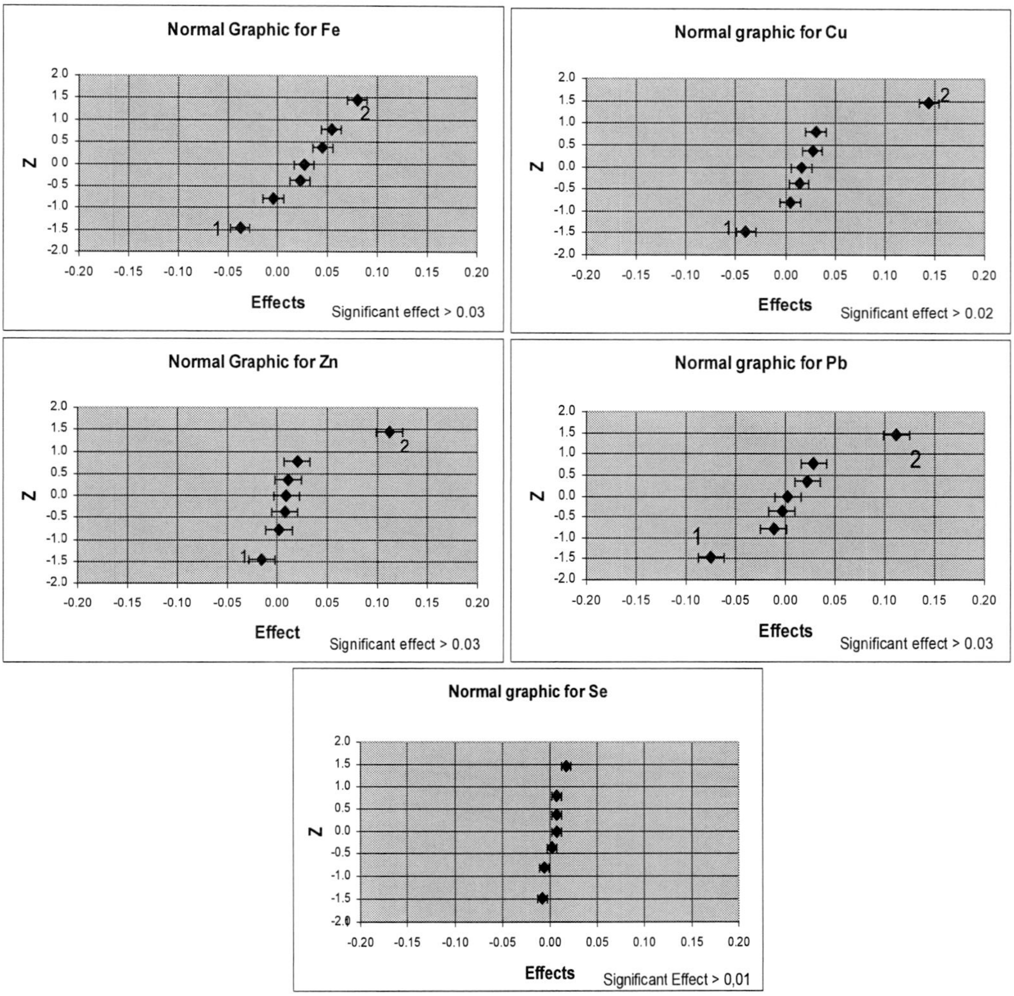

Figure 6. Normal graphic with 2^3 factorial design result.

Table 4. Results of concentrations for the 2^2 factorial design ($mg\ L^{-1}$)

Assay	*Factor*		*Fe (0.3)*[a]	*Cu (0.5)*[a]	*Zn (0.3)*[a]	*Se (0.3)*[a]	*Pb (0.5)*[a]
	pH	**V (mL)**					
I	5(+)	250(+)	0.32±0.03	0.46±0.01	0.27±0.01	0.02±0.01	0.47±0.02
J	4(-)	250(+)	0.26±0.01	0.41±0.01	0.19±0.01	0.27±0.01	0.38±0.01
K	5(+)	300(-)	0.25±0.02	0.49±0.01	0.24±0.01	0.01±0.01	0.44±0.01
L	4(-)	300(-)	0.25±0.01	0.46±0.04	0.25±0.02	0.32±0.03	0.42±0.05

[a] concentration in the multi-element solution

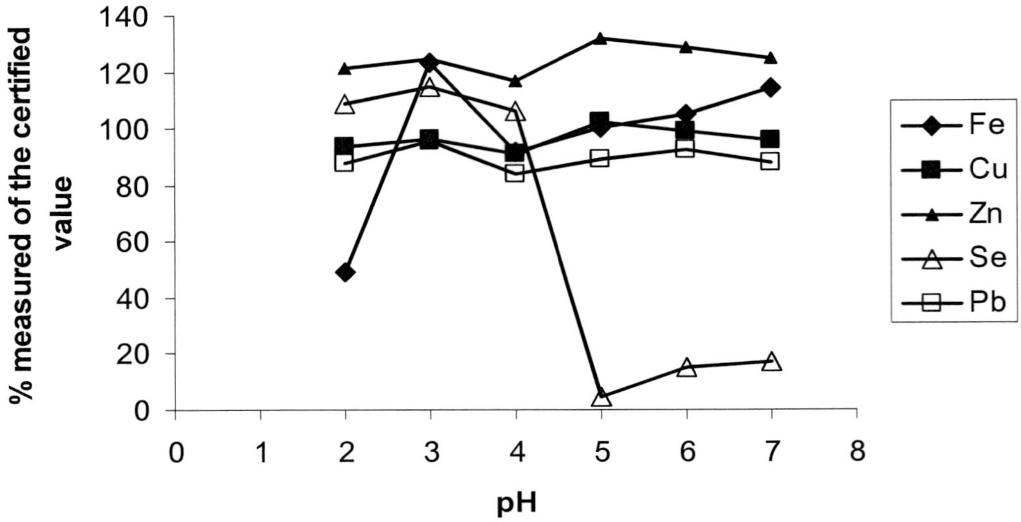

Figure 7. pH curve from the unvariate design.

Figure 7 shows the unvariate design results, in which a pH curve was plotted for each studied element. In the individual analyzes it was perceived that, for Fe and Cu, the best pH value was 5, for Zn and Se it was 4, and for Pb it was 3.

From the results of the three stages of factorial design and one unvariate design, it was concluded that the best optimized situation for Fe, Cu, Zn, Se and Pb preconcentration in a multi-element sample, using APDC precipitation methodology, is the following: 300 mL sample volume at pH 4, 1 mL APDC 2% solution with 10 min of stirring time and filtration in 0.45 μm pore size ester cellulose membranes.

2.4. In Situ Measurements Description

All the equipment necessary for *in situ* measurements can be put up in a car trunk. Two small bags (standard size for airplane hand luggage) are necessary, one for the EDXRF excitation/detection system and another for the equipments, reagents and materials for the pre-concentration procedure. A notebook, a portable electric generator and two portable table sets are, carried separately.

Initially is mounted the filtration and pre-concentration system. One liter of water is filtered, divided in two or three membranes depending on the suspended particulate matter (SPM) amount, since this filtration step can take up to 30 min. While is going on the filtration, the EDXRF system is mounted and calibrated.

As the membranes get ready, they are putted to dry at sun, when possible. It is necessary a minimum dry time of 30 min. (See Appendix II for details).

The normal measurement sequence is the following: measurement of distilled water blank, measurement of *in natura* samples in the an appropriated recipient, measurement of a membrane blank, measurement of the SPM membranes and measurements of the APDC pre-concentrated membranes.

2.5. Calibration

2.5.1. In Natura Standards

For calibration, *in natura* standards were measured. Mono-element spiked solutions were prepared with concentrations from 10 to 50 mg L^{-1} for Cr, Mn, Fe, Co, Ni, Cu, Zn, Se, Hg and Pb. A volume of 10 mL of the standard solution was placed in appropriated recipients for XRF analysis (Chemplex Inc.), covered with propylene film (Mylar, Chemplex Inc) for irradiation.

The validation was accomplished measuring three different multi-element standard solutions. Multi-element standard solution Sigma Aldrich 70002 (Ag, Al, B, Ba, Bi, Ca, Cd, Co, Cr, Cu, Fe, K, Li, Mg, Mn, Mo, Na, Ni, Pb, Sr, Tl, Zn), multi-element standard solution Sigma Aldrich 70006 (Al, As, B, Ba, Be, Cd, Co, Cr, Cu, Fe, Mn, Ni, Pb, Se, Tl, V, Zn) for analysis of drinking and sewage water and multi-element standard solution High Purity Inc.: Drinking Water Primary Standard (DWPS) (Ag, As, Ba, Cd, Cr, Pb, Hg, Se) and Drinking Water Secondary Standard (DWSS) (Cu, Fe, Mn, Zn) for the analysis of drinking water.

2.5.2. Membrane Standards

Elemental sensitivity values were determined irradiating mono-element standards deposited in polycarbonate thin films (MicroMatter Inc.). Were measured the following standards: K, Ca, Ti, Cr, Fe, Co, Ni, Cu, Zn, Ga, As, Se and Br, for K X-ray lines and Te, Cs, Ba, W, Au, Pb and Bi for L X-ray lines.

For validation purpose were used the standard reference materials SRM 1640 and SRM 1643e from NIST (USA National Institute of Standards and Technology), that were pre-concentrated as described at item 2.3. Also the three multi-element standard solutions (Sigma Aldrich 70002, Sigma Aldrich 70006 and DW), with different concentrations were prepared for methodology certification.

2.6. Quantification Procedure

The relationship between the fluorescence intensity of characteristic Kα or Lα lines and the concentration of an element in the sample is given by the fundamental parameters equation, Eq. (1) [47].

$$I_i = C_i .\ S_i .\ A \qquad (1)$$

where I_i is the net intensity of the characteristic X-ray (cps), C_i, represents the concentration (mg L^{-1}), S_i the elementary sensitivity (cps mg^{-1} L) of the analyzed element and A the absorption factor.

Detection limits (DL) and quantification limits (QL) were obtained using Eq. (2) and Eq. (3) [48]:

$$DL = \frac{3\sqrt{Bg_i}}{S_i\ t} \quad (2)$$

$$DQ = \frac{10\sqrt{Bg_i}}{S_i\ t} \quad (3)$$

where Bg_i is the background counts for the element and t is the measurement time.

3. RESULTS

3.1. *In Natura* Results

3.1.1. Calibration Curve, Detection and Quantification Limits

Calibration curves are presented in Figure 8 and Figure 9. The values for DL and QL, within 95% confidence level, are shown in Table 5.

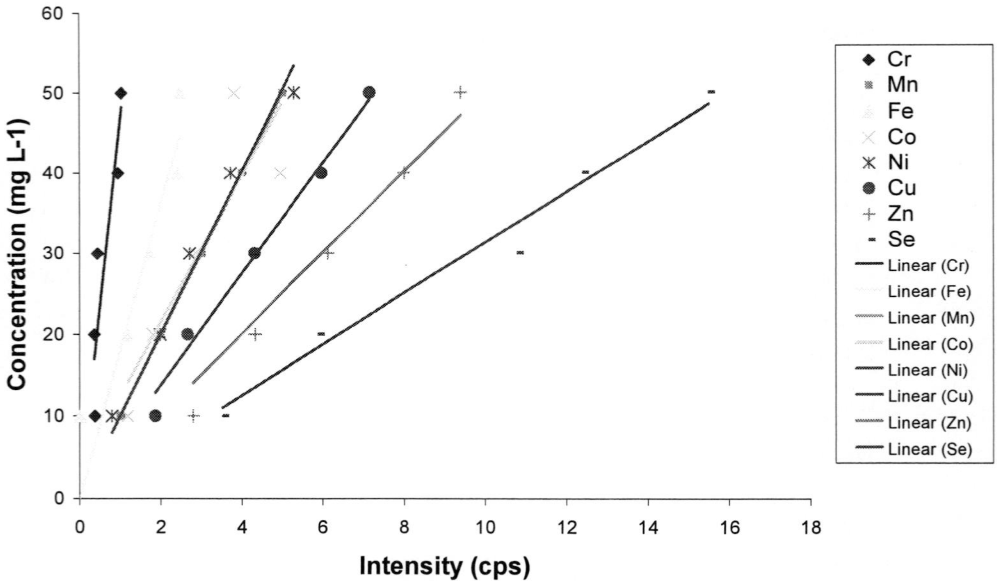

Figure 8. Sensitivity curve for Kα line elements. Standard deviation around 15%.

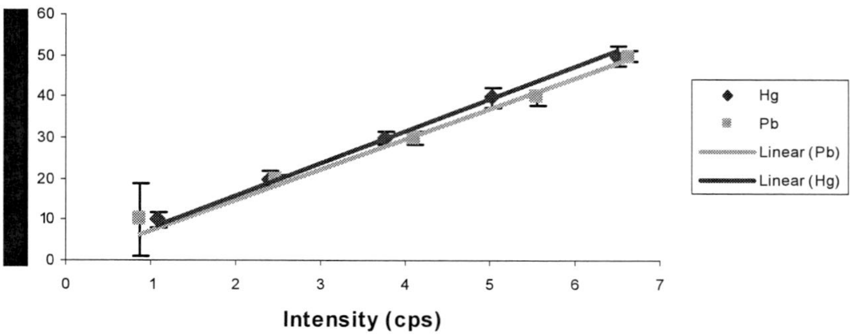

Figure 9. Sensitivity curve for Lα line elements.

Table 5. Detection and quantification limits for *in natura* water with a portable EDXRF system. Values, in mg L^{-1}, within 95% confidence level deviation

Element	Detection Limit	Quantification limit
Cr	12.5±1.5	41.8±4.9
Mn	7.5±0.8	24.8±2.7
Fe	4.9±0.5	16.3±1.8
Co	2.7±0.3	9.1±0.9
Ni	2.8±0.2	9.5±0.8
Cu	2.1±0.2	7.0±0.7
Zn	1.6±0.2	5.4±0.6
Se	1.9±0.1	6.2±0.4
Hg	3.4±0.2	11.5±0.8
Pb	3.7±0.4	12.4±1.2

3.1.2. Methodology Validation for In Natura Measurements

The results for the multi-element standards, in order to verify the accuracy of the methodology, are shown in Table 6.

Table 6. Results of the multi-element standards for methodology validation. Values in mg L^{-1}

	Standard 70002	
Element	**Certifed Concentration**	**Measured Concentration**
Cr	47.5 - 52.5	43.4 - 65.0
Mn	9.5 - 10.5	16.2 - 24.4
Fe	9.5 - 10.5	7.5 16.9
Co	9.5 - 10.5	5.2 - 14.8
Ni	47.5 - 52.5	45.3 - 60.1
Cu	9.5 - 10.5	9.8 - 14.4
Zn	9.5 - 10.5	11.3 - 12.3
Pb	95 - 105	91.1 - 99.3

Table 6.(Continued)

Standard 70006		
Element	**Certifed Concentration**	**Measured Concentration**
Mn	9.5 - 10.5	7.1 - 33.9
Fe	95 - 105	71.8 - 96.4
Co	9.5 - 10.5	6.8 - 11.8
Ni	19 - 21	17.7 - 24.1
Cu	19 - 21	19.1 - 22.5
Zn	95 - 105	75.8 - 85.4
Se	95 - 105	74.2 - 82.2
Pb	38 - 42	35.7 - 74.3

Standard DW		
Element	**Certifed Concentration**	**Measured Concentration**
Cr	95 - 105	79.3 - 111.5
Mn	47.5 - 52.5	41.6 - 51.0
Fe	91 - 105	75.2 - 102.0
Cu	47.5 - 52.5	44.7 - 48.4
Zn	47.5 - 52.5	75.8 - 85.4
Se	95 - 105	39.7 - 46.9
Hg	19 - 21	13.99 - 23.1
Pb	95 - 105	7.1 - 88.1

Some elements like Mn, Fe and Co in Sigma Aldrich standards are near to the quantification limit and were not determined with good accuracy. Quantification of Zn, Se and Pb was affected by system limitations and interference with other elements that turn difficult the deconvolution of peak areas with good precision. However, in Figure 10 and Figure 11 it can be seen that the qualitative identification is clearly possible.

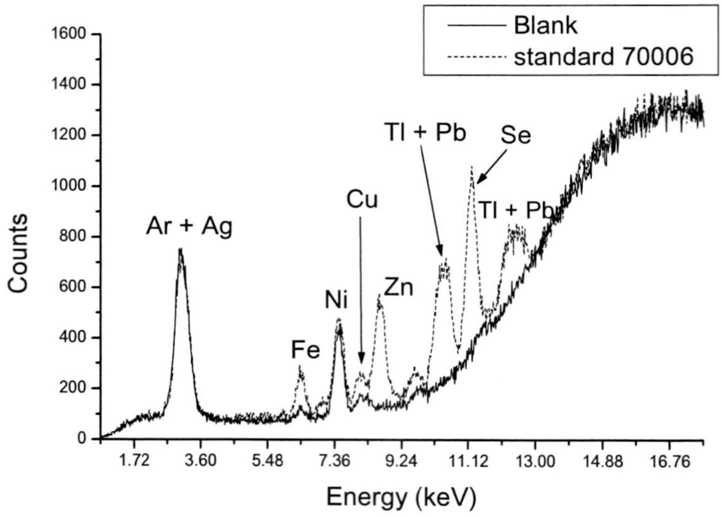

Figure 10. Spectra of the multi-element standard 70006.

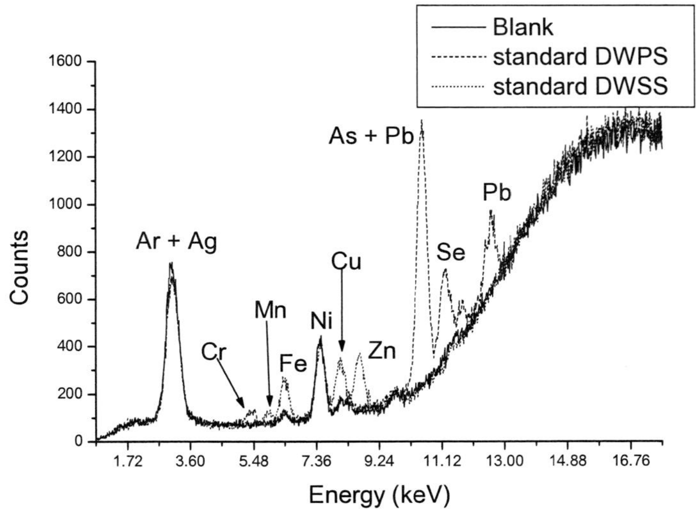

Figure 11. Spectra of the multi-element standard DWPS and DWSS.

As can be noted in all spectra, there is a Ni contamination because of an internal Ni rod that hold the Si PIN crystal[51], so nickel will be an element always present in the spectra. The Ar K line is due to air and Ag L lines are due to the Ag filter.

3.2. Membrane Results

3.2.1. Calibration Curve, Detection and Quantification Limit

Calibration curves are presented in Figure 12 and Figure 13. The values for DL and QL, within 95% confidence level, are shown in Table 7.

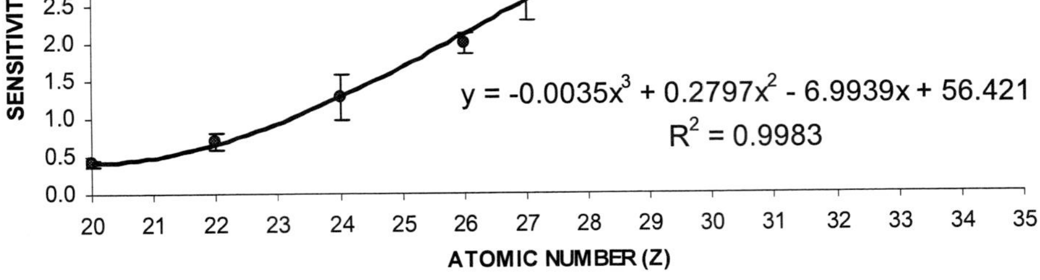

Figure 12. Elemental sensitivity curve for Kα X-rays using MicroMatter standards.

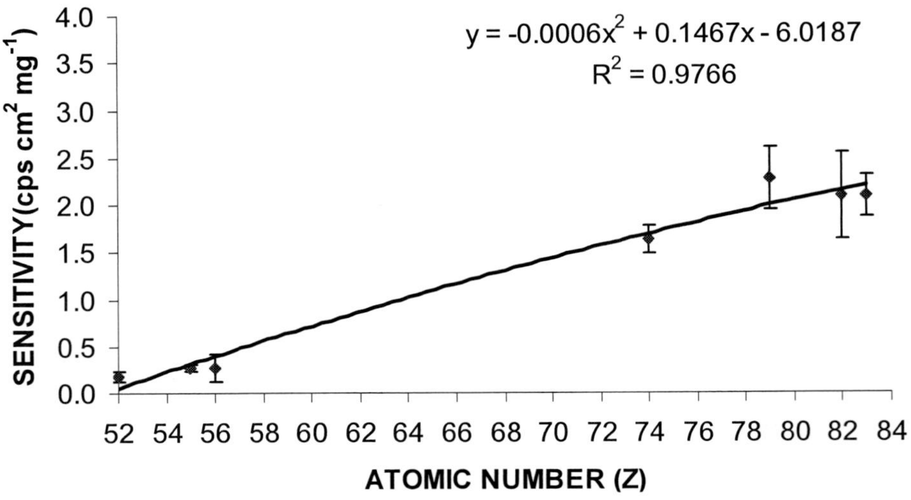

Figure 13. Elemental sensitivity curve for Lα X-rays using MicroMatter standards.

Table 7. Detection and Quantification limit for the MicroMatter standard membranes. Values in μg of the element

Element	DL	QL
Ca	1.32±0.42	4.39±1.41
Ti	1.66±0.47	5.53±1.57
Cr	1.46±0.22	4.85±0.75
Fe	0.69±0.13	2.28±0.45
Co	0.53±0.09	1.76±0.28
Cu	0.50±0.08	1.65±0.26
Zn	0.28±0.02	0.94±0.05
Ga	1.90±0.05	6.34±0.18
As	0.25±0.01	0.84±0.03
Se	0.52±0.01	1.74±0.01
Br	0.36±0.01	1.19±0.03
Te	3.66±1.02	12.22±3.41
Cs	3.27±0.40	10.89±1.34
Ba	2.27±0.14	7.58±0.45
W	0.92±0.01	3.06±0.03
Au	0.80±0.02	2.67±0.06
Pb	0.84±0.01	2.81±0.02
Bi	0.85±0.03	2.84±0.12

3.2.2. APDC Pre-Concentration Methodology Efficiency

To verify APDC chelating efficiency for pre-concentration of metals in water with posterior membrane filtering it was performed two tests using portable EDXRF and flame atomic absorption spectrometry (FAAS) methodologies.

Three multi-element standard solutions were analyzed for APDC efficiency verification. Multi-element standard solution Sigma Aldrich 70002 (at 300 μL/300 mL and 1000 μL/300 mL dilution), multi-element standard solution Sigma Aldrich 70006(at 300 μL/300 mL and 1000 μL/300 mL dilution) and multi-element standard solution DW (at 300 μL/300 mL and 500 μL/300 mL dilution).

Five membranes of each concentration listed above were prepared. For two membranes the percolate was filtered again in another membrane for EDXRF analysis. And for the other three membranes, 50 mL of the percolate was maintained refrigerated in a plastic flask for posterior FAAS analysis.

As for FAAS analysis it is necessary liquid samples, after the EDXRF analysis, three membranes were dissolved in 5 mL of concentrated HNO_3 at 50°C and completed with distilled water up to 25 mL.

Blanks were prepared for all cases.

3.2.2.1. Portable EDXRF Results

The spectra of the percolate membrane, compared with the standard sample and with the blank (distilled water with APDC) are presented in Figure 14.

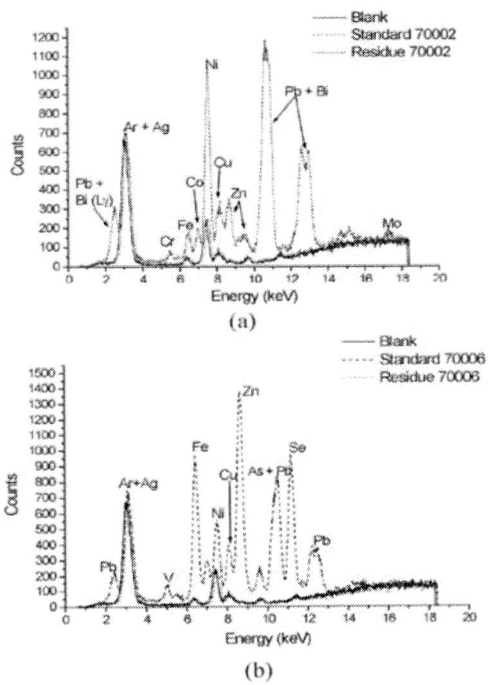

Figure 14. EDXRF spectra comparison of the Standard 70002(a) and Standard 70006 (b) with its respective percolate and blanks spectra.

3.2.2.2. FAAS Results

FAAS results are presented in Table 8. It can be noted that Mn concentration in the residue was always bigger than in the standard, indicating that Mn was not retained in the first filtering and that APDC is not a good chelating agent for this metal in the conditions used. For Fe the measured values are above the certified ones, demonstrating some contamination in the sample preparation or problems with the measurement equipment. For Cu and Zn the APDC was efficient and these elements were not found in the residue. For Cu the recovery was very good, but not so much for Zn. Pb was found in small concentration in the residues, conducting to partial chelating, therefore its recovery was satisfactory.

Table 8. FAAS measurements results for APDC efficiency verification. Values at 95% confidence level, in mg L^{-1}. Recovery values correspond to the percent of the measured concentration related to the certified ones

Sample	Measured Concentration	Concentration in the residue	Certified Value	Recovery
	Mn			
Standard 70006 300μL/300mL	0.0005±0.0001	0.0015±0.0002	0.010±0.001	5%
Standard 70006 1000μL/300mL	0.0026±0.0004	0.0046±0.0007	0.033±0.002	8%
Standard 70002 300μL/300mL	0.0017±0.0003	0.0026±0.0004	0.010±0.001	17%
Standard 70002 1000μL/300mL	0.0019±0.0003	0.0047±0.0007	0.033±0.002	6%
Standard DW 300μL/300mL	0.0028±0.0004	0.0069±0.0010	0.050±0.003	6%
Standard DW 500μL/300mL	0.0030±0.0004	0.0114±0.0017	0.083±0.004	4%
	Fe			
Standard 70006 300μL/300mL	0.176±0.026	0.020±0.003	0.100±0.005	176%
Standard 70006 1000μL/300mL	0.484±0.073	0.023±0.003	0.333±0.017	145%
Standard 70002 300μL/300mL	0.030±0.005	0.030±0.005	0.010±0.001	302%
Standard 70002 1000μL/300mL	0.055±0.008	0.046±0.007	0.033±0.002	166%
Standard DW 300μL/300mL	0.164±0.025	0.056±0.008	0.100±0.005	164%
Standard DW 500μL/300mL	0.246±0.037	0.085±0.013	0.167±0.008	147%
	Cu			
Standard 70006 300μL/300mL	0.022±0.003	0.0	0.020±0.001	109%
Standard 70006 1000μL/300mL	0.079±0.012	0.0	0.067±0.003	118%
Standard 70002 300μL/300mL	0.012±0.002	0.0	0.010±0.001	120%
Standard 70002 1000μL/300mL	0.037±0.006	0.0	0.033±0.002	111%
Standard DW 300μL/300mL	0.057±0.009	0.0	0.050±0.003	114%
Standard DW 500μL/300mL	0.099±0.015	0.0	0.083±0.004	119%
	Zn			
Standard 70006 300μL/300mL	0.059±0.009	0.0	0.100±0.005	59%
Standard 70006 1000μL/300mL	0.227±0.034	0.0	0.333±0.017	68%
Standard 70002 300μL/300mL	0.004±0.001	0.0	0.010±0.001	42%
Standard 70002 1000μL/300mL	0.021±0.003	0.0	0.033±0.002	64%
Standard DW 300μL/300mL	0.035±0.005	0.0	0.050±0.003	71%

Sample	Measured Concentration	Concentration in the residue	Certified Value	Recovery
	Mn			
Standard DW 500μL/300mL	0.069±0.010	0.0	0.083±0.004	83%
	Pb			
Standard 70006 300μL/300mL	0.052±0.008	0.007±0.001	0.012±0.001	434%
Standard 70006 1000μL/300mL	0.171±0.026	0.018±0.003	0.133±0.007	128%
Standard 70002 300μL/300mL	0.098±0.015	0.023±0.004	0.100±0.005	98%
Standard 70002 1000μL/300mL	0.378±0.057	0.030±0.004	0.333±0.017	113%
Standard DW 300μL/300mL	0.112±0.017	0.036±0.005	0.100±0.005	112%
Standard DW 500μL/300mL	0.213±0.032	0.060±0.009	0.167±0.008	128%

3.2.2.3. FAAS X Portable EDXRF Comparison

Comparing the results for the same membranes measured by EDXRF (triplicate measurements of two membranes) and, after dissolution, by FAAS, was obtained a good equivalence between the two techniques, since the values were coincident within the deviation values for the majority of the data. Table 9 and Figure 15 show the results.

Table 9. Results comparison for membranes prepared with multi-element standards and measured by FAAS and portable EDXRF. Values at 95% confidence level, in mg L^{-1}

Sample	Certified concentration	FAAS concentration	EDXRF concentration
	Fe		
Standard 70006 300μL/300mL	0.100±0.005	0.176±0.026	0.143±0.011
Standard 70006 1000μL/300mL	0.333±0.017	0.484±0.073	0.389±0.028
Standard 70002 300μL/300mL	0.010±0.001	0.030±0.005	<0.01[a]
Standard 70002 1000μL/300mL	0.033±0.002	0.055±0.008	0.040±0.005
Standard DW 300μL/300mL	0.100±0.005	0.164±0.025	0.129±0.005
Standard DW 500μL/300mL	0.167±0.008	0.246±0.037	0.200±0.027
	Ni		
Standard 70006 1000μL/300mL	0.067±0.003	0.092±0.014	0.074±0.003
Standard 70002 300μL/300mL	0.050±0.003	0.066±0.010	0.063±0.004
Standard 70002 1000μL/300mL	0.167±0.008	0.203±0.030	0.192±0.017
	Fe		
	Cu		
Standard 70006 300μL/300mL	0.020±0.001	0.022±0.003	0.025±0.002
Standard 70006 1000μL/300mL	0.067±0.003	0.079±0.012	0.071±0.004
Standard 70002 300μL/300mL	0.010±0.001	0.012±0.002	<0.01[a]
Standard 70002 1000μL/300mL	0.033±0.002	0.037±0.006	0.035±0.004

Table 9. (Continued)

Sample	Certified concentration	FAAS concentration	EDXRF concentration
Standard DW 300μL/300mL	0.050±0.003	0.057±0.009	0.062±0.003
Standard DW 500μL/300mL	0.083±0.004	0.099±0.015	0.102±0.017
	Zn		
Standard 70006 300μL/300mL	0.100±0.005	0.059±0.009	0.075±0.007
Standard 70006 1000μL/300mL	0.333±0.017	0.227±0.034	0.330±0.020
Standard 70002 300μL/300mL	0.010±0.001	0.004±0.001	<0.01[a]
Standard 70002 1000μL/300mL	0.033±0.002	0.021±0.003	<0.01[a]
Standard DW 300μL/300mL	0.050±0.003	0.035±0.005	0.041±0.004
Standard DW 500μL/300mL	0.083±0.004	0.069±0.010	0.079±0.009
	Pb		
Standard 70006 300μL/300mL	0.012±0.001	0.052±0.008	<0.02[a]
Standard 70006 1000μL/300mL	0.133±0.007	0.171±0.026	0.052±0.045
Standard 70002 300μL/300mL	0.100±0.005	0.098±0.015	0.087±0.009
Standard 70002 1000μL/300mL	0.333±0.017	0.378±0.057	0.338±0.012
Standard DW 300μL/300mL	0.100±0.005	0.112±0.017	0.105±0.022
Standard DW 500μL/300mL	0.167±0.008	0.213±0.032	0.164±0.043

[a] system quantification limit.

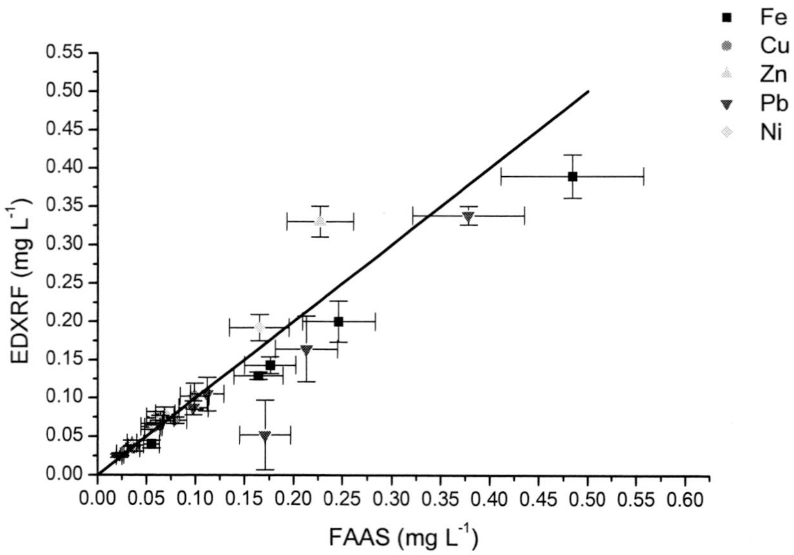

Figure 15. EDXRF and FAAS comparison results showing the equivalence of both methodologies. Values with 95% confidence level.

Table 10. Correction factor for the efficiency of APDC pre-concentration methodology

Element	Correction Factor (AAS)	Correction Factor (EDXRF)
Fe	0.670	0.836
Cu	0.857	0.866
Zn	1.474	1.035
Pb	0.866	0.998

With this analysis it was verified that manganese was not recovered, making it necessary to use other pre-concentration conditions or another chelating agent. For cupper the recovery values were down to 20%. For Fe, Ni, Zn and Pb the chelating was partially efficient. Therefore, in general the concentrations values for both techniques were equivalent considering the deviations. So APDC efficiency could be considered good for these metals.

In order to correct the sample concentration values, graphs of the measured concentrations versus the certified ones were constructed from Table 9 data. The results are presented at Table 10.

These correction factors were applied in all concentrations values of the samples and standards analyzed.

3.2.3. Methodology Validation for Membrane Measurements

The concentration values for the standard reference materials SRM 1640 and SRM 1643e, considering the sensitivity obtained from MicroMatter Standars and the correction factors related to APDC preconcentration efficiency (Table 10), with 95% confidence level, are presented in Table 11.

Table 11. Standard Reference Material results using portable EDXRF equipment. Values in mg L^{-1} within 95% confidence level

Element	Certified Concentration SRM 1640	Measured Concentration SRM 1640	Percent Variation (%)	Certified Concentration SRM 1643e	Measured Concentration SRM 1643e	Percent Variation (%)
V	0.0130±0.0004	0.006±0.002	-56	0.0216±0.0003	0.012±0.004	-46
Fe	0.0343±0.0016	0.047±0.003	36	0.056±0.001	0.110±0.008	96
Co	0.0203±0.0003	0.018±0.001	-13	0.0155±0.0002	0.013±0.003	-14
Ni	0.0274±0.0008	0.033±0.002	22	0.0357±0.0004	0.033±0.010	-7
Cu	0.0852±0.0012	0.069±0.003	-19	0.0130±0.0002	0.011±0.004	-19
Zn	0.0532±0.0011	0.014±0.001	-74	0.045±0.001	0.011±0.001	-76
Se	-	-		0.0068±0.0001	0.006±0.001	-9
Pb	0.0279±0.0001	0.038±0.012	37	0.0112±0.0001	0.027±0.016	141

It was verified that the percent variation related to the certified values were not satisfactory. First because the standard element concentration has magnitude of 10^1 µg L^{-1}, that is around the system quantification limit (Table 7) , making difficult its determination with accuracy. Another reason is that the MicroMatter standards used for calibration, has concentration values of two magnitude orders over the concentration values from the SRMs.

Still with validation purpose were analyzed three multi-element solutions, diluted in different concentrations. Table 12 present these values (equivalent to Table 9 with APDC pre-concentration efficiency correction)

Table 12. Results comparison for membranes prepared with multi-element standards and measured by FAAS and portable EDXRF. Values at 95% confidence level, in mg L^{-1}

	Fe				
Sample	Certified concentration	FAAS concentration	Recovery	EDXRF concentration	Recovery
Standard 70006 300μL/300mL	0.100±0.005	0.118±0.018	118%	0.120±0.009	120%
Standard 70006 1000μL/300mL	0.333±0.017	0.324±0.049	97%	0.325±0.023	98%
Standard 70002 300μL/300mL	0.010±0.001	0.020±0.003	202%	<0.01[a]	0%
Standard 70002 1000μL/300mL	0.033±0.002	0.037±0.006	111%	0.033±0.004	100%
Standard DW 300μL/300mL	0.100±0.005	0.110±0.016	110%	0.108±0.004	108%
Standard DW 500μL/300mL	0.167±0.008	0.165±0.025	97%	0.167±0.023	98%
	Ni				
Sample	Certified concentration	FAAS concentration	Recovery	EDXRF concentration	Recovery
Standard 70006 1000μL/300mL	0.067±0.003	0.074±0.011	111%	0.063±0.003	95%
Standard 70002 300μL/300mL	0.050±0.003	0.053±0.008	106%	0.054±0.004	108%
Standard 70002 1000μL/300mL	0.167±0.008	0.163±0.024	98%	0.164±0.017	98%
	Cu				
Sample	Certified concentration	FAAS concentration	Recovery	EDXRF concentration	Recovery
Standard 70006 300μL/300mL	0.020±0.001	0.018±0.003	94%	0.022±0.002	108%
Standard 70006 1000μL/300mL	0.067±0.003	0.068±0.010	101%	0.061±0.003	92%
Standard 70002 300μL/300mL	0.010±0.001	0.010±0.002	103%	<0.01[a]	
Standard 70002 1000μL/300mL	0.033±0.002	0.032±0.005	95%	0.030±0.003	91%
Standard DW 300μL/300mL	0.050±0.003	0.049±0.007	98%	0.054±0.003	107%
Standard DW 500μL/300mL	0.083±0.004	0.084±0.013	102%	0.088±0.015	106%
	Zn				
Sample	Certified concentration	FAAS concentration	Recovery	EDXRF concentration	Recovery
Standard 70006 300μL/300mL	0.100±0.005	0.088±0.013	88%	0.078±0.007	78%
Standard 70006 1000μL/300mL	0.333±0.017	0.335±0.050	100%	0.342±0.021	102%
Standard 70002 300μL/300mL	0.010±0.001	0.006±0.001	61%	<0.01[a]	
Standard 70002 1000μL/300mL	0.033±0.002	0.031±0.005	94%	<0.01[a]	
Standard DW 300μL/300mL	0.050±0.003	0.052±0.008	104%	0.042±0.004	85%
Standard DW 500μL/300mL	0.083±0.004	0.101±0.015	122%	0.082±0.009	99%

	Fe				
Sample	Certified concentration	FAAS concentration	Recovery	EDXRF concentration	Recovery
	Pb				
Sample	Certified concentration	FAAS concentration	Recovery	EDXRF concentration	Recovery
Standard 70006 300µL/300mL	0.012±0.001	0.045±0.007	376%	<0.01[a]	
Standard 70006 1000µL/300mL	0.133±0.007	0.148±0.022	111%	0.052±0.045	39%
Standard 70002 300µL/300mL	0.100±0.005	0.085±0.013	85%	0.087±0.009	87%
Standard 70002 1000µL/300mL	0.333±0.017	0.327±0.049	98%	0.337±0.012	101%
Standard DW 300µL/300mL	0.100±0.005	0.097±0.014	97%	0.105±0.022	105%
Standard DW 500µL/300mL	0.167±0.008	0.184±0.028	111%	0.164±0.043	98%

[a] system quantification limit

Analyzing these results is verified that there is a good concordance between the values from the two techniques, specially for Cu. In this case the recovery is better than the SRM results, due to the higher concentration values, below the quantification limits.

One of the factors that influenced the elements quantification was the matrix complexity, where interference and enhancement effects from the several elements present on the standards were relevant, since the solutions had between 12 and 29 elements in different concentrations.

3.3. Environmental Samples Results

3.3.1 First In Situ Measurements

The first in situ metal measurement, using the described portable EDXRF system, was realized in the Igapó Lake IV, near Baroré stream, at Londrina, Paraná, Brazil. It were analyzed samples from Igapó Lake IV and Baroré stream in a dry period. Figure 16 presents photographs of the sampling places.

Figure 16. Igapó Lake and Baroré stream.

Ten milliliters of *in natura* water were placed in a specific recipient (Chemplex Inc.) with Mylar film for direct EDXRF analysis.

From each samples 3 portions of 300 mL were filtered in field, using 0.45 μm cellulose esters membranes, for suspended particulate matter (SPM) retention. After this, the total volume was divided in three beakers, with 300 mL each one, for pre-concentration. A blank membrane was also prepared. So, it were prepared and measured six membranes of each point and the blank. The membranes were dried at sun for 30 min.

3.3.1.1. In natura results

In natura results are presented in Figure 17. In this case, it was not identified difference between samples and blank, evidencing that the elements are below the detection limit for liquid samples, presented at Table 5.

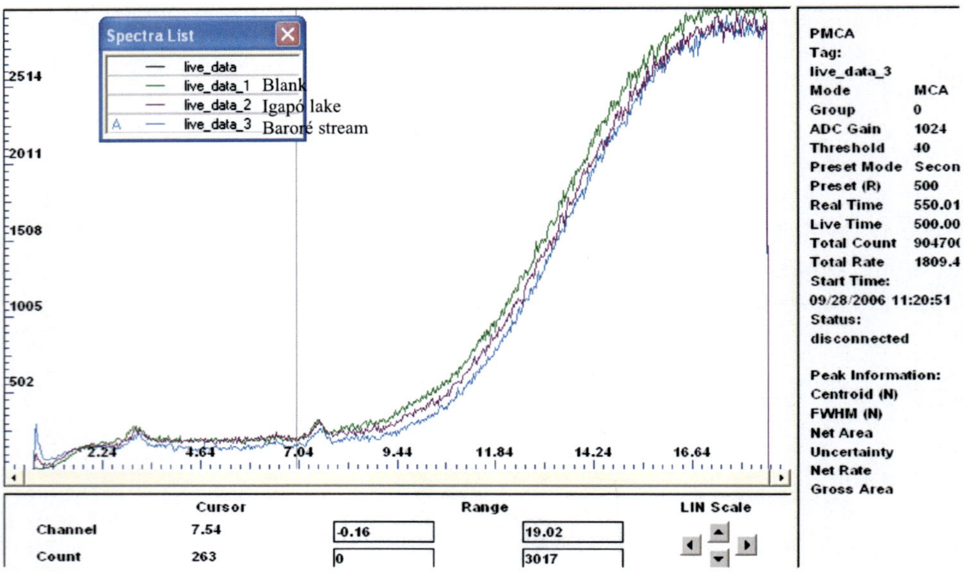

Figure 17. Igapó Lake IV, Baroré stream and blank comparison for in natura water measured in situ.

Table 13. Detection limit results, in mg L^{-1}, for *in natura* samples measured in the laboratory

Element	Igapó IV	Baroré stream
Cr	<20.5	<23.9
Mn	<12.4	<14.8
Fe	<8.0	<9.6
Co	<4.5	<5.4
Ni	<4.6	<5.5
Cu	<3.3	<3.9
Zn	<2.6	<3.0
Se	<2.8	<3.1
Hg	<5.3	<5.9
Pb	<5.7	<6.4

3.3.1.2. Membrane Results

For the samples analyzed in membranes, equivalent results, for the sample filtered on field and dried at sun for 30 min and the same membrane measured at laboratory after 24h from its preparation, was obtained. One of the results are at Figure 18. For APDC pre-concentrated membranes and for Igapó Lake IV membranes this behavior was reproducible.

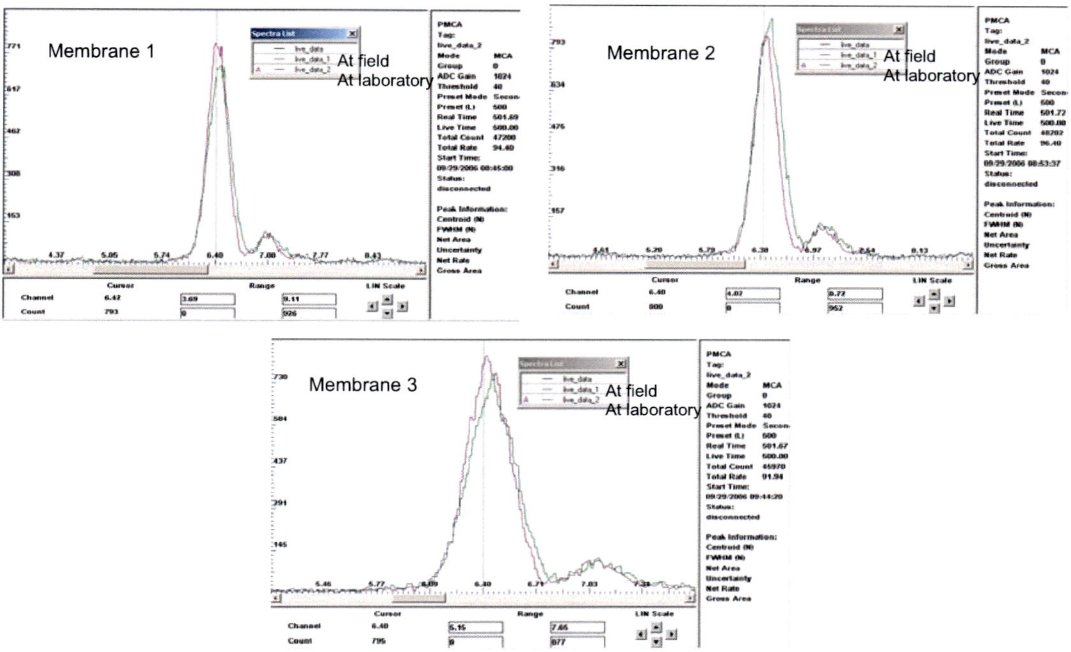

Figure 18. Comparison between the same membranes measured in situ and at the laboratory. Sample of suspended particulate matter retention from Baroré Stream.

The concentration values with its respective standard deviation are at Table 14 and Table 15. These results are the average of three measurements in each one of the three membranes from both samples.

The higher concentration verified for Fe in the suspended particulate matter could be related to soil particles. According to CONAMA (Brazilian National Environmental Commission) [49], it was not identified contamination in the samples, since the acceptable value is 0.3 mg L^{-1} for dissolved iron.

Table 14. Concentration values and standard deviation, in mg L^{-1}, for Baroré stream samples

Element	Water SPM	APDC pre-concentration
V	<0.045	<0.037
Cr	<0.003	<0.003
Mn	<0.002	<0.002
Fe	0.37±0.01	0.029±0.001
Co	<0.001	<0.001
Ni	<0.001	<0.001
Cu	<0.001	<0.001
Zn	<0.001	<0.001
As	<0.001	<0.001
Se	<0.001	<0.002
Hg	<0.002	<0.002
Pb	<0.002	<0.003

Table 15. Concentration values and standard deviation, in mg L^{-1}, for Igapó Lake IV samples

Element	Water SPM	APDC pre-concentration
Ca	0.034±0.003	
Ti	0.032±0.003	
V	<0.040	<0.039
Cr	<0.003	<0.003
Mn	<0.019	<0.002
Fe	0.70±0.02	0.024±0.002
Co	<0.001	<0.001
Ni	<0.001	<0.001
Cu	<0.001	<0.001
Zn	<0.001	<0.001
As	<0.001	<0.001
Se	<0.001	<0.002
Hg	<0.002	<0.002
Pb	<0.002	<0.003

3.3.2. Second In Situ Measurements

The second *in situ* measurement was realized in the Capivara stream, near a car battery plant, at Londrina, PR, Brazil. The stream is around 500 m from the highway and the rainwater flow into the stream.

It was analyzed three points: one in the Capivara stream, another in an well near the stream and one in the exit of the pipe which transport the rainwater from the highway. This sampling was performed in rainy period.

Were analyzed *in natura* water from the tree points and prepared 6 membranes of each point, three of suspended particulate matter and three applying the APDC pre-concentration methodology.

3.3.2.1. In Natura Results

In natura results presented divergences between *in situ* and laboratory measurements, as exemplify Figure 19. It was verified a difference in Fe concentration. As the sample was analyzed in rainy period, it was noticed visually an excess of suspended particulate matter, basically soil. In the field measurement these particles were suspended in the sample, since it was the first sample to be measured. After 24h at rest, the soil deposited in the film surface, resulting in higher concentration, as the effective measurement layer is around 1 mm deep in water.

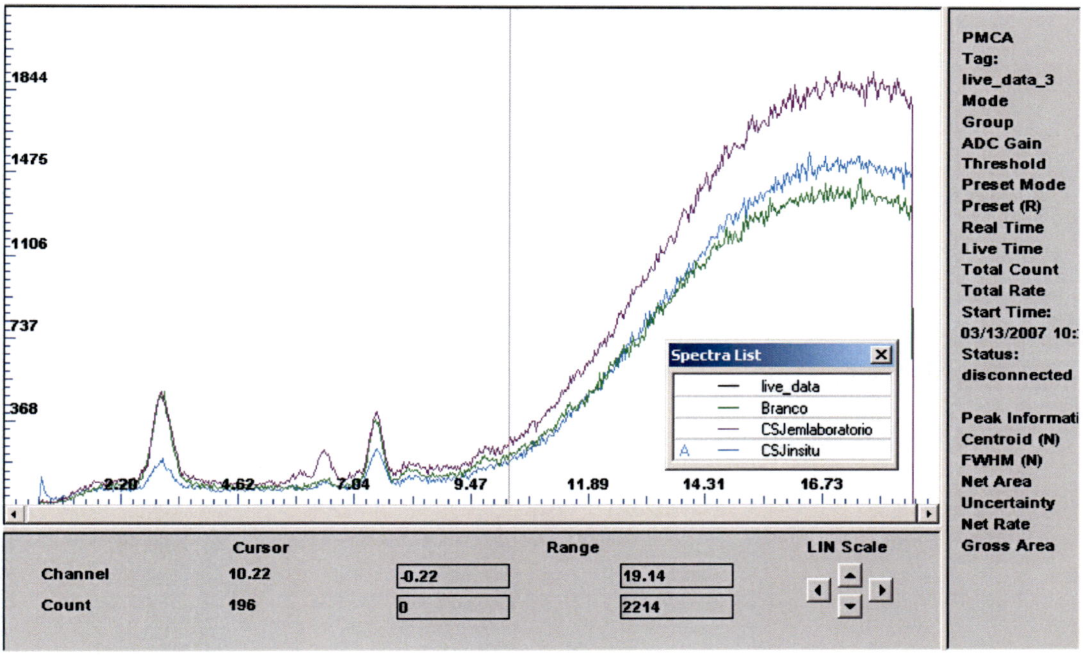

Figure 19. Blank, in situ and laboratory measurements comparison from Capivara stream in natura sample.

So, the field result for this particular situation migh be under estimated. Table 16 present the results for laboratory analysis.

3.3.2.2. Membrane Results

Table 17 shows the results comparison between *in situ* and laboratory measurements at Capivara stream. Table 18 presents the concentration values for this sampling point.

Table 16. Concentration values and standard deviation, in mg L^{-1}, for *in natura* samples measured in the laboratory

Element	Capivara Stream	Well	Pipe exit
Cr	<23.7	<25.6	<26.7
Mn	<14.0	<15.7	<16.3
Fe	94±5	595±8	80±6
Co	<5.1	<5.6	<5.8
Ni	<5.2	<5.6	<6.0
Cu	<3.6	<3.9	<4.1
Zn	<2.8	<2.9	<3.2
Se	<2.6	<2.6	<2.6
Hg	<5.2	<5.3	<5.5
Pb	<5.4	<5.4	<5.7

Table 17. Concentration values and standard deviation comparison , in mg L^{-1}, for Capivara Stream *in situ* and laboratory analysis

	Water SPM			APDC pre-concentration	
Element	in situ Concentration	Laboratory Concentration	Element	in situ Concentration	Laboratory Concentration
Ti	0.045±0.002	0.065±0.013	Ti	a	a
Fe	1.119±0.003	1.24±0.10	Fe	0.069±0.016	0.067±0.011
Zn	0.014±0.001	0.014±0.002	Zn	0.017±0.003	0.018±0.002

a – value below the system quantification limit.

Table 18. Concentration values and standard deviation comparison, in mg L-1, for Capivara Stream samples

Element	SPM concentration	APDC pre-concentration
Ti	0.065±0.013	<0.009
V	<0.007	<0.008
Cr	<0.005	<0.006
Mn	<0.003	<0.004
Fe	1.24±0.10	0.067±0.011
Co	<0.002	<0.003
Ni	<0.002	<0.002
Cu	<0.001	<0.002
Zn	0.014±0.002	0.018±0.002
As	<0.001	<0.002
Se	<0.002	<0.002
Hg	<0.002	<0.003
Pb	<0.002	<0.003

Table 19 shows the results comparison between *in situ* and laboratory measurements at the water well and Table 20 presents the concentration values with the detection limits.

Table 19. Concentration values and standard deviation comparison , in mg L^{-1}, for the water well *in situ* and laboratory analysis

	Water SPM			APDC pre-concentration	
Element	**in situ Concentration**	**Laboratory Concentration**	**Element**	**in situ Concentration**	**Laboratory Concentration**
Ti	0.19±0.12	0.23±0.16	Ti	A	a
Fe	1.52±0.85	1.7±1.1	Fe	A	a
Zn	a	a	Zn	A	a

a – value below the system quantification limit.

Table 20. Concentration values and standard deviation comparison, in mg L^{-1}, for water well samples

Element	SPM concentration	APDC pre-concentration
Ti	0.23±0.16	<0.011
V	<0.006	<0.008
Cr	<0.004	<0.006
Mn	<0.003	<0.004
Fe	1.7±1.1	<0.004
Co	<0.001	<0.003
Ni	<0.001	<0.002
Cu	<0.001	<0.002
Zn	<0.001	<0.002
As	<0.001	<0.002
Se	<0.001	<0.002
Hg	<0.001	<0.003
Pb	<0.002	<0.003

The measurements results for the pipe exit are in Table 21. It was identified the presence of Pb in the suspended particulate matter at higher levels than established in the legislation[49], which is 0,01 mg L^{-1}.

In general way, it was verified the presence of Ti and Fe coming from soil, which due to rain, concentrates in the streams. In the pipe exit sample it was identified Pb, which prove its antrophic origin in the environment.

3.3.3. Monitoring Results from Igapó Lakes

The focus of this monitoring study is to present results for metals in Igapó lakes water, at Londrina city. The Environment Institute of Paraná (EIP), that is the supervisory agency responsible for the region, makes the tracking twice a year in the lakes, collecting water samples that are analyzed by AAS as laboratory routine.

Table 21. Concentration values and standard deviation comparison, in mg L^{-1}, for pipe exit samples

Element	SPM concentration	APDC concentration
Ti	0.098±0.027	0.014±0.002
V	<0.010	<0.008
Cr	<0.006	<0.006
Mn	<0.005	<0.004
Fe	1.82±0.14	0.078±0.006
Co	<0.003	<0.003
Ni	<0.002	<0.002
Cu	<0.002	<0.002
Zn	<0.001	<0.002
As	<0.001	<0.002
Se	<0.001	<0.002
Hg	<0.002	<0.003
Pb	0.103±0.008	<0.004

Besides the monitoring results, another objective was to evaluate the potential of the portable EDXRF system for the determination of metal total concentration as an alternative analytical methodology. Then, researchers from the State University of Londrina accompanied the team of EIP in the samples collection, in order to evaluate the same points and the same samples by both techniques.

Ten sampling points from Igapó Lake and its effluents were monitored. The samples were collected in the summer of 2006 and 2007 and in the winter of 2006.

Two liters of superficial water were collected in polyethylene bottles from each one of the 10 sampling points. One liter was used for EDXRF analysis (in the university laboratories in Londrina) and the other for AAS (in the EIP laboratories). The samples were carried out to the laboratory and prepared in the same day for EDXRF analysis following the methodology presented on item 2.3. For AAS determination the samples were preserved with HNO_3 until its analysis. The preparation consists on the digestion of 100 mL of the sample with 5 mL of HNO_3 until almost total evaporation. Then, more 5 mL of HNO_3 is added, the solution is heated and 1 mL of it is completed with distilled water up to 25 mL.

The obtained values with EDXRF methodology are presented in Table 22. It was possible to quantify Ca, Ti, Mn, Fe, Ni, Cu and Zn. For Capivara stream, Água Fresca Stream, Rubi stream and Cambé river it was measured a small Zn contamination in the 2007 winter sampling, since the limit established by Brazilian National Environmental Commission, (CONAMA)[49], is 0.18 mg L^{-1} for this element. The concentrations for Mn and Ni were below the maximum permitted level of 0.1 mg L^{-1} and 0.025 mg L^{-1}, respectively.

Table 22. Total concentration and standard deviation values, in mg L^{-1}, for Igapó Lake samples and its effluents

IGAPÓ LAKE I

Element	March/06	June/06	May/07
Ca	<0.009	0.043±0.007	0.038±0.012
Ti	0.029±0.003	0.121±0.009	0.106
V	<0.004	<0.006	<0.004
Cr	<0.003	<0.004	<0.003
Mn	<0.002	0.052	<0.002
Fe	0.32±0.04	0.82±0.04	0.64±0.02
Co	<0.001	<0.001	<0.001
Ni	0.010±0.003	<0.001	<0.001
Cu	<0.001	<0.001	0.003±0.001
Zn	<0.001	0.003±0.001	0.071±0.010
As	<0.001	<0.001	<0.001
Se	<0.001	<0.001	<0.001
Hg	<0.001	<0.002	<0.002
Pb	<0.003	<0.003	<0.002

IGAPÓ LAKE II

Element	March/06	June/06	May/07
Ca	<0.009	0.032±0.005	<0.010
Ti	0.014±0.003	0.051±0.007	0.068±0.036
V	<0.004	<0.005	<0.005
Cr	<0.003	<0.003	<0.003
Mn	0.029±0.002	0.043±0.003	0.012±0.002
Fe	0.25±0.06	0.60±0.04	0.56±0.11
Co	<0.001	<0.001	<0.001
Ni	0.012±0.001	<0.001	<0.001
Cu	<0.001	<0.001	0.003±0.001
Zn	<0.001	<0.001	0.077±0.002
As	<0.001	<0.001	<0.001
Se	<0.001	<0.001	<0.001
Hg	<0.001	<0.002	<0.002
Pb	<0.003	<0.003	<0.002

IGAPÓ LAKE III

Element	March/06	June/06	May/07
Ca	<0.009	<0.010	0.020±0.012
Ti	0.023±0.004	0.033±0.004	0.095±0.007
V	<0.003	<0.004	<0.005
Cr	<0.002	<0.003	<0.004
Mn	0.039±0.003	0.008±0.001	0.016±0.002
Fe	0.37±0.03	0.57±0.04	0.92±0.09
Co	<0.001	<0.001	<0.001
Ni	0.011±0.003	<0.001	<0.001
Cu	<0.001	<0.001	0.005±0.001

IGAPÓ LAKE III (Continued)

Element	March/06	June/06	May/07
Zn	<0.001	<0.001	0.102±0.004
As	<0.001	<0.001	<0.001
Se	<0.001	<0.001	<0.001
Hg	<0.001	<0.002	<0.002
Pb	<0.002	<0.002	<0.002

IGAPÓ LAKE IV

Element	March/06	June/06	May/07
Ca	<0.009	<0.009	<0.010
Ti	0.027±0.015	0.023±0.002	0.024±0.005
V	<0.003	<0.004	<0.004
Cr	<0.002	<0.003	<0.003
Mn	<0.001	<0.002	0.005±0.001
Fe	0.66±0.14	0.65±0.12	0.62±0.04
Co	<0.001	<0.001	<0.001
Ni	0.010±0.003	<0.001	<0.001
Cu	<0.001	<0.001	0.009±0.001
Zn	<0.001	<0.001	0.135±0.011
As	<0.001	<0.001	<0.001
Se	<0.001	<0.001	<0.001
Hg	<0.001	<0.002	<0.002
Pb	<0.002	<0.002	<0.002

CAPIVARA STREAM

Element	March/06	June/06	May/07
Ca	<0.011	0.038±0.004	<0.012
Ti	0.043±0.001	0.136±0.002	0.112±0.041
V	<0.002	<0.006	<0.004
Cr	<0.001	<0.004	<0.003
Mn	<0.001	0.044±0.003	<0.002
Fe	0.28±0.02	0.89±0.04	0.71±0.10
Co	<0.001	<0.001	<0.001
Ni	0.009±0.005	<0.001	<0.001
Cu	<0.001	<0.001	0.004±0.001
Zn	<0.001	<0.001	0.19±0.02
As	<0.001	<0.001	<0.001
Se	<0.001	<0.001	<0.001
Hg	<0.001	<0.002	<0.002
Pb	<0.003	<0.003	<0.002

LEME STREAM

Element	March/06	June/06	May/07
Ca	<0.009	<0.010	0.020±0.011
Ti	0.015±0.003	<0.005	0.043±0.009
V	<0.001	<0.004	<0.003
Cr	<0.001	<0.002	<0.005

LEME STREAM (Continued)

Element	March/06	June/06	May/07
Mn	<0.001	<0.003	<0.002
Fe	0.18±0.02	0.12±0.01	0.36±0.06
Co	<0.001	<0.001	<0.001
Ni	0.005±0.003	<0.001	<0.001
Cu	<0.001	<0.001	0.003±0.001
Zn	<0.001	<0.001	0.055±0.005
As	<0.001	<0.001	<0.001
Se	<0.001	<0.001	<0.001
Hg	<0.001	<0.002	<0.002
Pb	<0.003	<0.003	<0.002

ÁGUA FRESCA STREAM

Element	March/06	June/06	May/07
Ca	<0.010	0.021±0.002	0.017±0.009
Ti	0.006±0.002	0.016±0.004	0.070±0.023
V	<0.001	<0.004	<0.005
Cr	<0.001	<0.003	<0.003
Mn	0.005±0.002	0.042±0.015	0.010±0.002
Fe	0.17±0.05	0.31±0.04	0.63±0.09
Co	<0.001	<0.001	<0.001
Ni	0.006±0.003	<0.001	<0.001
Cu	<0.001	<0.001	0.004±0.001
Zn	<0.001	<0.001	0.301±0.019
As	<0.001	<0.001	<0.001
Se	<0.001	<0.001	<0.001
Hg	<0.001	<0.002	<0.002
Pb	<0.003	<0.003	<0.003

RUBI STREAM

Element	March/06	June/06	May/07
Ca	<0.009	<0.009	<0.009
Ti	0.000	0.014±0.005	0.033±0.003
V	0.000	<0.004	<0.004
Cr	<0.02	<0.003	<0.003
Mn	0.000	<0.002	<0.002
Fe	0.25±0.01	0.18±0.03	0.24±0.01
Co	<0.001	<0.001	<0.001
Ni	<0.001	<0.001	<0.001
Cu	<0.001	<0.001	0.004±0.001
Zn	<0.001	<0.001	0.295±0.010
As	<0.001	<0.001	<0.001
Se	<0.001	<0.001	<0.001
Hg	<0.0001	<0.002	<0.002
Pb	<0.002	<0.003	<0.002

BARORÉ STREAM

Element	March/06	June/06	May/07
Ca	<0.010	<0.010	<0.009
Ti	0.023±0.011	<0.004	<0.004
V	<0.001	<0.003	<0.003
Cr	<0.001	<0.002	<0.002
Mn	<0.001	<0.002	<0.002
Fe	0.46±0.06	0.33±0.02	0.35±0.02
Co	<0.001	<0.001	<0.001
Ni	<0.001	<0.001	<0.001
Cu	<0.001	<0.001	0.004±0.001
Zn	<0.001	<0.001	0.167±0.008
As	<0.001	<0.001	<0.001
Se	<0.001	<0.001	<0.001
Hg	<0.001	<0.002	<0.002
Pb	<0.002	<0.002	<0.002

CAMBÉ RIVER

Element	March/06	June/06	May/07
Ca	0.000	<0.010	<0.019
Ti	0.006±0.001	0.023±0.006	0.289±0.086
V	<0.001	<0.004	<0.008
Cr	<0.001	<0.003	<0.004
Mn	<0.001	<0.002	<0.002
Fe	0.27±0.02	0.72±0.04	2.5±0.3
Co	<0.001	<0.001	<0.002
Ni	0.005±0.003	<0.001	<0.001
Cu	<0.001	<0.001	0.006±0.001
Zn	<0.001	<0.001	0.22±0.01
As	<0.001	<0.001	<0.001
Se	<0.001	<0.001	<0.001
Hg	<0.001	<0.002	<0.002
Pb	<0.002	<0.003	<0.002

For Fe and Cu, the concentration limits in the legislation are referred to the dissolved metal fraction, as 0.3 and 0.009 mg L^{-1}, respectively[49]. For Cu there was not identified contamination. In Fe case, the dissolved metal quantity ranged from 0.003±0.001 to 0.59±0.03. The contaminated points are identified at Figure 20, from which can be inferred the dissolved Fe dynamic in the evaluated points. Cambé river and Baroré stream encounter on the Igapó lake IV, which in turn supplies the other lakes. It can be noted that Fe dissolved concentration reduces from Igapó lake IV to Igapó lake I, suggesting that the iron is being diluted in the pathway.

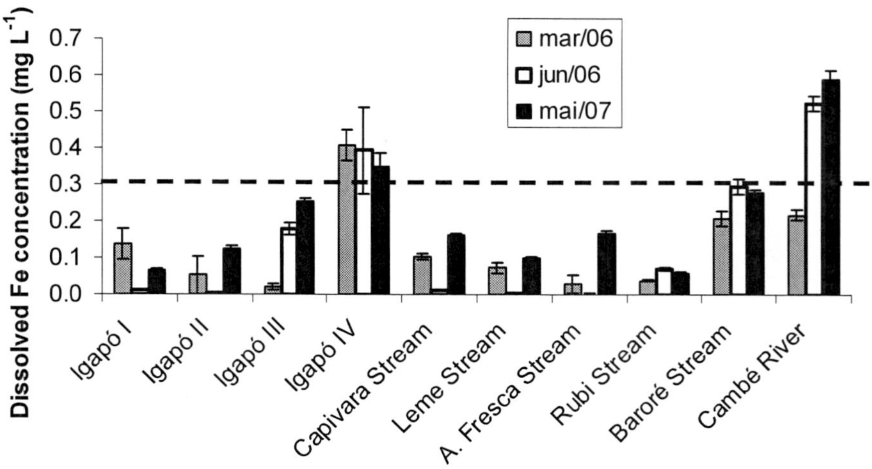

Figure 20. Concentration values for dissolved Fe in the Igapó Lake samples measured by portable EDXRF. The dashed line indicates the CONAMA maximum established value.

Environmental Institute of Paraná evaluates as a routine, the presence of Cr, Fe, Cd, Hg and Pb in same samples using AAS methodology. From this analysis, only Fe values were above the detection limits of the used system. So, Fe concentration was compared for both techniques. Figure 21 presents this comparison, showing that exist a good correlation between the results. The used AAS methodology determined total Fe concentration, however, CONAMA legislation consider contamination only for dissolved Fe, so this result can be considered inconclusive.

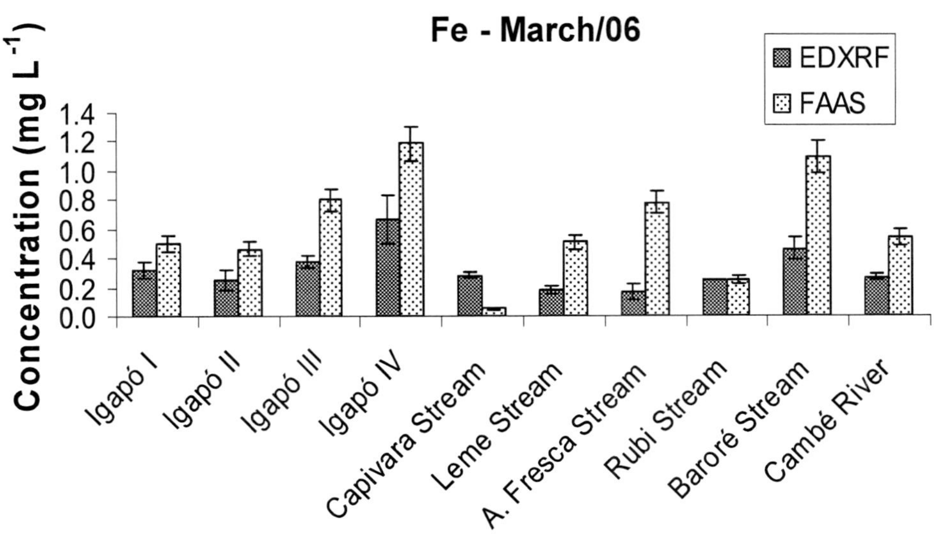

Figure 21. (Continued on next page.)

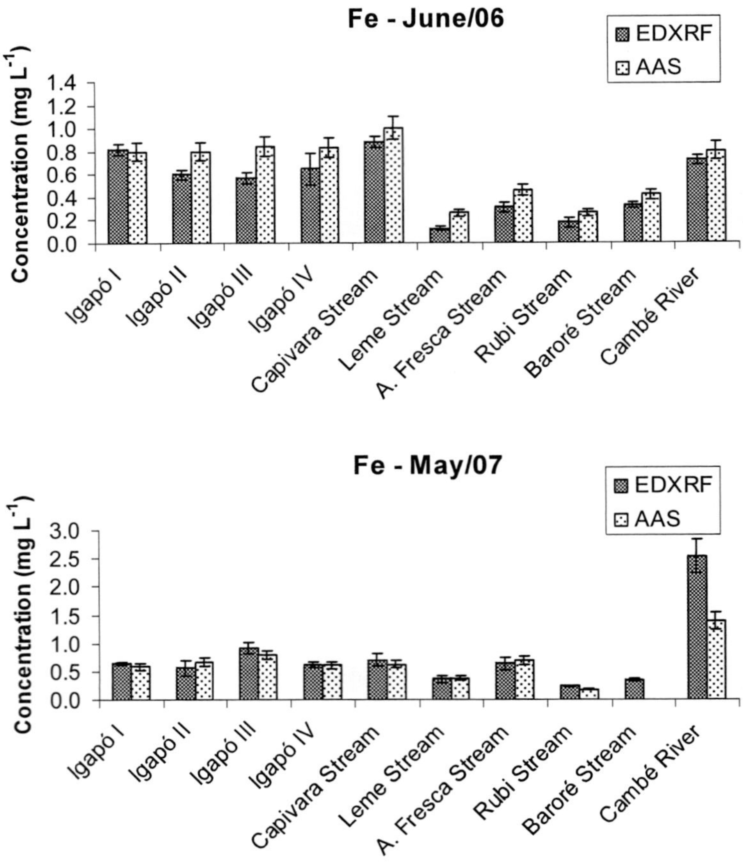

Figure 21. Comparison between EDXRF and AAS results for total Fe concentration values.

The monitoring result showed that there were no heavy metals in the water during the period. The presence of Ca, Ti, Mn, Fe, Ni, Cu and Zn was identified only by EDXRF methodology, due to its multi-element analysis facility, and not by AAS. The pre-concentration methodology enabled the quantification of the dissolved and non dissolved metals fraction without laborious steps.

The results of the two methodologies generated more detailed conclusions about the water conditions in the analyzed points, complementing the routinely analysis performed.

4. Conclusion

The performance of the equipment was very good and the results are satisfactory for *in situ* measurements employing a portable instrument, in which standard deviation around 30% are accepted (according to Report IAEA-TECDOC-1456, 2005) [50]. In this work, standard deviation within 95% confidence, for Fe and Ti, which were found in higher concentrations, ranged from 3% to 12%. For the concentration values around the system quantification limit, the deviation varied from 10% to 30%.

The analysis and quantification methodology presented good results, for in natura and membrane samples, considering the validation results for the SRMs and the different standards.

APDC pre-concentration methodology can be considered efficient for multi-element analysis of metals in water, especially for Fe, Cu, Zn, Se and Pb, for which were obtained recovery values with deviation below 20%.

In situ measurements is viable. In 4 h of work at field it was possible to prepare, measure and analyze around 14 membranes, presenting a partial report of the metal concentration in the sampling place. It was also verified that *in situ* and laboratory measurements conduct to equivalent results, considering the deviations, proving the robustness of the system in different use conditions.

The detection limits for *in natura* samples are above the national and international established values. In other hand, the analysis is fast and does not demand sample preparation. It is indicated for industries effluents or places with high pollution level.

Considering membrane measurements, the system detection limits are below the maximum concentration levels allowed, satisfying international and national legislation for potable water. The quantification limit is in the order of 0.01 mg L^{-1} for almost all elements, making viable the methodology application for water analysis.

Portable EDXRF showed to be a functional tool for environmental analysis, due to its fast and efficient way of analysis, with potential to replace analytical techniques more expensive and lengthy. The data obtained on field, allow to a inspection team, to take fundamental decisions related to the analyzed place. Besides this, the methodology is very useful for mapping regions and identification of contaminated areas.

ACKNOWLEDGMENTS

We are grateful to Fundação Araucária de Apoio ao Desenvolvimento Científico e Tecnológico do Paraná (009/2005-5402) and to the Conselho Nacional de Desenvolvimento Tecnológico, CNPq (470662/2004-2), for the financial support. We are also grateful to the Environmental Institute of Paraná staff, specially Leda Neiva Dias and Luis Zaransky for the collaboration in the sampling and in the discussion about FAAS results and methodology.

APPENDIX I – TESTS FOR GEOMETRY MEASUREMENTS DEFINITION

The objective of this appendix is to study the sensitivity variation for a portable excitation-detection system, employing a Si-PIN detector and a mini X ray tube, under different conditions. It was tested the use of Al, Ag or Pb collimator in the detector, the use of Ag or Mo filter in the X-ray tube and different values of high voltage and current.

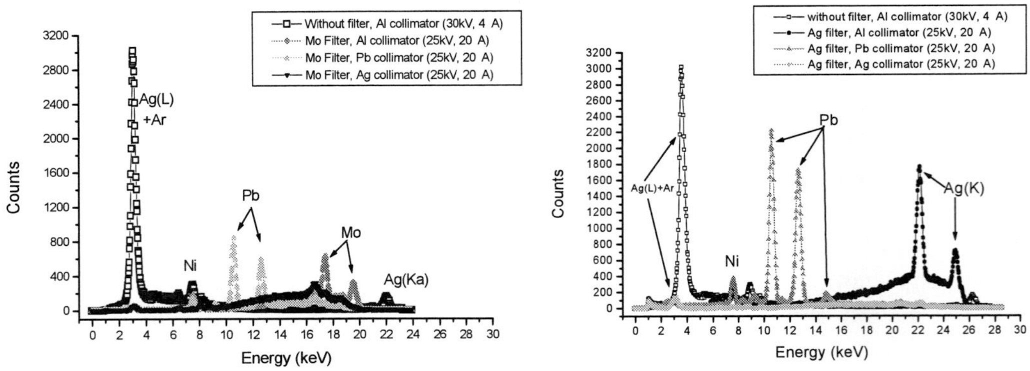

Figure 22. Blank spectra obtained with different geometry and excitation conditions.

I.1. Tests with Different Collimator and Filters

It was tested the following configurations: Al and Pb collimator without filter in the X-ray tube, Al, Ag and Pb collimator with Mo filter in the tube and Ag collimator with Ag filter.

Mo filter is made of 1 mm thickness resin. Ag filter are metallic with 50 μm and 10 μm thickness. Collimators made of Al, Pb and Ag have 3mm diameter aperture.

Individual thin film certified standards of K, Ca, Sc, Ti, Mn, Fe, Cu, Zn, Pb and a blank deposited in polycarbonate membranes from MicroMatter Inc were measured. Were also measured, for validation purpose, two certified reference material from NIST, SRM1832 and SRM1833. The measuring time was 200 s for each standard.

Comparative measurements of the blank standard spectra from MicroMatter, using different collimators, are shown at Figure 22. As can be seen there is a Ni contamination due to an internal rod that hold the Si PIN crystal[1], so that nickel will be an element always present in the spectra.

Even qualitatively, it is possible to verify that Ag collimator with Ag filter presents the smaller background. This result was quantitatively confirmed comparing the results at Table 23.

[1] Kump P, Necemer M, Rupnik Z. Development of the quantification procedures for in situ XRF analysis. IAEA/AL/130 REPORT, Vienna, 12-16 March, Attachment 5.10 - SLO-29042, 2001.

Table 23. Concentration results with the different tested conditions. In the last column, the marked values are related to the SRM1833 concentrations (C) and the others to the SRM1832

	Mo filter Pb Collimator 25 kV 40 μA	Mo filter Pb Collimator 25 kV 20 μ A	Mo filter Al Collimator 25 kV 40 μ A	Mo filter Al Collimator 25 kV 20 μ A	Mo filter Al Collimator 25 kV 10 μ A	Certified Value
Element (Atomic Num.)	C (μ g.cm^{-2})	C (μ g.cm^{-2})	C (μ g.cm^{-2})	C (μ g.cm^{-2})	C (μ g.cm^{-2})	C (μ g.cm^{-2})
K(19)	11,02	14,50	10,51	11,14	30,48	17,4 (1,6)

	Mo filter Pb Collimator	Mo filter Pb Collimator	Mo filter Al Collimator	Mo filter Al Collimator	Mo filter Al Collimator	Certified
Ca(20)	16,53	17,67	16,68	18,59	51,80	19,99 (1,30)
Ti(22)	11,07	11,10	9,91	10,06	12,20	12,9 (1,8)
V(23)	3,96	4,33	4,38	4,47	14,74	4,5 (0,5)
Mn(25)	4,69	5,03	5,58	5,37	31,47	4,5 (0,5)
Fe(26)	12,73	13,11	12,02	12,51	31,12	14,3 (0,5)
Co(27)	1,10	1,00	1,00	1,02	3,12	0,99 (0,1)
Cu(29)	3,20	2,58	2,54	2,43	7,43	2,52 (0,16)
Zn(30)	3,44	3,82	3,02	3,05	8,64	3,8 (0,2)
Pb(82)			12,79	12,65	39,98	15,9 (0,9)

	Mo filter Ag Collimator 25 kV 40 μ A	Mo filter Ag Collimator 25 kV 20 μ A	Mo filter Ag Collimator 25 kV 10 μ A	Ag-50 filter Ag Collimator 28 kV 20 μ A	Ag-10 Filter Ag Collimator 28 kV 20 μ A	Certified Value
Elemento (N. Atômico)	C (μ g.cm^{-2})	C (μ g.cm^{-2})	C (μ g.cm^{-2})	C (μ g.cm^{-2})	C (μ g.cm^{-2})	C (μ g.cm^{-2})
K(19)	6,24	9,74	22,64			17,4 (1,6)
Ca(20)	16,77	17,83	15,15	15,25	0,64	19,99 (1,30)
Ti(22)	11,40	10,98	11,90	9,38	0,36	12,9 (1,8)
V(23)	4,73	4,63	4,69	4,68	6,03	4,5 (0,5)
Mn(25)	5,17	5,31	5,38	3,65	0,19	4,5 (0,5)
Fe(26)	14,86	15,93	14,95	10,97	0,57	14,3 (0,5)
Co(27)	1,38	1,16	1,12	1,12	1,05	0,99 (0,1)
Cu(29)	3,01	2,69	3,02	2,03	0,50	2,52 (0,16)
Zn(30)	3,84	3,83	3,73	2,81	0,40	3,8 (0,2)
Pb(82)	15,47	14,67	14,57	8,83	4,07	15,9 (0,9)

I.2. Tests with Different Ag Filter Thickness

To verify the best Ag filter thickness, it was irradiated a Mylar film, using Ag collimator and 7 Ag filter of different thickness (20, 30, 50, 70, 100, 150 and 200 μm).

It was proved that the 20 and 30 μm filters do not stop the blank beam, as show Figure 23. It is also noted that the background is reduced beginning with the 70 μm filter, but with loss of beam intensity.

To fundament the choice, multi-element standards with: Cr, Fe, Co, Zn and Pb were prepared. The peak to background ratio and the percent deviation in the net peak area were calculated and the results are presented in Table 24.

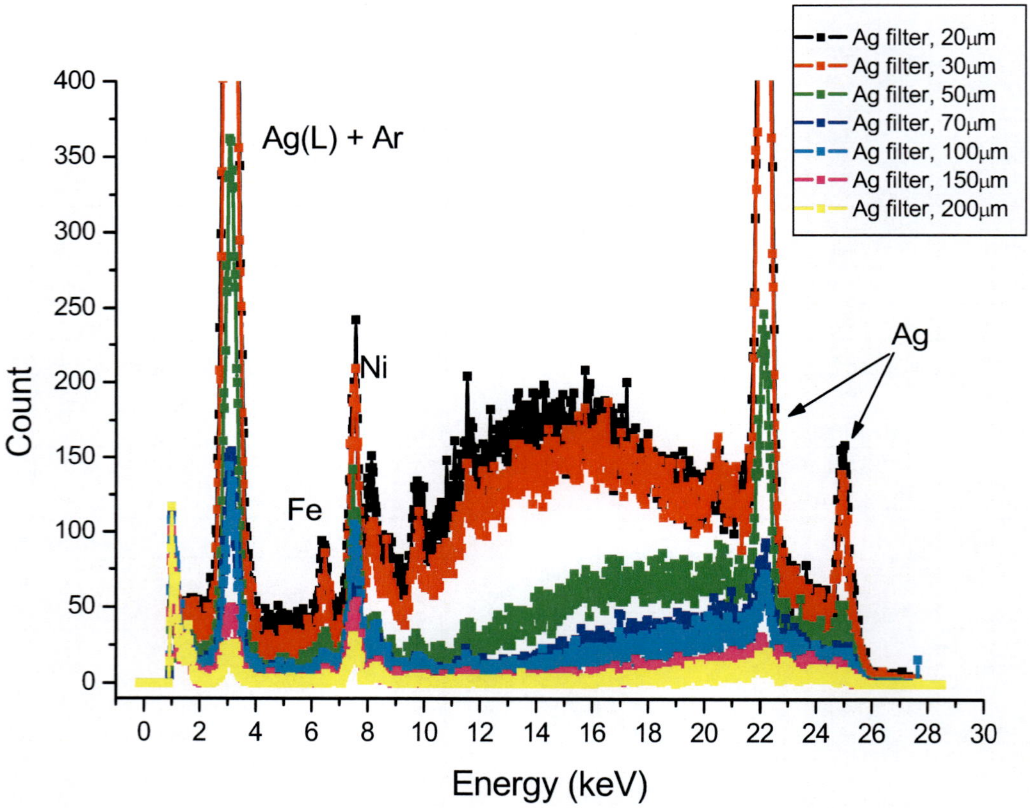

Figure 23. Espectro de um filme de Mylar, com filtros de Ag de diferentes espessuras, medidos com 28 kV, 10 μA, 200 s, com colimador de Ag no detector.

Table 24. Peak to background ratio and percent deviation in the net count from the elements determined with different thickness Ag filter for two membranes

Membrane A

	Peak net count Filter thickness (μm)					Background count Filter thickness (μm)			
Element	50	70	100	150	Element	50	70	100	150
Cr	41	15	5	12	Cr	208	112	85	20
Fe	1665	633	484	285	Fe	212	116	96	27
Co	1972	938	620	252	Co	267	125	102	33
Zn	1277	534	439	221	Zn	382	143	122	44
Pb	1572	771	547	273	Pb	483	132	134	51

	Peak/background ratio Filter thickness (μm)					Percent error Filter thickness (μm)			
Element	50	70	100	150	Element	50	70	100	150
Cr	0.2	0.1	0.1	0.6	Cr	51%	100%	260%	67%
Fe	7.9	5.5	5.0	10.6	Fe	3.0%	5.1%	6.0%	8.8%
Co	7.4	7.5	6.1	7.6	Co	4.6%	7.1%	8.5%	29.0%
Zn	3.3	3.7	3.6	5.0	Zn	4.6%	7.1%	8.0%	11.8%
Pb	3.3	5.8	4.1	5.4	Pb	4.4%	5.4%	6.9%	7.0%

Membrane B

	Peak net count Filter thickness (μm)					Background count Filter thickness (μm)			
Element	50	70	100	150	Element	50	70	100	150
Cr	18	14	17	12	Cr	242	93	66	38
Fe	2123	890	668	347	Fe	273	107	72	42
Co	2892	1055	709	464	Co	324	119	86	46
Zn	1687	646	557	321	Zn	409	161	114	54
Pb	2081	826	686	405	Pb	550	184	145	60

	Peak/background ratio Filter thickness (μm)					Percent error Filter thickness (μm)			
Element	50	70	100	150	Element	50	70	100	150
Cr	0.1	0.2	0.3	0.3	Cr	122%	100%	70.6%	66.7%
Fe	7.8	8.3	9.3	8.3	Fe	4.5%	4.0%	4.8%	7.2%
Co	8.9	8.9	8.2	10.1	Co	3.8%	6.6%	13.5%	15.7%
Zn	4.1	4.0	4.9	5.9	Zn	3.8%	6.3%	6.5%	8.1%
Pb	3.8	4.5	4.7	6.8	Pb	3.7%	5.6%	6.0%	4.7%

The bigger peak to background ratio is given for 150 μm Ag filter, therefore, for 50 μm filter is obtained the lower deviation. So, it was opted by 50 μm Ag filter, which makes possible higher precision.

Appendix II – Tests of Dry off Membranes

It was describe before that the separation process for suspended particulate matter or pre-concentration of water samples, consist on solution filtration to posterior EDXRF analysis. After filtration the membranes been wet, which generates a background increment due to bigger x-ray scattering from incident and emergent beam.

Tests with a membrane of suspended particulate matter from Capivara stream, filtered in ester cellulose membranes, were performed. The membranes were irradiated at different times just after its preparation. The evaluated times were: 5 min, 15 min, 30 min, 1 h, 2h, 3 h, 8 h, and 24 h. The membranes were dried at room temperature with around 65% air relative humidity.

Figure 24 presents a graph comparing the different dry off times evaluated.

It is verified that qualitative analysis is not prejudiced with the wet membrane. For low energy quantitative analysis, up to 8 KeV, the interference is minimal. It is suggested 1 h of minimum dry off time to analyze whole spectrum.

For *in situ* measurements the membrane was dried at sun for 30 min, since that 1 h is a relatively long wait time for field measurements. Figure 25 brings the spectra comparing *in situ* and laboratory measurements of the same membrane. For Fe peak the net peak area were 1179 ± 76, 1119 ± 55 and 1248 ± 59.

In this case, it was not verified difference between the *in situ* measurement and the laboratory one, showing that the membrane dry off at Sun is viable and efficient. This procedure reduces considerably the analysis time at field.

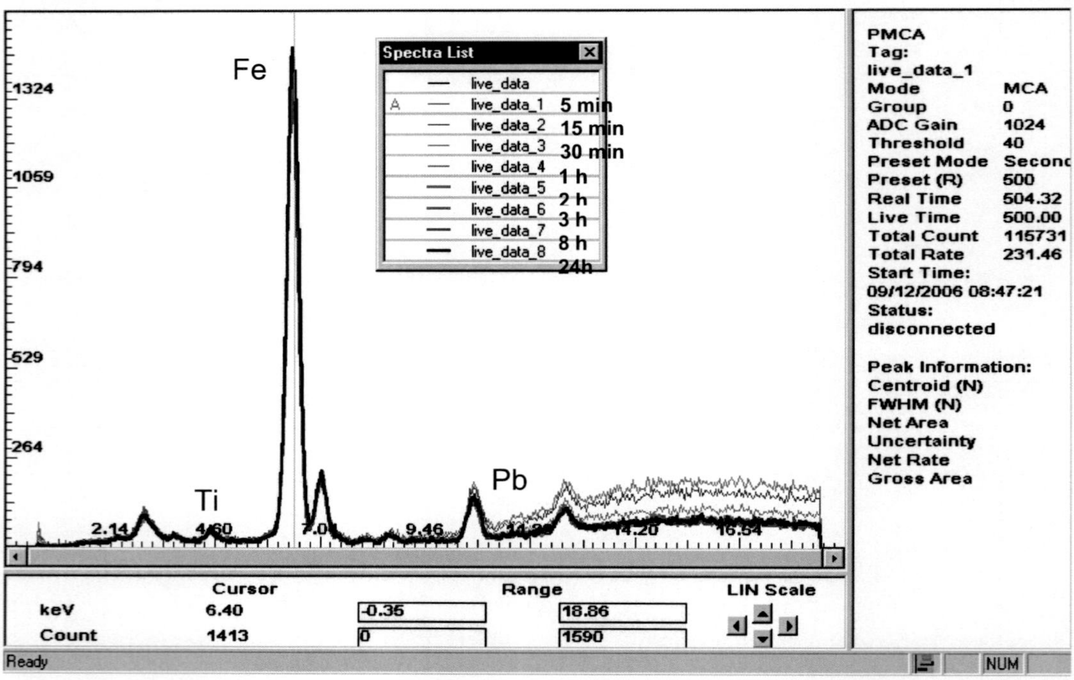

Figure 24. Different dry off time spectra. Sample from Capivara stream .

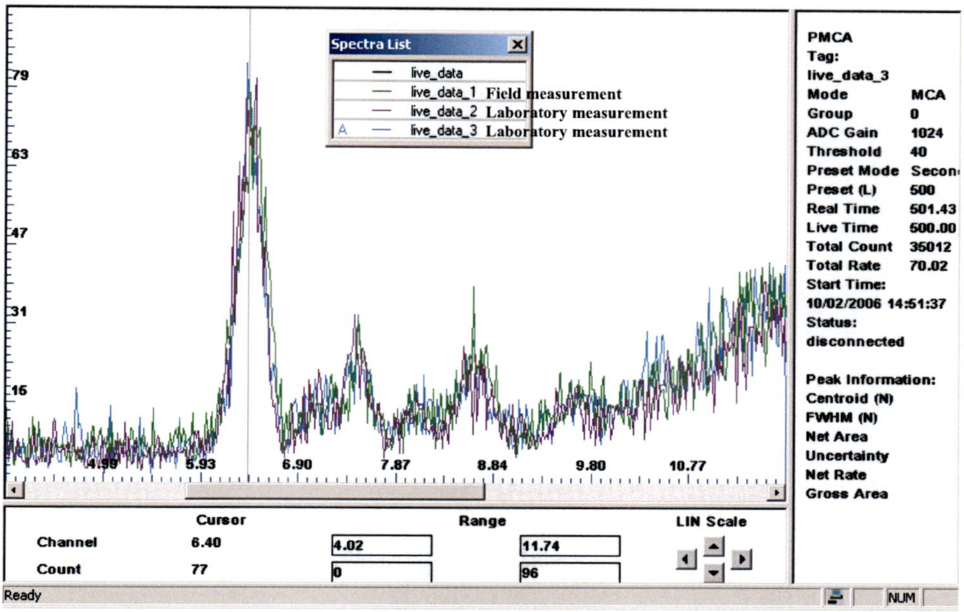

Figure 25. Field measurement (after 30 min. dry off at sun) and laboratory measurement (after 24 h dry off at room temperature) comparison of the same membrane.

REFERENCES

[1] Costa A. C. M., Anjos M. J., Lopes R. T., Perez C. A., Castro C. R. F. *X-Ray Spectrom.* 2005, 34, 183- 188.

[2] Daorattanachai P., Unob F.F., Imyim A. *Talanta* 2005, 67, 59-63.

[3] Balcerzak M. *Anal. Sci.* 2002, 18 ,737-750.

[4] Orescanin V., Mikelic L., Roje V., Lulic S. *Anal. Chim. Acta* 2006, 570 ,277-282.

[5] Hou X., Jones B.T. *Microchem. J.* 2000, 66 ,115-145.

[6] Hou X., He Y., Jones B.T. *Appl. Spectr. Reviews*, 2004, 39, 1-25.

[7] vanova J., Djingova, R., Kuleff, I. *J. Radioanal. Nucl. Chem.* 1998; 238, 29-32.

[8] Khusainov A.K., Antonova T.A., Bahlanov S.V., Derbin A.V., Ivanov V.V., Lysenco V.V., Morozov F., Mouratov V.G., Muratova V.N., Petukhov Y.A., Pirogov A.M., Polytsia O.P., Saveliev V.D., Solovei V.A., Yegorov K.A., Zhucov M.P. *Nucl. Instrum. Methods Phys. Res. Sect A.* 1999, 428, 223-231.

[9] Longoni A., Fiorini C., Leutenegger P., Sciuti S., Fronterotta G., Strüder L., Lechner P. *Nucl. Instrum. Methods Phys. Res. Sect A.* 1998, 409, 407-409.

[10] Ferretti M. *Nucl. Instrum. Methods Phys. Research B*, 2004, 226, 453-460.

[11] Van Grieken R., Araujo F., Rojas C., Veny P., XRF and PIXE applications in Life Sciences, Capri (Italy), 29-30 June 1989, edited by R. Moro and R. Cesareo, World Scientific, Singapore, 1990, p 79.

[12] Cesareo R. *Nuclear analytical techniques in medicine*, Elsevier, Amsterdam 1988; 19-121.

[13] Cesareo R., X-ray physics: Interaction with matter, production, detection, La Revista des Nuovo Cimento della Società Italiana di Fisica, Editrice Compositori, Bologna, 2000.

[14] Kalnicky J. D., Singhvi R. *J. Hazard. Mater.* 2001, 83, 93-122.

[15] Clement R.E., Yang P.W., Koester C.J. *Anal. Chem.* 1999; 71 ,257R-292R.

[16] Butler O. T., Cook J.M., Harrington C.F., Hill S.J.Riewters J., Miles D.L. *J. Anal. At. Spectrom.* 2005, 20, 301-330.

[17] Butler O. T., Cook J.M., Harrington C.F., Hill S.J.Riewters J., Miles D.L. *J. Anal. At. Spectrom.* 2007, 22, 187-221.

[18] Potts P.J., Ellis A.T., Kregsamer P., Streli C., Vanhoof C.,West M. Wobrauscheck P. *J. Anal. At. Spectrom.,* 2005, 20, 1124-1154.

[19] 1 Potts P.J., Ellis A.T., Kregsamer P., Streli C., Vanhoof C.,West M. Wobrauscheck P., *J. Anal. At. Spectrom.* 2006, 21, 1076-1107.

[20] West M., Ellis A.T., Kregsamer P., Potts, PJ; Streli C., Vanhoof C., Wobrauscheck P., *J. Anal. At. Spectrom.* 2007, 22, 1304-1332.

[21] Melquiades, FL, Appoloni, CR, *Radiat. Phys. Chem.*, 2004, 262, 533-541.

[22] Kirtay V.J., Kellum J.H., Apitz S.E., *Wat. Sci. Tech.* 1998, 37, 141-148.

[23] McDonald B.J., Unsell C.W., Elam W.T., Hudson K.R., Adams J.W. *Nucl. Instrum. Methods Phys. Res. Sect A.* 1999, 422, 805-808.

[24] Bernick M.B., Campagna P.R. *J. Hazard. Mater.* 1995, 43, 91-99.

[25] Economou T., *Radiat. Phys. Chem.* 2001, 61, 191-197.

[26] Rieder R., Economou T., Wänke H., Turkevich A., Crisp J., Brückner J., Dreibus G., McSween Jr. H. Y. *Science* 1997, 278, 1171-1174.

[27] Fiorini, C., Longoni A. *IEEE Transactions in Nuclear Science*, 1999, 46, 2011-2016.

[28] Mccomb M.E., Gesser H.D. *Talanta* 1999, 49, 869-879.

[29] Clark H.F., Brabander D.J., Erdil R.M. *J. Environ. Qual.* 2006, 35, 2066-2074.

[30] Kilbride C., Poole J., Hutchings T.R. *Environ. Pollut.* 2006, 143 ,16-23.

[31] Operating manual – XR100CR X ray detector system and PX2CR. Power supply, Amptek INC, 1998.

[32] Operating Manual – Miniature Bullet X ray Tube, Moxtek INC, 2003.

[33] Van Grieken R., *Anal. Chim. Acta.* 1982, 143, 3-34.

[34] Van Krieken, R, Bresseleers, K, Smiths, J, Vanderborht, B, Vanderstappen, M. *Adv. X-ray Anal.*, 1976, 19 ,435-447

[35] Ellis, AT, Leyden, DE, Wegscheider, W, Jablonski, BR, Bodnar, WB. *Chim. Acta* 1982, 142, 73.

[36] Kingston, H, Pella, PA. *Anal Chem*, 1981, 53, 223.

[37] Mccomb M.E., Gesser H.D., *Talanta* 1999, 49, 869-879

[38] Salvador, MJ, Lopes, G. N., Nascimento Filho, VF, Zucchi, O.D. *X-Ray Spectrom.* 2002, 31, 141-144.

[39] Alvarez, AM, Alvarez, JRE., Alvarez, RP, *J. Radioanal. Nucl. Chem.* 2000, 245 ,485-489.

[40] Narin I., Soylak M., *Anal. Chim. Acta* 2003, 493, 205-212

[41] Peralta-Zamora, P, Morais, JL, Nagata, N, *Eng. Sanit. Ambient.* 2005, 10, 106-110.

[42] Barros Neto, B, Scarmínio, IS, Burns, RE, Como fazer experimentos, 2nd ed., Editora da Unicamp: Campinas-SP, 2003.

[43] Melquiades, FL, Parreira, PS, Yabe, MJ, Corazza, MZ, Funfas, R, Appoloni, CR, *Talanta* 2007, 73, 121-126.

[44] Almeida, E, Nascimento Filho, VF, Valencia, EPE, Cunha, RM, *J. Radioanal. Nucl. Chem.* 2002, 252, 541-544.

[45] Nkono, NA, Asubiojo, OI, *J. Radioanal. Nucl. Chem.* 1998, 227, 117-119

[46] Necemer, M, Kump, P, *Spectrochim. Acta B* 1999, 54, 621-627

[47] Bertin, EP, Principles and practice of X-ray spectrometric analysis. London: Plenum Press, 1975. 1079p.

[48] Currie L.A. *Anal. Chem.* 1968, 40, 586-593.

[49] BRASIL. Ministério do Meio Ambiente, Conselho Nacional do Meio Ambiente (CONAMA). Resolução n 375, de 17 de março de 2005. Dispõe sobre a classificação dos corpos de água e diretrizes ambientais para o seu enquadramento, bem como estabelece as condições e padrões de lançamento de efluentes, e dá outras providências. Brasília, 2005.

[50] Kump P, Necemer M, Rupnik Z. *Report IAEA-TECDOC-1456*. Vienna; 2005.

In: Causes and Effects of Heavy Metal Pollution
Editor: Mikel L. Sanchez
ISBN 978-1-60456-900-1

Chapter 8

Biomonitoring of Heavy Metal Pollution in the Marine Environment Using Indicator Organisms

[1]*Joseph Selvin*, [2]S. Shanmugha Priya, [3]G. Seghal Kiran and [4]Saroj Bhosle

[1-2]Department of Microbiology, Bharathidasan University, Tiruchirappalli 620024, India
[3]Department of Biotechnology, Bharathidasan University, Tiruchirappalli 620024, India
[4]Department of Microbiology, Goa University, Goa 403 206, India

Abstract

Sustainable management of marine bioresources require an ecosystem perspective that includes recognition of natural and anthrapogenic disturbances on supporting food webs and resultant changes in community structure. The marine environment is continuously subjected to chemical pollution, which can have detrimental effect on aquatic organisms living in that environment. Each year several new synthetic chemicals enter the market, many of which are likely to reach and pose impacts on the marine environment. The concept to develop and use biological markers to monitor marine environments is at the base of recent studies. The biological marker reveals the status of environmental health condition by accumulation of metals in their tissues. The migrating organisms may be suitable indicators for a larger zone instead of confined areas. The selection of benthic and sedentary organisms could be ideal to select as bioindicators for notified regions. Among them, primitive multicellular organisms like sponges would be ideal. Sponges are filter feeders and are ideal for assessing the effect of silation and environmental contaminants on the primary food chain. Sponges seem to be filtering a large volume of water and accumulate heavy metals. Heavy metal contamination and sediment deposition had significant influence on the secondary metabolites synthesis of marine sponges. However, accumulation seems to depend on the metal and the species considered. Bacterial endosymbionts isolated from the sponges invariably showed resistance against a battery of heavy metals tested including copper, lead, mercury, cobalt and cadmium. Therefore, the bacteria associated with the sponges can be used as

---

* For Correspondence. Email: selvinj@rediffmail.com.

indicators of contamination in marine ecosystem. Seawater bacteria have already been established as biological indicators of contamination. Considering the overall complexity of ecological factors in the marine environment, developing a manageable set of bioindicators is a challenging task. The present report envisages the possibility of developing benthic nematodes as potential bioindicator model for monitoring the heavy metal pollution in the marine environment.

Keywords: *Heavy metal, marine-pollution, bioindicator, sponge-bacteria, pollution-monitoring*

Introduction

In the past few decades, uncontrolled urbanization and industrialization have caused a serious problem of pollution due to disposal of sewage and industrial effluents in most natural water bodies. The marine environment is continuously subjected to chemical pollution caused by discharge and disposal of untreated and partially treated domestic and industrial waste, discharge of industrial coolant waters, spilling of cargo such as chemicals and metal ores, fishing activities such as mechanized fishing vessels, draining of waste engine oils, painting of fishing vessels, shaping of metal lining of fishing boats, dumping of waste, thrash fish, etc. [Ramachandran *et al.,* 1989]. Heavy meals are one of the most serious pollutants in our natural environment due to their toxicity, persistence and bioaccumulation problems [Tam and Wong, 2000]. Trace metals in natural waters and their corresponding sediments have become a significant topic of concern for scientists, environmentalists and engineers in various fields associated with ecosystem management, as well as a concern of the general public. The marine environment is endlessly subjected to chemical pollution mainly by toxic heavy metals. The major sources of heavy metal pollution are the mining and smelting of metalliferous ores [Li and Thornton, 2001]. The presence of trace metals in aquatic environment originates from the natural interactions between the water, sediments and atmosphere with which the water is in contact. The sources of heavy metals contamination in the marine ecosystem can mostly be anthropogenic.

- Coastal / off-shore mining: Mining of coastal and marine waters might introduce heavy metals into the water from rocks and subsurface deposits.
- Industrial sources: Industrial processes, particularly those concerned with the mining and processing of metal ores, the finishing and plating of metals and the manufacture of metal objects.
- Maritime processes: Antifouling coatings / paints of marine structures lead to the leaching of metal ions into the water.
- Agricultural sources: Agricultural discharge containing residual pesticides and fertilizers which contains metals.
- Acid rains containing trace metals may cause the pollution of water with metals.

The concentrations fluctuate as a result of natural hydrodynamic chemical and biological forces. The continued influx of pollution load is aggravated in summers when the water evaporates increasing metal content. During this process, many bacteria acquire metal tolerance and the plasmid expression can lead to antibiotic resistance also. This may lead to new disease patterns and difficulties in management of infections. The activity of trace metals in aquatic systems and their impact on aquatic life vary depending upon the metal species. The most important heavy metals from the point of view of water pollution are Zn, Cu, Pb, Cd, Hg, Ni and Cr. Some of these metals (e.g. Cu, Ni, Cr and Zn) are essential trace metals to the living organisms, but become toxic at higher concentrations. Most of these metals particularly Pb and Cd have no known biological function but are toxic elements. Of major importance in this regard is the ability of metals to associate with other dissolved and suspended components. Most significant among these associations is the interaction between metals and organic compounds in water and sediment. This phenomenon would naturally alter the reactivity of metals in the aquatic environment. Presence of heavy metals in the marine environment is a matter of concern due to their nonbiodegradable nature and prolonged residence time, thus making them an important class of environmental pollutants. Direct toxicity to man and aquatic life and indirect toxicity through accumulations of metals in the aquatic food chain are the focus of this concern. WHO and other environmental agencies have specified the safe limit of these metals in drinking water as well as water used for other purposes. Heavy metal contamination has been described as a ticking environmental bomb. The heavy metal issue is believed to be a potential environmental disaster since the accumulation has been ongoing for decades and no one knows how long the pollutants will remain in the biosphere.

The quality of the marine environment is constantly being monitored by various national authoritative bodies analysing water, sediment and/or biota. The general methods used for checking the presence of metals in air, water and soil are chemical-based. These chemicals are not only costly and require exhaustive analysis in specialized laboratories, but to establish management priorities, frequent sampling is necessary, since under certain hydrodynamic conditions (for example, strong currents), the effects of a single waste dumping, could be detected for only a short time. Moreover chemical analysis can measure only a fraction of the contaminants but reveals nothing about the adverse effect. It has been previously demonstrated that a large number of errors may occur owing to the relatively low contents of pollutants [Quevauviller *et al.*, 1992]. For this reason, there is widespread recognition today that not only chemical analyses but biological techniques are required for assessment of marine ecosystem health. The use of bioindicators to check the presence of metals in the environment has thus gained significance. Further the bioindicators not only provide the level of heavy metal contamination, in addition, they provide the impact of these trace metals on the standing crop of the ecosystem. The bioindicator-based analysis involves both phenotypic and analytical parameters [Kahle *et al.,* 1999; Normandin *et al.*, 1999; Oliviera *et al.*, 2000]. Of the two, the former is quick, inexpensive and eco-friendly. Biological monitoring has been used for the control of anthropogenic pollution [Manning and Feder, 1980]. Biomonitoring has certain advantages in comparison to the direct measurement of industrial emission into the environment [Markert, 1996].

SIGNIFICANCE OF THE USE OF BIOINDICATORS

The ability of various pollutants (and their derivatives) mutually affect their toxic actions complicates the risk assessment based solely on environmental levels. Deleterious effects of contamination on populations are often difficult to detect in feral organisms since many of these effects tend to manifest only after longer periods of time. When the effect finally becomes clear, the destructive process may have gone beyond the point where it can be reversed by remedial actions or risk reduction. Such scenarios have triggered the research to establish early-warning signals, or biomarkers, reflecting the adverse biological responses towards anthropogenic environmental toxins [Bucheli and Fent, 1995].

- The most compelling reason for using biomarkers is that they can give information on the biological effects of pollutants rather than a mere quantification of their environmental levels.
- Biomarkers may provide insight into the potential mechanisms of contaminant effects.
- They detect the presence of both known and unknown contaminants.
- They provide a temporal and spatial measure of bioavailable pollutants.

BIOINDICATOR – DEFINITION

- Biomarkers are measurements in body fluids, cells or tissues indicating biochemical or cellular modifications due to the presence and magnitude of toxicants, or of host response [Bodin *et al.,* 2004].
- In an environmental context, biomarkers offer promise as sensitive indicators demonstrating that toxicants have entered organisms, have been distributed between tissues, and are eliciting a toxic effect at critical targets [McCarthy and Shugart, 1990].
- Several definitions have been given for the term 'biomarker', which is generally used in a broad sense to include almost any measurement reflecting an interaction between a biological system and a potential hazard, which may be chemical, physical or biological (WHO, 1993).
- A biomarker is defined as a change in a biological response (ranging from molecular through cellular and physiological responses to behavioral changes) which can be related to exposure or toxic effects of environmental chemicals [Peakall, 1994].
- Brummelen *et al.,* (1996) redefined the terms 'biomarker', 'bioindicator' and 'ecological indicator', linking them to different levels of biological organization. They considered a biomarker as any biological response to an environmental chemical at the subindividual level, measured inside an organism or in its products (urine, faeces, hair, feathers, etc.), indicating a deviation from the normal status that cannot be detected in the intact organism.
- A bioindicator is defined as an organism giving information on the environmental conditions of its habitat by its presence or absence or by its behavior, and an

ecological indicator is an ecosystem parameter, describing the structure and functioning of ecosystems.

Biomarkers- Subdivision

According to the NRC (1987), WHO (1993), biomarkers can be subdivided into three classes

- *Biomarkers of exposure*: covering the detection and measurement of an exogenous substance or its metabolite or the product of an interaction between a xenobiotic agent and some target molecule or cell that is measured in a compartment within an organism.
- *Biomarkers of effect*: including measurable biochemical, physiological or other alterations within tissues or body fluids of an organism that can be recognized as associated with an established or possible health impairment or disease
- *Biomarkers of susceptibility*: indicating the inherent or acquired ability of an organism to respond to the challenge of exposure to a specific xenobiotic substance, including genetic factors and changes in receptors which alter the susceptibility of an organism to that exposure.

Requirements for the Selection of a Bioindicator

The use of any organism or biomarker for environmental monitoring requires previous systematic studies to

- Establish the natural behavior of the organism in nature and in laboratory conditions
- Identify biomarkers altered in response to environmental conditions
- Establish the degree of susceptibility of the organism to specific agents [Chevre *et al.*, 2003].

Once established the organism's sensitivity to a given agent in concentration ranges just above the level found in natural environments, and identified one or more biomarkers useful for measuring the toxic effects of that agent, the organism can be considered potentially suitable for biomonitoring. The overview of various bioindicator models are presented in Figure 1.

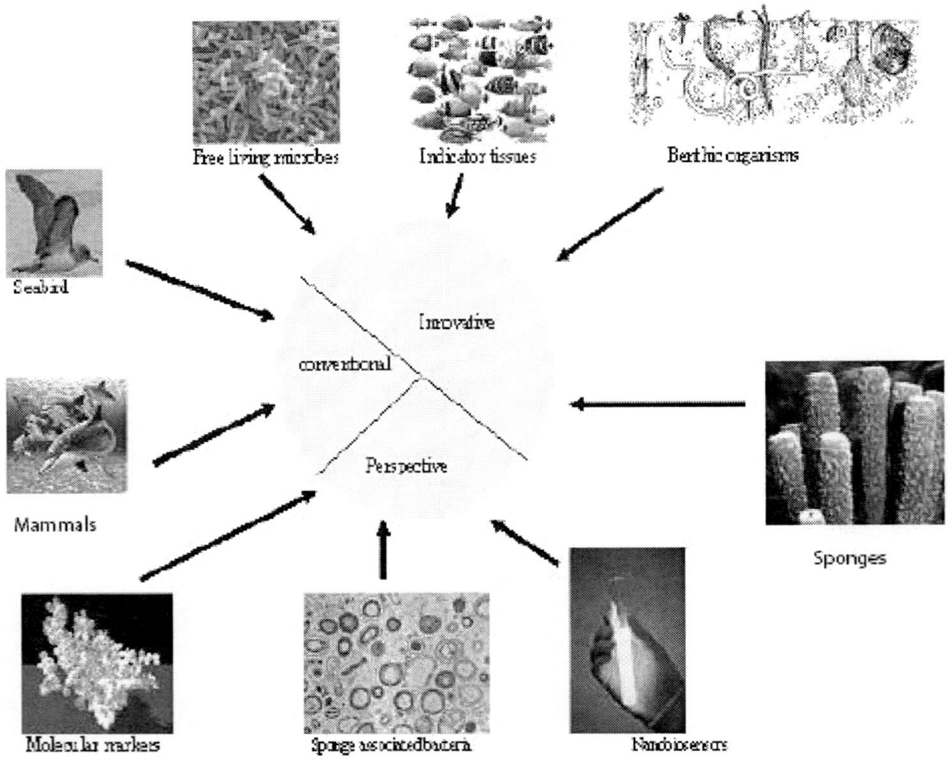

Figure 1. Marine bioindicators – an overview.

CONVENTIONAL MARINE BIOINDICATORS

1. Seabirds

Seabirds have been used extensively as monitors of oil spills and heavy metals. Though oil spills cause direct impact on the coastal avians, the impact heavy metals also considered to be critical. Concentrations of heavy metals are often reported for adult birds but less often for chicks or fledglings. However, chicks have been proposed as particularly useful indicators for both baseline pollution studies and monitoring programs, as they concentrate heavy metals during a specific period of time and from a local and definable foraging area. The Cory's shearwater, *Calonectris diomedea*, is a long- lived pelagic seabird found in warm marine waters from temperate to sub-tropical zones of the North Atlantic and Mediterranean [Cramp and Simmons, 1977]. High concentrations of heavy metals have been reported in tissues of adult Cory's shearwater from Mediterranean and Salvage islands which were attributed to accumulation from prey items [Renzoni *et al.*, 1986]. Therefore Cory's shearwater was used as biomarker for seawater polluted with heavy metals particularly Hg. Analysis of total Hg in the feather samples showed that Hg levels were independent of sex and age in adults. However, the reliability of avian models in heavy metal monitoring in the marine environments has not been established.

2. Marine Mammals

Marine mammals are sensitive to pollutants and frequent offshore landing of dolphins, porpoises and whales has been attributed with the heavy metal and other environmental poisoning. Many studies have been carried out concerning heavy metal accumulation in dolphins from different areas of the world, various dolphins from Mediterranean Sea, New Zealand [Koeman *et al.*,1973], the Japanese costs [Honda *et al.*, 1983; Itano *et al.*, 1984], Argentina [Marcovecchio *et al.*, 1990], French Atlantic coasts and French Mediterranean coasts [Andra *et al.*, 1991], British Isles [Law *et al.*,1991], Italian Mediterranean coasts (Tyrrhenian coasts) [Leonzio *et al.*, 1992], Apulian coasts and south Carolina coasts [Beck *et al.*, 1993]. In these studies, concentrations of Hg in dolphins from the Mediterranean are generally higher than those found in the same species from the Atlantic. Though the mammals are index of heavy metal pollution, they cannot be considered as a model organism for heavy metal monitoring. Because sampling of mammals for heavy metal analysis ultimately provokes wild life protection conflicts.

Newly Developed Bioindicator Models

1. Free Living Organisms as Bioindicators

Heavy metals are highly toxic pollutants of the marine (and terrestrial) environment. They produce considerable modifications of the microbial communities and their activities [Doelman *et al.,* 1994; Guzzo *et al.,* 1994; Hiroki, 1994; Starzecka and Bednarz, 1993]. Heavy metals generally exert an inhibitory action on microorganisms by blocking essential functional groups, displacing essential metal ions, or modifying the active conformations of biological molecules [Doelman *et al.,* 1994; Gadd and Griffiths, 1978; Li and Tan, 1994a; Wood and Wang, 1983]; however, at relatively low concentration some metals are essential for micoroorganisms (e.g. Co, Cu, Zn, Ni) since they provide vital co-factors for metallo-proteins and enzymes [Eiland, 1981; Doelman *et al.,* 1994]. The resistant of aquatic microorganisms to metals make their appearance as a result of exposure to aquatic environments contaminated with effluent from metal-associated industries. Bacteria have adapted to metals through a variety of chromosomal, transposon, and plasmid mediated resistance systems. Bacterial cells living in extreme metal concentrations possess an essential for their survival ability to resist such stress. The ability of microorganisms to grow in the presence of high metal concentrations results from specific mechanisms as: extracellular precipitation and exclusion of metal ions, binding of the metal ions to the outer surface of bacteria and its intracellular sequestration. Binding of metal cations on the surface of bacterial cells has become one of the most attractive means for metal biotranformation. Postulated mechanisms of metal resistance in microorganisms include

- Metal exclusion by permeability barrier (blockage at the level of cell wall)
- Active transport of the metal away from the cell/organism (systems of membrane transport)
- Intracellular sequestration of the metal by protein binding

- Extracellular sequestration
- Enzymatic detoxification of the metal to a less toxic form
- Reduction in metal sensitivity of cellular targets

Microbial resistance to heavy metals has been intensively studied over the past 25 years. However, the concentration of 'bioavailable' metal and resistance level of microbes to heavy metals is relevant to the microbial ecology of hazardous waste sites. Numerous toxicological studies have examined the heavy metal sensitivity and resistance of bacteria isolated from different habitats. In principle the free living bacteria are the pioneer organisms that are exposed immediately to any contamination in the environment. Thus the possibility of using metal resistant bacteria as bio-indicators of polluted environment has been shown to be a sensitive and reliable tool in detecting the sub-lethal toxicity of these polluting compounds. But the free living bacterial community is subjected to changes due to excess/ prolonged exposure of contaminants and external disturbances such as drifting behavior of migratory animals.

2. Indicator Tissues for Heavy Metal Monitoring

A combination of bioassays is increasingly recommended in the framework of integrated eco-toxicological approaches, in order to gain a better insight into the potential dangers associated with the disposal of complex industrial effluent in the environment. George and Olsson (1994) have suggested that fish species would be better candidates for monitoring heavy metal contamination. Additional attributes were added by Rayment and Barry (2000) where kidney tissues, visceral mass, adductor muscle, axial muscle of fishes act as indicator tissues for heavy metal monitoring.

3. Benthic Organisms as Sensitive Bioindicators

The use of benthic invertebrates as indicators of pollution dates from the work of Wilhelmi (1916).

- Polychaete *Capitella capitata* was a good indicator of sediments with high organic content. Since then, many other bioindicators have been used to indicate organic enrichment [Bellan *et al.,* 1981; Bellan, 1984] and the presence of heavy metals and other contaminants [Temara *et al.,* 1998]. Besides using specific species as bioindicators, Margalef (1975) suggested the use of ratios or indices to estimate the extent of pollution in coastal areas; these will vary as an ecosystem is undisturbed or under disturbance. Ratios were proposed as an extra ''tool'' to help in our understanding of effects of pollution on marine benthic communities. Indices were created as tools for assessing pollution, such as the one based on the ratio of ''pollution sentinel species'' versus ''pure water sentinel species,'' using polychaetes [Bellan, 1980; Bellan *et al.,* 1988; Quetin and Rouville, 1986], or the nematode–copepod ratio [Raffaelli and Mason, 1981].

- Unicellular algae are an ideal group to study responses to different environmental factors. One of the advantages of environmental study using algae is the possibility to achieve and observe many generations during relative short time period. Previous studies proved high sensitivity of the most algae towards changing of environmental conditions, especially as consequences of water pollution. Algae respond rapidly and predictably to a wide range of pollutants and provide potentially useful early warning signals of deteriorating conditions and possible causes.
- The blue crab, *Portunus pelagicus* indicated the contamination of heavy metals in the marine environment of Kuwait region by accumulating high concentrations of metals after a decade, in 1991 Gulf War.
- Mussels have been widely used for chemical biomonitoring in Mussel Watch Programs [NAS, 1980; RNO, 2000]. Specimens of mussel species *Anodonta anatina, Unio pictorum, U. tumidus* have been used as bioindicators of trace metal pollution [Jamil *et al.,* 1999].

Coull *et al.,* (1981) commented that the use of these taxa is not a valid tool because aspects like seasonal variation, and granulometry could alter the pollution level in the zone.

4. Sponges as Perceptive Bioindicator

Marine sponges are in most geographic regions an important component of marine benthic communities and in the past several years received increased attention from microbiologist as hotspots of unexplored diverse microbial communities. While sponges have long been exploited for the discovery of pharmacologically active natural products, little attention has been given to their potential role as bioindicator organisms for anthropogenic pollutants. Their potential as bioindicator organisms becomes apparent when one considers that they are filter feeders. Despite relatively low pumping rates, sponges can exhibit high retention rates as a function of well-developed aquiferous systems. This ability is significant in benthic-pelagic coupling and suggests the Porifera are able to influence or ameliorate microbial polluting assemblages associated with faecal contamination [Kefalas *et al.*, 2003].

In this sense, biological methods like the River Invertebrate Prediction and Classification System (RIVPACS) were developed to manage water quality based on macroinvertebrate communities [Armitage *et al.,* 1990; Wright *et al.,* 1993], which estimates the macroinvertebrate community that would exist in an area with no environmental stress, and vice versa. The use of macrobenthic taxa with relatively long life spans, i.e. sponges and ascidians, are much less influenced by transient changes in the environment, and the seasonal variations should not be a major problem. In the current scenario, bioaccumulation studies of pollutants in the tissues of sedentary animals would be more appropriate than periodically estimating concentrations in water and sediment or the use of free living bacteria to assess the environment health. Among them, primitive multicellular organisms like sponges would be ideal. High concentrations of pollutants including hydrocarbons [Zahn *et al.,* 1981], organochlorinated compounds [Perez *et al.,* 2003], and metals [Carballo *et al.,* 1996; Webster *et al.,* 2002; Perez *et al.,* 2004] have been reported in several sponge species. Sponges have some of the characteristics as good bioindicators and are convenient tools for characterizing the state of a marine ecosystem. They are attached to the substratum and have relatively long life span and are much less influenced by transient changes in the environment. Sponges are filter feeders frequently filtering large amounts of water and thus accumulate heavy metals from the surrounding medium. The accumulation of metals in sponges paralleled the concentrations of these contaminants in their medium. Sponges are the characteristic species of several communities, which are hotspots of biodiversity threatened by many perturbations and not regularly surveyed. Finally its very simple body plan allows a homogenous distribution of contaminants in the whole tissue and homogenous expression of biomarkers as well.

On the other hand, numerous natural environmental variables (e.g. water depth, sediment type, etc.) modify community structure, and it has not always been easy to separate such effects from anthropogenic consequences. The use of sponges and ascidians in monitoring studies is still limited by our relatively small knowledge about biotic and abiotic factors affecting the structure of these communities [Carballo *et al.,* 1994, 1996] and we are aware it may not be possible to use the same species or assemblages as indicators of polluted or semi-polluted conditions throughout the world. Many species have either limited geographical or ecological distribution; nevertheless, we think that lack of universal bioindicators of pollution must not be a deterrent to their use in specific monitoring studies. Though it has been reported that sponges can accumulate a wide range of pollutants from both suspensions and dissolved phases by principle the nonselective marine filter feeders obviously accumulate the suspended particles instead of dissolved compounds. Further it has been reported that the harmful effects of sublethal accumulation of heavy metals in sponges were not obvious and take time to appear.

Modern Gauge of Heavy Metal Contamination

Molecular Biomarkers as Pronounced Indicators

The molecular markers including cytochrome P4501A induction, DNA integrity, acetylcholine esterase activity and metallothionein induction have received special attention to evaluate the exposure of environment to pollution. Specific molecular biomarkers of biological effects of contaminants on marine organisms and their analytical technique of their measurements are presented in the following table.

Biomarker	Observation	Analytical technique
Cytochrome P4501A	Indicator of organic contaminants PAH, PCB, PCDD	Fluorometric, ELISA, spectrophotometric
DNA damage	Indicator of xenobiotics PAH, PCB, PCDD	DNA strand break measurement
AChE inhibition	Indicator of organophosphorous, carbamate, Cd, Pb, Cu	Measurement of enzyme activity
Metallothioneins	Indicator of Zn, Cu, Cd, Hg	Estimation of protein bound metal.

Microbial Endosymbionts of Marine Sponges as Promising Indicators

Most of the marine sponges harbor microorganisms that include bacteria, cyanobacteria and fungi within their tissues where they reside in the extra and intracellular space Bacteria can contribute up to 40% of the sponge biomass (equal to about 10^8–10^9 bacteria/ g of tissue) and are probably permanently associated with the host sponge unless they are disturbed by external stress factors. It has been reported that the bacterial load of sponge tissue exceeds the proximal load of seawater due to the sequestration of habitat bacteria in the mesohyl of marine sponges. Sponges have capability to filter the immediate vicinity within 24 h and to retain up to 80% of suspended particles including free-living bacteria. It was envisaged that the sponge-associated cultivable bacteria would be an index of normal bacterial flora in the immediate vicinity of sponges. Therefore, any consistent changes/influence of contaminants on the free-living bacteria could be estimated using the sponge-associated cultivable bacteria. It has been reported that the sublethal exposures to heavy metals may affect the physiological functions and behavior of organisms without killing them. Therefore, the effect of heavy metal pollution in the habitat could be detected directly from the sponge-associated bacteria instead of in situ monitoring of host sponges for behavioral changes. The sponge-associated bacteria might be an ideal bioindicator model for monitoring of heavy metal pollution in marine habitats.

A good bioindicator accumulates contaminants from the environment and accurately reflects environmental levels. Sponges invariably filter a large volume of seawater and potentially accumulate heavy metals and other contaminants from the environment. Sponge, being sessile marine invertebrates and modular in body organisation can live many years in the same location and therefore have the capability to accumulate anthropogenic pollutants such as metals over a long period. Almost all marine sponges harbour large numbers of microorganisms within their tissues where they reside in the extra and intra-cellular space. Bacteria in seawater has already been established as biological indicators of contamination. In a recent study, we established sponge associated bacteria as reliable biological indicators [Selvin *et al.,* 2008]. In this study, the bacteria associated with a marine sponge *Fasciospongia cavernosa* was evaluated as potential indicator organisms. The associated bacteria including *Steptomyces* sp. (MSI01), *Salinobacter* sp. (MSI06), *Roseobacter* sp. (MSI09), *Pseudomonas* sp. (MSI016), *Vibrio* sp. (MSI23), *Micromonospora* sp. (MSI28), *Saccharomonospora* sp. (MSI36) and *Alteromonas* sp. (MSI42) showed resistance against tested heavy metals. Literature evidenced that the sponge associated bacteria was seldom exploited for pollution monitoring though it has been extensively used for bioprospecting. In this background, the present findings come up with a new insight in the development of indicator models.

Application of Nanoparticles and Biosensors

A defence mechanism used by bacteria to avoid poisoning by heavy metals has been harnessed for the construction of atomic-scale ‘nano-clusters’. These tiny structures, made from atoms of platinum, palladium and gold, could be used to make more efficient catalysts or nano-electronic components. Researchers at the Rossendorf research centre in Dresden, Germany, used the bacteria *Bacillus sphaericus* to produce the nano-clusters. The organism can survive in soil contaminated with heavy metals due to a protective coating of proteins and sugars known as the S-layer. The bacteria's S-layer is covered with 8-nanometre-wide pores that allow nutrients through but prevent heavy metals from poisoning the cell beneath. Proteins inside the pores bind metal ions together which can later be assembled into nano-clusters of pure metal.

The chemical inertness and resistance to surface oxidation make gold an important material for use in nano-scale technologies and devices. Other materials that share similar corrosion resistance as gold are silver and platinum. However, silver is considered too reactive and platinum is significantly more expensive than gold [Corti and Holliday 2003]. The wide variation of optical properties of gold nanoparticles with particle size, particle-particle distance, and the dielectric properties of the surrounding media due to the phenomenon called surface plasmon resonance [Liz-Marzan, 2004] enables construction of simple but sensitive colorimetric sensors for various analytes. A recent report of fluorescence from strictly size controlled Au clusters, with the emission wavelength tuned by controlling the number of Au atoms in the cluster [Zheng *et al.,* 2004], demonstrates another potential application offered by controlling the size and shape of nano-sized gold particles. Recently, a novel strategy for detection of heavy metal ions in water has been developed employing 20 nm gold particles capped with a biopolymer called chitosan [Sugunan *et al.,* 2005]. Polymer capping of nanoparticles serves a two-fold purpose, that of stabilization and surface functionalization for application as sensors. Chitosan is widely used as a chelating agent for the removal of heavy metal contaminants in wastewater. Further modification of the attached chitosan molecules has the promise to achieve high-specificity sensors for various applications. Apart from applications in colorimetric pollution sensors, chitosan capped gold nanoparticles may have biology-oriented applications because it was found that chitosan shows selectivity in attachment to certain kinds of bacteria. Future experiments will be directed towards this aspect of chitosan capped gold nanoparticles.

Several analytical methods such as atomic absorption spectrometry, inductively coupled plasma with mass spectrometry as well as electrochemistry, have been developed for these purposes. Electrochemical biosensors have superior properties over the other existing measurement systems because they can provide rapid, simple and low-cost on-field determination of many biological active species and number of dangerous pollutants. In addition, biosensor technology is a powerful alternative to conventional analytical techniques, combining the specificity and sensitivity of biological systems in small devices. Recently, BIOMET®-sensor (based on light production by metal interacting bacteria) give a general interpretation of the availability of the metals (as a mixture) and of their risks compared to software models that try to calculate the speciation of the metals not taking into account unknown parameters or elements. The bioavailability of metals can change in function of time due to changing conditions as pH, redox, presence of complexing agents etc. A number of recently published papers describe the determination of heavy metals using electrochemical biosensors based on their interactions with DNA, enzymes, bacteria and proteins [Adam *et al.,* 2005]. Kansai Electric Power® has announced that, in collaboration with the Central Research Institute of Electric Power Industry, it has developed innovative antibodies that can identify heavy metals, including cadmium, mercury, and zinc. Using the antibody designed to identify cadmium, the company has developed a biosensor that can easily detect cadmium in soil as well as in foods, for the first time in the world. The biosensor enables researchers to detect heavy metals with the naked eye without having to use any special device. It takes 6-8 hours to detect a heavy metal compared to one week using conventional analyzer-based techniques. Torres *et al* (2005) described the design and properties of a heavy metal biosensor for the measurement of bio-available heavy metals in aqueous environments. The biosensor protein includes an N-terminal, cyan variant of the green fluorescent protein, an intervening metallothionein, and a C-terminal yellow variant of the green fluorescent protein from jellyfish. The genes encoding these proteins were codon-optimized for expression in Chlamydomonas and the modified green-fluorescent variants were engineered to reduce pH effects on fluorescence emission quantum yield and to reduce end-to-end dimerization. They experimentally proved the effects of Cd^{2+}, Cu^{2+} and Pb^{2+} ions on the fluorescence emission spectra of the biosensor. These results point microalgal biosensors like important organisms and could be used in biomonitoring of heavy metals in various water bodies. Though no commercial technology is available / developed for heavy metal monitoring in the marine environment, the biosensor / nanotechnology-based methods would be prospective approach.

REFERENCES

Adam V; Zehnalek J; Petrlova J; Potesil D; Sures B; Trnkova L; Jelen F; Vitecek J; Kizek R. Phytochelatin modified electrode surface as a sensitive heavy metal ion biosensor. *Sensors*, 2005, 5, 70-84.

Andre J; Boudou A; RibeyreF; Bernhard M. Comparative study of mercury accumulation in dolphins (Stenella coeruleoalba) from French Atlantic and Mediterranean coasts. *Sci.Total Environ.* 1991, 104: 191-209.

Armitage, PD; Pardo I; Furse MT; Wright JF. Assessment and predictions of biological quality. A demonstration of a British macroinvertebrate-based method in two Spanish rivers. *Limnetica*, 1990, 6,147–156.

BeckKM; FairP; McFeeW; Wolf D. Heavy metals in livers of bottlenose Dolphins stranded along the South Carolina coast. *Marine Pollution Bulletin*, 1993, 34, 59-63.

Bellan G. Indicateurs et indices biolgiques dans le domaine. *Marin. Bulletin Ecologie*, 1984, 15, 13–20.

Bellan G. Relationship of pollution to rocky substratum polychaetes on the French Mediterranean coast. *Marine Pollution Bulletin*, 1981, 11, 318–321.

Bellan G; Desrosiers G; Willsie A. Use of an annelid pollution index for monitoring a moderately polluted littoral zone. *Marine Pollution Bulletin*, 1988, 19, 662–665.

Bellan G; Olivari A; Picard J. Le peuplement des substrats meubles dans le couloir d'ecoulement des eaux usees de la ville de Marseille, In: Vemes Journees Etudes Pollutions Cagliari, 1980, 649–656.

Bodin N; Burgeot T. Seasonal variation of a battery of biomarkers and physiological indices for the mussel *Mytilus galloprovincialis* transplanted into the northwest Meditterranean Sea. *Comp. Biochem. Physiol. C Toxicol. Pharmacol.* 2004, 138, 411-427

Brummelen TC; Gestal CAM; Verweji RA. Long-term toxicity of five polycyclic aromatic hydrocarbons for the terrestrial isopods *Oniscus asellus* and *Porcellio scaber. Environmental Toxicology and Chemistry*, 1996, 15, 1199-1210

Bucheli TD; Fent K. Induction of cytochrome P450 as a biomarker for environmental contamination in aquatic ecosystems. *Crit. Rev. Environ. Sci. Technol,* 1995, 25, 201-268.

Carballo JL; Naranjo S; Gomez JC. The use of marine sponges at stress indicators in marine ecosystems at Algeciras Bay (southern Iberian Peninsula). *Mar. Ecol. Prog. Ser.*, 1996, 135, 109–122

Chèvre N; Gagné F; Blaise C. Development of a biomarker-based index for assessing the ecotoxic potential of aquatic sites. *Biomarkers*, 2003, 8, 287-298.

Coull BC; Hicks GR; Wells JBJ. Nematode/copepod ratios for monitoring pollution: a rebuttal. *Marine Pollution Bulletin*, 1981, 12, 378–381.

Cramp S; Simmons KEL. Handbook of the birds of Europe, the Middle East and North Africa. *The Birds of the Western Palearctic*, 1977, 1,136-140.

Doelman P; Jansen E; Michels M; Van Til M. Effects of heavy metals in soil on microbial diversity and activity as shown by the sensitivity-resistance index, an ecologically relevant parameter. *Biol. Fertil. Soil*, 1994, 17,177– 184

Eiland F. The effects of application of sewage sludge on microorganisms in soil. *Tidsskrift Planteavl,* 1981, 85, 39-46.

Gadd GM; Griffiths AJ. Microorganisms and heavy metal. *Microbiol. Ecol.,* 1978, 4, 303-317

Guzzo A; Bow DM; Bauda P. Identification and characterization of genetically programmed responses to toxic metal exposure in E. coli. Metals and microorganisms: relationships and application. *FEMS Microbiology Rev.*, 1994, 14, 369-374

Hiroki M. Populations of Cd-tolerant microorganisms in soil polluted with heavy metal. *Soil Sci. Plant Nutr.*, 1994, 40, 515-524

Honda K; Tatsukava R; Itano K; MiyazakiN; Fujiyama T. Heavy metals concentration in muscle,liver and kidney tissue of stripes dolphin Stenella coeruleaoalba and their variation with body length, weight, age and sex. *Agric. Biol.Chem.*, 1983, 47, 6, 1219-1228.

Itano K; Kawai S; Miyazaki N; Miyazaki N; Fujiyama T. Mercury and selenium levels in striped dolphins caught on the Pacific coast of Japan. *Agric. Biol.Chem.*, 1984, 48(5): 1109-1116.

Jamil A; Lajtha K; Radan S; Ruzsa G; Cristofor S; Postolache C. Mussels as bioindicators of trace metal pollution in the Danube Delta of Romania. *Hydrobiologia*, 1999, 392,143-158

Kahle S; Becker PH. Bird blood as bioindicator for mercury in the environment. *Chemosphere,* 1999, 39, 2415–2457.

Kefalas E; Castritsi-Catharios J ; Miliou H. Bacteria associated with the sponge *Spongia officinalis* as indicators of contamination. *Ecological Indicators* 2003, 2, 339-343.

Koeman JH; Peeters WHM; Koudstaa-HolCHM.; TjioePS; Goeij JJM. Mercury-selenium correlations in marine mammals. *Nature,* 1973, 245, 385-386.

Law RJ; Fileman CF; Hopkins AD; Baker JR; Harwood J; Jackson DB; Kennedy S; Martin AR; Morris RJ. Concentrations of trace metals in the livers of marine mammals (seals, porpoises and dolphins) from waters around the British Isles. *Marine Pollution Bulletin,* 1991, 22, 183-191.

Leonzio C; Focardi S; FossiC. Heavy metals and selenium in standard dolphins of the Northern Tyrrhenian (NW Mediterranean). *Sci. Total Environ.*, 1992, 119, 77-84.

Li F; Tan TC. Effect of heavy metal ions on the efficacy of a mixed bacilli BOD sensor. Biosens. *Bioelectron,* 1994, 9, 315-324.

Liz-Marzan LM. Nanometals: formation and color, Mater. *Today* 7, 2004, 26–31.

Manning W ; Feder W. Biomonitoring of Air Pollutants with Plants. London, New York: Applied Science Publishers, 1980.

Marcovecchio JE; MorenoVJ; BastidaRO; GerpeMS; Rodriguez DH. Tissue distribution of heavy metals in small cetaceans from the southwestern Atalantic ocean. *Marine Pollution Bulletin*, 1990, 21, 299-304.

Margalef R. Assessment of the effects on plankton. In: Pearson, E.A., de Frangipanr, E. (Eds.), Marine Pollution and Water Disposal, 1975, Pergamon Press, Oxford, pp. 301–306.

Markert B. Instrumental Element and Multi-element Analysis of Plant Samples. Methods and Applications. Chichester: John Wiley and Sons, 1996.

McCarthy JF; Shugart LR. Biomarkers of Environmental Contamination. Lewis Publishers, Chelsea, Mich, 1990, 457.

Normandin L; Kennedy G; Zayed J. Potential of dandelion (*Taraxacum officinale*) as a bioindicator of manganese arising from the use of methylcyclopentadienyl manganese in unleaded gasoline. *Sci. Total Environ.*, 1999, 239, 165–171.

Oliviera MH; Bonelli R; Aidoo KE; Batista CR. Microbiological quality of reconstituted internal formulations used in hospitals. *Nutrition.* 2000, 16, 729–733.

Peakall DW. Biomarkers: the way forward in environmental assessment. *Toxicol. Ecotoxicol. News,* 1994, 1, 55–60.

Perez T; Wafo E; Fourt M; Vacelet J. Marine sponges as biomonitor of polychlorobiphenyls contamination, concentration and fate of 24 congeners. *Environ. Sci. Technol.,* 2003, 37, 2152–2158

Quetin B; Rouville M. Submarine sewer outfalls. A design manual. *Marine Pollution Bulletin,* 1986, 17, 133–183

Raffaelli DG; Mason CF. Pollution monitoring with meiofauna, using the ratio of nematodes to copepods. *Marine Pollution Bulletin*, 1981, 1211, 158–163.

Ramachandran S; Sundaramoorthy S; Natarajan R. Status of coastal marine pollution in Tamil Nadu. In: Natarajan, R. (Ed.), Coastal Zone of Tamil Nadu. Status Reports, 1989, vol. III. Institute of Ocean Management – Center for Water Resources, Anna University, Madras – 25.

Renzoni A; Focardi S; FossiC; LeonzioC; Mayol J. Comparison between concentrations of mercury and other contaminants in eggs and tissues of Cory's shearwater islands. *Environ. Poll.,* 1986, 40, 17-35.

Starzecka A; Bednarz T. Comparison of development and metabolic activity of algae and bacteria in soil under the influence of short and long term contamination with metallurgic industrial dusts. *Arch. Hydrobiol.*, 1993, 98, 71–88.

Sugunan A; Thanachayanont C; Dutta J; Hilborn JG. Heavy-metal ion sensors using chitosan-capped gold nanoparticles. *Science and Technology of Advanced Materials*, 2005, 6, 335–340.

Tam NFY; Wong YS. Spatial variation of heavy metals in surface sediments of Hong Kong mangrove swamps. *Environmental Pollution*, 2000, 110, 195-205.

Temara A; Skei JM; Gillan D; Warnau M; Jangoux M; Dubois P. Validation of the Asteroid Asterias rubens (Echinodermata) as a bioindicator of spatial and temporal trends of Pb, Cd, and Zn contamination in field. *Marine Environmental Research*, 1998, 45, 341–356.

Torres MA; FalcaoVR; Colepicolo P; Rajamani S; Ewalt J; Sayre RT. Transgenic Microalgae as Heavy Metal Biosensors. T11 AM Aquatic Plants. Methods, Mechanisms and Markers, 2005, 343-344.

Webster NS; Negri AP; Wol CW; Maclean WJ; Munro MHG; Battershill CN. Human impacts and microbial ecology of antartic sponges. *Boll. Mus. Ist Boil. Univ. Genova*, 2002, 209, 66–67.

Wilhelmi J. Ubersicht uber die biologische Beurteilung des Wasers, Ges naturf Freunde, 1916, Berlin, 297–306.

Wood JM; Wang HK. Microbial resistance to heavy metals. *Environ. Sci. Technol,* 1983, 17, 582-590.

Wright JF; Furse MT; Armitage PD. RIVPACS a technique for evaluating the biological quality of rivers in the UK. *European Water Pollution Control*, 1993, 34, 15–25.

Zahn RK; Ahn G; Muller WEG; Kurelec B; Rijavec M; Batel R. Assessing consequences of marine pollution by hydrocarbons using sponges as model organisms. *Sci. Total Environ*, 1981, 20, 147–169

Zheng J; Zhang C; Dickson RM. Highly fluorescent, water-soluble, size-tunable gold quantum dots. *Phys. Rev. Lett.* 93, 2004, 077402.

In: Causes and Effects of Heavy Metal Pollution
Editor: Mikel L. Sanchez

ISBN 978-1-60456-900-1

Chapter 9

HEAVY METAL CONTAMINATION IN SELECTED URBAN COASTAL REGIONS IN US AND CHINA

***Huan Feng**[*1], **Weiguo Zhang**[2], **Luoping Zhang**[3], **Xu-Chen Wang**[4,5], **Lizhong Yu**[2] and **Danlin Yu**[1]*

1. Department of Earth and Environmental Studies, Montclair State University, New Jersey 07043, USA
2. State Key Laboratory of Estuarine and Coastal Research, East China Normal University, Shanghai 200062, PRC
3. State Key Laboratory of Marine Environmental Science, Environmental Science Research Center, Xiamen University, Xiamen, Fujian 361005, PRC
4. Department of Marine Ecology and Environmental Studies, Institute of Oceanography, Chinese Academy of Sciences, Qingdao, 266071, PRC
5. Department of Environmental, Earth and Ocean Sciences, University of Massachusetts at Boston, Boston, MA 02120-3393, USA

ABSTRACT

With urbanization and economic development in coastal area, metal pollution in coastal environment has been a problem. Estuaries and coastal intertidal zone are important habitats for aquatic and marine life. In the meantime, sediment is a repository of contaminants and records the pollution history. Sediment quality reflects the long-term health status of an estuarine or coastal system and can be evaluated by sediment quality guidelines/criteria, metal enrichment factors and other indicators. In this study, we summarize the results mainly from our previous studies in metal pollution in sediments in urban estuarine and coastal systems in US and China, including New York Harbor and Yangtze River estuary in metropolitan areas and Xiamen Bay and Jiaozhou Bay. The information from this study expands our knowledge in understanding metal pollution in urban coastal systems and assessing environmental quality impacted by industrialization and economic development.

---

* Corresponding author. Tel.: 973-655-7549; fax: 973-655-4072. E-mail: fengh@montclair.edu (H. Feng).

1. INTRODUCTION

Metals are naturally present in our environment and trace amount of metals is essential micronutrients for growth of many marine organisms. However, excessive amount of heavy metals can be toxic to marine life in the coastal environment (Underwood 1971; Langston 1990). With the urbanization and economic development, metal pollution in water and sediments in coastal areas has become a major environmental concern because the coastal area has a high population density and the estuaries and intertidal zone are important habitats for aquatic and marine life. Estuarine and coastal environment are considered as a complex system, where physical, chemical, and biological processes play major roles in behavior, transport and fate of metal contaminants. In coastal pollution study, anthropogenic impact is more apparent in estuarine and coastal area than that in open ocean because contaminants including toxic heavy metals are carried to the ocean by rivers via estuarine systems. Many heavy metals are transported predominantly in association with particulate matter that eventually settle down to the sediments, and consequently, high concentration of heavy metals are often detected in sediments in many industrialized harbors and coastal regions around the world because the major sources of anthropogenic metals to a coastal area are due to industrialization and urban development with population growth in the area (Forstner and Wittman 1983; Bothner et al. 1998; Feng et al. 2004; Santos et al. 2005). Therefore, estuarine and coastal sediments are repository for these pollutants that enter the sea and useful for environmental studies because it can provide time-integrated records (Goldberg et al. 1979; Trefry and Shokes 1981; Windom et al. 1989; Bopp and Simpson 1989; Valette-Silver 1993). This article summarizes our previous metal contamination studies in several coastal areas in the United States and China. It reflects the environmental concerns due to heavy metal contamination in these selected urban coastal regions in US and China with the urbanization and economic development.

2. STUDY AREAS

2.1. Lower Hudson River - New York Harbor, USA

The Hudson River estuary is classified as a partially mixed estuary and the water sources to the New York Harbor (km point -11 to 18, as defined by Olsen et al. 1984) are mainly Hudson River water, New York Bight water and New Jersey/New York waste water (Abood 1978; Klinkhammer and Bender 1981; Cooper et al. 1988) (Figure 1). With the early industrial development in this urban coastal area, pollution becomes a serious environmental problem. Sediments in the New York Harbor are found to be significantly contaminated by heavy metals, PCBs and other contaminants (Williams et al. 1978; Olsen et al. 1984; Bopp and Simpson 1989; Hunt et al. 1994; Gibbs 1994; Chillrud 1996; Feng et al. 1998a).

Sedimentation rates in the Hudson River estuary were estimated by Oslen et al. (1978) using reactor- and bomb-produced radionuclides ($^{239,240}Pu$, ^{137}Cs, ^{134}Cs and ^{60}Co) as time-stratigraphic markers for the sediment record.

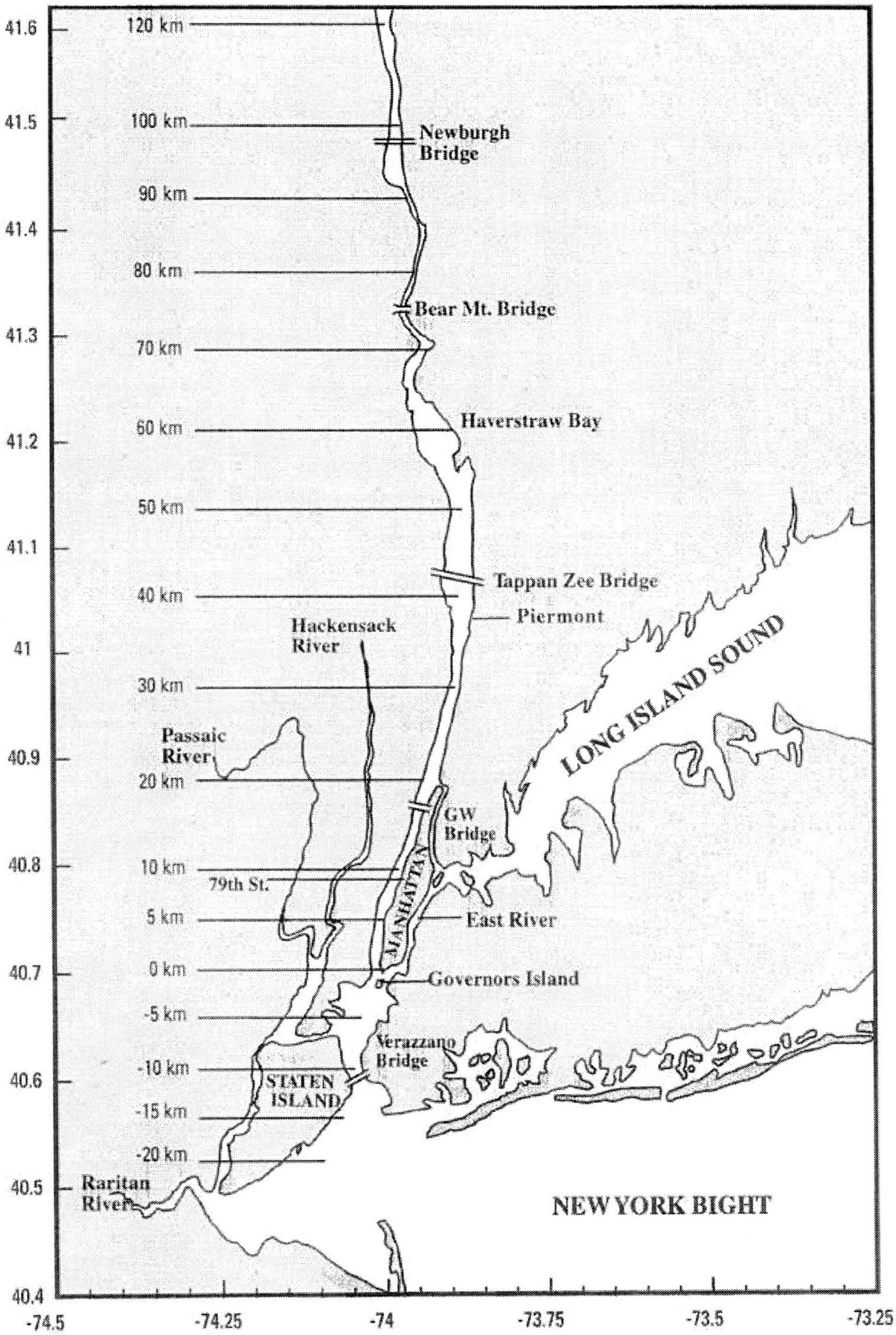

Figure 1. Map showing the lower Hudson River estuary, USA. (From Feng et al. 1998a).

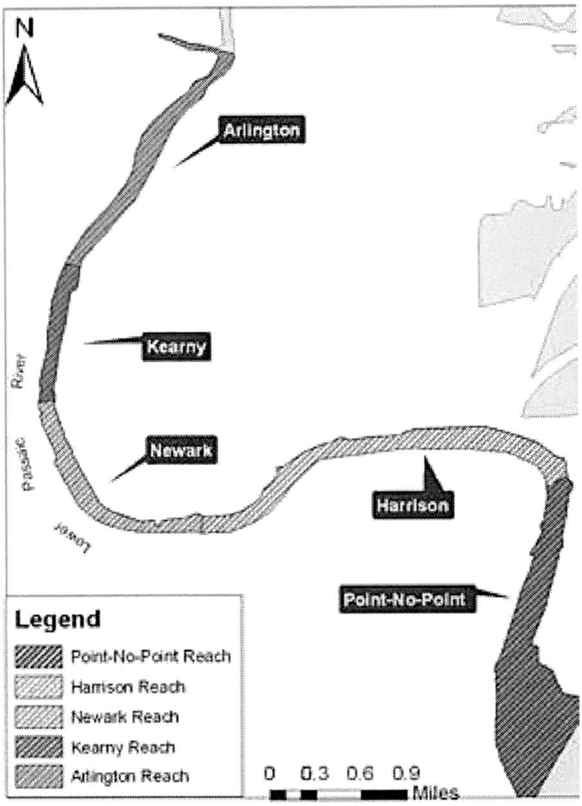

Figure 2. Map showing the lower Passaic River, New Jersey, USA. (From Onwueme and Feng, 2006).

Relating these radionuclide profiles in the Hudson sediments to the known input histories of these radionuclides to the estuary, they found that fine-grained sediment accumulation rates are between 5 and 20 cm y^{-1} in most depositional areas of the New York Harbor. They also found that the fine-grained particle accumulation rates in some areas adjacent to New York City are generally >3 cm y^{-1} and often as high as 10 to 20 cm y^{-1}, assuming that bioturbation in the Hudson River estuary is relatively unimportant (Olsen et al. 1981). Feng et al. (1998b) estimated that sedimentation rates in the western margin area at kmp ~8 varied from 6 to 26 cm y^{-1} based on ^{7}Be profiles.

2.2. Lower Passaic River, New Jersey, USA

The Passaic River is about 14 miles west to New York City, located in the New Jersey-New York metropolitan area. The Passaic River system consists of major and minor tributaries, and drains much of eight New Jersey counties as well portions of two New York counties and ultimately reaches Newark Bay at its confluence with the Hackensack River (Figures 1 and 2). The drainage basin of northern New Jersey's Passaic River system covers approximately 2200 km^2, about 11% of the state's surface area (NJDEP 1987). The basin contains some of the most densely populated land in the nation and is afflicted with numerous environmental problems, including major industrial pollution and suburban sprawl. This river has been heavily polluted by dioxins, PAHs, PCBs and heavy metals due to industrial activities and urbanization (Huntley et al. 1993, 1995, 1997; Iannuzzi et al., 2002; Onwueme and Feng 2006).

The Passaic River environmental status has been drawing much regional and national attention due to its urban environmental setting. The lower 10 km reach of the Passaic River is one of EPA Superfund Sites (Figure 2). In the lower Passaic River, fine-grained contaminated sediments have received considerable attention because these contaminants (metals, PCBs, PAHs, dioxins, etc.) have a potential to release into the aquatic system and air through diffusion, resuspension and/or evaporation, causing human health hazardous affect (Huntley et al. 1997; Walker et al. 1999; Wenning et al. 1994).

Since early 1980s, federal and New Jersey state agencies and private sectors have funded a series of remedial investigation of the lower 10 km of the Passaic River to characterize the horizontal and vertical distributions and concentrations of dioxins, PCBs, and metals in its sediment (Crawford et al. 1995; Finley et al. 1990; Hutley et al. 1993, 1995, 1997; Tierra Solutions, Inc. 2003; Wenning et al. 1993a, 1993b; Wolfskill and McNutt 2000). Numerous data were produced from these studies.

2.3. Yangtze River Estuary, China

The Yangtze River is the third largest river in the world and delivers about 4.7×10^8 tons of sediments to the sea annually (Chen 1998) (Figure 3). Therefore, intertidal zone in Yangtze River Estuary has been extensively well developed (Chen 1998). Previous studies show that heavy metal pollution has been found in some areas of the intertidal zone (Zhang et al. 2001; Feng et al. 2004). Shanghai, which is located along the Yangtze Estuary and one of the largest cities with most active industrial activities in China, has a direct impact on the coastal environment. It was reported that 5 million tons of industrial and domestic sewage were discharged into the Yangtze River estuary daily (Dai and Gu 1990). In addition to the local waste discharge, many areas in the intertidal zone have been used for harbors, iron and steel mills, and other heavy industries. Due to urbanization and economic development in past decades, metal contamination in the Yangtze River Estuary has received much attention (Xu et al. 1997; Chen et al. 2001; Feng et al. 2004; Zhang 1999; Zhang et al. 2001, 2007a).

2.4. Xiamen Bay, China

Western Xiamen Bay (24°29'N, 118°04'E) is a semi-enclosed bay in southeast China (Figure 4). The water depth ranges from 6-25 m with a deep-water coastline of about 30 km. Thus, it has an excellent natural condition for navigation and shipping activities. Although the economic development in Xiamen Economic Special Zone speeds up the urbanization and industrialization of the region in past 25 years, it has resulted in great environmental stress to western Xiamen Bay and its adjacent areas and brought environmental problems to the region.

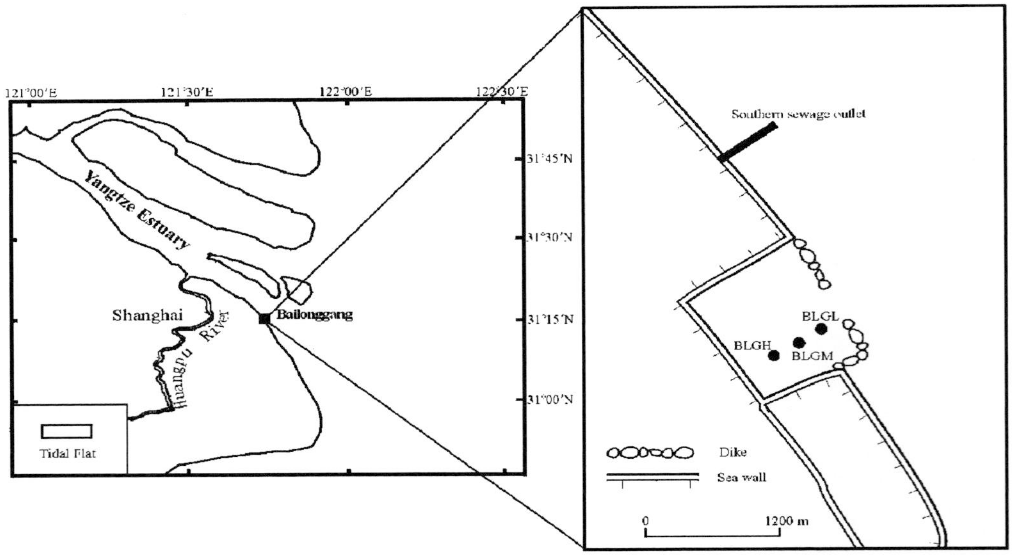

Figure 3. Map showing the study area in Yangtze River estuary intertidal zone, Shanghai, China. (From Feng et al. 2004).

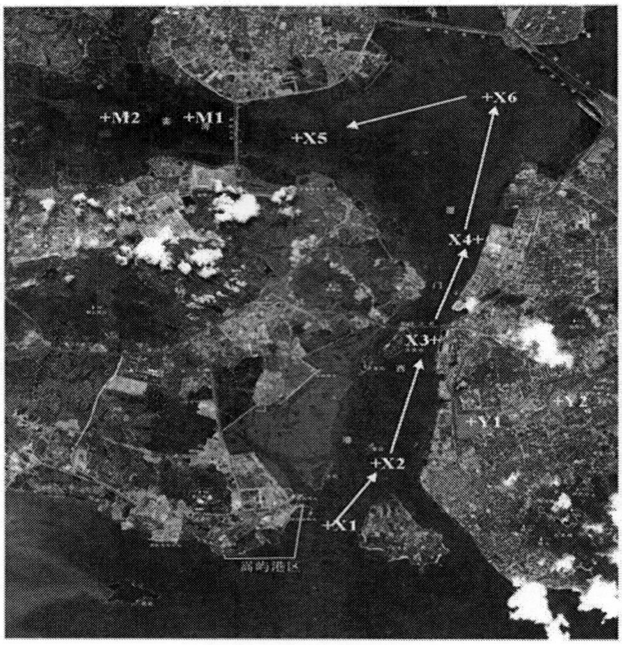

Figure 4. Map showing the study area in the western Xiamen Bay, Xiamen, China. (From Zhang et al. 2007b).

2.5. Jiaozhou Bay, China

Jiaozhou Bay is a semi-enclosed coastal embayment located on the east coast of China and adjacent to the Yellow Sea (Figure 5). The area of Jiaozhou Bay is about 400 km^2 with an average water depth of 7 m. Jiaozhou Bay is surrounded by three cities: Qingdao, Jiaozhou and Jiaonan that have a combined population of over 7 million and is one of the most populated coastal regions in the east coast of China. Over the years, Jiaozhou Bay has been the most important resource for civil, industrial and commercial development for the seaport city of Qingdao. Jiaozhou Bay is also the home of Qingdao Port, located on the east side of the bay and adjacent to the city. Qingdao Port is one of China's major trading ports, one of the top 7×10^8 ton ports in China, and one of the top 20 trading ports in the world. In the last ten years, the rapid urbanization along Jiaozhou Bay has affected the bay's ecosystem dramatically (Shen 2001; Wang et al. 2007). As a semi-enclosed system, the accumulation of pollutants discharged to the bay and deposition to the sediments has particular environmental concerns.

3. Result and Discussion

Table 1 summarizes metal contaminant concentrations in sediments in our selected areas in US and China. Significant spatial variations are seen in the study areas (Table 1). For comparison, natural levels of metal concentrations in rocks are also tabulated in Table 1. By comparison, it is seen that some metal concentrations in our study areas are above the natural levels, implying metal contamination. To properly evaluate anthropogenic heavy metal contamination in sediments, we can use enrichment factor (EF) of metals as an index. Mathematically, EF is a concentration ratio of observed metal to aluminum (or iron) in the sample of interest divided by the background metal/aluminum (or iron) concentration ratio and is expressed as (e.g., Covelli and Fontolan 1997; Feng et al. 1998a, 2004; Zhang and Liu 2002; Zhang et al. 2007b):

$$EF = \frac{\left(\frac{Me}{Al\,(\mathrm{Fe})}\right)_{Sample}}{\left(\frac{Me}{Al\,(\mathrm{Fe})}\right)_{Background}} \qquad (1)$$

where $(\mathrm{Me/Al(Fe)})_{\mathrm{Sample}}$ is the metal to Al (or Fe) ratio in the samples of interest; $(\mathrm{Me/Al(Fe)})_{\mathrm{Background}}$ is the natural background value of metal to Al (or Fe) ratio.

Advantage of using enrichment factor (EF) is that it can reflect the degree of metal contamination in the environment. This technique has been well applied in several studies to assess metal contamination in marine sediments (Windom et al. 1989, 1991; Zhang and Liu 2002; Feng et al. 1998a, 2004; Lu et al. 2005; Zhang et al. 2007b). The assessment criteria from Han et al. (2006) are listed in Table 2. In general, if an EF value is less than 2 (i.e., $EF < 2$), it suggests that the trace metals may be entirely from crustal materials or natural weathering processes, while a value of EF is greater than 2 (i.e., $EF>2$), it suggests that a significant portion of trace metal is delivered from non-crustal materials or non-natural weathering processes and a varying degrees of metal contamination exist depending on EF values (Sutherland 2000; Loska and Wiechuya 2003; Zhang and Liu 2002; Han et al. 2006).

Table 1. Heavy metal concentration in surface sediments (< 5cm) in selected urban estuaries and coastal areas in USA and China

Region		Ag (ppm)	Cd (ppm)	Cr (ppm)	Cu (ppm)	Hg (ppm)	Ni (ppm)	Pb (ppm)	Zn (ppm)	Reference
Hudson River Estuary, USA	mean±s.d	2.23 ± 1.49	1.00 ± 0.56		77.9 ± 31.7			81.5 ± 33.5	182 ± 41	Feng et al. 1998a
	range	0.20 - 6.38	0.18 - 2.29		17.5 - 149			23.7 - 177	101 - 257	
New York Harbor, USA	mean±s.d		4.6 ± 3.1		55 ± 30			92 ± 36	109 ± 62	Gibbs 1994
New York Harbor, USA	mean±s.d		1.6 ± 0.5		135 ± 27			150 ± 31	265 ± 64	Chillrud 1996
Lower Passaic River, USA	mean±s.d	6.37 ± 6.05	6.73 ± 4.11	198 ± 121		4.31 ± 2.33	61.5 ± 49.4	417 ± 338	681 ± 332	This Study
	range	0.62 - 42.3	0.32 - 29	8 - 860		0.1 - 12.4	7.1 - 369	4.4 - 2500	21 - 1900	
Yangtze River Intertidal Zone	mean±s.d				56.6 ± 12.1			31.7 ± 6.1	233 ± 131	Feng et al. 2004
(near a major sewage outlet)	range				40.6 - 84.1			23.9 - 44.7	109 - 550	
Yangtze River Intertidal Zone	range		0.025 - 0.925	58.43 - 567	29.5 - 226			9.35 - 582	58.2 - 842	Xu et al. 1997
Xiamen Bay, China	mean±s.d		0.33 ± 0.28	74.5 ± 25.5	44.0 ± 21.0		37.4 ± 12.1	54.0 ± 3.7	139 ± 39	Zhang et al. 2007b
	range		0.11 - 1.01	36.7 - 134	18.5 - 97.2		24.8 - 64.8	44.9 - 59.8	65 - 223	
Jiaozhou Bay, China	mean±s.d		0.059 ± 0.027	22.1 ± 14.9	12.0 ± 7.5		13.2 ± 7.5	10.3 ± 4.1	31 ± 15	Wang et al. 2007
	range		0.030 - 0.110	5.0 - 51.0	4.2 - 28.0		2.7 - 23.0	5.1 - 18.0	12 - 58	
Continental Rocks		0.07	0.2	71	32		49	16	127	Martin and Whitfield, 1983

3.1. Lower Hudson River - New York Harbor, USA

In the lower Hudson River estuary including New York Harbor, the concentrations of metal contaminants show a wide spatial variation. Feng et al. (1998a) reported that metal concentrations in the Hudson River estuary sediments range from 0.20 to 6.38 ppm for Ag, 0.18 to 2.29 ppm for Cd, 18 to 149 ppm for Cu, 24 to 177 ppm for Pb and Zn from 101 to 257 ppm for Zn. (Table 1). Gibbs (1994) found that metal concentrations in the New York Harbor sediments were 4.6±3.1 ppm for Cd, 55±30 ppm for Cu, 92±36 ppm for Pb and 109±62 ppm for Zn (Table 1). Chillrud (1996) also reported the concentrations of metal contaminants in the New York Harbor, which are 1.6±0.5 ppm for Cd, 135±37 ppm for Cu, 150±31 ppm for Pb and 265±64 ppm for Zn (Table 1).

Results from the previous studies indicate that mechanisms and processes governing the heavy metal contaminant input and distributions in the harbor could be explained by their chemically particle-reactive characteristics and estuarine hydrodynamics (Ullman and Wilson 1998; Feng et al. 2002). . When these metal contaminants were discharged into the Hudson River from the sources, they were scavenged onto the surface of suspended particles in water column, settled down to sediments along with the particles, and then resuspended and redistributed in the estuary (Hines et al. 1984; Windom et al. 1991; Feng et al. 1999a). Feng et al. (1999b) estimated that the residence times of ^{234}Th- and ^{7}Be-carrying particles in the Hudson estuary range from <1 to 10 d and the sediment could be transported up to 10-15 km up- and down-estuary during a tidal cycle.

Based on contamination categories by Han et al. (2006) (Table 2), metal enrichment factors in the Hudson River estuary sediment indicate that, the Hudson River estuary are generally contaminated by Ag, Cd, Cu, Pb and Zn and the contamination are more significant in some areas (Table 3).

Sediment contamination in the Hudson River estuary sediments has been studied by a number of workers since the 1970s (e.g., William et al. 1978; Olsen et al. 1978, 1981, 1984; Rohmann 1988; Bopp and Simpson 1989; Gibbs 1994; Hunt et al. 1994; Chillrud 1996; Hirschberg et al. 1996), in order to better understand the environmental quality of the Hudson River estuary, and the sources and factors controlling the sediments-associated contaminant distributions. Williams et al. (1978) found that Cu, Pb and Zn concentrations in the New York harbor sediments were high and anthropogenic metals discharged from dispersed contaminant sources were apparent to depths >60 cm in cores taken in this area. According to Olsen et al. (1984), deposition of suspended particles in the harbor removes virtually the entire riverborne load of particle-associated contaminants.

Along the Hudson River estuary, Gibbs (1994) found that the bottom sediment has a metal concentration maximum in the New York Harbor. Normalization of metal concentrations to Fe indicated a maximum level of pollution in the harbor (Gibbs 1994). Gibbs (1994) attributed the high metal concentrations in the harbor to local sources as well as sediment properties and hydrodynamics. Chillrud (1996) found that most metal contaminants such as Cd, Cu, Pb and Zn in New York Harbor sediments had declined by 50-90% from their maximum levels reached in the 1960s and 1970s, and the decreases seemed continuing during the early 1990s.

In an urban estuarine system, Ag can be referred to a good indicator of urban waster effluent input (Sañudo-Wilhelmy and Flegal 1992; Smith and Flegal 1993). In the Hudson

River estuary, Feng et al. (1998a) found that anthropogenic Cd, Cu, Pb and Zn showed linear correlations with anthropogenic Ag, suggesting that Cd, Cu, Pb and Zn were introduced to the Hudson River estuary along with Ag from the urban sources and distributed and accumulated in the area. It is very likely that that a certain amount of heavy metal contaminants are transported and redistributed with sediments due to estuarine circulation (Geyer 1993; Feng et al. 1999a, 1999b). These studies show that the Hudson River estuary is a heavily contaminated estuary.

3.2. Lower Passaic River, New Jersey, USA

The Passaic River basin contains some of the most densely populated land and the most polluted section of the river in the nation (Bonnevie et al. 1992, 1993, 1994; Gillis, et al. 1993 a, 1993b, 1995; Huntley et al. 1993, 1995, 1997; Iannuzzi and Wenning 1995; Iannuzzi et al. 1995, 1997, 2002; Wenning et al. 1993a, 1993b, 1994; Feng et al. 2005; Onwueme and Feng 2006). In this study, we analyzed the existing large dataset from the previous federal and New Jersey state sampling programs between 1990 and 2000 (e.g., R-EMAP 1998) and found that the contaminant concentrations in the lower Passaic River sediments are 6.37±6.06 ppm (ranging from 0.62 to 42.3 ppm) for Ag, 6.73±4.11 ppm (ranging 0.32-29 ppm) for Cd, 198±121 ppm (ranging 8-860 ppm) for Cr, 4.31±2.33 ppm (ranging 0.1-12.4 ppm) for Hg, 61.5±49.4 ppm (ranging 7.1-369 ppm) for Ni, 417±338 ppm (ranging 4.4-2500 ppm) for Pb, and 681±332 ppm (ranging 21-1900 ppm) for Zn (Table 1). We also calculated metal enrichment factors in this study and found that metal enrichment factor for Ag is 2.9±2.2 (ranging 0.2 - 15.9); Cd, 10.6±6.9 (0.4 - 51.2); Cr, 6.5±4.2 (1.0 - 32.4); Hg, 10.6±6.1 (0.5 – 37); Ni, 2.1±1.7 (0.7 – 12); Pb, 20±24 (0.2 – 207); and Zn, 8.2±4.1 (1.0 – 28) (Table 3). Based on the contamination categories (Han et al. 2006; Table 2), the metal enrichment factor indicate that the lower Passaic River is moderately to significantly contaminated by these metals as a whole and the contamination in some areas are very high.

Like other urban watershed and river-estuary systems, discharges of a variety of contaminants through runoff, river tributaries and industrial wastewater outfalls are problematic in coastal urban areas (Brosnan et al. 1994; Sañudo-Wilhelmy and Flegal 1994; Feng et al. 1998a, 2004; Zhang et al. 2007b). The lower Passaic River system is not an exception. In 1970, the United States Environmental Protection Agency (USEPA) declared the Passaic River the "second most polluted river in America," behind only the Cuyahoga, which had caught fire in 1969 (Peet and Johnson 1996), and the most chemically polluted estuary in the nation (NOAA 1984). In an earlier study, Feng et al. (2005) found that the Harrison Reach sediment in the lower Passaic River has the highest As, Cr, Hg, Pb and Zn concentrations. As there were and are numerous of industrial manufactures situated on the both side of the lower Passaic River, they could have been potential sources of these contaminants to the Passaic River (Bonnevie et al. 1992, 1993, 1994; Crawford et al. 1995; Gills et al. 1993a, 1993b; Huntley et al. 1997). This is especially true in the past when the discharge of various contaminants to the Passaic River system was uncontrolled and unlimited (Huntley et al. 1995; Iannuzzi and Wenning 1995; Iannuzzi et al. 1997; Meyserson et al. 1981).

Table 2. Contamination categories based on enrichment factor values (Han et al. 2006)

EF value	Category
EF<2	deficiency to minimal enrichment
EF=2–5	moderate enrichment
EF=5–20	significant enrichment
EF=20–40	very high enrichment
EF>40	extremely high enrichment

3.3. Yangtze River Estuary, China

In the Yangtze River intertidal zone, sediments near a point source of contaminants (e.g., a major sewage outlet) can preserve a record of contamination from the source and, therefore, metal concentration in the sediment could reflect an environmental impact due to urbanization. Feng et al. (2004) conducted a preliminary study in high, middle and low tidal flats of the Yangtze River estuary intertidal zone. Three short sediment cores (<20 cm) were collected in the high, middle and low tidal flats in the Yangtze River Estuary near the Southern (Nanqu) Sewage Outlet, one of the three largest sewage outlets in Shanghai, China. They found that metal concentrations in surface sediments (<5 cm) are 56.6±12.1 ppm (ranging 40.6 - 84.1 ppm) for Cu, 31.7±6.1 ppm (ranging 23.9 - 44.7 ppm) for Pb and 233±131 ppm (ranging 109 - 550 ppm) for Zn (Table 1). In an earlier study in the Yangtze River estuary intertidal zone, Xu et al. (1997) found that Cd concentration in the sediments ranges from 0.025 to 0.925 ppm; Cr, 58.4 - 567 ppm; Cu, 29.5 - 226 ppm; Pb, 9.35 – 582 ppm; and Zn 58.2 – 842 ppm. Feng et al. (2004) also calculated the metal enrichment factor based on their study, which are 3.2±0.6, 1.6±0.3 and 4.7±2.4 for Cu, Pb and Zn, respectively with a range of 2.5 – 4.6, 1.3 – 2.1 and 2.4 - 10.8, respectively (Table 3).

Although the sediment in Yangtze River intertidal zone is mainly from Yangtze River upper stream sediment discharge (4.7×10^{8} tons y^{-1}, Chen 1998), our early preliminary study (Feng et al. 2004) in a localized area indicate that the environmental impact due to urbanization and industrial development, as shown by this localized intertidal zone, is significant as reflected by their statistical analysis that time has a significant influence on metal concentrations whereas grain size has minor influence. The results from Feng et al. (2004) show that the intertidal zone near the sewage outlet is moderately contaminated by Cu and Zn, and, to less extent, contaminated by Pb based on Han et al. (2006) (Table 2). Feng et al. (2004) also indicate that the metal contamination is especially true in surface sediments due to environmental impact in recent years. The high EF values (EF>2, Table 3) support the conclusion that anthropogenic input of the metals is mainly due to urbanization and economic development in Shanghai metropolitan area happened in the past 10-20 years.

Table 3. Heavy metal enrichment factors (EF)in surface sediments (< 5cm) in selected urban estuaries and coastal areas in USA and China

Region		Ag	Cd	Cr	Cu	Hg	Ni	Pb	Zn	Reference
Hudson River Estuary, USA	mean±s.d.	9.4 ± 5.4	8.8 = 4.3		5.7 = 2.0			5.8 = 2.1	2.3 ± 0.3	Feng et al. 1998a
	range	0.8 - 23	2.0 - 18		1.5 - 10			1.9 - 12	1.5 - 3.1	
Lower Passaic River, USA	mean±s.d.	2.9 ± 2.2	10.6 = 6.9	6.5 ± 4.2		10.6 ± 6.1	2.1 ± 1.7	20 = 24	8.2 ± 4.1	This Study
	range	0.2 - 16	0.4 - 51	1.0 - 32		0.5 - 37	0.7 - 12	0.2 - 207	1.0 - 28	
Yangtze River Intertidal Zone	mean±s.d.				3.2 = 0.6			1.6 = 0.3	4.7 ± 2.4	Feng et al. 2004
	range				2.5 - 4.6			1.3 - 2.1	2.4 - 11	
Xiamen Bay, China	mean±s.d.		1.6 = 1.2	1.0 ± 0.5	1.3 = 0.6		0.7 ± 0.3	3.1 = 0.4	1.0 ± 0.2	Zhang et al. 2007b
	range		0.6 - 4.1	0.6 - 2.2	0.8 - 2.7		0.4 - 1.5	2.6 - 3.8	0.8 - 1.4	
Jiaozhou Bay, China	mean±s.d.		2.7 = 2.6	1.8 ± 0.4	2.4 = 1.0		1.6 ± 0.4	4.9 = 3.0	1.7 ± 0.9	Wang et al. 2007
	range		1.0 - 7.6	1.4 - 2.7	1.6 - 5.0		1.2 - 2.4	2.5 - 12	1.0 - 3.9	

3.4. Western Xiamen Bay, China

In western Xiamen Bay, field samplings for surface sediments (<5 cm) were conducted in December 2004 and July 2005, respectively, in order to examine if there is a significant temporal variation of metal concentrations and distribution. The sampling stations in western Xiamen Bay, Maluan Bay and Yuandang Lagoon are shown in Figure 4.

In a recent metal contamination study in western Xiamen Bay, China, Zhang et al. (2007b) found that metal concentrations in the surface sediments from the two samplings (December 2004 and July 2005, respectively) show a general consistency and the concentrations vary from 0.11-1.01 ppm for Cd, 37-134 ppm for Cr, 19-97 ppm for Cu, 25-65 ppm for Ni, 45-60 mg ppm for Pb and 65-223 ppm for Zn. They indicate that metal contaminant sources to the western Xiamen Bay and adjacent areas should include land-based point and non-point input, riverine/stream discharge and atmospheric fallout. Studies on atmospheric inputs to China sea have become a new focus of biogeochemical cycle in the past few years (e.g., Gao et al. 2002). However, this information of atmospheric fallout to the study area is missing based on our literature review. As the western Xiamen Bay area is within the coastal region, we expect the land-based influence should be more predominant.

The metal enrichment factors in Xiamen Bay sediment from Zhang et al. (2007) show that Cu enrichment factor, EF (Cu), ranges from 0.8 to 2.7, EF (Pb) from 2.6 to 3.8, and EF (Zn) from 0.8 to 1.4, EF (Cd) from 0.6 to 4.1, EF (Cr) from 0.6 to 2.2, and EF (Ni) from 0.4 to 1.5 (Table 3). Comparing with the metal contamination categories (Table 2), the metal enrichment factors (Table 3) indicate moderate Pb contamination in sediment appears everywhere in western Xiamen Bay, Ni and Zn contaminations are not significant at present, and moderate Cu, Cd, and Cr contaminations exist in some areas in western Xiamen Bay.

The results show that with the creation of Xiamen Economic Special Zone and establishment of industrial enterprises, the environmental impact has been seen as demonstrated by metal contamination in western Xiamen Bay and adjacent areas, which is becoming an environmental problem. Zhang et al. (2007b) indicated that although metal (Cu, Pb, Zn, Ni, Cd and Cr) concentrations in the surface sediments meet the Chinese National Standard of Marine Sediment Quality Criteria (CSTBS, 2002), metal enrichment factors (EF) and other index confirm the certain extent of metal contamination in the western Xiamen Bay and adjacent areas and that point source input of some specific metal contaminant(s) has caused significant sediment contamination in the local area.

3.5. Jiaozhou Bay, China

In our another study (Wang et al. 2007), sediments were collected from 10 sites in and outside Jiaozhou Bay (Figure 5) using an Ekman style stainless steel grab sampler (0.1 m^3). Wang et al. (2007) found that concentrations of toxic heavy metals ranged from 0.03 to 0.11 ppm for Cd, 5 to 51 ppm for Cr, 4.2 to 28 ppm for Cu, 2.7 to 23 ppm for Ni, 5.1 to 18 ppm for Pb, and 12 to 58 ppm for Zn. The distribution of heavy metal contaminants in the surface sediments showed significant spatial variations among the stations due to variability in metal source input.

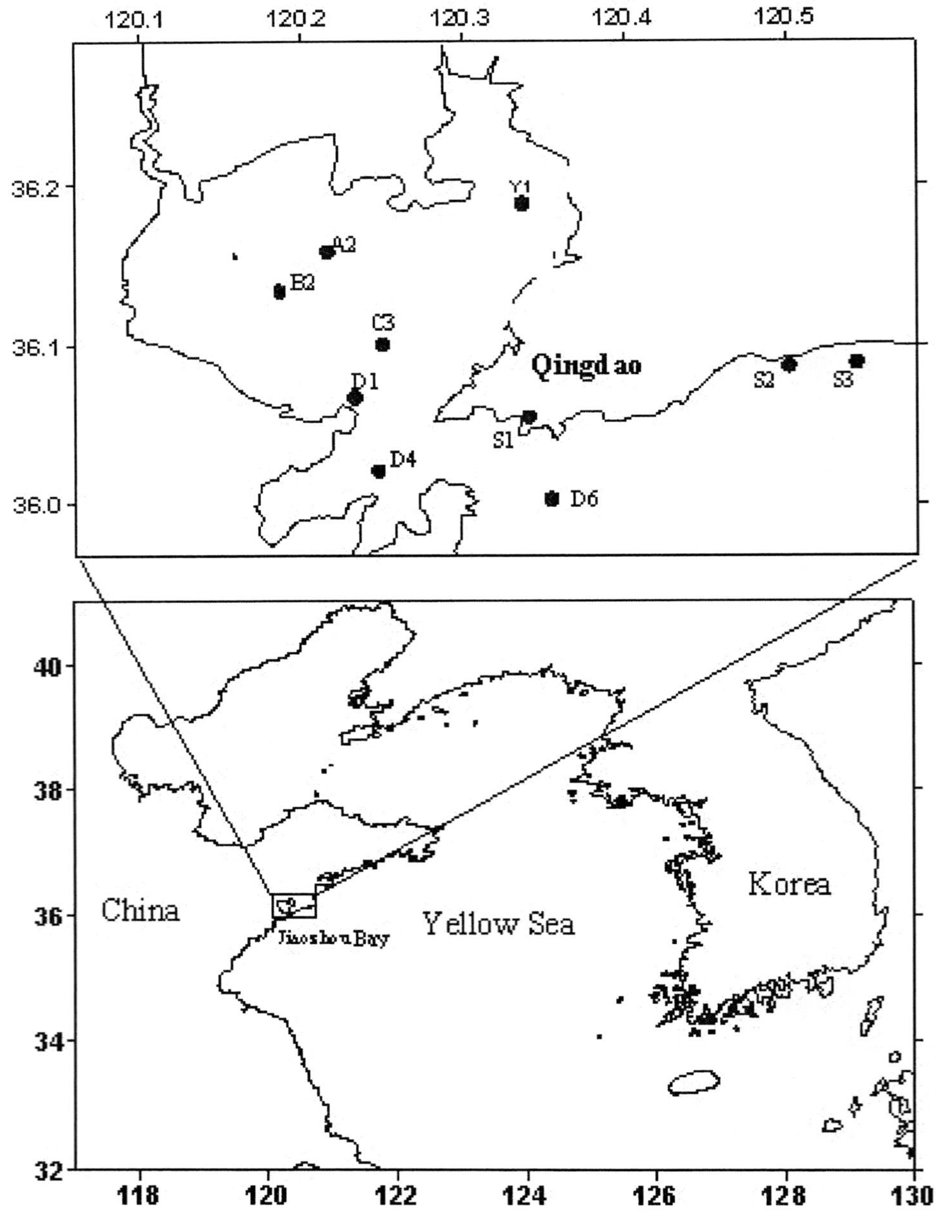

Figure 5. Map showing the study area in Jiaozhou Bay, China. (From Wang et al. 2007).

They found strong correlations between metal contaminants (Cu, Cr, Ni, Pb and Zn) and Al in the sediments (r^2 = 0.892 - 0.973), indicating that these heavy metals were strongly associated with the clay mineral phases in the sediments since Al is one of the most abundant elements in clay minerals and the variation of Al concentration usually suggests differences in particle grain size and Al-Si clay mineral content.

Wang et al. (2007) also used enrichment factor (EF) as an index to examine the extent of metal contamination in the sediments of Jiaozhou Bay. Based on the contamination categories (Han et al. 2006; Table 2), the metal EF values indicate that the sediments in Jiaozhou Bay have been contaminated by Cd, Cr, Cu, Ni, Pb and Zn to various degrees that, with a wide variation, Cd, Cu and Pb show a moderate enrichment in the sediments on average while Cr, Ni and Zn show a less contamination (Table 3). Lead (Pb) contamination is relatively more significant.

Although the sediment quality guidelines (SQDs) (Long et al. 1995, 1998) are useful indicators for assessing sediment contamination, Wang et al. (2007) show that the level of contamination may not be properly evaluated by comparing metal concentrations only without considering sedimentary mineralogy and geochemistry. For example, enrichment factor of metals is a useful indicator for assessment of metal contaminations in sediments. Based on their results, Wang et al. (2007) indicate that sediment grain size and organic matter play important roles in controlling the distribution of the heavy metals in surface sediments of Jiaozhou Bay and the major inputs of metal contamination to Jiaozhou Bay are from land-based anthropogenic sources such as discharges of industrial wastewater and municipal sewage and run-off. The elevated heavy metal concentrations in Jiaozhou Bay sediments could have significant impact on the bay's ecosystem. With the rapid economic development and urbanization around Jiaozhou Bay, coastal management and pollution control should focus on these contaminant sources, and continuing monitoring studies of heavy metal contamination within the bay is needed.

4. SUMMARY

In studying metal pollution in sediments, the extent of metal contamination can be categorized using the metal enrichment factor as suggested by Han et al. (2006). In this study, we compared the metal enrichment factors in different coastal regions (Figure 6). As shown in Figure 6, although the predominant metal contaminants vary from one place to another depending on the local industrial and economic setting, there is no exception to find that varying extent of metal contamination does exist in these regions. With the rapid industrialization and economic development in coastal regions, if lack of proper environmental protection procedures and regulations, heavy metals could be continuously introduced to estuarine and coastal environment through rivers, runoff and land-based point sources where metals are produced as a result of metal refinishing byproducts. Therefore, heavy metal contamination will have been still an environmental problem facing our society. The sediment quality guidelines, criteria, residential and non-residential standards, etc. established by federal and state environmental protection agencies in various countries are critical and useful for protecting human environment and assessing sediment/soil environmental quality. However, a multiple approach should be applied to properly evaluate and reflect the environmental status of an area and ensure the effectiveness in implementation of these regulatory requirements. As to the significantly polluted areas, remediation and restoration actions are needed to recover the system.

For the future work, we suggest that an in-depth understanding of the source, transport and fate of these contaminants be necessary for effective environmental management and ecosystem restoration. With the rapid urbanization and economic development in coastal areas, continuing monitoring and ecological studies of heavy metal pollution are certainly still needed.

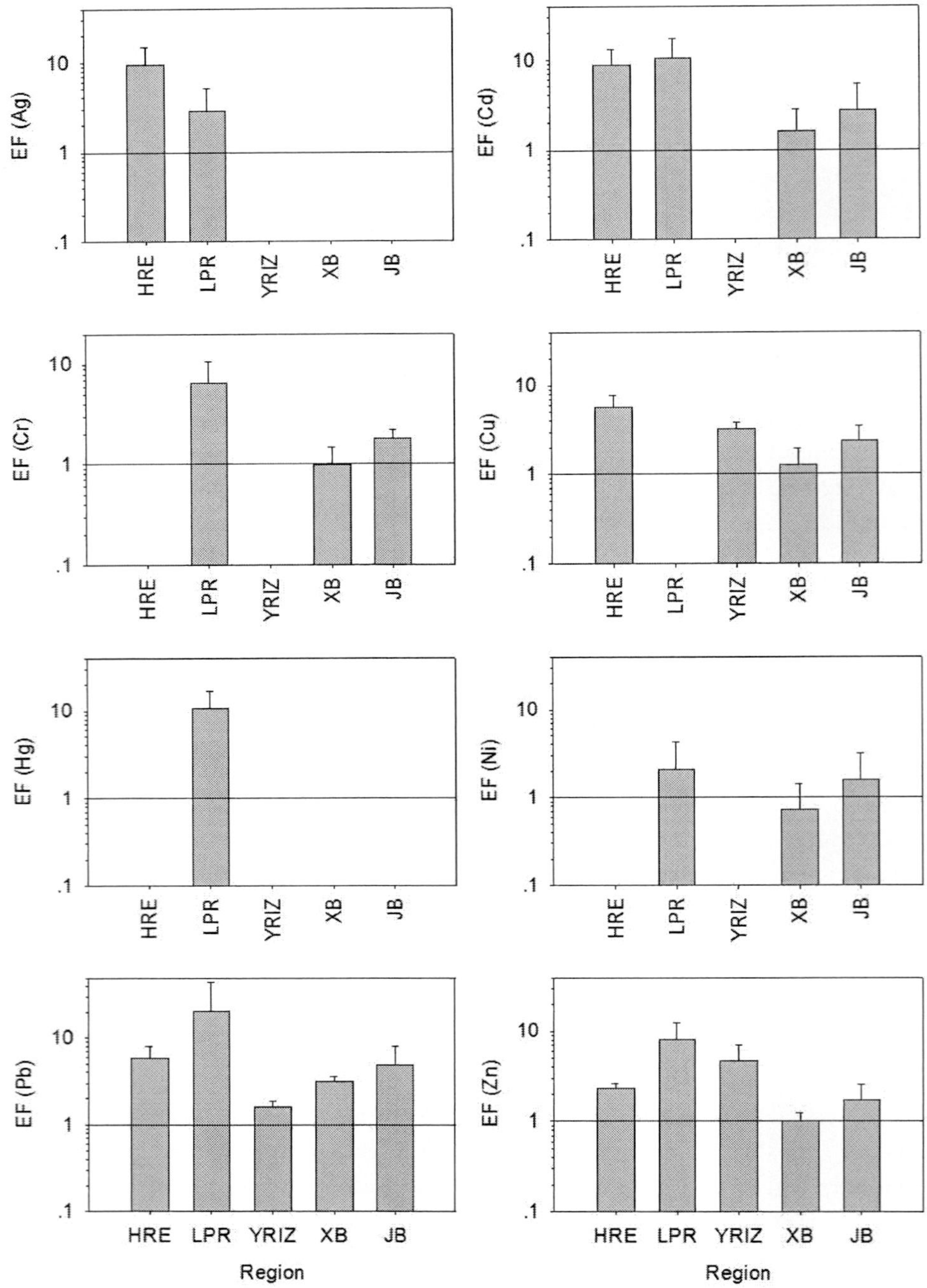

Figure 6. Comparison of metal enrichment factors (EF) in our selected areas. The predominant metal contaminants vary from one place to another due to the regional industrial and economic settings. However, varying extent of metal contamination are seen in these regions. The lower Hudson River estuary is designated as HRE, the lower Passaic River as LPR, Yangtze River intertidal zone as YRIZ, Xiamen Bay as XB, and Jiaozhou Bay as JB.

5. Acknowledgements

We would like to thank Libertad Urena and Jackie Kinney for their assistance in preparation of this manuscript.

6. Reference

Abood, K. A. (1978) Hudson River hydrodynamic and water quality characteristics. Prospectives on the New York Bight Symposium. Williamsburg, VA, 5-10 Nov. 1978.

Bonnevie N. L., Wenning R.J. and Gunster, G.D. (1992) Lead contamination in surficial sediments from Newark Bay, New Jersey. *Environmental International*, 18, 497-508

Bonnevie N.L., Wenning R.J. and Huntley S.L. (1993) Distribution of inorganic compounds in sediments from three waterways in Northern NJ. *Environmental Contamination and Toxicology*, 51, 672-680.

Bonnevie, N. L., Huntley, S. L., Found, B. W. and Wenning, R. J. (1994) Trace metal contamination in surficial sediments from Newark Bay, New Jersey. *Science of Total Environment*, 144, 1-16

Bopp, R. F. and Simpson, H. J. (1989) Contamination of the Hudson River - The sediment record. In: *Contaminated Marine Sediments Assessment and Remediation*. pp. 401-416.

Bothner, M.H., Buchhltz ten Brink, M. and Manheim, F.T. (1998) Metal concentrations in surface sediments of Boston Harbor – Changes with time. *Marine Environmental Research* 45, 127-155.

Brosnan, T. M., Stubin, A. I., Sapienza, V. and Ren, Y. G. (1994) Recent changes in metals loadings to New York Harbor from New York City water pollution control plants. In: *Hazardous and Industrial Wastes*. C. P. Huang (eds.). pp. 657-666.

Chen, X.Q. 1998. Changjiang (Yangtze) River Delta, China. *Journal of Coastal Research*, 14, 838-858.

Chen, Z., Kostaschuk, R. and Yang, M. (2001) Heavy metals on tidal flats in the Yangtze Estuary, China. *Environmental Geology* 40, 742-749.

Chillrud, S. N. (1996) Transport and fate of particle associated contaminants in the Hudson River basin. Ph. D. dissertation, Columbia University, Palisades, New York. 277 p.

Cooper, J. C., Cantelmo, F. R. and Newton, C. E. (1988) Overview of the Hudson River Estuary. *American Fisheries Society Monograph*, 4, 11-24.

Covelli, S. and Fontolan, G. (1997) Application of normalization procedure in determining regional geochemical baseline. *Environmental Geochemistry* 30, 34-45.

Crawford, D. W., Bonnevie, N. L. and Wenning, R. J. (1995) Sources of pollution and sediment contamination in Newark Bay, New Jersey, *Ecotoxicology and Environmental Safety*, 30, 85-100.

CSBTS (China State Bureau of Quality and Technical Supervision) (2002) The People's Republic of China National Standards GB 18668-2002 - Marine Sediment Quality. 10 pp. (In Chinese)

Dai, W.M. and Gu, Y.Z. (1990) Accumulation of heavy metal elements in the sediments near Xiqu and Nanqu sewage outlets, Shanghai. Shanghai *Environmental Science*, 9, 38–40 (in Chinese).

Feng, H., Cochran, J. K., Lwiza, H., Brownawell , B. and Hirschberg, D. J. (1998a) Distribution of Heavy Metal and PCB contaminants in the sediments of an urban estuary: the Hudson River. *Marine Environmental Research*, 45, 69-88.

Feng, H., Cochran, J. K., Hirschberg, D. J. andWilson, R. E. (1998b) Small-scale spatial variations of natural radionuclide and trace metal distributions in sediments from the Hudson River Estuary. Estuaries, 21, 263-280.

Feng H., Cochran, J.K. and Hirschberg, D.J. (1999a) ^{234}Th and ^{7}Be as tracers for the sources of particles to the turbidity maximum zone of the Hudson River estuary. *Estuarine, Coastal and Shelf Science,* 49, 629-645.

Feng, H., Cochran, J. K., and Hirschberg, D. J. (1999b) ^{234}Th and ^{7}Be as tracers for the transport and dynamics of suspended particles in a partially mixed estuary. *Geochimica et Cosmochimica Acta*, 63, 2487-2505.

Feng, H., Cochran, J.K. and Hirschberg, D.J. (2002) Mechanisms and behavior of metal contaminants over the course of a tidal cycle in the turbidity maximum zone of the Hudson River estuary. *Water Research*, 36, 733 –743.

Feng, H., Han X., Zhang, W. and Yu, L. (2004) A preliminary study of heavy metal contamination in Yangtze River intertidal zone due to urbanization. *Marine Pollution Bulletin*. 49, 910–915.

Feng, H., Onwueme, V., Jaslanek, W. J., Stern, E. A., and Jones, K.W. (2005) Application of geographical information system (GIS) to lower Passaic River sediment pollution study, New Jersey, USA. In: *Urban Dimensions of Environmental Change —— Science, Exposure, Policies, And Technologies*. (eds. H. Feng, L. Yu and W. Solecki). p p. 275-282. Science Press USA Inc., Monmouth Junction, New Jersey, USA.

Finley, B., Wenning, R. J., Ungs, M. J., Huntley, S.L. and Paustenbach, D. J. (1990) PCDDs and PCDFs in surficial sediments from the lower Passaic River and Newark Bay, p.409-414, In: Short Papers from the 10th International Meeting, Dioxin '90, Volume 1

Forstner, U. and Wittman, G.T.W., (1983). *Metal pollution in the aquatic environment*, 2nd. Ed., Verlag Publishers. New York. 18-19.

Gao, H.-W., Zhang, Y.-J, and Zhang, K. (2002) Atmospheric inputs of pollutants to the sea and their effects on marine envioronment and ecosystem. *Advance in Earth Sciences*, 17, 326-330. (In Chinese with English abstract.)

Geyer, R.W. (1993) The importance of suppression of turbulence by stratification on the estuarine turbidity maximum. *Estuaries*, 16, 113-125.

Gibbs, R. J. (1994) Metals in the sediments along the Hudson River Estuary. *Environmental International* 20, 507-516.

Gillis, C. A., Gunster, D. G., Bonnevie, N. L., Abel. T. B. andWenning, R. J. (1993a) Petroleum and hazardous chemical spills in Newark Bay, New Jersey, USA from 1982 to 1991. *Environmental Pollution*, 82, 245-253.

Gillis, C.A., Bonnevie, N.L. and Wenning, J. (1993b) Mercury contamination in the Newark Bay Estuary. *Ecotoxicology and Environmental Safety*, 25, 214-226

Gillis, C. A., Bonnevie, N. L., Su, S. H., Ducey, G. J., Huntley, S. L. and Wenning, R. J. (1995) DDT, DDD, DDE contamination of sediment in Newark Bay Estuary, New Jersey. *Archives of Environmental Contamination and Toxicology*, 28, 85-92.

Goldberg, E.D., Griffin, J.J., Hodge, V., Koid, M.and Windom, H.L. (1979) Pollution history of the Savannah River estuary. *Environmental Science and Technology*, 13, 588-594.

Han, Y.M., Du, P.X., Cao, J.J. and Posmentier E.S. (2006) Multivariate analysis of heavy metal contamination in urban dusts of Xi'an, Central China. *Science of the Total Environment*, 355, 176– 186.

Hines, M. E., Lynons, W., Amstrong, P. B., Orem, W. H., Spencer, M. J., Gaudette, H. E. and Jones, G. E. (1984) Seasonal metal remobilization in the sediments of Great Bay, New Hampshire. *Marine Chemistry*, 15, 173-187.

Hirschberg D. J., Chin P., Feng H. and Cochran J. K. (1996) Dynamics of sediment and contaminant transport in the Hudson River Estuary: evidence from sediment distributions of naturally occurring radionuclides. *Estuaries*, 19, 931-949.

Hunt, C. D., West, D. A.and Lewis, D. A. (1994) Trace metals concentrations in New York/New Jersey Harbor. In: *Hazardous and Industrial Wastes*. (ed. C. P. Huang). pp. 691-698.

Huntley S.L., Bonnevie N.L., Wenning R.J. and Bedbury H. (1993) Distribution of polycyclic aromatic hydrocarbons (PAHs) in three northern New Jersey waterways. *Bulletin of Environmental Contamination and Toxicology*, 51, 865-872.

Huntley S.L., Bonnevie N.L. and Wenning R.J. (1995) Distribution of polycyclic aromatic hydrocarbons and petroleum hydrocarbon contamination in sediment from Newark Bay estuary, New Jersey, *Archives of Environmental Contamination and Toxicology*, 28, 93-107.

Huntley, S. L., Iannuzzi, T. J., Avantaggio, J. D., Carlson -Lynch, H., Schmidt, C. W., and Finley, B. L. (1997) Combined Sewer Overflows as Sources of Sediment Contamination in the Lower Passaic River, New Jersey, II, Polychlorinated Dibenze-p-dioxins, Polychlorinated Dibenzofurans, and Polychlorinated Biphenyls, *Chemosphere* 34:233-250

Iannuzzi, T. J. and Wenning, R. J. (1995) Distribution and Possible Sources of Total Mercury in Sediments from the Newark Bay Estuary, New Jersey, *Bulletin of Environmental Contamination and Toxicology*, 55:901-908

Iannuzzi T. J., Huntley S. L., Finley B. L. and Wenning R. J. (1995) Distribution of possible sources of polychlorinated biphenyls in dated sediments from Newark Bay estuary, New Jersey. *Archives of Environmental Contamination and Toxicology*, 28: 108-117.

Iannuzzi, T. J., Huntley S.L., Schmidt, C. W., Finley, B. L., McNutt, R. P. and Burton, S. J. (1997) Combined Sewer Overflows (CSOs) as Source of Sediment Contamination in the Lower Passaic River, New Jersey. 1. Priority Pollutant and Inorganic Chemicals, *Chemosphere* 34:213-231

Iannuzzi, T. J., Ludwig, D. F., Kinnell, J. C., Wallin, J. M., Desvousges, W. H. and Dunford, R. W. (2002) *A Common Tragedy: History of an Urban River*, Amherst Scientific Publishers.

Klinkhammer, G. P. and Bender, M. L. (1981) Trace Metal Distributions in the Hudson River Estuary. Estuarine, *Coastal and Shelf Science*, 12, 629-643.

Langston, W.J. (1990) Toxic effects of metals and the incidence of metals pollution in marine ecosystems. In: Furness, R.W., Rainbow, P.S. (Eds.), *Heavy Metals in the Marine Environment*. CRC Press, Boca Raton.

Long, E.R., MacDonald, D.D., Smith, S.L. and Calder, F.D. (1995) Incidence of adverse biological effects within ranges of chemical concentrations in marine and estuarine sediments. *Environmental Management* 19, 81-97.

Long, E.R., Field, L.J. and MacDonald, D.D. (1998) Predicting toxicity in marine sediments with numerical sediment quality guidelines. *Environmental Toxicology and Chemistry* 17, 714-727.

Loska, K. and Wiechuya, D.(2003) Application of principle component analysis for the estimation of source of heavy metal contamination in surface sediments from the Rybnik Reservoir. *Chemosphere*, 51, 723– 33

Lu, X.Q., Werner, I. and Young, T.M. (2005) Geochemistry and bioavailability of metals in sediments from northern San Francisco Bay. *Environmental International* 31, 593-602.

Martin, J.-M. and Whitfield, M. (1983) The significance of the river input of chemical elements to the ocean. In: Trace Metals in Sea Water, C. S. Wong, E. Boyle, K. W. Bruland, J. D. Burton and E. D. Goldberg (Eds), Plenum Press, New York, pp.265-296.

Meyerson, A. L., Luther, G. W. III, Krajewski, J. and Hires, R. I. (1981) Heavy Metal Distribution in Newark Bay Sediments, *Marine Pollution Bulletin* 12(7): 244-250

NJDEP [New Jersey Department of Environmental Protection]. (1987) Passaic River Water Quality Management Study, NJDEP, Division of Water Resources

NOAA [National Oceanic and Atmospheric Administration]. (1984) A Geochemical Assessment of Sedimentation and Contaminant Distribution in the Hudson-Raritan Estuary, Rockville, MD: NOAA, National Ocean Service.

Olsen, C. R., Simpson, H. J., Bopp, R. F., Williams, S. C., Peng, T. H., and Deck, B. L. (1978) A geochemical analysis of the sediments and sedimentation in the Hudson Estuary. *J. Sediment. Petrol.*, 48, 401-418.

Olsen, C.R., Simpson, H.J. and Trier, R.M. (1981) Plutonium, radiocesium and radiocobalt in sediments of the Hudson river estuary. *Earth and Planetary science Letters*, 55, 377-392.

Olsen, C.R., Larsen, I.L., Brewster, R.H., Cutshall, N.H., Bopp, R.F. and Simpson, H.J. (1984) A geochemical assessment of sediment of sedimentation and contaminant distributions in the Hudson-Raritan Estuary. NOAA Technical Report NOS OMS 2. 101 pp.

Onwueme V. and Feng H. (2006) Risk characterization of contaminants in Passaic River sediments, New Jersey. *Middle States Geographer*, 39, 13-25.

Peet J., and Johnson T. (1996) The Forgotten River, Newark Star-Ledger, Three-Part Series. Property Owners. Chicago, IL, University of Chicago Press Publishing.

R-EMAP (Regional Environmental Monitoring and Assessment Program). (1998) Final Report: Sediment Quality of the NY/NJ Harbor System

Rohmann, S. O. (1988) Hudson River pollution: comparing point and nonpoint source emissions of hazardous chemicals. In: Symposium on Coastal Water Resources. American Water Resources Association. W. L. Lyke and T. J. Hoban (eds.). pp. 383-391.

Santos, I.R., Silva-Filho, E. V., Schaefer, C. E., Albuquerque-Filho, M. R. and Campos, L. S. (2005) Heavy metal contamination in coastal sediments and soil near the Brazilian Antarctic Station, King George Island. *Marine Pollution Bulletin* 50, 185-194.

Sañudo -Wilhelmy, S. and Flegal, A. R. (1992) Anthropogenic silver in the southern California Bight: A new tracer of sewage in coastal waters. *Environmental Science and Technology*, 26, 2147-2151.

Sañudo-Wilhelmy S.A. and Flegal A.R. (1994) The fate of metals in the Southern California Bight. *Environmental Science and Technology*, 26, 2147-2151

Shen, Z.-L. (2001) Historical changes in nutrient structure and its influence on phytoplankton composition in Jiaozhou Bay. Estuarine, *Coastal and Shelf Science* 52, 211-224.

Smith, G. J. and Flegal, A. R. (1993) Silver in San Francisco Bay estuarine waters. *Estuaries*, 16, 547-558.

Sutherland, R. A. (2000) Bed sediment-associated trace metals in an urban stream, Oahu, Hawaii. *Environmental Geology*, 39, 611– 27.

Tierra Solutions, Inc. (2003) Executive Summary: Passaic River Study Area Preliminary Findings

Trefry, J. H. and Shokes, R. F. (1981) History of heavy metal inputs to Mississippi Delta sediments. *Marine Environmental Pollution*, 2, 193-208.

Ullman, D. S. and Wilson, R. E. (1998) Model parameter estimation from data assimilation modeling: Temporal and spatial variability of the bottom drag coefficient. *Journal of Geophysical Research*, 103, 5531-5549.

Underwood, E.J. (1971) *Trace Elements in Human and Animal Nutrition* (3rd Editieon). Academic Press, Inc., New York.

Valette-Silver, N. J. (1993) The use of sediment cores to reconstruct historical trends in contamination of estuarine and coastal sediments. *Estuaries*, 16, 577-588.

Walker, W. J., McNutt, R. P. and Maslanka, C. K. (1999) The Potential Contribution of Urban Runoff to Surface Sediments if the Passaic River: Source and Chemical Characteristics, *Chemosphere* 38:363-377

Wang X-C., Feng H. and Ma, H-Q. (2007) An Assessment of Metal Contamination in Surface Sediments of Jiaozhou Bay, Qingdao, China. CLEAN - Soil, Air,Water. 35(1), 62-70.

Wenning J., Harris M., Paustenbach D.J. and Bedbury H. (1993a) Principal component analysis of potential sources of polychlorinated Dibenzo-p-Dioxin and Dibenzofuran residues in surficial sediments from Newark Bay, New Jersey, *Archives of Environmental Contamination and Toxicology*, 24: 271-289.

Wenning J., Harris M., Finley B. Paustenbach D.J. and Bedbury H. (1993b) Application of pattern recognition technique to evaluate polychlorinated Dibenzo-p-Dioxin and Dibenzofuran distributions in surficial sediments from the Lower Passaic River and Newark Bay. *Ecotoxicology and Environmental Safety*, 25, 103-125.

Wenning R.J., Bonnievie N.L. and Huntley S.L. (1994) Accumulation of metals, polychlorinated biphenyls, and plocyclic hydrocarbons in sediment from lower Passaic River, New Jersey. Archives of Environmental Contamination and Toxicology, 27:64-81.

Williams, S. C., Simpson, H. J., Olsen, C. R. and Bopp, R. F. (1978) Sources of heavy metals in sediments of the Hudson River Estuary. *Marine Chemistry*, 6, 195-213.

Windom, H. L., Schropp, S. J., Calder, F. D., Ryan, J. D., Smith, R. G. Jr., Burney, L.C., Lewis, F. G. and Rawlinson, C. H. (1989) Natural trace metal concentrations in estuarine and coastal marine sediments of the southeastern United States. *Environmental Sciences and Technology*, 23, 314-320.

Windom, H., Byrd, J., Smith, R. Jr., Hungspreugs, M., Dharmvanij, S., Thumtrakul, W. and Yeats, P. (1991) Trace metal-nutrient relationships in estuaries. *Marine Chemistry*, 32, 177-194.

Wolfskill L. A. and McNutt R.P. (2000) An Environmental Study of the Passaic River and its Estuary, *Steton Hall Law Review*, 29:37-46

Xu, S. Y., Tao, J., Chen, Z. L., Chen, Z. Y. and Lü, Q. R. (1997) Dynamic accumulation of heavy metals in tidal flat sediments of Shanghai. *Oceanologia et Limnologia Sinica*, 28, 509-515 (In Chinese with English abstract).

Zhang J. (1999) Heavy metal compositions of suspended sediments in the Changjiang (Yangtze River) estuary: Significance of riverine transport to the ocean. *Continental Shelf Research* 19, 1521-1543.

Zhang, J. and Liu, C. L. (2002) Riverine composition and estuarine geochemistry of particulate metals in China - Weathering features, anthropogenic impact and chemical fluxes. Estuarine, *Coastal and Shelf Science*, 54, 1051-1070.

Zhang L., Ye X., Feng H., Jing Y., Ouyang T., Yu X., Liang R. and Chen W. (2007) Heavy Metal Contamination in Western Xiamen Bay Sediments and Its Vicinity, *China. Marine Pollution Bulletin*, 54, 974–982

Zhang, W., Yu, L., Hutchinson, S. M., Xu, S., Chen, Z. and Gao, X. (2001) China's Yangtze Estuary: I. Geomorphic influence on heavy metal accumulation in intertidal sediments. *Geomorphology*, 41, 195-205.

Zhang W., Yu L., Lu M., Hutchinson S. M. and Feng H. (2007) Magnetic approach to normalizing heavy metal concentrations for particle size effects in intertidal sediments in the Yangtze Estuary, China. *Environmental Pollution*, 147, 238-244

In: Causes and Effects of Heavy Metal Pollution
Editor: Mikel L. Sanchez
ISBN 978-1-60456-900-1

Chapter 10

MONITORING HEAVY METAL POLLUTION WITH TRANSGENIC PLANTS

Igor Kovalchuk [*] ***and Olga Kovalchuk***
Department of Biological Sciences, University of Lethbridge, Lethbridge, AB. T1K 3M4, Canada

ABSTRACT

Heavy metals are metallic elements with high atomic weights. They tend to accumulate in the food chain and can be toxic and mutagenic. Such elements like mercury, chromium, cadmium, arsenic, and lead, when they are in their ionic and complexed forms, pollute soil, water and even air. Conventional methods of identification of polluted environment are laborious and costly. The presence of contamination is very often difficult to detect. It is even more difficult to evaluate its potential danger to living organisms. In this review, we will discuss the use of transgenic plants for the detection of heavy metal pollution and for the evaluation of its potential toxicity and mutagenicity. The greatest advantage of transgenic plants is that they can be made to be more sensitive to a particular pollutant. Plants are an excellent alternative to conventional methods, since they can be planted and grown at the site of pollution. In this chapter, we will describe transgenic plants that have already been successfully used for biomonitoring heavy metal pollution, and will also present novel ideas for generating efficient transgenic phytosensors.

INTRODUCTION

Heavy metals contamination of the environment can result from anthropogenic as well as natural causes. Activities such as mining, smelting and agricultural practices represent an anthropogenic factor which contributes to the increased levels of heavy metals such as Cd, Pb, Ni and alike in soil and water (Sharma and Agrawal, 2005). The persistence of heavy

[*] to whom correspondence should be addressed; Department of Biological Sciences, University of Lethbridge, Lethbridge, AB. T1K 3M4, Canada; tel: 403 394 3916; e-mail: igor.kovalchuk@uleth.ca.

metal salts in the environment allows them to be accumulated in soil, water and living organisms. Their accumulation in plants is particularly dangerous, since plants are at the bottom of the food chain and are consumed by animals and humans (Sharma and Agrawal, 2005; Vij and Tyagi, 2007). Consumption of food and water contaminated by heavy metals has substantial and long lasting detrimental effects on human health (Järup, 2003; Islam et al., 2007).

In plants, heavy metals disrupt various physiological activities such as photosynthesis, gas exchange and absorption of nutrients. They inhibit growth, decrease yield and diminish the nutritional dietary value of plant food products. The exact mode of action of various heavy metals and their effect on living organisms is beyond the scope of this review, and we refer readers to comprehensive reviews on the subject (Filipic *et al.*, 2006; Clemens, 2006; Sharma and Dietz, 2006; Vij and Tyagi, 2007).

Monitoring heavy metal pollution will allow avoiding unnecessary exposures. Numerous bacteria-, plant- and animal-based assays exist for biomonitoring the presence, toxicity and mutagenicity of metal pollution. All of them have their advantages and disadvantages. Monitoring the environment with plants is called phytosensoring. The main advantage of using phytosensors is that plants can be grown at the polluted areas, where they naturally take up metal ions through roots or even leaves. Therefore, it makes possible to correlate metal concentration with toxicity or mutagenicity in cells.

In this chapter, we will discuss the use of transgenic plants as sensors of heavy metal pollution. We will also describe plants that have previously been used for this purpose, as well as plants that can specifically be generated.

Why Plants?

Bacterial and animal models were among the first ones used for the detection of potential toxicity or mutagenicity of environmental pollution, various food additives and work-related exposures, etc. They were further modified, many of them via transgenesis, to become sensitive to a particular contaminant (Miller, 1985; Green *et al.*, 1986; Levin and Ames, 1986; Schaaper and Dunn, 1991; Murti *et al.*, 1994; Sacco *et al.*, 1997; Mayer *et al.*, 1998; Amanuma *et al.*, 2000; Hendricks and Engelward, 2004; Rugo *et al.*, 2005). Despite being found useful for environmental biomonitoring, animal and bacterial models unfortunately cannot be used for studying the influence of complex patterns of soil pollution. This issue becomes especially challenging when monitoring needs to be done in the open environment at a given place. Not every organism is suitable for biomonitoring especially in the field (Sheehan and Loucks, 1994). To be field biomonitors, test-organisms should absorb and integrate doses of toxicants from polluted air, water or/and soil. The use of animal-based models is problematic due to their non-settled life style. As plants are sedentary organisms, they completely suit that very purpose. They can be grown at a certain polluted site or at a potentially dangerous site, and thus they can monitor pollution. Of a particular importance for sensoring heavy metal pollution is the fact that plants possess the mechanisms of an active up-taking metals from the environment and distributing them throughout an entire plant. It is considered to be more or less uniform exposure. On top of it, plants are a better and more convenient alternative, in terms of ethics and aesthetics.

Requirements for a Plant to Be a Transgenic Phytosensor

Biomonitoring the environment using phytosensors is a well documented process. It is, however, outside of the scope of the current review, as the information about plant biomonitors can be found in numerous assays (Fiskesjo, 1988; Ichikawa, 1992; Grant, 1994; Kanaya *et al.*, 1994; Fiskesjo, 1995; Bolle *et al.*, 2004; Ma *et al.*, 2005; Grant and Owens, 2006). Here, we attempted to cover all known transgenic phytosensors.

Transgenic phytosensors are plants that were intentionally modified to become better sensors. It could be done via introduction of either a marker that would be easy to visualize or genetic elements that are more sensitive to the impact of a mutagen.

Not every plant species can be used for biomonitoring. As mentioned above, an efficient phytosensor should be able to absorb heavy metal salts and make them available for plant metabolic processes. If an intended phytosensor monitors genotoxicity of an agent, this agent should be able to penetrate into nucleus and damage DNA directly, or should be able to influence the processes that might disturb either chromatin structure or DNA repair. The additional essential factors justifying that a plant is a good biomonitor are short growth seasons, high biomass, good uptake rate and relative “toughness”. In this respect, Arabidopsis (*Arabidopsis thaliana*) is an ideal plant: it has the short growth season – 2-3 months, it is tolerant to high levels of pollution, and it is relatively easy to transform (Somerville and Koornneef, 2002; Koornneef *et al.*, 2004).

The generation of transgenic biomonitors requires a transformation step, therefore it puts a limit on plant choices. Not every plant species can be transformed to become a transgenic one (Altpeter, 2006; Sparrow *et al.*, 2006). Moreover, it is a well-known fact that many transgenic lines intended for a direct use as food and feed were shown to be unstable in the open environment as opposed to phytotrons and greenhouses. This is yet another restriction on the use of plants as biomonitors. A transgenic marker should be stably integrated, and it should not be silenced or overreact to normal environmental cues such as temperature changes, wind, water availability, etc. (Filipecki and Malepszy, 2006). As these factors could become synergistic with or antagonistic to heavy metal pollutants, they can potentially skew the results and lead to the false positive or negative outcomes.

Marker Genes to be Used for Mutagenesis Assays

A marker or a reporter gene represents a gene which activity can be easily and cost-effectively monitored in plants. These genes include, but are not limited to, β-glucuronidase (*uid*A or GUS), luciferase and GFP (YEFP, RFP and other modifications) (Stewart, 2001, 2006). All of these genes have different sensitivity of detection in plants. Moreover, different procedures are used to visualize them. To the best of our knowledge, the GFP gene-marker is the cheapest and the easiest to visualize. It just requires a blue lamp to see the GFP expression as green on red background. The main disadvantage of this marker is that it has the lowest sensitivity which makes it the least feasible to be applied in biomonitoring. The GUS gene-marker is the second cheapest and easiest to use. Visualization of the GUS gene requires histochemical staining using X-glu as a substrate for the β-glucuronidase enzyme. The reaction results in a blue precipitate accumulating in cells. The LUC-gene marker is the most expensive and the most difficult of these three to visualize, but it is by far the most sensitive one (Ilnytskyy *et al.*, 2005). Plants to be analyzed are sprayed with the substrate luciferine; the cleavage of luciferine by luciferase results in the fluorescence emission that is detected in a special custom-build LN CCD cooled luciferase camera. The major advantage of the luciferase reporter gene is *in vivo* detection of luciferase activity. One of the GUS marker disadvantages is that it requires destructive non-vital histochemical staining for the visualization of the events, and this can be done only once, at a specific time point during plant growth. In contrast, detection of LUC and GFP markers does not kill plants and can be done multiple times during plant growth.

GUS and LUC-based reporter assays for the detection of potential mutagenicity of pollutants have been used before (Kovalchuk *et al.*, 1998, 1999, 2001; Ries *et al.*, 2000; Besplug *et al.*, 2004; Boyko *et al.*, 2006). A GFP-based reporter system was never used due to its low sensitivity (Jan Lucht, personal communication). However, it has been successfully used in animals. Using this system for the detection of DNA damage requires a complex procedure that includes fluorescent cells sorting to be conducted by flow cytometry (Kovalchuk *et al.*, 2004). Additionally, reporter genes based on an antibiotic resistance marker have been also generated and successfully used in the lab, and yet they require various steps of tissue culturing for isolation of damaged cells and characterization of recombination events (Lebel *et al.*, 1993).

Transgenic Phytosensors Used for Mutagenicity Detection of Various Agents Including Heavy Metals

The only transgenic phytosensors that have been previously used for the detection of harmful influence of pollutants, including heavy metals, were plants that allow to detect DNA damage at a transgene locus. All of these plants carry in their genome an active copy of a transgenic marker gene. Various types of DNA damage, including strand breaks, or nucleotide damage are processed via various types of repair and lead to the restoration of transgene activity. Kanamycin has been the first transgene used for sensing the genotoxicity (Lebel *et al.*, 1993). The GUS transgene, however, has been the most frequently used. The generated phytosensors include a homologous recombination substrate (Lebel *et al.*, 1993; Kovalchuk *et al.*, 1998), a point mutation substrate (Kovalchuk *et al.*, 2000; Van der Auwera *et al.*, 2008), a frame shift mutation substrate (Leonard *et al.*, 2003) and a microsatellite stability substrate (Azaiez *et al.*, 2006). All of these systems score one single endpoint – the number of blue spots on plants after histochemical staining.

Recombination Reporter Assay

The first recombination reporter assay used for detection of mutagenicity is based on truncated overlapping copies of kanamycin transgene. Protoplasts of transgenic tobacco (*Nicotiana tabacum*) plants carrying such a transgene have exhibited 2-10-fold increase in recombination frequency while exposed to X-rays, mitomycin C and heat shock (Lebel *et al.*,

1993). Due to the major disadvantage involving extensive and laborious tissue culturing, this system has not been used for biomonitoring.

Transgenic Arabidopsis and tobacco plants transformed with the overlapping truncated versions of a β-glucuronidase marker gene have been originally used for detection and analysis of homologous recombination events (Figure 1A; Swoboda *et al.*, 1994). Breaks at the regions of homology can be restored via recombination events and visualized via hisotchemical staining. The procedure is based on the ability of the active version of the GUS (*uidA*) coding enzyme β-glucuronidase to cleave the provided substrate, X-glu. The reaction results in the formation of a blue precipitate. The cells where homologous recombination events occur can be precisely localized as blue sectors on transparent plants enabling the quantitative assay (Figure 1B).

The recombination assay, mentioned above, does not allow to trace the recombination frequency in exposed plants before and after exposure to mutagens. To overcome this obstacle, we generated other recombination reporter constructs that utilize the luciferase gene. The similar recombination substrate was generated, and transgenic Arabidopsis and tobacco plants carrying the construct were obtained (Ilnytskyy *et al.*, 2005). Frequent recombination events were observed using an *in vivo* imaging system (Figure 2).

We have also tested a different transgenic visual marker frequently used in mammalian research, green fluorescent protein (GFP). This marker gene appeared to be not suitable for recombination and mutation assays. Although the whole GFP+ plants are easily detectable, single cells with recombination or mutation events that lead to restoration of the GFP function are hardly detectable and require a laborious microscopic analysis.

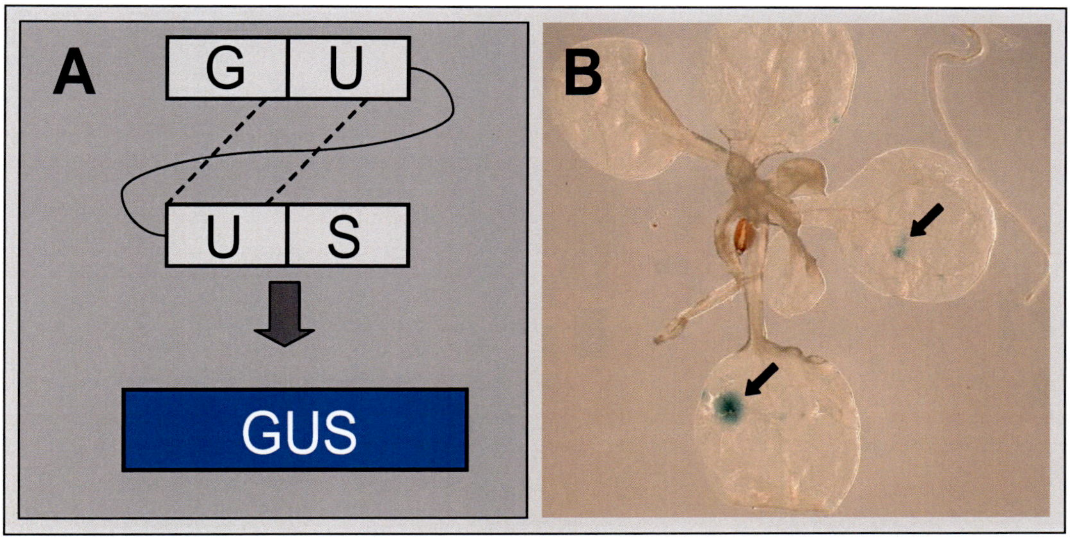

Figure 1. Transgenic GUS-based "recombination" system for the detection of environmental mutagens. A. Recombination construct consists of two truncated (one without the 3' end and one without the 5' end and the promoter) non-functional copies of GUS gene depicted as "GU" and "US". Damage to either region of homology (U) can be repaired by homologous recombination, restoring transgene integrity. B. Recombination events are visualized as blue sectors in plants.

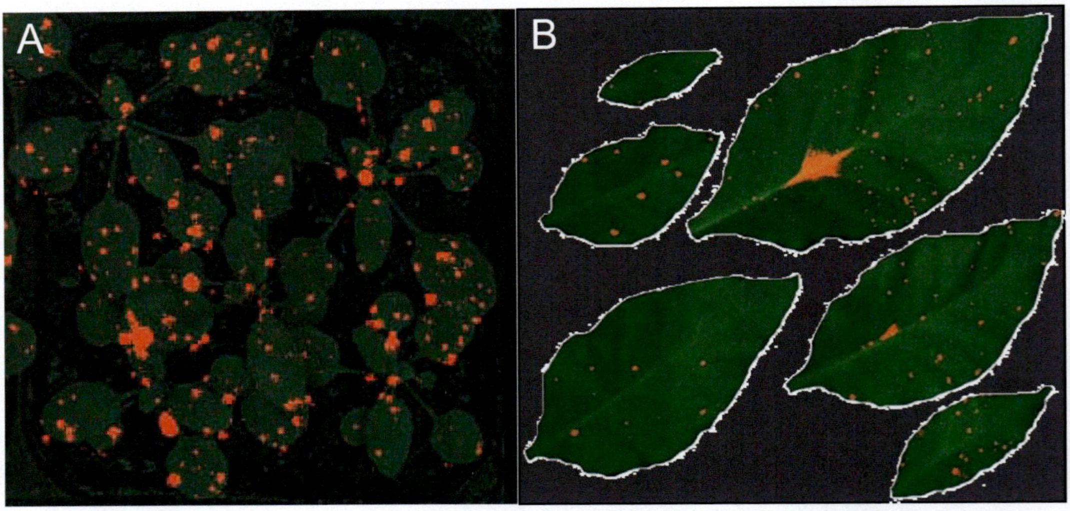

Figure 2. Luciferase recombination system A. Visualization of luciferase activity in Arabidopsis. Plants were sprayed with luciferine.B. Visualization of luciferase activity in tobacco: individual leaves were submerged in luciferine solution.

Point Mutation Reporter Assay

Point mutations are perhaps the most common type of mutations in cells (Kovalchuk *et al.*, 2000; Dollé *et al.*, 2002). Since plants do not have a predetermined germ line, any somatic mutation in the meristem can potentially be inherited. Thus, it is important that phytosensors that are measuring point mutation frequency are being used for genotoxicity studies.

A system for the detection of such events has been developed (Kovalchuk *et al.*, 2000). Using site-directed mutagenesis, a single point mutation creating the stop codon sequence at the 5' end of the GUS gene has been generated (Kovalchuk *et al.*, 2000). Several independent lines carrying various stop codons at different positions have been made. These mutations resulted in a complete inactivation of the transgene. Transgenic plants carrying such constructs exhibit occasional sectors of blue upon histochemical staining. These sectors represent the spontaneous restoration of *uidA* activity due to reversion of stop codons to original codons. These lines allow detecting the following reversions: T → C, T → A and T →G.

Recently, another set of lines allowing the detection of the most frequently occurring C → T mutations has been generated. The authors reported a substantially higher mutation frequency than the one observed by Kovalchuk *et al.* (2000). This can be attributed to the fact that C → T mutations scored by Van der Auwera *et al.* (2008) are more frequent than the ones scored by Kovalchuk *et al.* (2000). Point mutation reporter plants responded strongly to various mutagens such as UV-C, X-rays, methyl methanesulfonate (MMS) and ethyl methane sulfonate (EMS) by increasing mutation frequency at the transgene (Kovalchuk *et al.*, 2000). However, the authors reported that using these lines, they found no increase in mutation frequency upon heavy metals exposure, and they did not describe conditions under which heavy metals exposure occurred.

Frame Shift Mutation/ Microsatellite Stability Assays

None of the above-mentioned systems can detect various small deletions and insertions occurring frequently upon DNA damage repair. Leonard *et al.* (2003) and Azaiez *et al.* (2006) developed assays in which a frameshift was introduced into the GUS gene. Insertion of (G)7 (Leonard *et al.*, 2003) or (G)16 (Azaiez *et al.*, 2006) creates a frameshift in the GUS sequence. These sequences represent a small sequence repeat called a microsatellite, and they are known to be highly unstable and respond to various stresses with expansion or retraction (Schmidt and Anderson, 2006). Insertion of (G)2 or deletion of a single G restores the frame, and thus activates the transgene. As in the above-mentioned assays, these events are scored as blue spots on a transparent background (Figure 1B). These assays have not been tested for biosensoring, although UVC treatment of plants carrying (G)7 reporter showed a 5-fold increase in the number of blues spots.

TRANSGENIC PLANTS USED FOR GENOTOXICITY ANALYSIS OF HEAVY METALS

The first use of transgenic "recombination plants" was for testing soil and water radioactive pollution. Large-scale environmental monitoring experiments with the use of transgenic Arabidopsis and tobacco plants showed substantial genotoxicity of soils from different contaminated areas of Ukraine (Kovalchuk *et al.*, 1998; Kovalchuk *et al.*, 1999). The assay made it possible to monitor the frequency of homologous recombination in plants grown in soil types affected by different levels of pollution and to correlate it with an absorbed dose of Cs, Sr and other radioactive elements in plants. This dose consisted of an external and internal dose. The sensitivity of these plants allowed to detect a difference in the frequency of HR between plants grown in the clean soil (22 Bq/kg) and plants grown in the soil with a contamination level as low as 1.5-3.3 Ci/km^2 (188-575 Bq/kg) (Kovalchuk *et al.*, 1999).

The further work confirmed the usefulness of these plants for bimonitoring. Plants germinated at various concentrations of heavy metal salts such as Cd^{2+}, Pb^{2+}, Ni^{2+}, Zn^{2+}, Cu^{2+} showed a substantial dose- and uptake-dependent increase in point mutations frequencies. It is noteworthy that the mutation frequency increase correlates strongly and positively with the HR frequency increase in "recombination" plants grown in the presence of the same concentrations of aforementioned salts (Kovalchuk *et al.*, 2001).

Transgenic Arabidopsis and tobacco plants were also used for the analysis of genotoxicity of radioactively polluted water. We sampled water from private wells in the villages located in the inhabited areas contaminated after the Chernobyl accident. The radiological analysis of water samples did not reveal any ^{137}Cs or ^{90}Sr activity, because concentrations of these nuclides were below detectable limits. Despite this fact, we found an increase of the HR frequency in Arabidopsis plants grown in the media prepared from this contaminated water (Kovalchuk *et al.*, 2003a).

Transgenic Plant Systems Can Be Efficiently Used to Monitor other Types of Environmental Pollutants

Transgenic phytosensors were used for the analysis of water contaminated with different concentrations of various herbicides. It was found that herbicides 2,4-D and dicamba could increase both the HR and point mutations frequencies, whereas herbicide atrazine increased only the HR frequency but not the point mutations frequency (Besplag *et al.*, 2004).

Transgenic recombination lines were also used for the analysis of potential mutagenic influence of a high level UV-B radiation. This system showed a UV-B dose-dependent increase of the HRF (Ries *et al.*, 2000), and therefore, it was used for the analysis of heritable changes that could be detected as totally blue stained plants. These plants inherited the restoration of a marker gene as a result of germline recombination events (Ries *et al.*, 2000).

Recently, we have also showed that exposure to viral infection results in an increase of the HRF in somatic (Kovalchuk *et al.*, 2003b) and meiotic cells (Boyko *et al.*, 2007). Our unpublished data also suggest that bacterial infection results in a similar response. These experiments confirm that transgenic phytosensors can detect genome destabilization caused by a variety of mutagens in the environment.

Transgenic Plant Systems that Can Be Generated for Biosensoring

The transgenic plant systems used to date do not allow to score all kinds of mutations in one single assay. In order to check whether a particular heavy metal is able to generate double strand breaks, damage to nucleotides, deletions, or insertions, we have to use all three phytosensor systems mentioned above. One way to overcome this obstacle is to increase a target size and to design a marker that functions in the form of a repressor-gene combination. A tetracycline repressor is an ideal system that can potentially be used for the development of a new biosensor. The system would include a reporter gene, GUS or LUC, a tetracycline promoter and a repressor (Figure 3). Under normal conditions, an active repressor binds the promoter and prevents transgene expression. Any mutation that disrupts the repressor sequence restores the promoter function and activates the reporter transgene. Transgenic plants carrying the repressor/promoter/reporter construct should have a substantial number of sectors expressing the transgene. Ideally, such sectors should have a discrete character and be similar to those mentioned above. It is, however, possible that repressor binding will not repress the transgene activity completely and will have a certain level of "leakage". In this case, plants potentially will have a very low level of transgene expression and will have occasional sectors of high activity that still can be scored. The additional downside of using such plants is that they have to be heterozygous for such a construct to ensure inactivation of a single copy repressor element by a mutation event.

Another way of designing a good transgenic biosensor would be to generate plants that carry an inactive version of the antibiotic (or herbicide) resistance gene. Such a system will make it possible to analyze heritable changes occurring in plants exposed to heavy metals. The progeny of exposed plants should be grown in the presence of a selective agent, and resistant plants are scored.

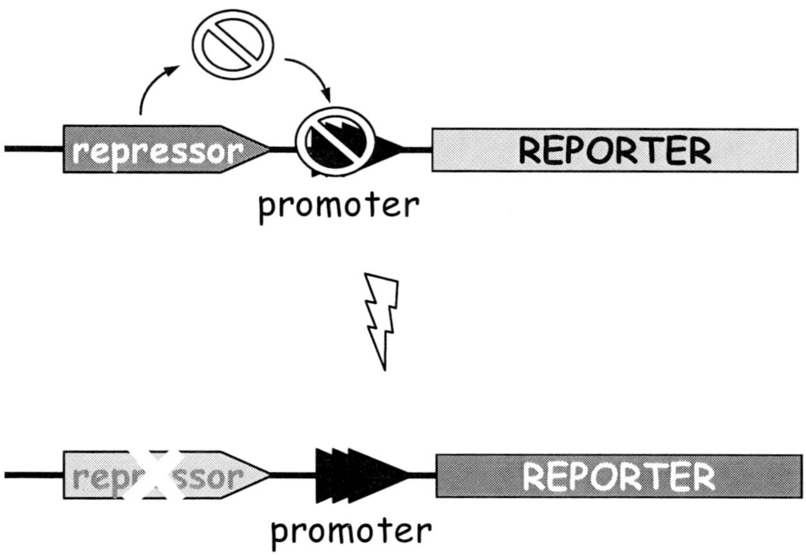

Figure 3. Inducible repressor/promoter/marker biosensor. Marker gene placed under the tetracycline inducible promoter is normally inactive due to the binding of the repressor. Any mutation that inactivates the repressor releases activity of the promoter and thus activates the marker gene expression.

This approach, despite its attractiveness, requires much longer time for the evaluation of the environmental influence as well as a larger number of progeny plants to be scored. Our lab has recently generated another transgenic biosensor that carries a dual recombination marker, a visible marker based on the luciferase transgene and antibiotic-based marker sulfonamide. This system should allow the analysis of somatic recombination events via a luciferase marker and meiotic events via a sulfonamide marker. The system has not been tested yet.

Production of a Transgenic Phytosensor on Demand

One of the important advantages that transgenic biosensors can offer is the ability to customize an assay in accordance with monitoring needs. This approach, however, requires substantial knowledge of changes that occur in a cell upon the influence of a particular heavy metal salt. The first step would be analysis of specific changes in the transcriptome. Identification of genes that are induced only by exposure to a particular heavy metal salt should allow the generation of a specific biosensor. This biosensor would carry in its genome a transgene driven by the promoter of a specific inducible gene. Activation of the promoter upon exposure to heavy metals could be visualized by an increase in marker gene expression. A similar system has been described for bacterial biosensors (Harms *et al.*, 2006). Moreover, such systems have been produced in various animals. Transgenic fish, worm and mice have been generated that carry in their genome GFP under the control of stress-responsive promoters such heat shock protein 70, metallothionein 2, cytochrome P450 family protein 35A2, glutathione-S-transferase 4, superoxide dismutase 1, catalase 2, C. elegans p53-like protein 1, and apoptosis enhancer 1 genes (Seok *et al.*, 2006; Roh *et al.*, 2006). Activation of these promoters by exposure to heavy metal stress has been tested in fish and warm. Recently, such a system has been generated in *Physcomitrella patens*; a heat shock response promoter has been shown to be activated by exposure to organic contaminants (Saidi *et al.*, 2007).

An alternative strategy may involve the analysis of genes induced by a particular heavy metal and identification of common specific regulatory elements. In yeast and animals, the heavy metal-responsive elements (HMREs) and the metal-induced transcription factors have been characterized in details (Stuart *et al.*, 1985). The metal-regulatory elements MREs consist of a highly conserved heptanucleotide core (5′-TGCRCNC-3′) and less conserved flanking nucleotides (Searle *et al.*, 1987). A metal response element-binding transcription factor-1 (MTF-1) binds to MREs and regulates metallothioneins gene transcription. The MTF-1 is a zinc-dependent, zinc-responsive transcription factor (Westin and Schaffner, 1988; Heuchel *et al.*, 1994). In yeast, for example, it has been demonstrated that a copper metallothionein gene is induced through a copper-responsive element, and this process is mediated by the copper-binding transcription factor ACEl (Fürst *et al.*, 1988). The ACE1 binds to the MRE upstream promoter of a target gene. The consensus sequence of this MRE is 5′-HTHNNGCTGD-3′ (D = A, G, or T; H = A, C, or T; N = any residue; Dixon *et al.*, 1996). As soon as such elements are identified, one would use a visible marker gene regulated by the minimal promoter that would include these regulatory elements (Figure 4A). Qi *et al.* (2007) has recently reported that the bean (*Phaseolus vulgaris*) stress-related gene (PvSR2) responds only to heavy metal stress. This gene promoter contains MRE elements. The authors fused the promoter to the GUS gene and showed the activation of the transgene reporter by exposure to heavy metals (Qi *et al.*, 2007). These plants, however, had a certain level of background expression. The ideal system would include a minimal promoter that contains metal-specific MREs. Such system will respond to specific heavy metals and will have a very low background of GUS expression.

Several other models for biosensoring can be produced. One of the examples can measure senescence activity occurring naturally in plants and enhanced dramatically by exposure to stress. Recently, Zhang *et al.* (2007) have reported the system that allows the measurement of delayed fluorescence (DF) as the reflection of age-dependent senescence. They detected DF emissions from ageing or stressed leaves with a home-made DF biosensor (Zhang *et al.*, 2007). Having proper calibration and enhancement via transgenesis, such system can potentially be used for phytosensoring heavy metal pollution.

Yet another example proposes a more complex detection system. Lalonde *et al.* (2005) suggest that the generation of a fusion tag encoding fluorescent protein (FP) allows visualization of protein/protein interactions. The *in vivo* combination of FPs with a second fluorophore provides novel applications that rely on fluorescence resonance energy transfer (FRET). FRET detects rearrangements of the relative orientation and distance of two fluorophores within the 1–10 nm range, and thus provides extraordinary spatial resolution, vast possibilities for measuring protein–protein interactions or the creation of small molecule biosensors. First, the generation of a specific heavy-metal inducible biosensor makes it possible to analyze a specific proteome induced by exposure. Next, one would analyze heavy-metal response specific protein-protein interactions via the yeast two-hybrid system. When a pair of proteins is identified, fusion of FPs to both proteins can be generated (Figure 4B). Such transgenic plant would respond to heavy metal exposure by the production of transgenic proteins carrying two fluorophores; the interaction of these proteins could be then detected by measuring the level of fluorescence (Figure 4B) (Lalonde *et al.* (2005).

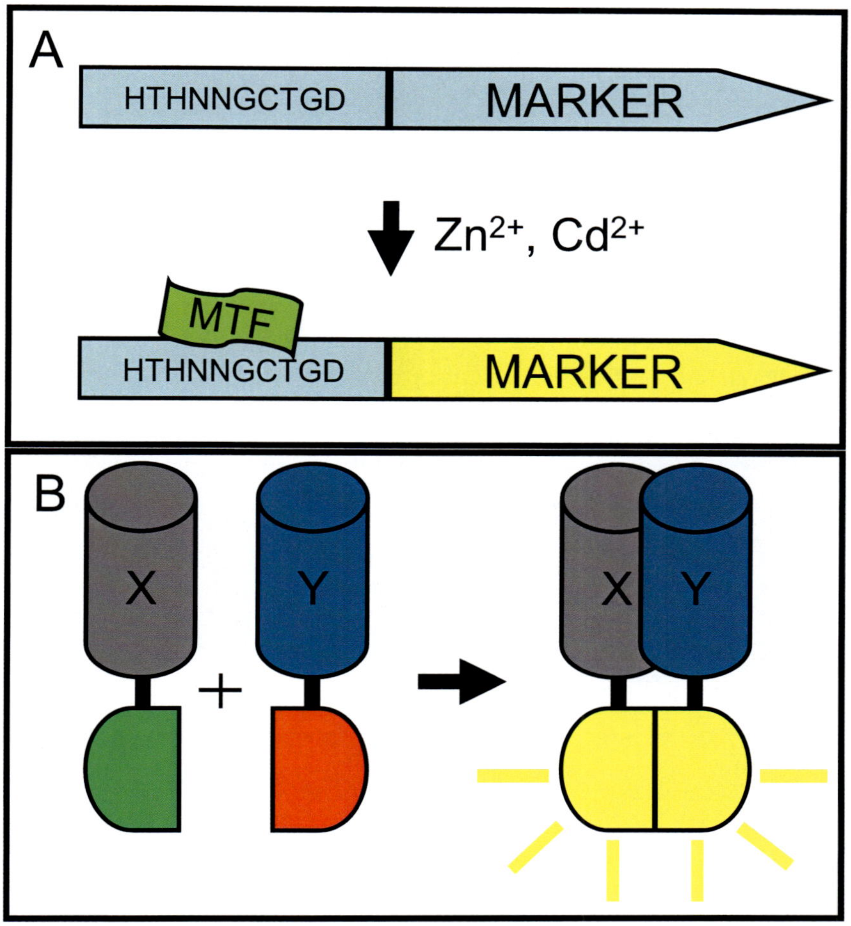

Figure 4. Metal-specific biosensors. The logistics of these two biosensors is based on the ability of heavy metals to induce the expression of specific genes. A. Use the specific promoter that carries the metal responsive elements (MRE) – depicted as HTHNNGCTGD sequence. Specific metal response element-binding transcription factors (MTFs) would bind the promoter upon the exposure to specific metal and activate the expression of the marker gene.B. First, heavy-metal induced specific protein-protein interaction has to be identified – depicted as X and Y. Fusions with individual fluorophores (fusion tag) has to be made. Exposure to specific heavy metal should activate the interaction of X and Y proteins, resulting in interaction of two fluorophores. Emitted fluorescence can be detected by custom build detectors.

CHANGES IN THE EPIGENOME

Multiple reports suggest that exposure to stress, including heavy metals, frequently results in genome modifications that do not alter genome sequence (Zoroddu *et al.*, 2000; Oruambo *et al.*, 2007; Schnekenburger *et al.*, 2007). Such changes, named epigenetic, involve DNA methylation and histone modifications. Epigenetic changes are frequent signatures of genome instabilities leading to cancer. Exposure to heavy metals, especially to nickel, is known to lead to severe changes in histone modifications and cancer (Lu *et al.*, 2005; Ke *et al.*, 2006; Kondo *et al.*, 2006; Schnekenburger *et al.*, 2007).

Changes in DNA methylation patterns can serve as a hallmark of exposure to stress. We have recently reported the substantially changes in DNA methylation in the progeny of pine trees exposed to Chernobyl fall-out (Kovalchuk *et al.*, 2003c), and in the progeny of Arabidopsis and tobacco plants exposed to various stresses, including heavy metals, salt and even pathogens (Boyko *et al.*, 2007; unpublished data).

It is possible to generate a model transgenic phytosensor capable of responding to heavy metal stress by changes in methylation. One would need to find a tissue-specific promoter or a promoter of poorly expressed genes and fuse them to a marker gene such as LUC or GUS. Transgenic plants with such constructs can be generated, and the marker expression can be checked. Plants with low activity of the marker gene can be used for the analysis of promoter or/and gene methylation levels. Plants with heavily methylated promoters can be used for biosensoring. It should be checked, however, whether this system will respond to heavy metal stress by promoter demethylation, and thus by activation of a transgene (Figure 5). These constructs can potentially include flower- and root-specific promoters. One can also use the Sadhu6-1 transposon element promoter with the strongest CpHpG methylation (Rangwala and Richards, 2007). Transposons are known to be activated by stress (Grandbastien et al., 2005), and their promoters should be well suited for the generation of transgenic biosensors.

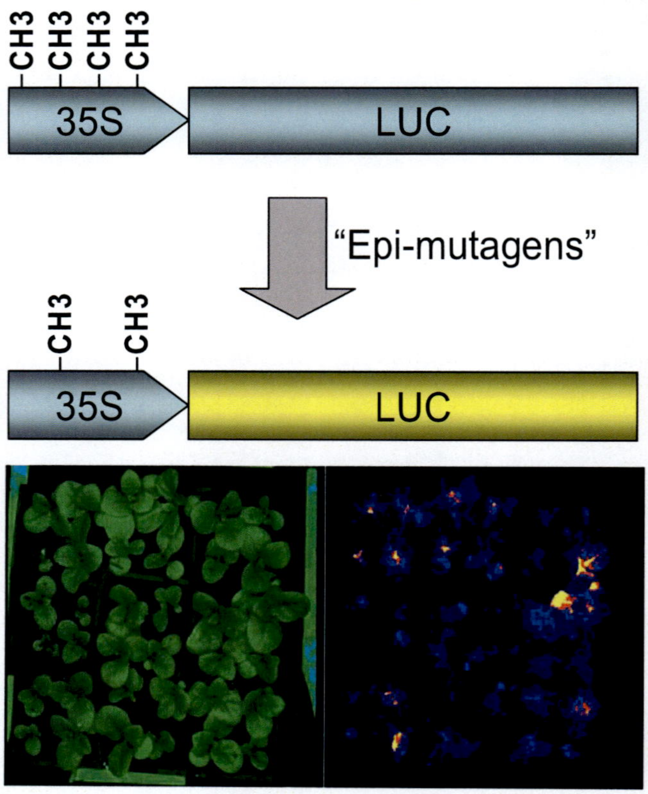

Figure 5. Sensor of "epi-mutagens" Heavy methylated 35S promoter can be designed either by incorporation of additional CGs or can be selected from independent transgenic plants that carry 35S:LUC but do not express it. "Epi-mutagens" that result in decrease in methylation level should activate the promoter and result in LUC expression. Plants carrying such a construct should have spontaneous or induced expression of luciferase.

Conclusion

Biosensoring environmental pollution requires careful consideration of constantly changing patterns of water, soil and air pollution. Transgenic phytosensors are ideal organisms to fulfill this task – they can be generated to respond only to specific heavy metal pollution. Such plants are assumed to be modified in such a way that they will be able not only to sense heavy metal contamination but also to effectively decontaminate any pollution. It is also conceivable that plants described in this review can be used as a means of remediation quality control used for testing and evaluating genotoxicity of contaminated soil and water before and after remediation.

Acknowledgement

We are grateful to Valentina Titova and Andrey Golubov for critical reading of the manuscript. The original work presented in this review was supported by the Swiss National Science Foundation, NSERC operating grant to O.K. as well as EMBO fellowship to I.K

References

Altpeter, F. (2006). Rye (*Secale cereale L.*). *Methods Mol. Biol., 343*, 223-31.

Amanuma, K., Takeda, H., Amanuma, H., Aoki, Y. (2000). Transgenic zebrafish for detecting mutations caused by compounds in aquatic environments. *Nat. Biotechnol., 18*, 62-65.

Azaiez, A., Bouchard, E.F., Jean, M., Belzile, F.J. (2006). Length, orientation, and plant host influence the mutation frequency in microsatellites. *Genome, 49(11)*, 1366-73.

Besplug, J., Filkowski, J., Burke, P., Kovalchuk, I., Kovalchuk, O. (2004). Atrazine induces homologous recombination but not point mutation in the transgenic plant-based biomonitoring assay. *Arch. Environ. Contam. Toxicol., 46(3)*, 296-300.

Bolle, P., Mastrangelo, S., Tucci, P., Evandri, M.G. (2004). Clastogenicity of atrazine assessed with the *Allium cepa* test. *Environ. Mol. Mutagen., 43(2)*, 137-41.

Boyko, A., Greer, M., Kovalchuk, I. (2006). Acute exposure to UVB has a more profound effect on plant genome stability than chronic exposure. *Mutat. Res., 602(1-2),* 100-9.

Boyko, A., Kathiria, P., Zemp, F.J., Yao, Y., Pogribny, I., Kovalchuk, I. (2007). Transgenerational changes in the genome stability and methylation in pathogen-infected plants: (virus-induced plant genome instability). *Nucleic Acids Res., 35*, 1714-25.

Clemens, S. (2006). Toxic metal accumulation, responses to exposure and mechanisms of tolerance in plants. *Biochimie, 88(11),* 1707-19.

Dixon, W.J., Inouye, C., Karin, M., Tullius, T.D. (1996). CUP2 binds in a bipartite manner to upstream activation sequence c in the promoter of the yeast copper metallothionein gene. *J. Biol. Inorg. Chem., 1,* 451–459.

Dollé, M.E., Snyder, W.K., Dunson, D.B., Vijg, J. (2002). Mutational fingerprints of aging. *Nucleic Acids Res., 30(2),* 545-9.

Filipecki, M., Malepszy, S. (2006). Unintended consequences of plant transformation: a molecular insight. *J. Appl. Genet., 47(4)*, 277-86.

Filipic, M., Fatur, T., Vudrag, M. (2006). Molecular mechanisms of cadmium induced mutagenicity. *Hum. Exp. Toxicol., 25(2),*67-77.

Fiskesjo, G. (1995). *Allium* test. *Methods Mol. Biol., 43,* 119-27.

Fiskesjo, G. (1988). The *Allium* test - an alternative in environmental studies: the relative toxicity of metal ions. *Mutat. Res., 197*, 243-260.

Fürst, P., Hu, S., Hackett, R., Hamer, D. (1988). Copper activates metallothionein gene transcription by altering the conformation of a specific DNA binding protein. *Cell, 55,* 705–717

Grandbastien, M.A., Audeon, C., Bonnivard, E., Casacuberta, J.M., Chalhoub, B., Costa, A.P., Le, Q.H., Melayah, D., Petit, M., Poncet, C., Tam, S.M., Van Sluys, M.A., Mhiri, C. (2005). Stress activation and genomic impact of Tnt1 retrotransposons in *Solanaceae. Cytogenet. Genome Res., 110(1-4),* 229-41.

Grant, W.F., Owens, E.T. (2006). Zea mays assays of chemical/radiation genotoxicity for the study of environmental mutagens. *Mutat. Res., 613(1),* 17-64.

Grant, W. F. (1994). The present status of higher plant bioassays for detection of environmental mutagens. *Mutat. Res., 310*, 175-185.

Green, M., Todo, T., Ryo, H. and Fujikawa, K. (1986). Genetic-molecular basis for a simple Drosophila melanogaster somatic system that detects environmental mutagens. *Proc. Nat. Acad. Sci. U S A, 83*, 6667-6671.

Harms, H., Wells, M.C., van der Meer, J.R. (2006). Whole-cell living biosensors--are they ready for environmental application? *Appl. Microbiol. Biotechnol., 70(3),* 273-80.

Hendricks, C.A., Engelward, B.P. (2004). "Recombomice": the past, present, and future of recombination-detection in mice. *DNA Repair (Amst)., 3(10),*1255-61.

Heuchel, R., Radtke, F., Georgiev, O., Stark, G., Aguet, M., Schaffner, W. (1994) The transcription factor MTF-1 is essential for basal and heavy metal-induced metallothionein gene expression. *EMBO J., 13,* 2870–2875.

Ichikawa, S. (1992). Tradescantia stamen-hair system as an excellent botanical tester of mutagenicity: its responses to ionizing radiations and chemical mutagens, and some synergistic effects found. *Mutat. Res., 270*, 3-22.

Ilnytskyy, Y., Yao, Y., Kovalchuk, I. (2005). Double-strand break repair machinery is sensitive to UV radiation. *J. Mol. Biol., 345(4),* 707-15.

Islam, E., Yang, X.E., He, Z.L., Mahmood, Q. (2007). Assessing potential dietary toxicity of heavy metals in selected vegetables and food crops. *J. Zhejiang Univ. Sci. B., 8(1),* 1-13.

Järup, L. (2003). Hazards of heavy metal contamination. *Br. Med. Bull., 68,* 167-82.

Kanaya, N., Gill, B., Grover, I., Murin, A., Osiecka, R., Sandhu, S. and Andersson, H. (1994). *Vicia faba* chromosomal aberration assay. *Mutat. Res., 310*, 231-247.

Ke, Q., Davidson, T., Chen, H., Kluz, T., Costa, M. (2006). Alterations of histone modifications and transgene silencing by nickel chloride. *Carcinogenesis., 27(7),* 1481-8.

Kondo, K., Takahashi, Y., Hirose, Y., Nagao, T., Tsuyuguchi, M., Hashimoto, M., Ochiai, A., Monden, Y., Tangoku, A. (2006). The reduced expression and aberrant methylation of p16(INK4a) in chromate workers with lung cancer. *Lung Cancer., 53(3),* 295-302.

Koornneef, M., Alonso-Blanco, C., Vreugdenhil, D. (2004). Naturally occurring genetic variation in Arabidopsis thaliana. *Annu. Rev. Plant. Biol.*, *55*, 141-72.

Kovalchuk, I., Kovalchuk, O., Kalck, V., Boyko, V., Filkowski, J., Heinlein, M., Hohn, B. (2003b). Pathogen-induced systemic plant signal triggers DNA rearrangements. *Nature, 423,* 760-762.

Kovalchuk, O., Burke, P., Arkhipov, A., Kuchma, N., James, S.J., Kovalchuk, I., Pogribny, I. (2003c). Genome hypermethylation in Pinus silvestris of Chernobyl--a mechanism for radiation adaptation? *Mutat Res., 529(1-2),* 13-20.

Kovalchuk, O., Hendricks, C.A., Cassie, S., Engelward, A.J., Engelward, B.P. (2004). In vivo recombination after chronic damage exposure falls to below spontaneous levels in "recombomice". *Mol. Cancer Res., 2*, 567-73.

Kovalchuk, O., Telyuk, P., Kovalchuk, L., Kovalchuk, I., Titov, V. (2003a). Novel plant bioassays for monitoring the genotoxicity of drinking water from the inhabited areas of the Ukraine affected by the Chernobyl accident. *Bull. Environ. Contam. Toxicol., 70(5),* 847-53.

Kovalchuk, I., Kovalchuk, O. and Hohn B. (2000). Genome-wide variation of the somatic mutation frequency in transgenic plants. *EMBO J., 19*, 4431-4438.

Kovalchuk, I., Kovalchuk, O., Arkhipov, A. and Hohn, B. (1998). Transgenic Plants are Sensitive Bioindicators of Nuclear Pollution Caused by the Chernobyl Accident. *Nat. Biotechnol., 16*, 1054-1057.

Kovalchuk, O., Kovalchuk, I., Titov, V., Arkhipov, A. and Hohn, B. (1999). Radiation hazard caused by the Chernobyl accident in inhabited areas of Ukraine can be monitored by transgenic plants. *Mutat. Res., 446,* 49-55.

Kovalchuk, O., Titov, V., Hohn, B. and Kovalchuk, I. (2001). A sensitive transgenic plant system to detect toxic inorganic compounds in the environment. *Nat. Biotechnol., 19,* 568-72.

Lalonde, S., Ehrhardt, D.W., Frommer, W.B. (2005). Shining light on signaling and metabolic networks by genetically encoded biosensors. *Curr. Opin. Plant Biol., 8(6),* 574-81.

Lebel, E.G., Masson, J., Bogucki, A., Paszkowski, J. (1993). Stress-induced intrachromosomal recombination in plant somatic cells. *Proc. Nat. Acad. Sci. U S A, 90(2),* 422-6.

Leonard, J.M., Bollmann, S.R., Hays, J.B. (2003). Reduction of stability of arabidopsis genomic and transgenic DNA-repeat sequences (microsatellites) by inactivation of AtMSH2 mismatch-repair function. *Plant Physiol., 133(1)*, 328-38.

Levin, D.E., and Ames B.N. (1986). Classifying mutagens as to their possible specificity in causing the six possible transitions and transversions: a simple analysis using the Salmonella mutagenicity assay. *Environ. Mutagen., 8*, 9-28.

Lu, H., Shi, X., Costa, M., Huang, C. (2005). Carcinogenic effect of nickel compounds. *Mol. Cell Biochem., 279(1-2),* 45-67.

Ma, T.H., Cabrera, G.L., Owens, E. (2005). Genotoxic agents detected by plant bioassays. *Rev. Environ. Health, 20(1)*, 1-13.

Mayer, C., Klein, R.G., Wesch, H. and Schmezer, P. (1998). Nickel subsulfide is genotoxic in vitro but shows no mutagenic potential in respiratory tract tissues of BigBlue rats and Muta Mouse mice *in vivo* after inhalation. *Mutat. Res., 420*, 85-98.

Miller, J.H. (1985). Mutagenic specificity of ultraviolet light. *J. Mol. Biol., 182*, 45-68.

Murti, R., Schimenti, K. and Schimenti, J. (1994). A recombination-based transgenic mouse system for genotoxicity testing. *Mutat. Res., 307*, 583-595.

Oruambo, I.F., Kachikwu, S., Idabor, L. (2007). Dose-related Increased Binding of Nickel to Chromatin Proteins; and Changes to DNA Concentration in the Liver of Guinea Pigs Treated with Nigerian Light Crude Oil. *Int. J. Environ. Res. Public Health., 4(3)*, 211-5.

Qi, X., Zhang, Y., Chai, T. (2007). Characterization of a novel plant promoter specifically induced by heavy metal and identification of the promoter regions conferring heavy metal responsiveness. *Plant Physiol., 143(1),* 50-9.

Rangwala, S.H., Richards, E.J. (2007). Differential epigenetic regulation within an Arabidopsis retroposon family. *Genetics, 176(1),* 151-60.

Ries, G. *et al.* (2000). Elevated UV-B radiation reduces genome stability in plants. *Nature, 406*, 98-101.

Roh, J.Y., Lee, J., Choi, J. (2006). Assessment of stress-related gene expression in the heavy metal-exposed nematode *Caenorhabditis elegans*: a potential biomarker for metal-induced toxicity monitoring and environmental risk assessment. *Environ. Toxicol. Chem., 25(11),* 2946-56.

Rugo, R.E., Almeida, K.H., Hendricks, C.A., Jonnalagadda, V.S., Engelward, B.P. (2005). A single acute exposure to a chemotherapeutic agent induces hyper-recombination in distantly descendant cells and in their neighbors. *Oncogene, 24(32),* 5016-25.

Sacco, M.G. et al. (1997). A transgenic mouse model for the detection of cellular stress induced by toxic inorganic compounds. *Nat. Biotech., 15*, 1392-1397.

Saidi, Y., Domini, M., Choy, F., Zryd, J.P., Schwitzguebel, J.P., Goloubinoff, P. (2007). Activation of the heat shock response in plants by chlorophenols: transgenic *Physcomitrella patens* as a sensitive biosensor for organic pollutants. *Plant Cell Environ., 30(6),* 753-63.

Schaaper, R.M. and Dunn, R.L. (1991). Spontaneous mutation in the *Escherichia coli. Carcinogenesis, 11*, 1087-1095.

Schmidt, A.L., Anderson, L.M. (2006). Repetitive DNA elements as mediators of genomic change in response to environmental cues. *Biol. Rev. Camb. Philos. Soc., 81(4)*, 531-43.

Schnekenburger, M., Talaska, G., Puga, A. (2007). Chromium cross-links histone deacetylase 1-DNA methyltransferase 1 complexes to chromatin, inhibiting histone-remodeling marks critical for transcriptional activation. *Mol Cell Biol., 27(20),* 7089-101.

Searle, P.F., Stuart, G.W., Palmiter, R.D. (1987). Metal regulatory elements of the mouse metallothionein-I gene. *Experientia Suppl., 52,* 407–414.

Seok, S.H., Park, J.H., Baek, M.W., Lee, H.Y., Kim, D.J., Uhm, H.M., Hong, J.J., Na, Y.R., Jin, B.H., Ryu, D.Y., Park, J.H. (2006). Specific activation of the human HSP70 promoter by copper sulfate in mosaic transgenic zebrafish. *J. Biotechnol., 126(3),* 406-13.

Sharma, R.K., Agrawal, M. (2005). Biological effects of heavy metals: an overview. *J. Environ. Biol., 26(2 Suppl),* 301-13.

Sharma, S.S., Dietz, K.J. (2006). The significance of amino acids and amino acid-derived molecules in plant responses and adaptation to heavy metal stress. *J. Exp. Bot., 57(4),* 711-26.

Sheehan, P.J. and Loucks, O.L. (1994). Issue paper on effects characterization. *Ecological Risk Assessment issue Papers*, EPA/630/R-94/009, U.S. Environmental Protection Agency, Washington, D.C.

Somerville, C., Koornneef, M. (2002). A fortunate choice: the history of Arabidopsis as a model plant. *Nat. Rev. Genet., 3(11)*, 883-9.

Sparrow, P.A., Dale, P.J., Irwin, J.A. (2006). *Brassica oleracea. Methods Mol. Biol., 343*, 417-26.

Stewart, C.N. Jr. (2001). The utility of green fluorescent protein in transgenic plants. *Plant Cell Rep., 20(5)*, 376-82.

Stewart, C.N. Jr. (2006). Go with the glow: fluorescent proteins to light transgenic organisms. *Trends Biotechnol., 24(4)*, 155-62.

Stuart, G.W., Searle, P.F., Palmiter, R.D. (1985) Identification of multiple metal regulatory elements in mouse metallothionein-I promoter by assaying synthetic sequences. *Nature, 317,* 828–831.

Swoboda. P., Gal, S., Hohn, B. and Puchta, H. (1994). Intrachromosomal homologous recombination in whole plants. *EMBO J., 13,* 484-489.

Van der Auwera, G., Baute, J., Bauwens, M., Peck, I., Piette, D., Pycke, M., Asselman, P., Depicker, A. (2008). Development and Application of Novel Constructs to Score C:G-to-T:A Transitions and Homologous Recombination in Arabidopsis. *Plant Physiol., 146(1),* 22-31.

Vij, S., Tyagi, A.K. (2007). Emerging trends in the functional genomics of the abiotic stress response in crop plants. *Plant Biotechnol. J., 5(3),* 361-80.

Westin, G., Schaffner, W. (1988). A zinc-responsive factor interacts with a metal-regulated enhancer element (MRE) of the mouse metallothionein-I gene. *EMBO J., 7,* 3763–3770.

Zhang, L., Xing, D., Wang, J., Li, L. (2007). Rapid and non-invasive detection of plants senescence using a delayed fluorescence technique. *Photochem. Photobiol. Sci., 6(6),* 635-41.

Zoroddu, M.A., Kowalik-Jankowska, T., Kozlowski, H., Molinari, H., Salnikow, K., Broday, L., Costa, M. (2000). Interaction of Ni(II) and Cu(II) with a metal binding sequence of histone H4: AKRHRK, a model of the H4 tail. *Biochim. Biophys. Acta., 1475(2)*, 163-8.

In: Causes and Effects of Heavy Metal Pollution
Editor: Mikel L. Sanchez
ISBN 978-1-60456-900-1

Chapter 11

GEOCHEMISTRY OF MAJOR AND TRACE ELEMENTS IN CORE SEDIMENTS OF SUNDERBAN DELTA, INDIA: AN ASSESSMENT OF METAL POLLUTION USING ATOMIC ABSORPTION SPECTROMETER AND INDUCTIVELY COUPLED PLASMA MASS SPECTROMETRY

A. Bhattacharya*, K.K. Satpathy**, M.V.R. Prasad**, J. Canario**, M. Chatterjee*, S.K Sarkar*, V. Branco**, B. Bhattacharya*, A. K. Bandyopadhyay * and Md. Aftab Alam****

*Department of Marine Science, University of Calcutta, 35, Ballygunge Circular Road, Calcutta-700019, India

** Indira Gandhi Centre for Atomic Research, Environmental and Industrial Safety Section, Safety Group, Kalpakkam 603102, Tamil Nadu, India.

***IPIMAR- National Institute of Biological Resources, Av. Brasilia, 1449-006 Lisboa, Portugal.

**** Department of Statistics, University of Calcutta, 35, Ballygunge Circular Road, Calcutta-700019, India

ABSTRACT

The paper documents a detailed account of spatial distribution and possible sources of major elements along with heavy metals in fine grained fractions (<63 µm) of core sediments collected from seven sites in Sunderban wetland, northeastern part of Bay of Bengal, India. This work aims to evaluate the fluvio-marine and geochemical processes influencing the trace element distribution and to check the suitability of employing heavy metals data in evaluating biological effects on the basis of sediment quality guidelines. Both Inductively Coupled Plasma Mass Spectrophotometer (ICPMS) and Atomic Absorption Spectrometry (AAS) were employed to determine the elemental concentration in acid-digested sediment samples. Trace element concentrations were available at different core depths with an erratic pattern of distribution. An overall enrichment of majority of the elements has been recorded at the site Kakdwip, located along the main stream of Ganges, and this can be attributed to domestic and industrial effluent discharge, intensive fishing and boating activities coupled with use of antifouling paints. In contrast, the site Canning, located further east in the mudflat of Matla River is characterized by minimum trace element content. An abrupt variations of Mn, V, Cd and U were encountered at the site Jambu Island- an offshore island facing Bay of Bengal. For Cu, Ni and As, a smaller proportion of the samples had exceeded the effects range-low (ER-L) concentrations indicating that the dataset would be suitable for future use in evaluating predictive abilities of sediment quality guidelines

INTRODUCTION

Trace metal pollution in the marine environment has long been recognized as a matter of serious environmental concern (Balkas et al., 1982). Hence, the knowledge of trace elements levels and distribution in the marine environment, leads to a better understanding of their behavior in aquatic systems and allows the detection of sources of pollution (Forstner and Wittman, 1979). Heavy metals, in particular, are likely to insinuate potentially hazardous conditions to estuarine and marine organisms, due to their persistence in the environment and toxicity above threshold levels, They may be subdivided into two categories: (i) transition metals (e.g., cobalt, copper, iron, manganese, etc) which are essential to metabolism at low concentration but may be toxic at high concentrations and (ii) metalloids (e.g., arsenic, cadmium, lead, mercury, selenium, tin) which generally are not required for metabolic function but are toxic at low concentrations.

Due to their environmental persistence, biogeochemical recycling and ecological risks, heavy metals are of particular concern worldwide (Liu et al., 2003; Gonzalez-Macias et al., 2006). Metal contaminated environments may release metals to surrounding waters in three ways: (a) by desorption from suspended particles upon contact with seawater, (b) by desorption from bottom sediments and (c) by diffusion from interstitial water subsequent to diagenetic alteration of sediments (Duinker, 1980). However, if the equilibrium between marine sediments and the overlying water body is broken, marine sediments would transfer most pollutants into seawater (Valdes et al., 2005). As metal concentrations in marine sediments increase, more heavy metals will return to water bodies via chemical and biological processes (Sin et al., 2001).

The spatial distribution of heavy metals in marine sediments is of major importance in clarifying the pollution history of aquatic systems (Rubio et al., 2001; Birch et al., 2001; Liu et al., 2003). The process is impacted by natural and anthropogenic factors, such as climate, parent rock weathering, industrial and domestic wastewater, aquaculture, and agriculture (Morillo et al., 2004; El Nemr et al., 2006; Luo et al., 2006). It is essential to distinguish between natural and human impacts on heavy metals in marine sediments. Eventually, geoaccumulation index, enrichment factor and factor analysis have been applied to indicate the degree of contamination by heavy metals from lithogenic and anthropogenic sources (Rubio et al., 2001; Simeonov et al., 2003; Santos et al., 2005; El Nemr et al., 2006).

The present study has been undertaken in the Sunderban delta of West Bengal, northeast India at the northernmost interface of the Bay of Bengal. It is formed by the alluvial and tidal network systems of Hugli, Saptamukhi, Jamira, Matla and Gosaba rivers, all of which have funnel shaped openings to the south before meeting the Bay of Bengal. The other tidal distributory drainage network includes, among others, the Bidya and Bidyadhari rivers, which are connected to the major river systems by interlacing creeks. The lower delta plain exhibits a fascinating network of channels and interchannel areas with dynamically changing tidal shoals and islands of variable dimensions and shapes (Bhattachacharya and Das, 2002). The seaface is characterized by macro to mesotidal amplitude and moderate to moderately high wave climate. The wave-tidal energy wanes out farther inland toward north.

MATERIALS AND METHOD

Sampling Locations

Seven sampling stations from S_1 to S_7 have been chosen along a SW-NE transect covering an approximate distance of 120 km from high-energy macrotidal mouth of the Hugli river to low-energy delta plain about 90 km inland from the sea face where Bidya and Bidyadhari distributory systems form a complex network with the upstream stretch of the Matla River (Figure 1). The alluvial sediments of the delta around these distributory systems are reworked and manipulated both by tidal hydrodynamics and bioturbational activities, which involve both lateral and vertical transport of material. As a result, the intertidal mudflats of this mangrove-infested Sunderban delta, also recognized as a 'Biodiversity Hotspot', undergo constant changes in terms of their topographical, textural and structural attributes. The sampling stations S_1, S_2 and S_3 occur along the main channel of the Hugli River which has a 16 km-wide funnel shaped opening with bifurcating arms around Sagar Island at the mouth of the estuary (Figure 1).

The intertidal flats of these three stations on the Hugli estuary are dynamically active by ebb and flood tidal currents, tidal bores, wind driven waves and seasonal cyclonic storms, all of which act as potential natural agents for deposition, erosion, suspension and resuspension of sediments. Being located along the Hugli navigation channel, the intertidal sediments of all these stations are modified by dredged out sediments. Sampling station S_4 is positioned on the foreshore sandy beach of the Jambu Island - an offshore island in the open sea environment. Sampling station S_5 is situated on the point bar near Canning at the meander bend of Matla River. Sediments at this point bar are recycled by tidal flushing of the Matla River where both longitudinal and helical flows of this meandering river control the accretionery point bar deposits. Sampling stations S_6 and S_7 belong to the mudflats deposited on both sides of the Bidyadhari River of which the former is located to the erosional site and the latter to the depositional site of a sharp meander bend.

The sediment properties of each sampling station vary in accordance with the hydrodynamic and sedimentological setting of each area (See also Rubio et al., 2001 Bhattachacharya and Das, 2002). The upland sediment discharge by the Hugli and its tributaries, to-and-fro movement of tidal currents, shoreline and river bank erosion, bioturbational churning by benthic animals together with human activities involving agriculture, fisheries, dredging, navigation, deforestation and reclamation, all contribute to the sediment characteristic of each sampling station.

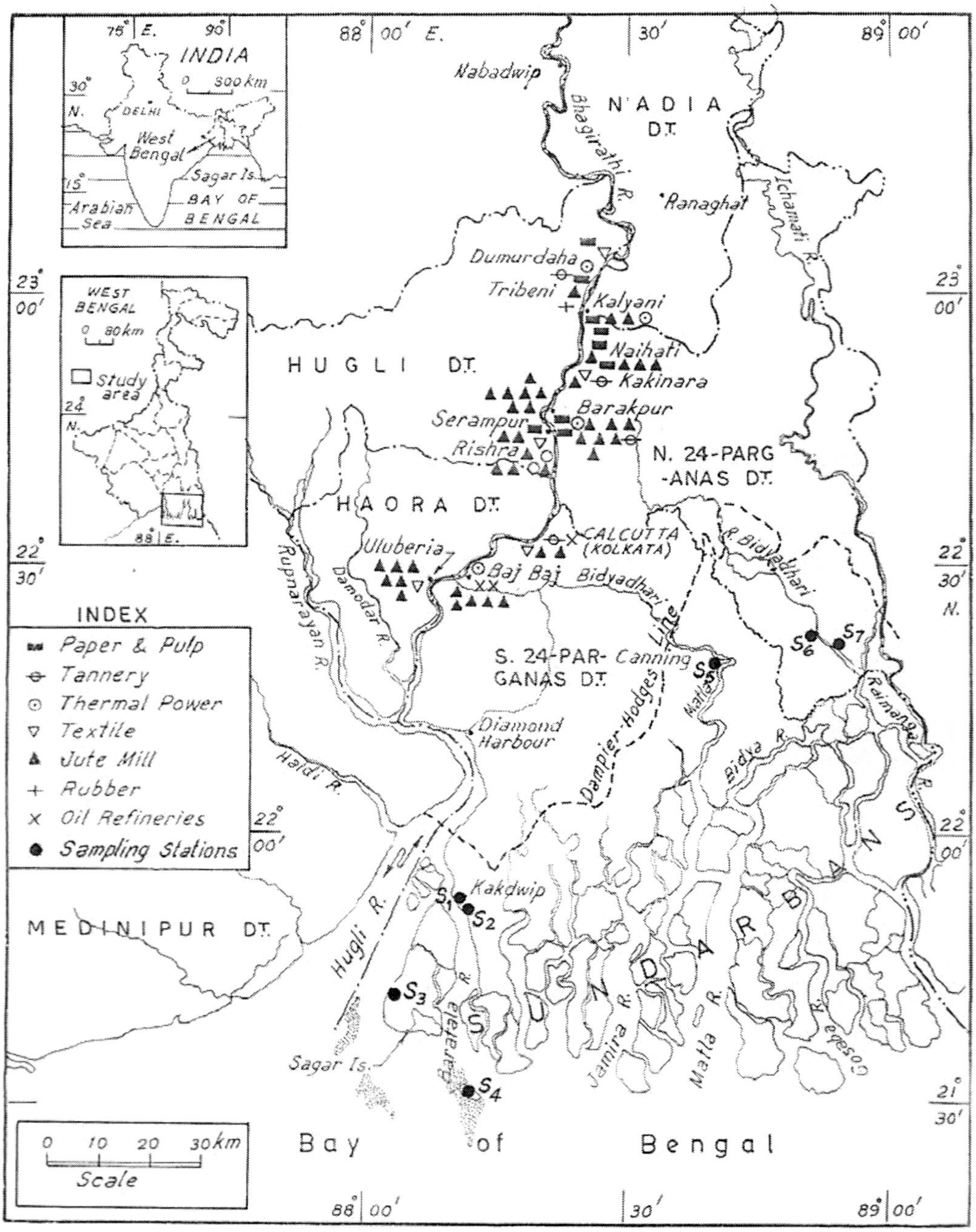

Figure 1. Map showing the locations of the seven sampling stations located in Sunderban wetland.

Collection and Preservation of Sediment Samples

During winter months (January - February, 2006), cores of 30 cm length were collected with a steel corer (40 cm length and 5 cm diameter) by gently pushing it into the sediment. Cores were capped and frozen and transported to the laboratory. The upper 40 cm of each core were sliced into 4 cm fractions (subsamples) with the help of PVC spatula. Prior to sample collection, all glasswares for the collection and storage of sediment samples were thoroughly cleaned with acid (10% HNO_3), and then rinsed in double-distilled water before use.

The core length differed between stations due to variations in the nature of the substratum. Core fractions were stored in labelled polyethylene bags and kept in iceboxes until reaching the laboratory where they were frozen at -20 ^{0} C. Within two days, a portion of each sample was placed in a ventilated oven at a very mild temperature (max. 40^{0}C). Dried samples were then disaggregated using an agate mortar and pestle, sieved through a 63μm metallic sieve and stored in hermetic plastic bags until analysis. All visible marine organisms and coarse shell fragments, sea grass leaves and roots were removed manually when present. Sediments were analysed for quality parameters (organic carbon, pH, % of silt, clay and sand) and trace elements' content. Organic carbon (C_{org}) content of sediments was determined following a rapid titration method (Walkey and Black, 1934) and pH with the help of a deluxe pH meter (model no. 101E) using a combination glass electrode manufactured by M.S. Electronics (India) Pvt. Ltd. Mechanical analyses of sediment were done by sieving in a Ro-Tap Shaker (Krumbein and Pettijohn, 1938) manufactured by W.S. Tyler Company, Cleveland, Ohio, and statistical computation of textural parameters was done by using formulae of Folk and Ward (1957).

Geochemical Analyses of Elements

In the laboratory, total determinations of Al, Si, Ca, Mg, Fe and Mn were performed by mineralization of the sediments with a mixture of acids (HF, HNO_3 and HCl) according to the method described by Rantala and Loring (1975). Element concentrations were determined by flame-AAS (Perkin-Elmer Annalist 100) using direct aspiration into N_2O-acetylene flame (Al, Ca, Mg and Si) or air-acetylene flame (Fe and Mn). The V, Cr, Co, Ni, Cu, As, Se, Mo, Cd, Pb and U levels were determined using the sample digestion described above in a Inductively Coupled Plasma Mass Spectrometer Thermo Elemental - X Series. The precision expressed as relative standard deviation was less then 4% ($p<0.05$). International certified standards (BCSS-1, MESS-2, MESS-3, PACS-2,) were used to ensure the accuracy of the procedure. For all metals investigated, obtained and certified values were not statistically different ($p<0.05$).

Statistical Analyses

The raw data were classified objectively using factor analysis. The purpose of factor analysis is to reduce the complexity within the similarity matrix of a multivariate data collection and transform it into a simpler and easier to interpret factor matrix. Regression analyses and Pearson product correlation coefficients between trace metals and organic carbon and mud (silt and clay %) in sediment were also worked out.

RESULT AND DISCUSSION

The core samples of different depths have been analyzed for major and trace elements (Al, Si, Ca, Mg, Fe, Mn, Zn, V, Cr, Co, Ni, Cu, As, Mo, Cd, Pb and U). Table 2 shows pooled mean values (along with standard deviation) of pH, organic carbon (%), percentages of carbonate, sand, silt and clay, concentrations of Mn, Zn, V, Cr, Cu, Ni, Co, As, Mo, Cd, Pb, U (in $\mu g\ g^{-1}$) and Al, Fe, Mg, Si (%) obtained in sediment profiles of the seven sampling stations. Variations of metal concentrations both spatially and in sediment columns reflected variations in sources, hydrodynamic conditions, together with variations in textural (sand, silt and clay %) and/or carbonate and organic matter content. Zn, Pb, Cr, Ni, Cu and As contents in sediments are relatively high at S_1 and S_3 compared to other stations (Table 2. The contents of almost all metals (Fe, Mn, Zn, V, Cr, Co, Ni, As and Pb) have a decreasing trend from SW to NE i.e., from the mudflats adjoining Hugli estuary to the inner part of the delta indicating a lesser dispersal opportunity for the metals from the main source of their origin. It is thus evident that the two stations S_1 and S_3 lying along the main course of the Hugli estuary receive a larger amount of metals from anthropogenic inputs from nearby Haldia port, as well as from the cluster of industries (e.g.,paper and pulp, textile, tannery etc.,) further upstream. The sedimentological settings of these two stations favour settling and concentration of metals in the intertidal sediments, and are controlled by the behaviour of the Hugli estuary. The Hugli, presently acts as a seasonal river with dominating flow only during four months of the year (Goodbred, 2000). During the remaining months it is the tidal flux from the Bay of Bengal that keeps the estuary active. The heavy metals together with tidal sediments are recirculated in the lower stretch of the estuary and undergo burial during the monsoon months with high discharge of sediments.

For Mn, Fe, V, Cd and U, maximum concentrations are recorded at S_4, the sandy beach area of Jambu offshore island. The characteristic of these metal concentrations is probably attributable to Fe and Mn oxide coatings on the sandy grains of this station. Other metals, particularly, V and Cd show relatively higher values probably due to their association with ferrous alloys that are used in the floating buoys along the shipping line close to this station. The rusty barges may also have some contributions. The high Mn and Fe contents in the sandy sediments have also been noticed by Shrader et al., (1977) from a fluvial system in eastern Tennessee. Uranium content in sediments is very consistent, ranging from 2.1-2.96$\pm$ $\mu g\ g^{-1}$ in all sampling stations except in station S_4, where the value goes up to 12$\pm$1.09 $\mu g\ g^{-1}$. The source of U, however, is not clearly known. One possible nearby source may be due to alongshore drift of U from the Subarnarekha River delta occurring at a distance of 45 km west of this station (S_4). The Subarnarekha River drains through the Uranium mining sites of Jaduguda, Jharkhand State, adjoining to the state of West Bengal. In the more mangrove-infested stations like S_5, S_6 and S_7 (Figure 1) show decreasing grain size with increasing depth up to 28 cm. The finer grained sediment in these stations tends to have slightly higher contents of Fe, Cr, V, Ni, and Cd. This trend may be attributed to higher surface absorption on the higher specific surface of the smaller particles aided by more ionic attraction Horowty and Elrick, 1987).

Since the alluvial and tidal sediments are primarily generated from the weathering of crustal rocks, analyses of some common crustal elements have been taken into consideration in the present study.

Table 1. Results of certified Reference Materials as well as the observed values

a.

		BCSS-1		MESS-3		PACS-2	
	Element	Certified	Obtained	Certified	Obtained	Certified	Obtained
CRM	Al (%)	6.26±0.22	6.10	8.59±0.23	8.31	6.62±0.32	5.84
	Si (%)	30.9±0.47	29.6	27*	24	28*	22
	Ca (%)	0.54±0.05	0.80	1.47±0.06	1.20	1.96±0.18	1.78
	Mg (%)	1.47±0.14	2.03	1.6*	1.85	1.47±0.13	1.41
	Fe (%)	3.29±0.09	3.91	4.34±0.11.	4.84	4.09±0.06	4.15
	Mn (μg/g)	229±15	249	324±12	325	440±19	423
	Zn (μg/g)	119±12	112	159±8	146	364±23	366

b.

		BCSS-1		MESS-3		PACS-2	
	Element	Certified	Obtained	Certified	Obtained	Certified	Obtained
CRM	V (μg/g)	93.4 ± 4.9	100	243 ± 10	242	133 ± 5	129
	Cr (μg/g)	123 ± 14	110	105 ± 4	108	90.7 ± 4.6	84
	Co (μg/g)	11.4 ± 2.1	12.0	14.4 ± 2.0	13.1	11.5 ± 0.3	10.9
	Ni (μg/g)	55.3 ± 3.6	55.3	46.9 ± 2.2	40.9	39.5 ± 2.3	36.5
	Cu (μg/g)	18.5 ± 2.7	16.8	33.9 ± 1.6	35.0	310 ± 12	280
	As (μg/g)	11.1 ± 1.4	12.4	21.2 ± 1.1	21.9	26.2 ± 1.5	27.3
	Mo (μg/g)	-	112	2.78 ± 0.07	2.75	5.43 ± 0.28	366
	Cd (μg/g)	0.25 ± 0.04	0.27	0.24 ± 0.01	0.22	2.11 ± 0.15	5.09
	Pb (μg/g)	22.7 ± 3.4	24.8	21.1 ± 0.7	23.8	183 ± 8	175
	U (μg/g)	-	2.3	4*	4	3*	2

Table 2. Pooled mean values (X±SD) of pH, organic carbon (%); carbonate (%), percentages of sand, silt and clay and concentrations of Al(%), Si(%), Ca(%), Mg(%), Fe(%), Zn, Mn, V, Cr, Co, Ni, Cu, As, Mo, Cd, Pb and U ($\mu g\ g^{-1}$) in sediment profiles of the 7sampling stations in Sunderban mangrove environment

Parameter	S_1	S_2	S_3	S_4	S_5	S_6	S_7
pH	8.6 ± 0.17	8.4 ± 0.21	8.5 ± 0.07	8.1 ± 0.33	6.9 ± 0.22	8.5 ± 0.16	7.9 ± 0.18
Organic Carbon%	0.52 ±0.09	0.74 ± 0.62	0.56 ± 0.02	0.15 ± 0.16	0.73 ±0.14	0.78 ± 0.13	0.94 ± 0.12
Carbonate %	3.7 ± 0.35	2.7 ± 0.52	5.5 ± 0.55	3.7 ± 0.41	7.6 ± 0.65	6.93 ± 0.60	6.4 ± 1.07
Sand%	2.24 ± 0.51	18.25 ± 7.26	3.15 ± 0.62	99.74 ± 0.55	27.5 ± 6.11	16.67 ± 5.67	10.05 ± 3.45
Silt%	41.97 ± 15.75	47.42 ± 21.67	41.13 ± 11.67	0.26 ± 0.55	32.54 ± 11.19	37.58 ± 9.19	33.15 ± 9.74
Clay%	55.79 ± 15.73	34.33 ± 17.84	55.71 ± 11.44	-	39.95 ± 11.41	45.75 ± 9.62	56.79 ± 10.21

Table 2. (Continued)

Parameter	S_1	S_2	S_3	S_4	S_5	S_6	S_7
Texture	Silty clay – clayey very fine	Silty-Clayey fine	Silty clay – clayey very fine	Sandy	Fine loam – Clayey fine	Fine loam – Clayey fine	Silty clay – clayey very fine
Al%	7.4 ± 0.36	5.4 ± 0.71	7.53 ± 0.27	4.81 ± 0.44	4.99 ± 0.41	6.14 ± 0.33	6.95 ± 0.59
Si%	28.8 ± 1.86	32.13 ± 3.92	26.2 ± 3.55	29.07 ± 4.93	29.8 ± 1.88	30.9 ± 2.1	30.4 ± 4.75
Ca%	1.6 ± 0.22	1.97 ± 0.15	1.69 ± 0.13	3.47 ± 0.4	2.86 ± 0.24	2.55 ± 0.21	1.99 ± 0.19
Mg%	1.56 ± 0.15	1.09 ± 0.16	1.5 ± 0.08	1.24 ± 0.07	1.02 ± 0.1	1.23 ± 0.11	1.34 ± 0.11
Fe%	4.46 ± 0.28	2.84 ± 0.48	4.18 ± 0.20	7.12 ± 0.69	2.19 ± 0.2	2.76 ± 0.22	3.43 ± 0.54
Mn	690.8 ± 30.6	529.5 ± 46.24	686.86 ± 27.9	3411.3 ± 589.08	491.5 ± 68.73	470.5 ± 24.6	505 ± 25.5
Zn	92.5 ± 15.03	56.67 ± 8.98	72.28 ± 12.63	58.17 ± 9.02	48.25 ± 5.56	68 ± 7.74	54 ± 16.06
V	86.3 ± 5.71	56 ± 6.96	82.43 ± 4.68	93.5 ± 5.01	43 ± 4.24	54.5 ± 15.4	69.28 ± 9.18
Cr	80.8 ± 5.98	49.67 ± 7.68	78.86 ± 6.07	67.5 ± 4.81	35.75 ± 3.77	49 ± 14.8	68.43 ± 13.18
Co	13.5 ± 1.05	8.47 ±	13.14 ± 0.69	10.07 ± 0.83	6 ± 0.83	7.73 ± 2.28	10.11 ± 1.3
Ni	36.5 ± 3.27	20.67 ± 3.88	35.43 ± 2.51	11.97 ± 3.54	14.5 ± 2.38	19.27 ± 5.96	25.43 ± 3.91
Cu	47.3 ± 6.89	27.3 ± 4.18	49. 14 ± 3.76	16.33 ± 3.08	17.5 ± 3.11	22.3 ± 6.67	23.71 ± 6.72
As	9.18 ± 0.78	6.13 ± 1.66	8.78 ± 0.54	4.1 ± 0.81	4.25 ± 0,87	5.98 ± 1.84	7.08 ± 1.19
Mo	0.27 ± 0.04	0.15 ± 0.08	0.28 ± 0.05	0.31 ± 0.09	0.13 ± 0.05	0.18 ± 0.1	0.24 ± 0.13
Cd	0.09 ± 0.007	0.08 ± 0.02	0.07 ± 0.005	0.26 ± 0.02	0.05 ± 0.006	0.07 ± 0.02	0.13 ± 0.03
Pb	25.5 ± 1.87	20 ± 1.55	30.28 ± 14.43	14.5 ± 1.38	18.75 ± 0.96	18 ± 4.47	22.86 ± 2.48
U	2.25 ± 0.14	2.45 ± 0.4	2.1 ± 0.08	12 ± 1.09	2.47 ± 0.17	2.33 ± 0.55	2.96 ± 0.40

It is seen that average values of concentrations of Si, Al, Ca, Mg and Fe of all 7 stations show a close resemblance with that of crustal rocks (Table 4). The slightly higher concentration of Si in the samples of the studied area is a reflection of greater maturity of the sediments deposited, after long transportation and reworking. The slightly low content of Al, on the other hand, may be the result of weathering of feldspar, on the source rocks, into clay minerals.

Carbonate Contents

The relatively high mean carbonate contents (6.4 - 7.6%) in the mudflat sediments of stations S_5, S_6 and S_7 is a reflection of the presence of more bioclasts produced by fragmentary shells of benthic organisms (mainly molluscs and crustaceans). The mudflat of these areas belongs is less wave-tide dominated which favors a good habitat for the benthic organisms. In addition to benthic organisms, foraminifera and other shelly fauna from the sea floor may be swept shoreward by wave action to yield some more carbonate fractions in the intertidal sediments. The lower carbonate content (2.7-5.5 %) at stations S_1, S_2, S_3 and S_4 reflects a lesser importance of bioclasts, as these stations are located close to the seaface in a high energy environment, less suitable for survival of the benthic fauna.

Organic Matter Content

Fine sediments with more mud (silt and clay) are relatively rich in organic carbon (0.52 to 0.94%) with an increasing trend with depth as seen in all stations, except S_4, which is sandy and has a very low (0.15%) organic carbon content that does not vary significantly with core depth (Figure 2). Mean values obtained for organic carbon (0.15-0.94%) are comparatively lower than the typical values found in sediments of other mangrove areas (5 DW% Corg) (Gonneea et al., 2004; Bouillon et al., 2000). With exception of S_1 and S_4, the highest percentages of organic carbon contents (0.61-1.87) are found in the deeper layers. This is due to the fact that the increase in organic carbon with depth is dominantly controlled by the increasing proportion of mud fraction. The organic matter accumulated in these estuarine sediments is a result of high biological productivity associated with upwelling events (Rodriguez et al., 1991; Maríne and Olivares, 1999). A variety of marine sources contribute to the organic matter and these include phytoplankton from overlying water column, faecal pellets from zooplankton and filter feeding organisms (mussels), benthic animals, microalgal populations dominated by diatoms, bacteria and other heteromorphs in the sediment. Burial of macroalgae such as *Ulva* and *Enteromorpha* may cause anoxia in sediments (Volkman et al., 2000). Mangrove macrophytes, sea grass and algal macrophytes, contribute organic carbon in these intertidal mud depending on their opportunities of burial. A relatively lower percentage of organic carbon in the top soil (0-5 cm) than in the deeper layers could be attributed to the constant flushing activities by tides along with the impact of waves which remove the finer fractions of the sediments from the fringing areas. In contrast, tidal influence and wave actions are comparatively weaker at the other two stations (S_6 and S_7) situated far east of the Hugli estuary.

Sediments of the studied stations are characterized by pH varying from slightly acidic to basic (6.6-8.8). This is probably due to oxidation of FeS_2 and FeS to Fe_2SO_4 in part resulting from the decomposition of mangrove litter and hydrolysis of tannin in mangrove plants, that release various kinds of organic acids (Liao, 1990). A pH above 7 results in the precipitation of the major dissolved metallic species with a concomitant scavenging of many trace metals by the hydroxide floccules of the precipitating iron, aluminum and manganese; dissolved carbon dioxide, chlorine and areas rich in organic matter (mangroves) could have a reverse effect: keeping some of the cations in free forms (see, for example, the modules of Turner et al., 1981).

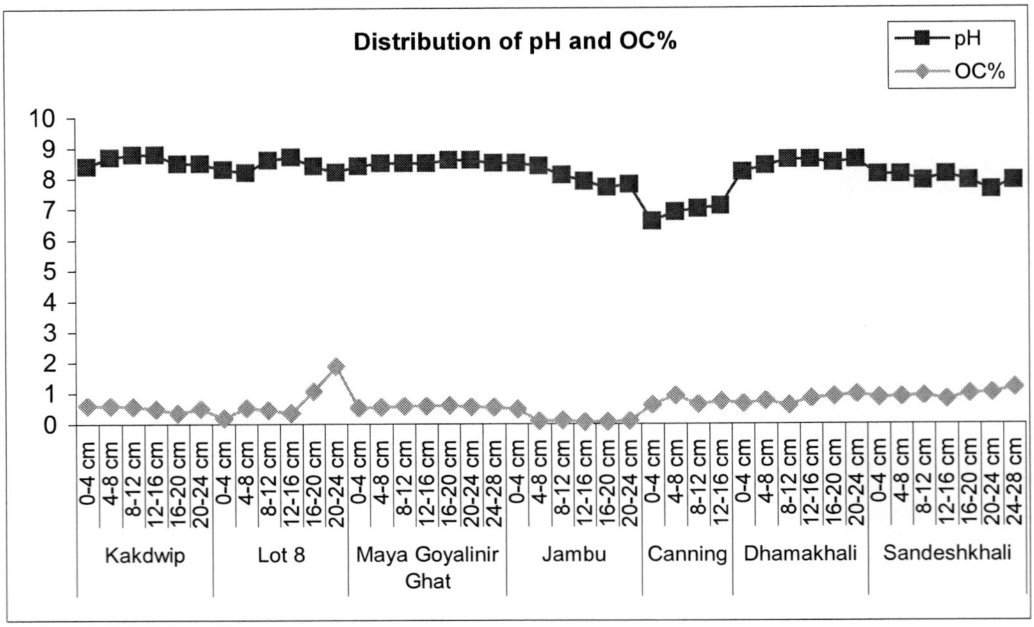

Figure 2A. Distribution of pH and % organic carbon (OC) at different depth profiles of seven sampling sites.

I would make secondary axis the the OC values and would put concentrations in Y axis. The profiles represented this way are quite confusing. I really don't know the solution since we have many paramenters but for sure you should not connect the profiles in the diferetn stations. Probably the metal concentrations should be divided by Al as you mentioned in the discussion,

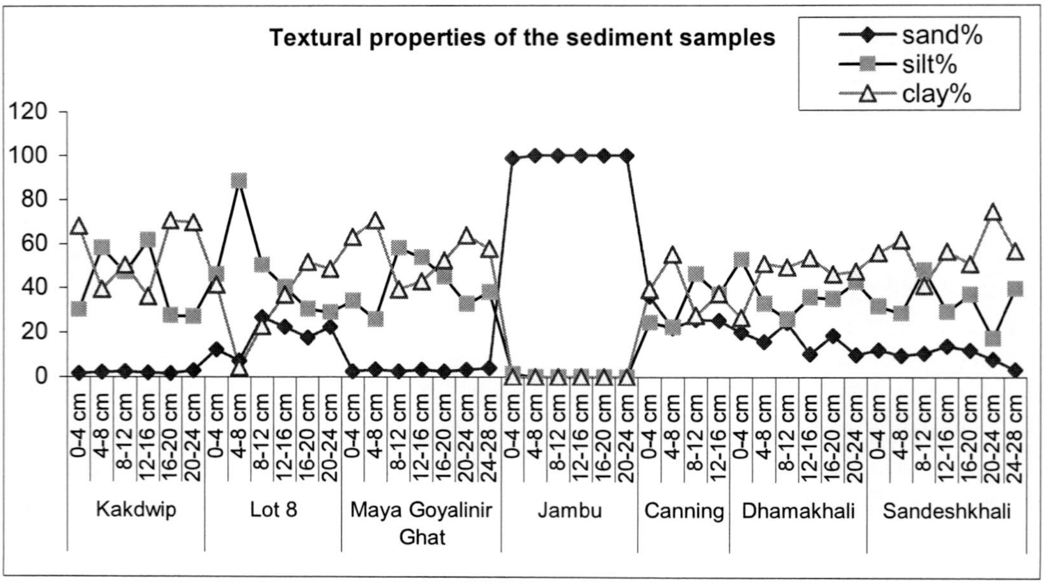

Figure 2B. Distribution of % sand, silt and clay at different depth profiles of seven sampling sites.

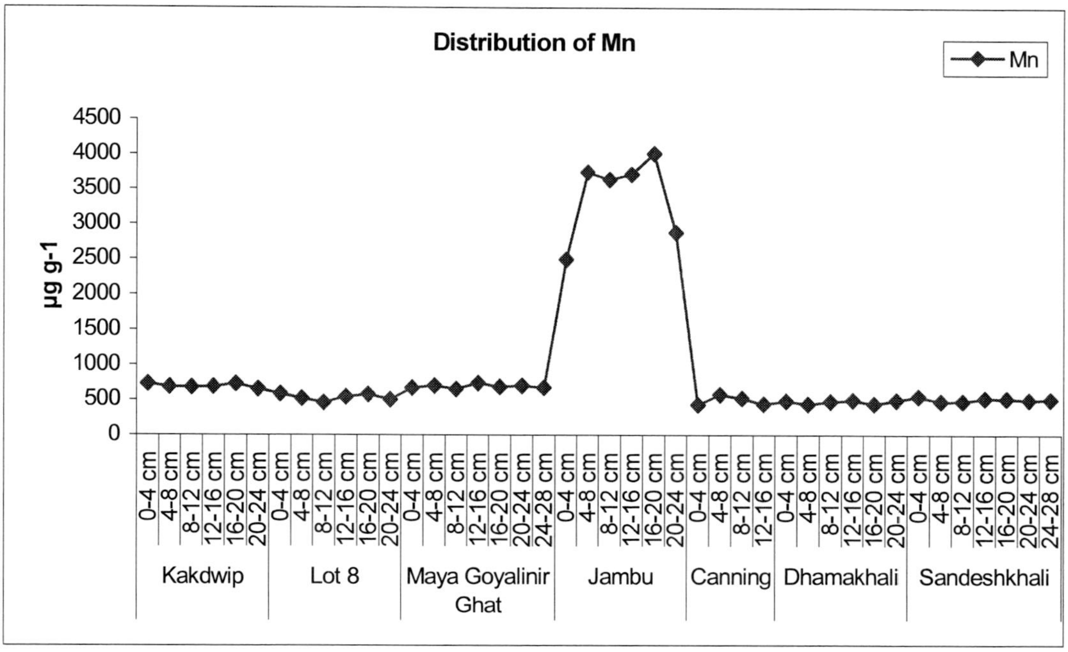

Figure 2C. Distribution of Mn at different depth profiles of the seven sampling sites.

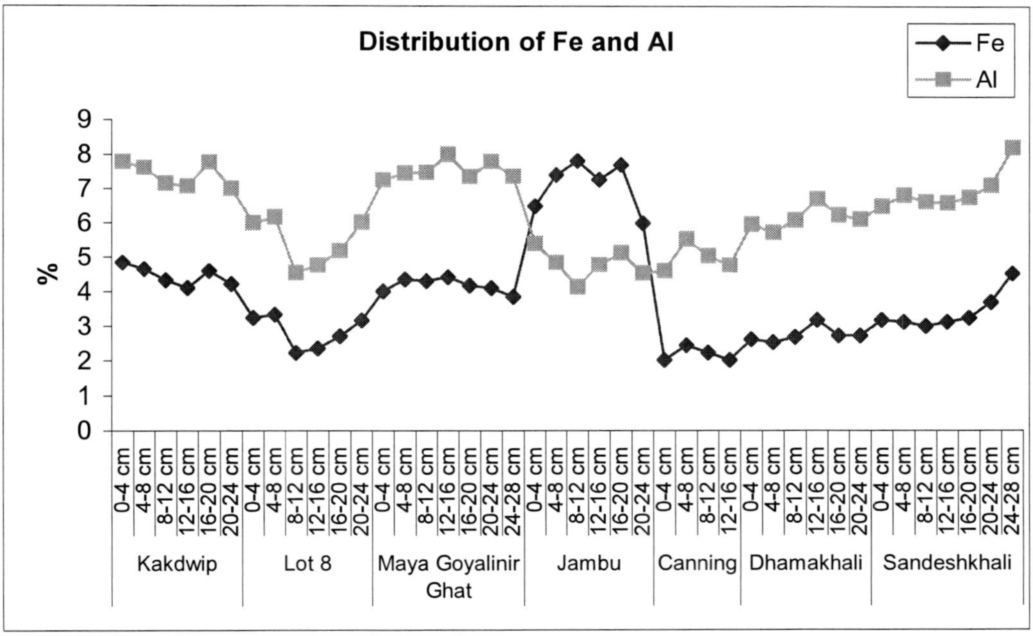

Figure 2D. Distribution of % of Fe and Al at different depth profiles of seven sampling sites.

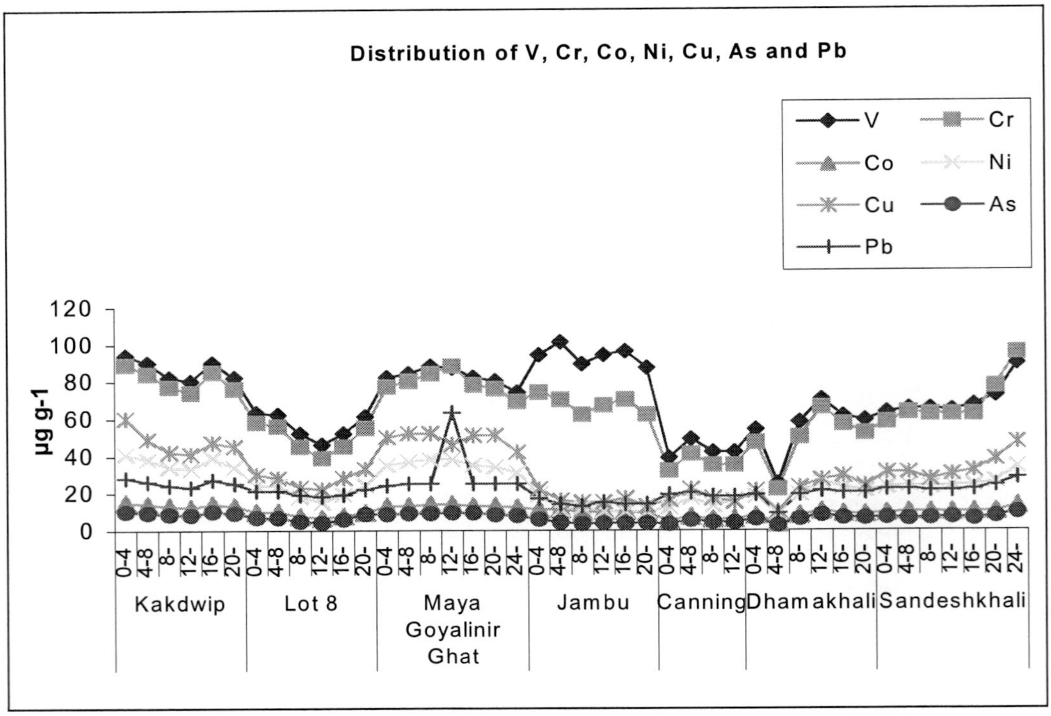

Figure 2E. Distribution of V, Co, Cu, Pb, Cr, Ni and As at different depth profiles of seven sampling sites.

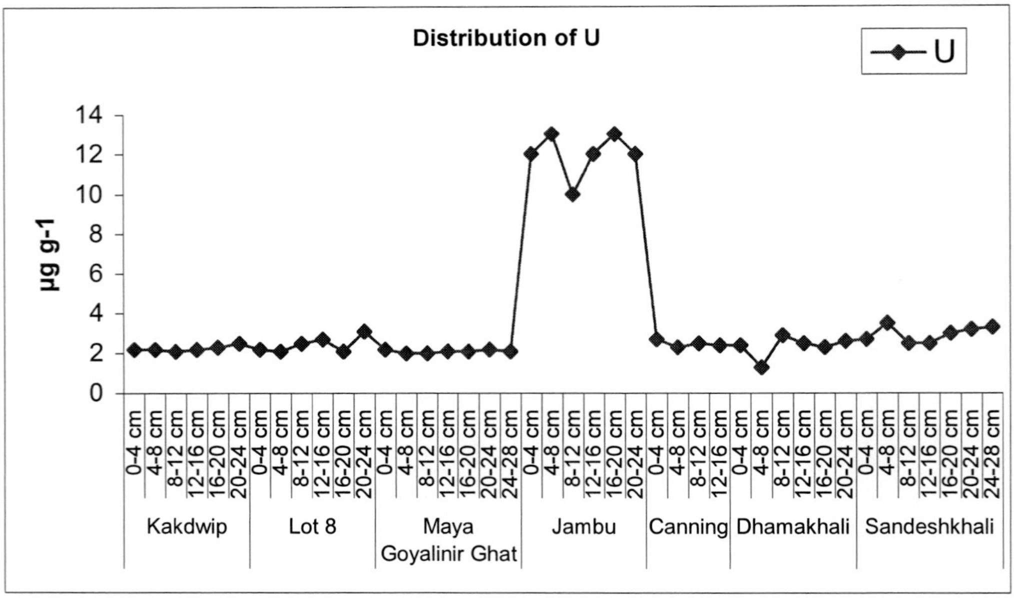

Figure 2F. Distribution of U at different depth profiles of seven sampling sites.

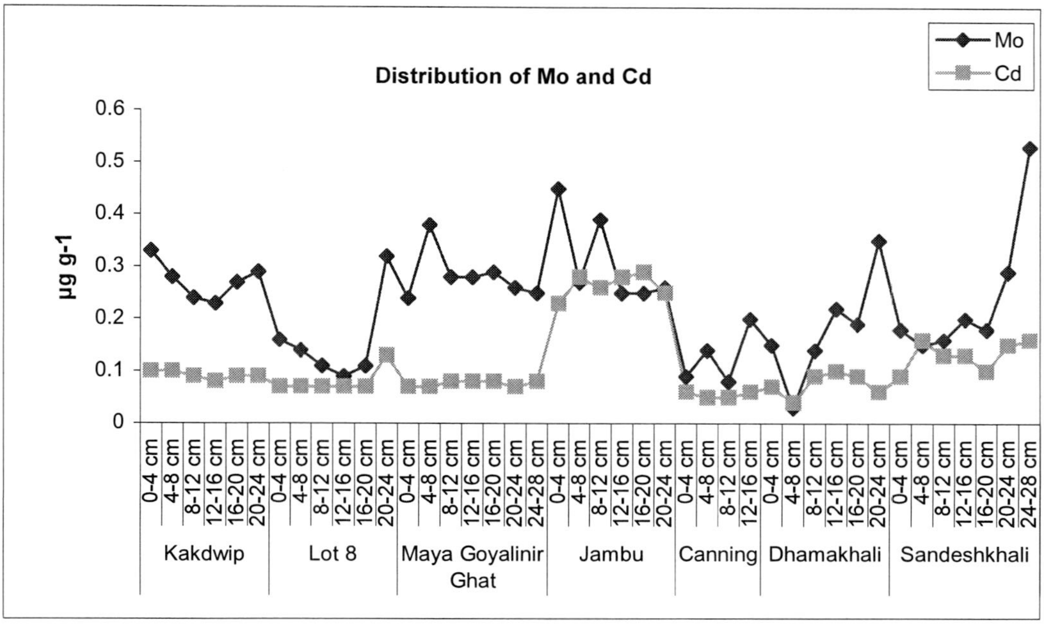

Figure 2G. Distribution of Mo and Cd at different depth profiles of seven sampling sites.

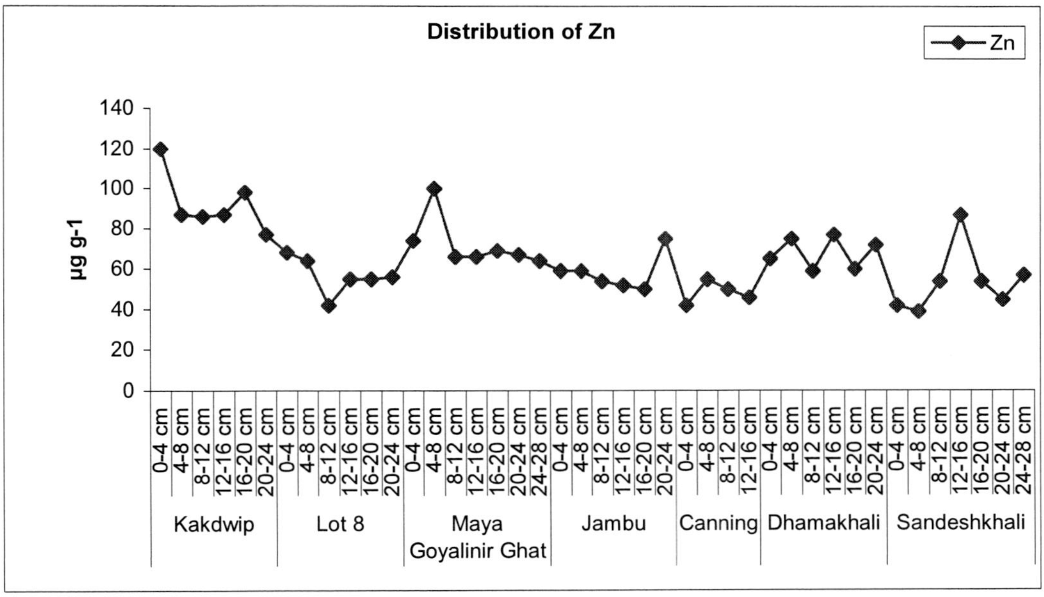

Figure 2H. Distribution of Zn at different sediment profiles of seven sampling sites.

The content of organic matter and carbonates are important controlling factors in the accumulation of trace metals in sediments. Fine-grained sediments are more likely to have relatively high metal contents because of the larger surface area of the small sediment particles. In this study the concentration of various metals in core sediments varied greatly with location as well as with depth. Higher concentration of most of the elements (except Mo,

Mn, U and V) was observed in S_1 (Kakdwip) located near the estuary. Although Mo and Pb concentrations showed only a narrow range of variation Cd and Ni varied widely from 0.04 to 0.29 μg g^{-1} and 8.6 to 41 μg g^{-1} respectively. With the exception of S_5, S_6 and S_7, all studied stations show elevated concentration of V, Cr, Co, Ni, Cu, As and Mo at surface/near surface sediment profiles. Pb and U, on the other hand, showed elevated concentrations at or near deeper layers in all the seven studied cores.

A high enrichment of Mn was detected at S_4, Jambu Island (2492-4009 μg g^{-1}). However, the reason for this four/five-fold increase in the concentration of this particular element at this pristine site is unknown. The variation in the elemental concentrations in the study area could be at least partly associated with sediment type and sediment grain size. This relationship between grain size and trace metal concentrations has been reported elsewhere (Rubio et al., 2000; Cho et al., 1999).

Heavy metals can exist in several phases that are associated with organic matter, carbonates, oxyhydroxides, and aluminosilicates forms in the total sediment matrix. The portion of the total heavy metal concentration caused by anthropogenic contamination is also believed to be associated with organic matter or adsorption on Fe/Mn hydrous oxides. Therefore, before the degree of pollution is evaluated, the association between organic matter and hydrous oxide forms of Fe/Mn and total metal concentration should be determined. Because several elements can easily accumulate under anoxic conditions, the portions of organic matter related to metal concentration could be evaluated indirectly by comparing the behaviour of redox-sensitive elements. Mobile elements, such as Mo, Mn and V, generally accumulate in anoxic environments (Morford et al., 2001), while Cu concentrations can either decrease or increase in anoxic environments (Jung et al., 1996). It is therefore possible to evaluate benthic environments on the basis of elemental behavior and their associations. Local boating activity in the middle and outer estuary likely contributes significant quantities of metals from antifouling paints to this estuarine delta. Arsenium, Cd, Cr, Cu, Pb and Zn have been used in antifouling paints in recent years as an alternative to tributyltin (Bryan and Langston, 1992). In particular As and Zn concentrations were elevated at S_1 and S_3 and are likely to be associated with antifouling paint residues arising from boat maintenance and cleaning in the sites in close vicinity.

Vertical profiles of Mn (Fig 2C) and Fe (Fig 2D) show decrease in their concentrations from the surface suggesting a diagenetic enrichment (Santschi et al., 1990 and Shaw et al., 1990) during which Fe-Mn oxyhydroxides dissolve in the partly reduced sediment layer producing Fe^{2+} and Mn^{2+} species, which migrate upward in the sediment column and get precipitated near the oxic-suboxic interface. The above inference is well supported by high concentration of Fe (S_1, 0-4 cm, 4.85%; S_2, 4-8 cm, 3.34% and S_5, 4-8 cm, 2.46%) and Mn (S_1, 0-4 cm, 727 μg g^{-1}; S_2, 4-8 cm, 3.34% and S_5, 4-8 cm, 574 μg g^{-1}) and low values of Fe at 12-16 cm in S_1 (4.11%), 8-12 cm in S_2 (2.23%) and at 0-4 and 12-16 cm at S_5 (2.03%). On the other hand, the higher values of Fe in the deeper half of S_3, S_4, S_6 and S_7 are due to stable reduced layers existing in the layer below 8-12 cm in S_3 and S_4, 12-16 cm in S_6 and 24-28 cm in S_7. So sediment layers of 8-12 cm, 12-16 cm and 24-28 cm in S_3, S_4, S_6 and S_7 respectively act as a transition zone between oxygenated water and the reduced zone in the sediments (see also Janaki-Raman et al., 2007). Vertical distribution of Mn (Fig 2 C) shows variation with depth and also with respect to sampling location. The low values of Mn at 8-12 cm (461 μg g^{-1} and 651 μg g^{-1}at S_2) and 4-8 cm (438 μg g^{-1} at S_6 and 478 at S_7) indicate that dissolved Mn ions, with greater mobility, are easily removed from the pore water to surface sediments and then to the upper water column through diffusion, advection processes. In addition, tidal currents have played a vital role in their mobility (Jung et al., 1996).

Vertical profiles of Cu indicate high values in the reduced layers at 16-28 cm depth in S_2, S_6 and S_7, which are due to its absorption and scavenging by Fe and Mn- oxides and hydroxides (Prohic and Kniewald, 1987). The profiles of Ni and Co resemble each other (Fig 2 E). The increase of Ni content at 0-4 cm in S_1 (41 $\mu g\ g^{-1}$) and S_2 (24 $\mu g\ g^{-1}$) is probably due to Ni sorption on to Mn (oxy) hydroxides. The redox status of the sediments determines the extent to which mobilization of Mn and associated Ni will occur, i.e., the presence of a distinct sub-oxic-reduced condition leading to loss of (absorbed) Ni from the surface sediments (Froelich et al., 1979). The surface layer at S_1, S_4 and surface layer at S5 reveal an increase in concentration of Pb, which is attributed to the exchangeable fraction. The higher Pb values in S_3 (63 $\mu g\ g^{-1}$) may be attributable to the atmospheric inputs, operation of a large number of mechanized fishing boats in this area. The low values of Pb at S_6 (9 $\mu g\ g^{-1}$) and S_4 (13 $\mu g\ g^{-1}$) at 8-16 cm depth may indicate re-suspension processes. The maximum concentration of vanadium at station S_4 (values) followed by S_3 (values) may be ascribed to discharge of crude oils in considerable quantities (Nriagu, 1989) as these two stations are located near an international shipping line.

Increasing urbanization, resulting in rising solid waste and wastewater production, increasing harbour activities and, to a lesser extent, recent industrialization has apparently maintained Zn emission in the local environment.

Interpretation of Statistical Data

The inter-element relationship between various sedimentological parameters and other elements are presented in Table 3. Three common factors were derived from the 42 samples describing 78.1% of the total variance (Table 3). The three factors provide additional support (grouping of elements) in determining the enrichment and association among the analyzed parameters. In the Factor analysis, Factors 1, 2 and 3 account for 38.3, 33.9 and 5.9 % of the total variance.

Factor analysis was used to identify the parameters that control metal distributions in the studied area. Table 4 shows the Varimax rotation factor matrix results. Three factors explain 78.1% of the total variance in 42 sediment samples analyzed.

Table 3. Factor analyses (after Varimax rotation) showing the contribution of statistically dominant variables measured in this study

Variable	Factor1	Factor2	Factor3
OC	-0.003	-0.447	0.065
Mud%	0.372	-0.895	0.027
V	0.749	0.616	-0.112
Cr	0.926	0.252	-0.121
Co	0.959	0.130	-0.099
Ni	0.877	-0.399	-0.085
Cu	0.894	-0.338	-0.135

Table 3. (Continued)

Variable	Factor1	Factor2	Factor3
As	0.868	-0.344	-0.088
Mo	0.586	0.406	-0.104
Cd	0.038	0.966	0.047
Pb	0.508	-0.260	-0.218
U	-0.142	0.964	-0.027
Fe%	0.336	0.894	-0.052
Mn	-0.081	0.959	-0.047
Si%	-0.230	-0.063	0.961
Ca%	-0.609	0.647	0.010
Mg%	0.846	-0.005	-0.070
Zn	0.468	-0.114	-0.028
% Var	38.3	33.9	5.9
Cumul var%	38.3	72.2	78.1

Table 4. A comparison table to show the values of major elements in the studied samples and that of the crust

	Average of all seven stations (%)	earth's crust
Si	29.54	27.69
Al	6.2	8.07
Ca	2.30	3.65
Mg	1.28	2.08
Fe	3.85	5.05

Factor 1 account for 38.3% of the total variance and groups the majority of the variables with high positive loadings except OC, U, Mn, Si and Ca. This indicates that the organic flux through the water column is not an efficient mechanism for transport of metals. The metals are presumably bound either in clays via adsorption and/or exchange. Factor 2 explains 33.9% of total variance and is characterized by negative loading of OC, Mud, Ni, Cu, As, Pb, Si, Mg and Zn indicating some degree of continental influence. Factor 3 accounts for 5.9% of the total variance and shows a strong positive loading of Si. So this factor can be termed as 'silicon factor'.

In the case of trace metals and the relationships between them, the strong association between most of them indicates that they are primarily controlled by Fe-Mn oxyhydroxides (Buckley et al., 1995; Zwolsman et al., 1993; 1996). The relationship by significant correlation ($p<0.05$) of Fe versus Mn, V, Cr, Ni, Cu, Mo, Cd and U ($r=0.89$, 0.85, 0.56, 0.85, 0.56, 0.61, 0.85 and 0.83 respectively). Mn versus Ca, V, Ni, Cd and U (0.73, 0.56, 0.56, 0.91 and 0.96 respectively). The relationships of these metals indicate that they are closely associated with Fe and Mn oxides and that Fe is available in the form of oxide coating on sediment particles as evidenced from other studies in southeast coast of India (Rao and Murthy, 1990; Subramanian and Mohanachandran, 1990).

Correlation coefficient analysis shows strong positive relation ($p<0.05$) of Al with Mg, Zn, Cr, Co, Cu and As. Strong positive correlations are also observed between heavy metals as well as other major elements studied. The positive correlation coefficient is indicative either of the discharges having a common source, such as special chemical industries, fertilizers and petrochemical complexes, etc., or that the associated metallic cations have isogeochemical processes (Bhosale and Sahu, 1991)

ERL and ERM (Effect Range Low and Effect Range Medium) Values

To estimate the possible environmental consequences of metals analyzed, our results were also compared to US NOAA's sediment quality guidelines. In this study the effects range-low (ER - L) and Effects range-median (ER - M) concentrations are considered. The ER - L represents chemical concentrations below which adverse biological effects were rarely observed, while the ER - M represents concentrations above which effects were more frequently observed (Long et al., 1995, 1997). Computing the mean values of the elements in each studied station, majority of the elements revealed lower values excepting Ni (at S_1, S_3 and S_7), Cu (at S_1 and S_3) and As (at S_1 and S_3).

Index of Geoaccumulation (I_{geo})

Possible sediment enrichment of metals was evaluated in terms of the I_{geo} of Muller (1979). The formula used for the calculation of I_{geo} is: $\log_2 (C_n/1.5\ B_n)$, where C_n is the measured content of element "n", and B_n the element's content in "average shale" (Turekian and Wedepohl, 1961). The geoaccumulation Index (I_{geo}) in the present work showed very low values indicating that sediments are uncontaminated (Muller, 1979), similar to previous observation reported by Chatterjee et al., 2006 from the core sediment samples in Sundarban environment. The observed I_{geo} values are in contrast as worked out by diverse authors in Indian estuarine sediments (Biksham and Subramanian, 1988; Ramesh et al., 1990; Subramanian et al., 1988; Ramanathan et al., 1993).

Enrichment Factor (EF) Values

Differentiating the elements originating from human activities and those from natural weathering is an essential part of geochemical studies. One such technique largely applied is 'normalization' where metal concentrations were normalized to a textural or compositional characteristic of sediments. Normalizing elements relative to Al is widely used to compensate for variations in both grain size and composition, since it represents the quantity of aluminosilicates, which is the predominant carrier phase for adsorbed elements in coastal sediments. According to Nolting et al. (1999), this method is also a powerful tool for the regional comparison of trace metal content in sediments and can also be applied to determine enrichment factors (EFs).

The values of EFs can be obtained using the equation $EF=(\text{metal/Al})_{\text{sediment}}/(\text{metal/Al})_{\text{shale}}$. EFs close to unity point indicate crustal origins while those greater than 10 are considered to be non-crustal source (Nolting et al., 1999). The minimum EFs obtained for many elements (Mn, V, Co, Ni, As, Mo and Cd) are less than unity implying that these elements are depleted in some of the phases relative to crustal abundance in the study area. However, it is evident that all the elements with a EF value greater than unity reveals sediment contamination, for example the higher EF values for Mg, Fe, Zn, Cu, Pb and U. An overall higher EF values for these elements in all the seven stations suggest the presence of contaminated sediments derived from multifarious sources like domestic sewage, power-plant operation, major storm events or dumping of sediments. In addition, the importance of atmospheric deposition of Pb from industry and vehicle emissions is also of great importance. Emissions from automobiles consuming leaded petrol all over India are the major source of atmospheric Pb (Huntzicker et al., 1975). A high proportion of the Pb from vehicle emissions may also be transferred to the estuary by the way of road runoff through storm water drains.

Interpretative Problems

In coastal ocean and estuarine systems, bioturbation and physical mixing of the sediments can occur (Benninger et al., 1979; Nittrouer et al., 1983/1984) and these are very much pronounced at S_3 where the mudflat is populated with the tube-dwelling polychaete *Diopatra cuprea* which stabilize the bottom against erosion and also enhance sediment accumulation. The processes may obliterate the information contained in the top part of the sedimentary column. Another typical problem is concerned with the role of diagenesis, particularly for trace metal remobilization as pointed out by Valette-Silver (1993). The vertical distribution of some metals (Fe, Mn etc.,) can be affected after sediment deposition by diagenetic processes that solubilize the elements in the anoxic portions of the core and redeposit them in the upper oxic layers. Based on studies conducted off the California coast, Shaw et al., (1990) described a variety of chemical behavior for the transition metals depending on the form under which they are transported to the sediments (e.g., scavenged by the Mn-Fe oxides, biogenic material, detrital material etc.,). Transport, burial and diagenesis play a key role in the preservation of historical records for metal contamination (Valette-Silver, 1993; Bhattacharya et al., 2000). Hence for reliable results an excellent knowledge of the environment and of the problems specific to each area is to be taken into consideration.

Conclusion

The spatial distribution of the heavy metals in the Hugli estuarine complex is controlled by both natural and anthropogenic processes. The natural processes involve macro to meso tidal to-and-fro flushing, ebb and flood flows along flood and ebb channels, a seasonal moderate to high tidal bore, estuarine circulation, edge currents, suspension-resuspension process under helical flow, and bioturbational activities. The anthropogenic sediment input is primarily linked to catchment processes linked to urbanization, agriculture, industrialization, deforestation and reclamation activities.

Acknowledgement

The work was financially supported by a research project entitled "Concentration of heavy metals in sediment profiles of Sunderban mangrove environment, northeast India" (Sanction No: 24(0276)/05/EMR-II) funded by Council of Scientific and Industrial Research (CSIR), New Delhi, India. One of the authors (M. Chatterjee) is grateful to CSIR for awarding her SRF.

References

Balkas, IT., Tugrel, S. and Salhogin, I., (1982). Trace metal levels in fish and Crustaceans from Northeastern Mediterranean coastal waters. *Marine Environmental Research*, 6: 281-289.

Benninger, LKRC., Aller, JK., Cochran. and Turekian, KK., (1979). Effects of biological sediment mixing on the ^{210}Pb chronology and trace metal distribution in a Long Island Sound sediment core. *Earth and Planetary Science Letters,* 43: 241-259.

Bhattacharya, A. and Das, GK., (2002). Dynamic geomorphic environment of Indian Sunderbans. In: Changing Environmental scenario of the Indian subcontinent. S. P Basu (ed.) Acb Publication, Kolkata, 284-298.

Bhattacharya, A., Sarkar, SK., Das, GK., Giri, S., Bose, G., Mitra, K. and Dey, D., (2000). Holocene tidal flat sedimentation in the Sundarban Biosphere Reserve, northeast India. *S.D.M.C.E.T; Dharwad and IGCP-367, Spl. Pub*. 2: 51-55.

Biksham, G. and Subramanian, V., (1988). Elemental composition of Godavari sediments (Central and Southern Indian Subcontinent). *Chemical Geology*, 70: 275-286.

Bouillon, S., Mohan, PC., Sreenivas, N. and Dehairs, F., (2000). Sources of suspended organic matter and selective feeding by ooplankton in an estuarine mangrove ecosystem as traced by stable isotopes. *Marine Ecology Progress Series*, 208: 70-92.

Birch, GF., Taylor, SE. and Matthai, C., (2001) Small-scale spatial and temporal variance in the concentration of heavy metals in aquatic sediments: a review and some new concepts. *Environ Pollution*, 113: 357-372.

Buckley, DE., Smith, JN. and Winters, GV., (1995). Accumulation of contaminant metals in marine sediments of Halifax harbour, Nova Scotia: environmental factors and historical trends. *Applied Geochemistry* 10: 175-195.

Chatterjee, M., Silva-Filho EV., Sarkar, SK., Sella, SM., Bhattacharya, A., Satpathy, KK., Prasad, MVR ., Chakraborty, S. and Bhattacharya, BD., (2007). Distribution and possible source of trace elements in the sediment cores of a tropical macrotidal estuary and their ecotoxicological significance. *Environment International*, 33: 346-356.

Cho, Y., Lee, C. and Choi, M ., (1999). Geochemistry of surface sediments off the southern and western coast of Korea. *Marine Geology*, 159: 111-129.

Duinker, JC., (1980) Suspended matter in estuaries: adsorption and desorption processes, in *Chemistry and Biogeochemistry of Estuaries*, Olausson, E., and Cato, I (Eds.). John Wiley and sons, Chichester: 121.

El Nemr, A., Khaled, A. and El Sikaily, A., (2006). Distribution and statistical analysis of leachable and total heavy metals in the sediments of the Suez Gulf. *Environmental Monitoring and Assessment*,118: 89-112.

Folk, RL. and Ward, WC., (1957). Brazos River bar, a study of the significance of grain size parameters. *Journal of Sedimentary Petrology*, 27: 3-26.

Forstner, U. and Wittman, GTW., (1979). Metal pollution in the aquatic environment. *Berlin Heidelberg*: *Springer-Verlag;* pp. 486.

Gonneea, ME., Paytan, A. and Herrera-Silveira, JA., (2004). *Estuarine, Coastal and Shelf Science*, 61: 211-227

Gonzalez-macias, C., Schifter, I., Liuch-Cota, DB., Mendez-Rodriguez, L. and Hernandez-Vazquez, S., (2006). Distribution, enrichment and accumulation of heavy metals in coastal sediments of Salina Cruz Bay, Mexico. *Environmental Monitoring and Assessment* 118: 211-230.

Huntzicker, JJ., Friedlander, SK. and Davidson, CI., (1975). Material balance for automobile-emitted lead in Los Angels Basin. *Environmental Science and Technology* 9: 448-457.

Janaki-Raman, D., Jonathan, MP., Srinivasalu, S., Armstron-Altrin, JS., Mohan, SP. and Ram-Mohan, V., (2007). Trace metal enrichments in core sediments in Muthupet mangroves, SE coast of India: Application of Acid leachable technique. *Environmental Pollution* 145(1): 245-257.

Jung, HJ., Lee, CB., Cho, YG. and Kong, JK., (1996). A mechanism for the enrichment of Cu and depletion of Mn in anoxic marine sediments, Banweol Intertidal flat, Korea, *Marine Pollution Bulletin* 32(11): 782-787.

Krumbein, WC. and Pettijohn, FJ., (1938). *Manual of Sedimentary Petrology*. New York: Plenum: p.549.

Liao, JF., (1990). The chemical properties of the mangrove Solonchak in the northeast part of Hainan Island. *The Acta Scientiarum Naturalium Universities Sunyatseni*. 9(4): 67-72.

Liu, WX., Li, XD., Shen, ZG., Wang, DC., Wai, OWH. and Li, YS., (2003). Multivariate statistical study of heavy metal enrichment in sediments of the Pearl River Estuary. *Environmental Pollution,*121: 377-388.

Long, ER., Field, LJ. and MacDonald, DD., (1997). Predicting toxicity in marine sediments with numerical sediment quality guidelines. *Environmental Toxicology and Chemistry,* 17(4): 714-727.

Long, ER., MacDonald, DD., Smith, SC. and Calder, FD., (1995). Incidence of adverse biological effects within ranges of chemical concentrations in marine and estuarine sediments. *Environmental Management*, 19(1): 81-97.

Luo, W., Wang, TY., Lu, YL., Giesy, JP., Shi, YJ., Zheng, YM., Xing, Y. and Wu, GH., (2006). Landscape ecology of the Guanting Reservoir, Beijing, China: multivariate and geostatistical analyses of metals in soils. *Environmental Pollution*, 146: 567-576.

Marín, V. and Olivares, G., (1999). Estacionalidad de la productividad primaria en bahía Mejillones del Sur (Chile): una aproximación proceso-functional. *Revista Chilena de Historia Natural*, 72: 629-641.

Morillo, J., Usero, J. and Gracia, I., (2004). Heavy metal distribution in marine sediments from the southwest coast of Spain. *Chemosphere*, 55: 431-442.

Morford, J. and Emerson, S., (1999). The geochemistry of redox-sensitive trace metals in sediments. *Geochimica et Cosmochimica Acta*, 63(11-12): 1735-1750.

Muller, G., (1979). Schermetalle in den sedimenten des Rheins-Veran-derungen seitt, 1971. *Umschan*, 79: 778-783.

Nittrouer, CA., Demaster, DJ., McKee, BA., Cutshall, BA. and Larsen, IL., (1983/1984). The effect of sediment mixing on ^{210}Pb accumulation rates for the Washington continental shelf. *Marine Geology*: 54, 201-221.

Nolting, RF., Ramkema, A. and Everaarts, JM., (1999). The geochemistry of Cu, Cd, Zn, Ni and Pb in sediment cores from the continental slope of Banc d' Arquin (Mauritani). *Continental Shelf Research,* 19: 665-691.

Nriagu, JO., (1989). A global assessment of natural sources of atmospheric trace metals. *Nature*,338: 47-49.

Ramanathan, AL., Vaithiyanathan, P., Subramanian, V. and Das, BK., (1993). Geochemistry of the Cauvery estuary, East Coast of India. *Estuaries*, 16: 459-474.

Ramesh, R., Subramanian, V. and Van Grieken, R., (1990). Heavy metal distribution in sediments of Krishna river basin, India. *Environmental Geology Water Science*, 15: 207-216.

Rantala, RTT. and Loring, DH., (1975). Multi-element analysis of silicate rocks and marine sediments by atomic absorption spectrophotometry. *Atom Absorption Newsletter*, 14: 117-120.

Rao, CHM. and Murty, PSN. (1990).Geochemistry of the continental margin sediments of the central west coast of India. *Journal of Geological Society of India*, 35: 19-37.

Rodriguez, L., Marin, V., Farias, M. and Oyarce, E. (1991) Identification of an upwelling zone by remote sensing and *in situ* measurement. Mejillones del Sur Bay (Antofagasta-Chile). *Scientia Marina*, 55(3): 467-473.

Rubio, B., Nombela, MA. and Vilas, F.(2000). Geochemistry of major and trace elements in sediments of the Ria de Vigo (NW Spain): an assessment of metal pollution. 40: 968-980.

Rubio, B., Pye, K., Rae, JE. and Rey, D. (2001). Sedimentological characteristics, heavy metal distribution and magnetic properties in subtidal sediments, Ria de Pontevedra, NW Spain. *Sedimentology*, 48: 1277-1296.

Salomons, W. and Forstner, U.(1984) *Metal in the Hydrocycle.* Berlin: Springer-Verlag: 349 pp.

Santos, IR., Silva, EV., Schaefer, CEGR., Albuquerque, MR. and Campos, LS. (2005). Heavy metal contamination in coastal sediments and soils near the Brazilian Antarctic Station, King George Island. *Marine Pollution Bulletin*, 50: 185-194.

Santschi, PH., Hohener, P., Benoit, G. and Brink, B. (1990). Chemical processes at the sediment water interface. *Marine Chem*istry 30: 269-315.

Shaw, TJ., Gieskes, JM. and Jahnke, RA. (1990). Early digenesis in differing depositional environments. The response of transition metals in pore water. *Geochim. Cosmochim. Acta*, 54: 1233-1246.

Simenov, V., Stratis, JA., Samara, C., Zachariadis, G., Voutsa, D., Anthemidis, A., Sofoniou, M and Kouimtzis, T. (2003). (Assessment of the surface water quality in Northern Greece. *Water Research*, 37: 4119-4124.

Sin, SN., Chua, H., Lo, WW. and Ng, LM. (2001). Assessment of heavy metal cations in sediments of Shing Mun River, Hong Kong. *Environment International*, 26: 297-301.

Subramanian, V. and Mohanachandran, G. (1990). Heavy metals distribution and enrichment in the sediments of southern east coast of India. *Marine Pollution Bulletin*, 21: 324-330.

Subramanian, V., Jha, PK. and Van Grieken, R. (1988). Heavy metals in the Ganges estuary. *Marine Pollution Bulletin*, 19: 290-293.

Turkian, KK. and Wedephol, KH.(1961) Distribution of the elements in some major units of the earth crust. *Bulletin of Geological Society of America*, 72, 175– 92.

Walkey, A. and Black, TA. (1934). An examination of the Dugtijaraff method for determining soil organic matter and proposed modification of the chronic and titration method. *Soil Sci*ence, 37: 23-38.

Valdes, J., Vargas, G., Sifeddine, A., Ortlieb, L. and Guinez, M. (2005). Distribution and enrichment evaluation of heavy metals in Mejillones Bay (230 S), Northern Chile: geochemical and statistical approach. *Marine Pollution Bulletin*, 50: 1558-1568.

Valelte-Silver, HJ. (1993). The use of sediment cores to reconstruct historical trends in contamination of estuarine and coastal sediments. *Estuaries*, 16(3B): 577-588

Volkman, JK., Rohjans, J., Rullkotter, J., Scholz-Bottcher. and Liebezeit, G. (2000). Sources and diagenesis of organic matter in tidal flat sediments from the German Wadden Sea. *Continental Shelf Research*, 1139-1158.

Zwolsman, JJG., Berger, GW. and Van Eck, GTM. (1993). Sediment accumulation rates, historical input, post depositional mobility and retention of major elements and trace metals in salt marsh sediments of the Scheldt estuary, SW Netherlands. *Marine Chemistry*, 44: 73-94.

Zwolsmam, JJG., Van Eck, GTM. and Berger, GW. (1996). Spatial and temporal distribution of trace metals in sediments from the Scheldt estuary, south-west Netherlands. *Estuarine Coastal and Shelf Science,* 43: 55-79.

Correlation Matrix of all the Major Elements, Heavy Metals and Sediment Quality Parameters of the Seven Sampling Stations

S1 (KAKDWIP)

	Al	Si	Ca	Mg	Fe	Mn	Zn	V
Si	-0.358							
	0.486							
Ca	-0.719	0.745						
	0.107	0.089						
Mg	0.142	-0.320	0.160					
	0.789	0.537	0.763					
Fe	0.940	-0.501	-0.813	0.015				
	0.005	0.311	0.049	0.978				
Mn	0.858	-0.334	-0.519	0.384	0.724			
	0.029	0.517	0.291	0.452	0.104			

	Al	Si	Ca	Mg	Fe	Mn	Zn	V
Zn	0.768	-0.664	-0.694	0.326	0.792	0.877		
	0.075	0.150	0.126	0.528	0.060	0.022		
V	0.962	-0.533	-0.859	0.041	0.986	0.760	0.794	
	0.002	0.276	0.029	0.938	0.000	0.080	0.059	
Cr	0.966	-0.503	-0.840	0.028	0.990	0.787	0.818	0.997
	0.002	0.309	0.036	0.958	0.000	0.063	0.047	0.000
Co	0.887	-0.518	-0.880	-0.156	0.978	0.658	0.768	0.968
	0.019	0.292	0.021	0.768	0.001	0.156	0.075	0.002
Ni	0.968	-0.522	-0.858	0.042	0.976	0.808	0.828	0.995
	0.001	0.288	0.029	0.937	0.001	0.052	0.042	0.000
Cu	0.755	-0.816	-0.935	0.038	0.886	0.620	0.856	0.891
	0.082	0.048	0.006	0.943	0.019	0.189	0.030	0.017
As	0.870	-0.371	-0.881	-0.291	0.869	0.658	0.640	0.910
	0.024	0.468	0.020	0.576	0.025	0.155	0.171	0.012
Mo	0.582	-0.751	-0.971	-0.290	0.747	0.362	0.626	0.768
	0.226	0.086	0.001	0.578	0.088	0.480	0.184	0.075
Cd	0.681	-0.479	-0.759	-0.242	0.868	0.288	0.486	0.821
	0.137	0.336	0.080	0.644	0.025	0.580	0.328	0.045
Pb	0.892	-0.498	-0.921	-0.197	0.928	0.701	0.750	0.954
	0.017	0.315	0.009	0.708	0.008	0.121	0.086	0.003
U	-0.218	-0.027	-0.290	-0.614	-0.229	-0.330	-0.333	-0.127
	0.678	0.959	0.577	0.195	0.662	0.523	0.519	0.811
oc%	-0.334	0.319	-0.044	-0.789	-0.349	-0.476	-0.548	-0.282
	0.517	0.537	0.934	0.062	0.498	0.340	0.260	0.588
sand%	-0.760	0.568	0.720	-0.217	-0.781	-0.427	-0.456	-0.818
	0.080	0.240	0.107	0.679	0.067	0.399	0.363	0.047
silt%	0.273	-0.130	-0.056	0.269	0.369	-0.084	-0.030	0.306
	0.601	0.807	0.916	0.606	0.471	0.874	0.955	0.555
clay%	-0.023	-0.073	-0.225	-0.240	-0.130	0.275	0.222	-0.039
	0.966	0.891	0.668	0.646	0.805	0.598	0.672	0.942

	Cr	Co	Ni	Cu	As	Mo	Cd	Pb
Co	0.973							
	0.001							
Ni	0.997	0.962						
	0.000	0.002						
Cu	0.885	0.913	0.887					
	0.019	0.011	0.018					
As	0.910	0.919	0.923	0.788				
	0.012	0.010	0.009	0.062				
Mo	0.752	0.844	0.761	0.918	0.806			
	0.084	0.035	0.079	0.010	0.053			

	Cr	Co	Ni	Cu	As	Mo	Cd	Pb
Cd	0.807	0.887	0.772	0.797	0.723	0.784		
	0.052	0.019	0.072	0.058	0.105	0.065		
Pb	0.956	0.968	0.964	0.884	0.982	0.858	0.781	
	0.003	0.001	0.002	0.019	0.000	0.029	0.067	
U	-0.158	-0.069	-0.111	-0.042	0.233	0.321	-0.096	0.116
	0.765	0.896	0.834	0.937	0.657	0.535	0.856	0.826
oc%	-0.298	-0.189	-0.270	-0.284	0.116	0.100	-0.163	-0.031
	0.566	0.720	0.605	0.586	0.827	0.850	0.758	0.954
sand%	-0.771	-0.720	-0.768	-0.702	-0.655	-0.618	-0.772	-0.688
	0.073	0.107	0.075	0.120	0.158	0.191	0.072	0.131
silt%	0.269	0.250	0.216	0.173	0.014	0.041	0.571	0.071
	0.607	0.633	0.681	0.743	0.978	0.938	0.236	0.894
clay%	-0.013	-0.010	0.050	0.075	0.249	0.202	-0.379	0.194
	0.981	0.986	0.925	0.887	0.634	0.701	0.458	0.712

	U	oc%	sand%	silt%				
oc%	0.926							
	0.008							
sand%	0.069	0.260						
	0.897	0.618						
silt%	-0.501	-0.474	-0.649					
	0.312	0.342	0.163					
clay%	0.581	0.471	0.382	-0.951				
	0.227	0.346	0.455	0.004				

Cell Contents: Pearson correlation P-Value

S2 (LOT 8)

	Al	Si	Ca	Mg	Fe	Mn	Zn	V
Si	0.690							
	0.129							
Ca	-0.148	-0.080						
	0.779	0.880						
Mg	0.864	0.628	0.322					
	0.026	0.182	0.534					
Fe	0.998	0.700	-0.124	0.874				
	0.000	0.121	0.815	0.023				
Mn	0.293	0.227	0.773	0.561	0.335			
	0.573	0.665	0.071	0.247	0.516			
Zn	0.818	0.658	0.422	0.979	0.837	0.698		
	0.047	0.156	0.404	0.001	0.038	0.123		

	Al	Si	Ca	Mg	Fe	Mn	Zn	V
V	0.916	0.487	-0.346	0.732	0.918	0.096	0.647	
	0.010	0.327	0.502	0.098	0.010	0.857	0.165	
Cr	0.917	0.475	-0.303	0.756	0.920	0.125	0.670	0.999
	0.010	0.341	0.559	0.082	0.009	0.814	0.145	0.000
Co	0.936	0.529	-0.355	0.739	0.937	0.101	0.657	0.997
	0.006	0.281	0.490	0.094	0.006	0.850	0.156	0.000
Ni	0.937	0.519	-0.376	0.715	0.937	0.105	0.639	0.992
	0.006	0.292	0.462	0.110	0.006	0.843	0.172	0.000
Cu	0.852	0.337	-0.279	0.579	0.844	0.260	0.547	0.819
	0.031	0.513	0.592	0.229	0.035	0.619	0.261	0.046
As	0.850	0.344	-0.447	0.524	0.830	0.032	0.451	0.846
	0.032	0.505	0.374	0.285	0.041	0.952	0.369	0.034
Mo	0.612	0.045	-0.457	0.287	0.567	-0.187	0.179	0.605
	0.197	0.933	0.363	0.581	0.240	0.722	0.734	0.203
Cd	0.396	-0.079	-0.456	0.060	0.342	-0.291	-0.036	0.352
	0.437	0.882	0.363	0.909	0.507	0.575	0.945	0.494
Pb	0.902	0.408	-0.445	0.637	0.885	-0.031	0.532	0.946
	0.014	0.422	0.377	0.174	0.019	0.954	0.277	0.004
U	-0.103	-0.443	-0.215	-0.244	-0.169	-0.468	-0.357	-0.137
	0.846	0.380	0.682	0.641	0.749	0.349	0.487	0.796
oc%	0.248	0.096	-0.191	0.343	0.241	-0.376	0.188	0.465
	0.636	0.857	0.718	0.506	0.646	0.462	0.721	0.353
sand%	0.074	-0.120	-0.817	-0.313	0.017	-0.846	-0.460	0.178
	0.889	0.821	0.047	0.546	0.975	0.034	0.359	0.736
silt%	-0.316	0.249	0.137	-0.094	-0.328	-0.338	-0.123	-0.399
	0.542	0.634	0.796	0.860	0.526	0.512	0.816	0.433
clay%	0.312	-0.247	-0.112	0.102	0.326	0.363	0.136	0.393
	0.547	0.637	0.833	0.848	0.529	0.479	0.797	0.441

	Cr	Co	Ni	Cu	As	Mo	Cd	Pb
Co	0.995							
	0.000							
Ni	0.988	0.998						
	0.000	0.000						
Cu	0.814	0.839	0.871					
	0.049	0.037	0.024					
As	0.836	0.868	0.896	0.969				
	0.038	0.025	0.016	0.001				
Mo	0.597	0.626	0.658	0.821	0.900			
	0.211	0.184	0.155	0.045	0.014			
Cd	0.340	0.381	0.421	0.664	0.757	0.956		
	0.510	0.457	0.406	0.150	0.081	0.003		

	Cr	Co	Ni	Cu	As	Mo	Cd	Pb
Pb	0.941	0.956	0.965	0.896	0.956	0.824	0.632	
	0.005	0.003	0.002	0.016	0.003	0.044	0.178	
U	-0.137	-0.124	-0.103	0.120	0.248	0.644	0.799	0.162
	0.796	0.815	0.845	0.821	0.636	0.168	0.057	0.759
oc%	0.477	0.424	0.367	-0.076	0.062	0.021	-0.145	0.343
	0.339	0.402	0.474	0.886	0.907	0.969	0.784	0.505
sand%	0.149	0.194	0.213	0.223	0.434	0.666	0.750	0.413
	0.778	0.713	0.686	0.671	0.390	0.149	0.086	0.416
silt%	-0.398	-0.395	-0.444	-0.732	-0.616	-0.541	-0.460	-0.452
	0.435	0.438	0.378	0.098	0.193	0.268	0.359	0.368
clay%	0.393	0.388	0.436	0.723	0.601	0.519	0.435	0.439
	0.441	0.447	0.387	0.104	0.207	0.291	0.388	0.384

	U	oc%	sand%	silt%				
oc%	-0.056							
	0.917							
sand%	0.721	0.189						
	0.106	0.720						
silt%	-0.068	0.389	-0.013					
	0.899	0.446	0.981					
clay%	0.045	-0.394	-0.018	-1.000				
	0.933	0.440	0.973	0.000				

Cell Contents: Pearson correlation P-Value

Station S3 (MG GHAT)

	Al	Si	Ca	Mg	Fe	Mn	Zn	V
Si	-0.593							
	0.160							
Ca	-0.173	0.124						
	0.710	0.791						
Mg	0.391	-0.810	-0.052					
	0.386	0.027	0.911					
Fe	0.534	-0.821	-0.548	0.726				
	0.217	0.024	0.202	0.064				
Mn	0.796	-0.497	-0.390	0.014	0.501			
	0.032	0.256	0.387	0.976	0.252			
Zn	-0.230	0.085	-0.515	-0.223	0.357	0.146		
	0.619	0.856	0.236	0.631	0.432	0.755		
V	0.362	-0.681	-0.608	0.799	0.897	0.217	0.178	
	0.426	0.092	0.147	0.031	0.006	0.640	0.703	

	Al	Si	Ca	Mg	Fe	Mn	Zn	V
Cr	0.566	-0.789	-0.561	0.778	0.923	0.447	0.083	0.958
	0.186	0.035	0.190	0.039	0.003	0.315	0.859	0.001
Co	0.532	-0.715	-0.497	0.850	0.834	0.269	-0.082	0.957
	0.219	0.071	0.256	0.015	0.020	0.559	0.861	0.001
Ni	0.378	-0.686	-0.637	0.750	0.938	0.291	0.295	0.989
	0.404	0.089	0.124	0.052	0.002	0.526	0.520	0.000
Cu	-0.119	-0.150	-0.645	0.358	0.477	-0.168	0.441	0.573
	0.799	0.748	0.118	0.430	0.280	0.719	0.322	0.179
As	0.342	-0.864	-0.526	0.696	0.880	0.423	0.194	0.833
	0.453	0.012	0.226	0.082	0.009	0.344	0.677	0.020
Mo	0.018	-0.372	-0.330	0.179	0.635	0.270	0.859	0.361
	0.969	0.411	0.470	0.701	0.125	0.558	0.013	0.426
Cd	0.101	-0.558	0.619	0.530	0.078	-0.083	-0.596	0.086
	0.830	0.193	0.139	0.221	0.868	0.860	0.158	0.855
Pb	0.783	-0.649	-0.201	0.344	0.536	0.794	-0.220	0.429
	0.037	0.115	0.665	0.450	0.215	0.033	0.636	0.336
U	0.083	0.495	-0.160	-0.564	-0.551	0.139	-0.404	-0.436
	0.859	0.259	0.731	0.187	0.200	0.767	0.369	0.329
oc%	0.086	-0.736	0.174	0.569	0.347	0.078	-0.250	0.260
	0.855	0.059	0.709	0.183	0.445	0.869	0.588	0.574
sand%	0.224	0.230	0.570	-0.464	-0.356	0.274	0.006	-0.651
	0.629	0.620	0.182	0.294	0.433	0.552	0.989	0.113
silt%	0.313	-0.662	0.218	0.827	0.371	-0.057	-0.634	0.537
	0.494	0.106	0.639	0.022	0.413	0.904	0.126	0.214
clay%	-0.329	0.659	-0.254	-0.815	-0.357	0.045	0.644	-0.510
	0.472	0.107	0.582	0.025	0.432	0.924	0.118	0.242

	Cr	Co	Ni	Cu	As	Mo	Cd	Pb
Co	0.961							
	0.001							
Ni	0.958	0.922						
	0.001	0.003						
Cu	0.374	0.440	0.558					
	0.409	0.323	0.193					
As	0.859	0.767	0.843	0.461				
	0.013	0.044	0.017	0.298				
Mo	0.315	0.141	0.474	0.427	0.481			
	0.491	0.763	0.282	0.340	0.275			
Cd	0.184	0.194	0.036	-0.462	0.264	-0.211		
	0.694	0.677	0.940	0.297	0.567	0.650		
Pb	0.665	0.548	0.452	-0.369	0.506	-0.016	0.364	
	0.103	0.203	0.308	0.415	0.247	0.972	0.422	

	Cr	Co	Ni	Cu	As	Mo	Cd	Pb
U	-0.370	-0.296	-0.489	-0.163	-0.416	-0.703	-0.382	-0.014
	0.414	0.520	0.266	0.727	0.353	0.078	0.398	0.976
oc%	0.293	0.271	0.224	0.154	0.645	0.165	0.665	0.162
	0.524	0.556	0.629	0.741	0.118	0.723	0.103	0.728
sand%	-0.492	-0.587	-0.573	-0.729	-0.579	0.029	0.037	0.115
	0.262	0.166	0.178	0.063	0.173	0.951	0.937	0.806
silt%	0.586	0.677	0.451	-0.099	0.478	-0.289	0.810	0.480
	0.166	0.095	0.310	0.833	0.278	0.530	0.027	0.275
clay%	-0.569	-0.656	-0.427	0.143	-0.453	0.293	-0.827	-0.495
	0.182	0.110	0.339	0.760	0.308	0.524	0.022	0.259

	U	oc%	sand%	silt%				
oc%	-0.321							
	0.483							
sand%	-0.000	-0.334						
	1.000	0.464						
silt%	-0.298	0.540	-0.355					
	0.517	0.211	0.434					
clay%	0.305	-0.529	0.306	-0.999				
	0.507	0.222	0.505	0.000				

Cell Contents: Pearson correlation P-Value

Station S4 (JAMBU ISLAND)

	Al	Si	Ca	Mg	Fe	Mn	Zn	V
Si	-0.436							
	0.388							
Ca	-0.504	0.109						
	0.308	0.837						
Mg	-0.458	0.884	0.235					
	0.361	0.019	0.654					
Fe	-0.213	0.213	0.682	0.369				
	0.685	0.686	0.136	0.471				
Mn	-0.272	-0.006	0.923	0.210	0.858			
	0.602	0.990	0.009	0.689	0.029			
Zn	-0.184	0.157	-0.367	0.064	-0.857	-0.644		
	0.728	0.766	0.474	0.905	0.029	0.167		
V	0.556	0.087	0.288	0.016	0.416	0.414	-0.476	
	0.252	0.869	0.581	0.976	0.412	0.415	0.340	
Cr	0.935	-0.184	-0.391	-0.274	0.022	-0.149	-0.367	0.735
	0.006	0.727	0.444	0.600	0.968	0.778	0.475	0.096

	Al	Si	Ca	Mg	Fe	Mn	Zn	V
Co	0.691	0.175	-0.200	-0.041	0.183	-0.048	-0.395	0.838
	0.129	0.741	0.704	0.938	0.729	0.928	0.438	0.037
Ni	0.517	0.156	-0.855	-0.060	-0.300	-0.719	-0.005	0.038
	0.294	0.768	0.030	0.910	0.564	0.108	0.992	0.943
Cu	0.792	-0.048	-0.689	-0.151	-0.056	-0.424	-0.312	0.376
	0.061	0.928	0.130	0.775	0.916	0.402	0.547	0.462
As	0.782	-0.071	-0.849	-0.290	-0.468	-0.724	0.095	0.245
	0.066	0.893	0.033	0.577	0.349	0.104	0.858	0.640
Mo	0.123	0.384	-0.704	0.216	-0.098	-0.594	-0.059	-0.206
	0.816	0.453	0.118	0.682	0.853	0.213	0.911	0.696
Cd	-0.102	-0.185	0.896	0.024	0.693	0.954	-0.535	0.504
	0.847	0.725	0.016	0.964	0.127	0.003	0.274	0.308
Pb	0.795	-0.389	-0.778	-0.623	-0.487	-0.661	0.008	0.188
	0.059	0.446	0.068	0.186	0.327	0.153	0.988	0.721
U	0.703	-0.359	0.077	-0.294	-0.134	0.148	0.020	0.692
	0.119	0.484	0.884	0.571	0.800	0.779	0.970	0.127
oc%	0.587	0.054	-0.915	-0.134	-0.398	-0.773	0.064	0.005
	0.220	0.920	0.011	0.800	0.435	0.072	0.904	0.993
sand%	-0.664	0.091	0.901	0.300	0.419	0.756	-0.022	-0.053
	0.150	0.865	0.014	0.564	0.408	0.082	0.968	0.921
silt%	0.664	-0.065	-0.898	-0.265	-0.397	-0.744	0.008	0.063
	0.150	0.902	0.015	0.612	0.436	0.090	0.988	0.905
clay%	*	*	*	*	*	*	*	*
	*	*	*	*	*	*	*	*

	Cr	Co	Ni	Cu	As	Mo	Cd	Pb
Co	0.894							
	0.016							
Ni	0.591	0.567						
	0.217	0.241						
Cu	0.866	0.758	0.885					
	0.026	0.081	0.019					
As	0.781	0.675	0.904	0.893				
	0.067	0.141	0.013	0.016				
Mo	0.253	0.320	0.896	0.683	0.622			
	0.629	0.537	0.016	0.135	0.187			
Cd	-0.046	-0.011	-0.777	-0.432	-0.652	-0.774		
	0.931	0.984	0.069	0.393	0.161	0.071		
Pb	0.740	0.578	0.782	0.802	0.926	0.462	-0.546	
	0.093	0.230	0.066	0.055	0.008	0.356	0.262	
U	0.608	0.441	-0.160	0.178	0.246	-0.550	0.404	0.265
	0.201	0.382	0.763	0.736	0.638	0.258	0.427	0.612

	Cr	Co	Ni	Cu	As	Mo	Cd	Pb
oc%	0.609	0.524	0.988	0.888	0.936	0.844	-0.797	0.828
	0.199	0.286	0.000	0.018	0.006	0.034	0.058	0.042
sand%	-0.667	-0.556	-0.964	-0.898	-0.960	-0.769	0.745	-0.912
	0.148	0.252	0.002	0.015	0.002	0.074	0.089	0.011
silt%	0.674	0.565	0.971	0.910	0.959	0.783	-0.741	0.898
	0.142	0.242	0.001	0.012	0.002	0.066	0.092	0.015
clay%	*	*	*	*	*	*	*	*
	*	*	*	*	*	*	*	*

	U	oc%	sand%	silt%				
oc%	-0.079							
	0.881							
sand%	0.000	-0.984						
	1.000	0.000						
silt%	-0.003	0.989	-0.999					
	0.995	0.000	0.000					
clay%	*	*	*	*				
	*	*	*	*				

Cell Contents: Pearson correlation P-Value

NOTE * All values in column are identical.
Station S5 (CANNING)

	Al	Si	Ca	Mg	Fe	Mn	Zn	V
Si	-0.605							
	0.395							
Ca	-0.056	-0.298						
	0.944	0.702						
Mg	0.949	-0.376	-0.334					
	0.051	0.624	0.666					
Fe	0.985	-0.544	0.064	0.919				
	0.015	0.456	0.936	0.081				
Mn	0.974	-0.649	0.171	0.864	0.989			
	0.026	0.351	0.829	0.136	0.011			
Zn	0.983	-0.740	0.009	0.890	0.954	0.969		
	0.017	0.260	0.991	0.110	0.046	0.031		
V	0.959	-0.518	-0.332	0.983	0.906	0.871	0.933	
	0.041	0.482	0.668	0.017	0.094	0.129	0.067	
Cr	0.960	-0.552	-0.321	0.974	0.903	0.873	0.941	0.999
	0.040	0.448	0.679	0.026	0.097	0.127	0.059	0.001
Co	0.922	-0.588	-0.384	0.941	0.844	0.819	0.919	0.986
	0.078	0.412	0.616	0.059	0.156	0.181	0.081	0.014

	Al	Si	Ca	Mg	Fe	Mn	Zn	V
Ni	0.654	-0.458	-0.686	0.749	0.518	0.483	0.667	0.825
	0.346	0.542	0.314	0.251	0.482	0.517	0.333	0.175
Cu	0.937	-0.293	-0.140	0.962	0.955	0.900	0.858	0.910
	0.063	0.707	0.860	0.038	0.045	0.100	0.142	0.090
As	0.984	-0.662	-0.163	0.944	0.938	0.931	0.985	0.980
	0.016	0.338	0.837	0.056	0.062	0.069	0.015	0.020
Mo	0.035	-0.160	-0.822	0.182	-0.136	-0.163	0.079	0.286
	0.965	0.840	0.178	0.818	0.864	0.837	0.921	0.714
Cd	-0.857	0.721	-0.466	-0.662	-0.899	-0.949	-0.883	-0.680
	0.143	0.279	0.534	0.338	0.101	0.051	0.117	0.320
Pb	0.608	0.264	-0.400	0.781	0.644	0.524	0.454	0.656
	0.392	0.736	0.600	0.219	0.356	0.476	0.546	0.344
U	-0.784	0.718	0.406	-0.773	-0.665	-0.672	-0.834	-0.874
	0.216	0.282	0.594	0.227	0.335	0.328	0.166	0.126
oc%	0.802	-0.281	-0.641	0.930	0.720	0.647	0.749	0.936
	0.198	0.719	0.359	0.070	0.280	0.353	0.251	0.064
sand%	-0.787	0.872	0.167	-0.695	-0.684	-0.727	-0.869	-0.815
	0.213	0.128	0.833	0.305	0.316	0.273	0.131	0.185
silt%	-0.196	-0.640	0.621	-0.468	-0.217	-0.069	-0.022	-0.326
	0.804	0.360	0.379	0.532	0.783	0.931	0.978	0.674
clay%	0.611	0.162	-0.699	0.828	0.576	0.454	0.484	0.754
	0.389	0.838	0.301	0.172	0.424	0.546	0.516	0.246

	Cr	Co	Ni	Cu	As	Mo	Cd	Pb
Co	0.991							
	0.009							
Ni	0.835	0.896						
	0.165	0.104						
Cu	0.895	0.828	0.540					
	0.105	0.172	0.460					
As	0.985	0.971	0.769	0.884				
	0.015	0.029	0.231	0.116				
Mo	0.301	0.417	0.777	-0.088	0.205			
	0.699	0.583	0.223	0.912	0.795			
Cd	-0.688	-0.627	-0.243	-0.743	-0.793	0.367		
	0.312	0.373	0.757	0.257	0.207	0.633		
Pb	0.623	0.546	0.366	0.840	0.538	-0.079	-0.302	
	0.377	0.454	0.634	0.160	0.462	0.921	0.698	
U	-0.892	-0.942	-0.943	-0.596	-0.882	-0.612	0.507	-0.255
	0.108	0.058	0.057	0.404	0.118	0.388	0.493	0.745
oc%	0.929	0.938	0.911	0.807	0.853	0.514	-0.380	0.712
	0.071	0.062	0.089	0.193	0.147	0.486	0.620	0.288

	Cr	Co	Ni	Cu	As	Mo	Cd	Pb
sand%	-0.839	-0.885	-0.828	-0.544	-0.874	-0.473	0.637	-0.098
	0.161	0.115	0.172	0.456	0.126	0.527	0.363	0.902
silt%	-0.287	-0.235	-0.231	-0.496	-0.150	-0.072	-0.180	-0.883
	0.713	0.765	0.769	0.504	0.850	0.928	0.820	0.117
clay%	0.728	0.702	0.668	0.776	0.612	0.323	-0.162	0.917
	0.272	0.298	0.332	0.224	0.388	0.677	0.838	0.083

	U	oc%	sand%	silt%				
oc%	-0.841							
	0.159							
sand%	0.966	-0.700						
	0.034	0.300						
silt%	-0.011	-0.528	-0.234					
	0.989	0.472	0.766					
clay%	-0.504	0.891	-0.303	-0.855				
	0.496	0.109	0.697	0.145				

Cell Contents: Pearson correlation P-Value

Station S6 (DHAMAKHALI)

	Al	Si	Ca	Mg	Fe	Mn	Zn	V
Si	-0.305							
	0.556							
Ca	-0.535	-0.383						
	0.274	0.453						
Mg	0.517	-0.738	0.291					
	0.294	0.094	0.576					
Fe	0.970	-0.234	-0.606	0.365				
	0.001	0.655	0.202	0.477				
Mn	0.477	-0.366	-0.563	0.322	0.500			
	0.339	0.476	0.245	0.534	0.313			
Zn	0.185	0.393	-0.693	-0.667	0.394	0.217		
	0.726	0.441	0.127	0.148	0.440	0.680		
V	0.850	-0.491	-0.376	0.755	0.725	0.634	-0.220	
	0.032	0.323	0.463	0.082	0.103	0.177	0.676	
Cr	0.900	-0.426	-0.422	0.712	0.782	0.573	-0.159	0.991
	0.014	0.400	0.405	0.113	0.066	0.235	0.764	0.000
Co	0.853	-0.454	-0.404	0.724	0.725	0.610	-0.206	0.998
	0.031	0.365	0.428	0.104	0.103	0.199	0.695	0.000
Ni	0.900	-0.397	-0.450	0.668	0.781	0.546	-0.133	0.982
	0.014	0.435	0.371	0.147	0.067	0.262	0.801	0.000

	Al	Si	Ca	Mg	Fe	Mn	Zn	V
Cu	0.783	-0.424	-0.265	0.736	0.612	0.396	-0.363	0.952
	0.065	0.402	0.612	0.095	0.197	0.437	0.480	0.003
As	0.898	-0.427	-0.471	0.691	0.797	0.656	-0.104	0.992
	0.015	0.398	0.346	0.128	0.058	0.157	0.845	0.000
Mo	0.516	-0.347	-0.569	0.231	0.436	0.682	0.088	0.708
	0.295	0.501	0.238	0.660	0.388	0.136	0.868	0.115
Cd	0.826	-0.510	-0.053	0.880	0.707	0.307	-0.366	0.866
	0.043	0.301	0.920	0.021	0.117	0.554	0.476	0.026
Pb	0.730	-0.468	-0.341	0.748	0.582	0.655	-0.323	0.979
	0.099	0.349	0.508	0.087	0.226	0.158	0.532	0.001
U	0.526	-0.762	-0.027	0.858	0.395	0.700	-0.458	0.856
	0.283	0.078	0.960	0.029	0.439	0.122	0.361	0.030
oc%	0.366	0.129	-0.588	-0.277	0.329	0.072	0.362	0.282
	0.475	0.808	0.220	0.595	0.524	0.892	0.481	0.588
sand%	-0.421	-0.253	0.853	0.486	-0.528	-0.345	-0.840	-0.152
	0.406	0.629	0.031	0.328	0.281	0.503	0.036	0.774
silt%	-0.096	0.580	-0.603	-0.281	-0.107	0.411	0.141	0.094
	0.856	0.228	0.205	0.589	0.839	0.419	0.790	0.859
clay%	0.338	-0.406	0.076	-0.019	0.411	-0.191	0.359	-0.002
	0.513	0.425	0.886	0.971	0.418	0.717	0.484	0.997

	Cr	Co	Ni	Cu	As	Mo	Cd	Pb
Co	0.994							
	0.000							
Ni	0.996	0.990						
	0.000	0.000						
Cu	0.958	0.964	0.965					
	0.003	0.002	0.002					
As	0.994	0.991	0.987	0.928				
	0.000	0.000	0.000	0.008				
Mo	0.684	0.729	0.716	0.672	0.707			
	0.134	0.100	0.110	0.143	0.116			
Cd	0.879	0.851	0.853	0.853	0.852	0.286		
	0.021	0.032	0.031	0.031	0.031	0.583		
Pb	0.951	0.978	0.942	0.942	0.955	0.746	0.792	
	0.003	0.001	0.005	0.005	0.003	0.088	0.060	
U	0.784	0.829	0.752	0.767	0.802	0.632	0.713	0.892
	0.065	0.041	0.085	0.075	0.055	0.179	0.112	0.017
oc%	0.341	0.339	0.417	0.392	0.318	0.701	-0.023	0.264
	0.509	0.511	0.411	0.443	0.540	0.121	0.966	0.613
sand%	-0.213	-0.187	-0.265	-0.085	-0.247	-0.557	0.155	-0.080
	0.686	0.722	0.612	0.873	0.637	0.251	0.769	0.881

	Cr	Co	Ni	Cu	As	Mo	Cd	Pb
silt%	0.070	0.112	0.074	0.036	0.119	0.293	-0.233	0.217
	0.895	0.833	0.889	0.946	0.822	0.573	0.657	0.679
clay%	0.057	0.002	0.083	0.015	0.029	0.049	0.129	-0.162
	0.915	0.997	0.875	0.978	0.956	0.927	0.808	0.759

	U	oc%	sand%	silt%				
oc%	-0.028							
	0.958							
sand%	0.166	-0.783						
	0.753	0.066						
silt%	0.011	0.132	-0.234					
	0.983	0.803	0.655					
clay%	-0.109	0.337	-0.365	-0.819				
	0.837	0.514	0.476	0.046				

Cell Contents: Pearson correlation P-Value

Station S7 (SANDESHKHALI)

	Al	Si	Ca	Mg	Fe	Mn	Zn	V
Si	-0.100							
	0.831							
Ca	-0.774	0.497						
	0.041	0.257						
Mg	0.911	-0.236	-0.629					
	0.004	0.611	0.131					
Fe	0.974	-0.244	-0.851	0.865				
	0.000	0.598	0.015	0.012				
Mn	-0.182	-0.562	-0.039	-0.068	-0.004			
	0.697	0.189	0.934	0.885	0.994			
Zn	-0.033	-0.114	-0.095	-0.055	-0.030	0.271		
	0.943	0.808	0.840	0.907	0.950	0.557		
V	0.994	-0.210	-0.815	0.920	0.982	-0.119	0.000	
	0.000	0.652	0.026	0.003	0.000	0.799	1.000	
Cr	0.981	-0.155	-0.864	0.836	0.978	-0.185	-0.002	0.981
	0.000	0.741	0.012	0.019	0.000	0.691	0.996	0.000
Co	0.981	-0.141	-0.732	0.957	0.937	-0.164	0.062	0.982
	0.000	0.763	0.062	0.001	0.002	0.726	0.895	0.000
Ni	0.990	-0.151	-0.790	0.918	0.977	-0.070	0.064	0.990
	0.000	0.746	0.035	0.004	0.000	0.881	0.892	0.000
Cu	0.956	-0.187	-0.814	0.819	0.992	0.019	-0.074	0.957
	0.001	0.688	0.026	0.024	0.000	0.967	0.875	0.001

	Al	Si	Ca	Mg	Fe	Mn	Zn	V
As	0.910	-0.336	-0.903	0.841	0.945	0.096	0.174	0.931
	0.004	0.461	0.005	0.018	0.001	0.837	0.709	0.002
Mo	0.966	-0.281	-0.877	0.879	0.988	0.019	0.090	0.980
	0.000	0.542	0.010	0.009	0.000	0.967	0.848	0.000
Cd	0.615	0.562	-0.399	0.367	0.498	-0.729	-0.022	0.544
	0.142	0.189	0.375	0.418	0.255	0.063	0.962	0.207
Pb	0.976	-0.198	-0.812	0.884	0.992	-0.042	-0.138	0.977
	0.000	0.670	0.026	0.008	0.000	0.929	0.768	0.000
U	0.575	0.512	-0.133	0.401	0.517	-0.405	-0.553	0.499
	0.177	0.240	0.776	0.372	0.235	0.367	0.198	0.254
oc%	0.906	-0.288	-0.708	0.864	0.888	-0.222	-0.235	0.922
	0.005	0.532	0.075	0.012	0.008	0.632	0.611	0.003
sand%	-0.939	0.042	0.749	-0.836	-0.902	0.368	0.320	-0.921
	0.002	0.930	0.053	0.019	0.005	0.417	0.483	0.003
silt%	0.075	-0.298	0.071	0.423	-0.039	-0.137	0.130	0.114
	0.873	0.516	0.880	0.344	0.933	0.769	0.781	0.808
clay%	0.249	0.272	-0.322	-0.119	0.345	0.003	-0.232	0.206
	0.591	0.555	0.481	0.799	0.449	0.994	0.617	0.658

	Cr	Co	Ni	Cu	As	Mo	Cd	Pb
Co	0.943							
	0.001							
Ni	0.970	0.983						
	0.000	0.000						
Cu	0.961	0.900	0.956					
	0.001	0.006	0.001					
As	0.928	0.907	0.948	0.910				
	0.003	0.005	0.001	0.004				
Mo	0.973	0.951	0.985	0.964	0.981			
	0.000	0.001	0.000	0.000	0.000			
Cd	0.641	0.558	0.558	0.499	0.449	0.493		
	0.121	0.193	0.193	0.254	0.312	0.261		
Pb	0.967	0.935	0.971	0.987	0.920	0.970	0.509	
	0.000	0.002	0.000	0.000	0.003	0.000	0.244	
U	0.523	0.466	0.494	0.590	0.260	0.402	0.604	0.582
	0.228	0.292	0.260	0.163	0.574	0.371	0.151	0.170
oc%	0.888	0.886	0.863	0.866	0.751	0.849	0.417	0.899
	0.008	0.008	0.012	0.012	0.052	0.016	0.352	0.006
sand%	-0.934	-0.890	-0.892	-0.887	-0.805	-0.870	-0.675	-0.934
	0.002	0.007	0.007	0.008	0.029	0.011	0.096	0.002
silt%	-0.025	0.247	0.089	-0.148	0.064	0.040	-0.194	-0.024
	0.957	0.593	0.850	0.751	0.891	0.932	0.677	0.958

	Cr	Co	Ni	Cu	As	Mo	Cd	Pb
clay%	0.343	0.068	0.220	0.443	0.213	0.258	0.417	0.341
	0.452	0.885	0.636	0.319	0.647	0.576	0.352	0.454

	U	oc%	sand%	silt%				
oc%	0.549							
	0.202							
sand%	-0.661	-0.902						
	0.106	0.005						
silt%	-0.373	0.189	-0.040					
	0.410	0.685	0.933					
clay%	0.581	0.128	-0.303	-0.940				
	0.171	0.785	0.509	0.002				

In: Causes and Effects of Heavy Metal Pollution
Editor: Mikel L. Sanchez
ISBN 978-1-60456-900-1

Chapter 12

HEAVY METAL POLLUTION, RISK ASSESSMENT AND REMEDIATION IN PADDY SOIL ENVIRONMENT: RESEARCH EXPERIENCES AND PERSPECTIVES IN KOREA

Jae E. Yang*, Yong Sik Ok, Won-Il Kim[1] and Jin-Soo Lee[2]
Department of Biological Environment, Kangwon National University, Chuncheon, Korea,
[1]National Institute of Agricultural Science and Technology (NIAST), Suwon, Korea and
[2]Mine Reclamation Corp. (MIRECO), Seoul, Korea.

ABSTRACT

This invited paper reviews the status of heavy metal pollution in the paddy soils and rice in Korea, the human health risk assessment for heavy metals, and the remediation approaches to reduce the metal translocation from soil to rice grain. The Soil Environment Conservation Law (SECL) designates the soil pollution standard for As, Cd, Cu, Hg, Cr^{6+}, Pb, Ni, Zn and F, and these are used as both the maximum permissible levels of such metals in agricultural soil. The extensive monitoring for heavy metals in soil and crop revealed that concentrations of metals in paddy soils, without an evident anthropogenic source of contaminants, were mostly below the threshold levels designated by the SECL. Major sources of metal pollution in paddy soils were however related with mining activities. The increased level of Cd and Cu in soil increased the activities of cations (Ca>Mg>K) temporally, decreased the level of exchangeable cations, altered the supply mechanisms and decreased the nutrient buffering capacity of soil. The most widely described effects of metal toxicity in plants were the stunted growth, leaf epinasty and chlorosis. The Korean Government implements various countermeasures to prevent the soil pollution by metals through legislation, monitoring networks, risk assessment and remediation. The potential risk of the adverse effects of metals on human health was assessed based on the human exposure pathways to rice, groundwater and soil in three abandoned mines where metal contents in soil and rice exceeded the safety guidelines.

* Tel: +82-33-250-6446, Fax:+82-33-241-6640, Email: yangjay@kangwon.ac.kr.

The hazard index (HI) values for As and Cd exceeded 1, representing a potential toxic risk of As and Cd to the human health. The cancer risk for As via the rice and groundwater consumptions exceeded one cancer case in ten thousand. Health risk assessment indicated that a long term exposure to rice grown in the metal contaminated paddy soils could pose a potential health threat. The soil and plant management options have been considered to prevent the heavy metal transfer to rice from the contaminated paddy soils. The soil management options include the uses of soil ameliorants, fertilizers and irrigation control, soil covering/dressing, reversing and soil layer mixing methods. In the plant management options, the 24 rice cultivars were screened to find the accumulating or excluding variety. The Japonica cultivars were considerably low accumulating rice for Cd. These cultivars might be screened to cultivate in the contaminated soil environment to have the metal concentration at a low enough for the safe consumption. The continuous submersion of the soil was interacted better with fertilizer than the intermittent irrigation to retard the Cd uptake by rice. Based on the regulatory criteria of Cd for soil pollution and food safety, the quantity of Cd which should be remediate at most was estimated to be only 0.04% of Cd in the contaminated soils. Is it worthwhile to remove such a small quantity of Cd with effort and budget which may be greater than land price? Are those criteria the risk-based or the concentration-based? At least limiting to rice, we need to devote to the development of protocols for pollution monitoring, risk assessment and remediation to cope with such dilemmas in the paddy soil environment.

Keywords: heavy metal pollution, paddy soil, rice, risk assessment, remediation, Cd.

INTRODUCTION

Recently in Korea, conservation of soil quality has been stimulated by recognizing that the soil resources are the critically important components of the earth's biosphere and directly related to human health. Soil plays an important role, not only in the production of food and fiber but also in the maintenance of our environmental quality. Thus, soil contamination by toxic chemicals such as heavy metals, their translocation into crops and threat to human health through the food chain are one of the major environmental concerns in Korea.

Heavy metals are natural constituents of rocks, soils, sediments, and water. Some of the heavy metals are essential for plant growth, but most of them are not (Kabata-Pendias, 2001). When heavy metals are accumulated in soil beyond the self purification capacity, they may inhibit the growth of crops or accumulate in the crops which may be harmful to human health. For the last several decades, heavy metal concentrations in some soils have been increased due to heavy industrialization. Sources of heavy metals causing soil contamination in Korea are mostly derived, directly or indirectly, from the mining sites, industrial or domestic wastewater, solid wastes, and sewage sludge. Also smeltering industry and combustion of fossil fuel constitute a significant contribution to the heavy metal contamination in the soil (Yang et al., 2000).

Data from the extensive monitoring research have shown that the levels of trace metals in soil have increased in last several decades; but current concentrations of heavy metals are, except in a few extreme cases, lower than the regulatory levels (SGIS, 2007; Yang et al., 2000; Kim, 1993; Lim, 1994). Critical impacts of heavy metals on human health, such as

Minamata and Itai-Itai diseases as reported in Japan (Salomons and Forstner, 1984) have not been reported in Korea.

Rice is the staple crop in the Southeast and East Asia. In Korea, accumulations of heavy metal in rice become major environmental and health concerns. Previous works have shown that metal contents in rice grown nearby the abandoned mines are higher than those in unpolluted areas (Kang, 2007; Yang et al., 2000). Among various anthropogenic sources for soil pollution, the heavy metals from the mining activities are known to be the major contribution to the accumulated metals in crops in Korea. Mining operation, tailings and wastewater provide obvious sources of metal contamination (Adriano 1986; Lee et al. 2001). In South Korea, over 1,000 metal mines have been closed or abandoned due to the depression of mining industry since the late 1980s (Lee et al. 2001; Yang et al. 2000). In addition, most of mine tailings were abandoned on slopes and had been discharged by rains directly to the streams and agricultural lands adjacent to the closed or abandoned metal, causing soil and water contaminations (Yang et al., 2006).

The elevated concentrations of heavy metals have been reported in the agricultural soils and crops in the vicinity of the abandoned mines (Lee et al., 2005a). Through the exposure pathways, such as groundwater, rice and soil inhalation, the high level of metals can pose a potential health risk to the residents in the mining territories. The quantitative prediction of the toxicological risk to human health in the metal-contaminated area has not been scrutinized.

In Korea, realizing the importance of preventing soil from heavy metal contamination, a number of pollution control measures are being taken to conserve, prevent and remediate the soil contamination by heavy metals. The Soil Environment Conservation Law (SECL) becomes stricter for regulation of metal pollution and requires the corrective action assessment, risk assessment and remediation for soils exceeding the standard guidelines. Recently, the Mine Hazard Prevention Law (MHPL) was promulgated by the Ministry of Commerce, Industry and Energy (MOCIE) for the integrated management of soil remediation in the mining areas. This article will be devoted to the discussion and perspectives of the present status, impacts, management strategies, risk assessment and remediation of heavy metal pollution in the paddy soil environment in Korea.

I. Present Status, Environmental Impacts and Management Strategies of Heavy Metals in Korean Soils

A. Sources of Trace Metal Contamination in Paddy Soils

Any metal of either naturally occurring or artificial origin can be considered to be a pollutant when it exists in excess concentrations in the wrong place (Adriano, 1986; Kabata-Pendias, 2001). Human activities often mobilize and redistribute metals such that they can cause adverse effects. Certain metal is accumulated in soil to concentrations that are toxic to plants of which may pose a health hazard to domestic animals and humans (Yang et al., 2000).

Heavy metals of environmental concern in Korea, in terms of the degree of phytotoxicity or undesirable entrance into the food chain, are As, Cd, Cr, Cu, Hg, Pb, Ni and Zn. Table 1 summarizes major sources of metals and their pollution pathways into the soil (Yang et al., 2000; Kim, 1989). Metal accumulation in soils is closely connected to the specific local sources such as discharges from smelters, metal-based industries, chemical manufacturing industries, active, inactive, and abandoned mining sites and irrigation water. Among these sources, effluents from the abandoned mines and erosion of mine tailings are crucial factors for threatening soil and crop contaminations by metals in Korea in view of agricultural production.

Numbers of the inactive and abandoned mines have been increasing due to the depressed mining industries over the last two decades (Ministry of Environment, 1993; Park, 1995). The probability for the influx of metals into soils and streams from these mining sites is high. The Ministries of Environment (MOE), Agriculture and Forestry (MAF), and Commerce, Industry and Energy (MOCIE) aware of these potential environmental problems and continues to make all possible efforts to preserve soils from pollution sources, by enacting the Soil Environment Preservation Law (SEPL) and Mine Hazard Prevention Law (MHPL), operating the soil quality monitoring networks, mandatory metal analysis for soil quality and its impact assessment, and setting up the standard for soil quality assessment and crops.

Table 1. Selected hazardous metal sources and major contamination pathways in Korean soil (Kim, 1989; Yang et al., 2000)

Metals	Pollution Source Industries	Pollution Pathways
As	Mining, Agrochemicals	Irrigation water
Cd	Mining Smelter Electroplating / Finishing	Irrigation water Smoke Suspended particles
Cr	Mining, Smelter, Electroplating / Finishing, and Leather	Irrigation water
Cu	Mining Smelter Agrochemicals	Irrigation water Smoke Suspended particles
Hg	Chemical manufacturing, Agrochemicals, Battery, Mining, Metallurgy, Paints, Dyes, Pulp, and Paper	Irrigation water
Ni	Electroplating / Finishing	Irrigation water
Pb	Automobile exhaust, Mining, Metallurgy, Smelter, and Agrochemicals	Irrigation water, Smoke, and Suspended particles
Zn	Mining and Metallurgy Smelters	Irrigation water Suspended particles

B. Monitoring of Heavy Metals in Paddy Soils

1. Natural Abundance of Trace Metals in Korean Soils

Table 2 shows the natural abundance of heavy metals in Korean paddy and upland field soils and selected crops collected from the unpolluted areas. Similar metal contents exist between paddy and upland soils, however, Zn contents in upland field soils for vegetables and fruits are higher than those in paddy soils, possibly due to the use of fertilizers.

Of the total national land areas in Korea (about 9.93 million ha), 65% is forest area and 21% is cultivated lands. A significant feature of farmland use in Korea is the abundance of paddy soils, composing of 63% of total cultivated areas. A large majority (74%) of these paddy fields are irrigated for rice growth. As shown in Table 1, irrigation is a major pathway for metal contamination in soil. Not surprisingly, concentrations of metals in these irrigated paddy soils, especially near the industrialized and mining areas, are usually greater than those in other sites (Yang et al., 2000).

Average concentrations of metals in major crops grown in uncontaminated soils are shown in Table 2. These data have been used as the natural background concentration of metals in crops. Concentrations of Cu and Zn, which are essential plant nutrients, were greater than those of non-essential nutrients of Cd, Cr, Pb, Hg and As.

2. Distribution of Trace Metals in Korean Paddy Soils

More than 60% of farming lands in Korea is used for rice production. Table 3 shows examples of metal distributions in paddy soils and rice as influenced by the various metal pollution sources. Metal concentrations were varied with sampling locations and pollution sources, but those in soils and rice were much higher than the natural occurrence levels (Table 2), indicating soil contamination is closely related to the mining activities.

Table 2. Natural abundance of trace metals in paddy and upland soils, and crops grown in the unpolluted soils

Classification		Sample Number	As	Cd	Cr	Cu	Hg	Pb	Zn	References
			mg/kg							
Soil¶	Paddy(Rice)	407	n.a.†	0.13	n.a.	4.15	n.a.	4.67	3.95	Cited in Yang et al., 2000
	Paddy(Rice)	330	0.56	0.14	0.49	4.00	0.09	5.38	4.36	
	Upland(Barley	105	0.62	0.16	0.28	4.00	0.09	5.49	9.94	
	Upland (Vegetables)	420	n.a.	0.13	0.57	3.78	n.a.	1.93	16.23	
	Upland(Fruits)	325	n.a.	0.22	0.63	3.59	n.a.	1.81	24.61	
Crop ‡	Brown Rice	407	n.a.	0.05	n.a.	3.31	n.a.	0.44	20.55	
	Barley grain	108	0.12	0.05	0.21	4.49	0.04	0.54	19.35	
	Vegetables	420	n.a.	0.15	0.57	5.50	n.a.	2.63	50.79	
	Fruits	374	n.a.	0.07	1.59	3.79	n.a.	2.50	20.70	

¶ Top soils (0~15 cm) were collected at harvesting periods from the unpolluted areas.

† Not analyzed.

‡ Crops were collected from the unpolluted fields.

Table 3. Average metal contents in paddy soil (0~15 cm) and brown rice collected from the selected areas influenced by different pollution sources (Yang et al., 2000)

Samples	Locations	Pollution sources	Sample No.	Cd	Cu	Pb	Zn
				---------------- mg/kg			----------------
Soil	(12)[§]	Pb and Zn mining waste	796	1.11	8.54	27.40	30.30
Rice	(12)	Pb and Zn mining waste	679	0.22	4.07	0.58	4.94
Soil	Siheung	Zn mining	6	7.68	–[¶]	–	938.00
	Changwon		8	1.25	–	–	60.90
	Seongju		12	1.53	–	–	105.80
	Uljin		9	2.16	–	–	104.60
	Chilgok		10	1.79	–	–	17.80
Rice	Siheung	Zn mining	6	0.87	–	–	34.70
	Changwon		8	1.57	–	–	29.61
	Seongju		12	0.55	–	–	28.66
	Uljin		9	0.43	–	–	29.33
	Chilgok		10	1.11	–	–	24.11
Soil	Janghang	Metal smeltering	30	1.92	106.28	123.1	120.13
Rice	Janghang	Metal smeltering	30	0.71	4.44	6.33	30.34
Soil	Kangwon	Metal mines	28	7.35	35.83	98.86	118.77
Rice	Kangwon	Metal mines	28	0.38	2.38	1.31	22.31

[§] Samples from twelve different locations.
[¶] Not analyzed.

Table 4 shows the extensive monitoring results for heavy metal contaminations in the paddy soils nearby the closed mines and industrial sites. Concentrations of all metals were greater than the natural abundances and highest concentrations were even exceeding the Korea soil pollution criteria. These results demonstrate that the major sources of metals in paddy soils in Korea are directly related to the wastes from mines and industrial sites.

Table 4. Mean and ranges of As, Cd, Cu, Ni, Pb, and Zn contents in paddy soils collected near the closed mines and industrial areas[¶]

Year	Samples	Locations	As	Cd	Cu	Ni	Pb	Zn
						mg/kg		
2000	600	Closed mines	3.68 (tr-62.0)	0.59 (tr-8.60)	17.88 (tr-306.0)	0.99 (tr-55.5)	22.61 (tr-5573)	34.6 (tr-429)
2004	600	Closed mines	1.88 (tr-43.4)	0.46 (tr-5.55)	13.88 (tr-292.0)	30.5 (2.9-2073)	14.81 (tr-403)	161.4 (11-1789)
2001	600	Industrial	1.02	0.28	9.16	0.92	9.03	8.30

Year	Samples	Locations	As	Cd	Cu	Ni	Pb	Zn
			mg/kg					
		area	(tr-22.7)	(tr-25.9)	(tr-324.8)	(tr-49.5)	(tr-149.1)	(0.4-244)
2005	600	Industrial area	0.44 (tr-7.18)	0.25 (tr-17.9)	6.68 (tr-164.8)	18.8 (1.0-168)	7.04 (tr-97.9)	78.1 (15-1105)
Natural Abundance			0.56	0.13	4.08	0.82	5.03	4.16
Korea Soil Pollution Criteria (Threshold of Danger Level)			6	1.5	50	40	100	300
Korea Soil Pollution Criteria (Corrective Action Level)			15	4	125	100	300	700

¶Cd, Cu and Pb were extracted by 0.1M HCl and As by 1.0M HCl. Concentrations of Ni and Zn were total content basis after aqua regia extraction. Korean regulation of soil contamination was changed to total content basis for Ni and Zn since 2003 (Soil Environmental Conservation Act, MOE). Data of 2000 and 2001 for Ni and Zn were extracted by 0.1M HCl.

Table 5 shows the mean and ranges of As, Cd, Cu, Ni and Zn contents in rice grown in the contaminated and non-contaminated paddy soils. Those metal contents grown in the contaminated paddy soil as influenced by mining and municipal activities were higher than those in the reference areas. Contents of Cd and Pb in the closed mining areas were exceeding the safety guidelines

Other chemical factors in the pollution sources can cause the problems such as soil acidification and crop growth retardation, although metal contents in mining wastes are critical for the environmental concerns. Kim (1993) reported the rice damages from the mining wastes, which unfortunately were also partly used as irrigation source. The pH and sulfate concentrations of the irrigated soil were 3.6 and 1860 mg/kg, and the respective values of the paddy water were 2.8 and 2850 mg/L.

Table 5. Mean and ranges of As, Cd, Cu, Ni, Pb and Zn contents in rice

Year	Samples	Sites	As	Cd	Cu	Ni	Pb	Zn
			mg/kg					
1999†	500	Non-contaminated	0.07 (tr-0.27)	0.035 (tr-0.46)	2.74 (0.20-20.8)	0.29 (tr-5.27)	0.28 (tr-1.80)	18.3 (3.1-31.3)
1999†	300	Contaminated (municipal)	0.08 (tr-0.43)	0.050 (tr-0.95)	3.59 (0.23-14.8)	0.38 (tr-3.52)	0.29 (tr-2.86)	20.1 (4.1-32.6)
2000‡	300	Contaminated (closed mine)	0.10 (tr-1.38)	0.105 (tr-2.13)	3.35 (0.50-8.92)	0.18 (tr-1.61)	0.22 (tr-5.05)	16.3 (5.6-57.3)
2001‡	30	Non-contaminated	0.06 (tr-0.17)	0.011 (tr-0.07)	2.20 (0.80-3.93)	0.12 (0.4-0.49)	0.04 (tr-0.15)	13.9 (6.8-19.3)
Safety Guideline§			-	0.2‡ (1.0†)	-	-	0.2	-

† Brown rice; ‡ Policed rice

§ Maximum permissible concentration by Korean Food and Drug Administration (KFDA) was 1.0 mg/kg of Cd in brown rice before 2000 and 0.2 mg/kg in polished rice after 2000 in Korea. In case of Pb, 0.2 mg/kg in polished rice was set in 2006.

The pH of the irrigation water was 2.8. These results demonstrate that careful monitoring of the degree of contamination in the soils may be the simplest way of reducing human exposure to toxic chemicals.

Yoo (1990) also reported the influence of mining wastes on the chemical characteristics of the soil. The pH of the paddy water influenced by mining waste was the lowest, as compared to other sources. Metal contents and chemical oxygen demand (COD) in wastes of mines were much higher than those in non-contaminated irrigation water. As expected, metal concentrations in soils irrigated with wastewater were greater than those in unpolluted soils. Better management of wastewater before being discharged into the paddy field is still required to prevent soil from pollution. The MOE requires the mandatory installment of the wastewater treatment facilities at industrial plants. All expenses shall be borne by the producers of pollutants, based on the principle of "Polluters Pay." However, many acid mine drainage (AMD) from the abandoned mines is still discharging into environment without a proper wastewater treatment facilities (Yang et al., 2006).

Heavy metals are natural elements, ubiquitous in trace concentrations in the different components of the ecosystem. High natural soil contents are occasionally found originating from the geological processes. Generally, however, high metal concentrations are mostly the results of industrial and agricultural activities. Metalliferous mining and industrial activities are important sources of metal dispersion and enrichment in the environment in Korea. Therefore, soil contamination by heavy metals is not a recent phenomenon but its intensity and frequency highly increased during the last decades. Total numbers of metalliferous mines are estimated as 2006 places, of which 730 are active mines and 1276 are inactive and closed mines (MIRECO, 2007). Erosion of solid wastes and tailings, and waste water discharges from the abandoned tailings piles, waste rocks and abandoned portal are the major sources of soil pollution in Korea (Yang, 2007).

C. Impacts of Trace Metal Contamination in Soil and Crops

Metals and metalloids are found associated with a wide variety of cellular activities in plants, either as structural components of key molecules or they are involved in a wide range of metabolic processes such as enzymes (Kabata-Pendias, 2001; Farago, 1994). All plants need a variety of metal ions for the optimum integration of growth and metabolic activity. Excessive concentrations of some metals however may produce toxic symptoms in plants. Thus, metal ions can be grouped as essential and non-essential elements. Among metals, the essential elements such as Zn and Cu show the general yield response curve but the non-essential elements such as Cd and Pb show the yield plateau over which elements are neither toxic nor deficient, and the biomass is independent of the tissue concentration of the element. However, there is an upper critical tissue concentration and above this concentration the metal is toxic and the yield is reduced. Metals in the plant operate as stress factors in that they cause physiological reaction change and in so doing can reduce vigor, or in the extreme, totally inhibit plant growth. Sensitivity describes the effects of a stress which result in injury or death of the plant. The most widely described effects of metal toxicity in plants are the stunted growth, leaf epinasty and chlorosis (Farago, 1994). At lower degrees of soil pollution, however, these visible symptoms are less pronounced or can even be absent, whereas at the cellular level several physiological processes are affected due to increased local metal concentrations, such as alteration of the plasma membrane permeability and inhibition of enzymes.

Table 6 summarizes the impacts of heavy metals on growth and yield of crops observed in Korea. Generally small amounts of metals as compared to the total levels were translocated into the plants and most of metals were found in roots (Yang et al., 2000; 2007). Heavy metals caused decreases in several growth parameters such as germination, height, leaf growth, root growth, and number of tillers. Nutrient uptake and crop yield were also decreased with increasing metal concentration. Decreases in some of the physiological parameters such chlorophyll content and transpiration rate were also reported.

Cases for resolution and petition against the environmental pollutions in the specific industrial areas have been increased. Most of them were related to the pollutions of air, water, noise and vibration, and offensive odor (Ministry of Environment, 1993a). The petition case for soil contamination was limited to the crop damages by metals and other substances due to the inlet of wastewater into the farmland. Soil contamination problems specifically arising from the accumulation of trace metals have not been filed. Serious metal impacts on human health through the food chains, as the cases of Minamata and Itai-Itai diseases, have not been reported in Korea.

The increased concentrations of toxic metals in the environment cause degradation of soil chemical and biological qualities (Lothenbach et al. 1999; Yang and Skogley, 1989, 1990) and exhibit toxicity to most plants and human health through food web (Chandrajith et al. 1995; Chen et al. 2000). Research concerning metal speciation, transformation, and bioavailability in Korean soils has been much more limited than that on metal distribution. Because it is well known that metal speciation and transformation in the environment are directly connected to its bioavailability and biotoxicity, more vigorous research activity in these fields is necessary in Korea to assess the impact of trace metals on the soil-plant system.

Contaminations with Cu or Cd changed the ionic distributions between soil and solution, resulting in the temporary increases of cations such as Ca, Mg and K in soil solution and decreases in nutrient buffering capacity, due to the metal adsorption on the soil particles (Yang and Skogley, 1989, 1990).

Table 6. Summary for the impacts of heavy metals on plant growth

Metals	Crops	Impacts
As	Rice	Yield decrease; higher [As] in root; sterilized grain increase; root growth retardation; abnormal cells in roots; decrease in transpiration, height, tillers, and leaf growth
Cd, Cu, Ni Zn	Radish, Cabbage	Germination inhibited; yield decrease; stunted leaf and root growth
Cd	Rice	Decrease in height, tillering, and yield
As, Cd, Cu, Pb, Zn	Corn	Decrease in dry weight, height, peroxidase activity, chlorophyll content, nutrient uptake, stunted growth, ear length and yields
Cr	Cabbage	Chlorosis and necrosis; root growth retardation; decrease in nutrient uptake and yield

References are cited in Yang et al. (2000)

Magnesium was a more responsible cation than Ca for changing ionic distribution induced by metal pollution (Lee, 1995). These changes were considered to degrade the function of the soil to supply nutrients to plant roots over the long term. Most of metals discharged into soil were later adsorbed by the soil, and low concentrations of metals remained in the soil solution.

The primary species of Cu in soil solutions, as predicted by the GEOCHEM modeling (Parker et al., 1987), were free ionic forms (> 99 molar %) and the remaining of Cu was complexed with Cl^-, SO_4^{2-} and OH^-, strongly depending on the soil pH. In case of Cd, Cd^{2+} and $CdCl^+$ forms were the primary species existed in the soil solution, followed by the secondary species such as $CdSO_4$, $CdHPO_4$, and $CdCl_2$ complexes. More than 73 molar % of Cd in soil solution were existed as free Cd species at <100mg/kg additions. Increasing Cd additions increased Cd concentration in soil solution but Cd species complexed with Cl were increased in compensation for the decreases of free Cd species (Lee, 1995). Additions of trace metal such as Cu, Cd, Ni and Zn decreased soil urease activities and the orders of metals reducing 50% of the urease activity were Cu > Zn > Cd > Ni (Im, 1994). These differences in speciation between Cu and Cd influence the amounts of metal taken up by crops showing more Cu transfers into crops than Cd.

Intake of heavy metals from foods is of primary impact of soil pollution. This necessitates the risk assessment to human health. The major food consumption in Korea is through the food crops such as rice, vegetative and fruits. Thus the Korean government awakes of these concerns and implements various countermeasures to prevent metal transfer into crop through legislation, technical remediation, and risk assessment.

II. Management Strategies of Heavy Metals in Korean Soil and Crop

A. Regulatory Strategies for Soil Pollution Prevention

In Korea, the laws relating to the conservation and control measures of the soil pollution are the Soil Environment Conservation Law, the Mine Hazard Prevention Law, the Agricultural Land Law, the Basic Environmental Policy Law, the Water Quality Conservation Law, the Air Quality Conservation Law, and Mining Safety Law, etc.

In recent years, there have been increasing occurrences soil contaminations having metals discharging from the mining, urban and industrial areas. It is believed that this increasing trend has been partly triggered by recent urban development as well as waste disposal problems that are now becoming serious. As the awareness of metal impacts on the ecosystem and human health is increasing, stricter measures to control soil pollution are implemented in Korea.

A separated Law dealing with soil pollutions has not been enacted in Korea by 1994, and rather few articles were included in the Water Quality Preservation Law. Under this condition, proper control and management of soil pollution had been limited. The Soil Environment Conservation Law was promulgated in 1994 by MOE and was effective from 1995, in which soils of agricultural fields, forest, and urban areas have been included for application of the pollution standard. Since after, the MOE has been revised the soil environmental quality standards for additional toxic substances which may affect the crop growth and human health. Based on the continuous measurements of pollutants in soils through the monitoring networks, the soil pollution policy area will be designated and an effective control measures have been established such as risk assessment and remediation protocol etc. The cost of soil pollution policy projects is to be placed on any person responsible for such pollution. A stricter administrative measure will be taken to any to cause such pollution.

B. Designation Of The Specific Harmful Substances for Soil Pollution Assessment Criteria

Heavy metals of As, Cd, Cu, Cr^{6+}, Hg, Pb, Ni and Zn, which are likely to cause soil contamination and restrict crop growth, were designated as the specific harmful substances under the Articles 4 and 16 of the SECL, MOE (1996, 2007) (Table 7). The SECL designates the pollution standard for concentrations of As, Cd, Cu, Hg, Cr^{6+}, Pb, Ni, Zn and F (Table 7). The standards are based on threshold values for limited crop growth and are used as both the maximum permissible levels of such metals in agricultural soil and as the critical values for assessing the environmental impact on the soil.

C. Monitoring Network of Heavy Metal Concentrations in Soils

The MOE installed the nationwide monitoring network of soil quality assessment for heavy metals, according to the Article 5 and 6 of the SECL. The objectives of installing the monitoring network are to (1) observe the changes of metal concentration in soils, (2) use these data for soil contamination control countermeasures cooperating with regional governments, and (3) prevent soil contamination in advance. The number of monitoring locations of the network for surveying metal concentrations in soils has been increased from 780 locations in 1996 to 1,500 locations in 2005. Among them, 25.2% is the soils from agricultural areas.

Table 7. The threshold of danger level and corrective action level for soil pollution criteria in Korea

Substances	Threshold of Danger Level (mg/kg)		Corrective Action Level (mg/kg)	
	Agricultural Area*	Factory/Industrial Area**	Agricultural Area	Factory/Industrial Area
Cd†	1.5	12	4	30
Cu†	50	200	125	500
As‡	6	20	15	50
Hg‡	4	16	10	40
Pb†	100	400	300	1,000
Cr^{6+}†	4	12	10	30
Zn¶	300	800	700	2,000

Table 7. (Continued)

Substances	Threshold of Danger Level (mg/kg)		Corrective Action Level (mg/kg)	
	Agricultural Area*	Factory/Industrial Area**	Agricultural Area	Factory/Industrial Area
Ni¶	40	160	100	400
F	400	800	800	2,000
Organic Phosphates	10	30	-	-
PCB	-	12	-	30
CN	2	120	5	300
Phenol	4	20	10	50
BTEX	-	80	-	200
TPH	500	2,000	1,200	5,000
TCE	8	40	20	100
PCE	4	24	10	60

*Agricultural lands include soil uses for upland, paddy, orchard, pasture, forest, park, etc.
** Factory/Industry lands include soil uses for factory, road, railroad land etc.
† 0.1M HCl extraction
‡ 1.0M HCl extraction
¶ Total content extracted by aqua regia

The monitoring networks are controlled by the MOE and seven regional offices are in charge of sampling and conducting analyses. The monitoring network covers soil samples from agricultural upland fields, paddy fields, urban residences, industrial sites, etc. The analytical results have been reported annually and compiled into the Soil Groundwater Information System (SGIS: http://www.sgis.or.kr; sgisinfo@emc.or.kr). The SGIS provides archives for soil and groundwater monitoring networks results.

Through the monitoring networks, a series of soil pollution survey and management actions are followed depending on metal concentrations in soils based on the soil pollution criteria.

- Site characterization activities at the monitoring sites
- Corrective action assessment: The detailed survey for soil pollution at locations where metal concentration exceed the corrective action level (Table 7)
- Soil environmental impact assessment at sites where the potential soil pollution is plausible due to the presence of the pollution-causing facilities
- Risk assessment application to contaminated sites
- Soil remediation plan for contaminated sites

The monitoring networks data showed that metal distributions in soils from the agricultural lands were similar to those of natural contents, except for the mining and industrial sites. Metal concentrations in soils are fairly constant with time; however, concentrations of some metal such as Cu, Hg and Pb tended to decrease with time (SGIS, 2007). Metal concentrations in soils, collected from the mining areas, the waste landfill areas, and the lower reaches of a stream that was influenced by the type of the industrial wastes discharged into it, were in general higher than those in soils from farming lands. Metals in soils from these farm areas were tended to decrease with time. Concentrations of metals in soils from the residence areas were comparatively low and similar to the natural occurrence.

These results indicate that metal contents in Korean soils were influenced by several contamination sources such as waste water and mining discharges, but the average values of the metal concentrations were in the similar ranges to the natural abundances and were seem to be safe for crop cultivation, except for a few numbers of contaminated areas. Measurement of metal contents in soils from the monitoring networks is continuing for the purposes of soil quality assessment and environmental safety.

D. Designation of the Soil Pollution Policy Area

Based on the analytical results on metal concentrations from the monitoring networks, the MOE may designate areas, corresponding to the requirements for the pollution standard criteria (Table 7), as agricultural soil pollution policy areas for the more intensive but detailed survey for metals. Metal mining and smeltering areas are cases in point. The National Institute of Environmental Research (NIER) is in charge of metal analyses in soils, irrigation water, sediments, slags, and rice.

Metal concentrations in soils from the metal mining areas are much higher than those in other areas. Mining industries in Korea are declining, and the corresponding number of inactive and abandoned mines is increasing since the late 1980s. Therefore metal contaminations of soils in these areas are becoming increasingly serious. The MOE, MAF, MOCIE and Provincial Governments provide guidance, inspection, subsidies and other assistances; make an attempt to prevent soil contamination and crop damage; and establish the soil pollution control measures in these areas.

E. Establishment of Soil Pollution Policy Projects

After the designation of a policy area, the Provincial Governor and Mayor are authorized to design the soil pollution policy projects for soil pollution prevention and removal of soil pollution, and the rationalized use of contaminated agricultural land. When a soil is judged to be contaminated by the specific harmful metals (Table 7), with the approval of the Minister of Environment and help from the Minister of Agriculture and Forestry, the Governor and Mayor can limit the crop growth in these contaminated soils, and can collect and remove the agricultural products from these soils. The cost is to be wholly borne by the person responsible for such contamination. The Governor and Mayor implement the remedial work in the soil pollution policy areas with the physical, chemical and biological remedial methods. For preventing the agricultural soil pollution, a Provincial Governor or Mayor, after designating a policy area, may set a stricter standard than effluent water standards stipulated in accordance the Water Quality Conservation Law.

F. Irrigation Water Quality for Paddy Fields

River water quality standards to be used as a source of irrigation water for rice growth are pH 6 to 8.5, BOD <8 mg/L, COD <8 mg/L, dissolved oxygen >2 mg/L, suspended solid <100 mg/L, EC <1 dS/m, and total nitrogen <5 mg/L. Permissible wastewater discharge standards for the irrigation water are pH 5 to 9, Cr <2 mg/L, Fe <10 mg/L, Zn <5 mg/L, Cu <3 mg/L, Cd <0.1 mg/L, Hg <0.005mg/L, As <0.5 mg/L, Pb <1 mg/L, Cr^{6+} < 0.5 mg/L and soluble Mn <10 mg/L (Ministry of Environment, 1993a). Based on the inspection of the wastewater discharge facilities, the Minister of Environment may take the administrative measures to pollution producers, such as ordering improvement and repair, temporary suspension, license withdraw, accusation, transfer, and others. Also the Minister of Environment may impose the pollution charges to the producers.

G. Safety Criteria of Crops for Cd And Pb

The MOE, MAF and KFDA conducted the extensive survey nationwide to determine the heavy metal contents in 10 major crops in last 10 years. The survey locations divide into non-contaminated areas and contaminated areas such as mining sites. After the risk assessment of these crops to human health, the KFDA revised the food safety standards for Cd and Pb in Dec. 21, 2006. Previously, Cd and Pb standards in rice were available but the new regulation extended to 10 crops (Table 8). Table 9 shows the heavy metal standards for agricultural products designated by EU and FAO/WHO Codex Alimentarius Commission.

Table 8. The maximum permissible Cd and Pb contents in 10 crops designated by KFDA (2006)

Crops	Cd (mg/kg)	Pb (mg/kg)
Rice (peeled)	<0.2	<0.2
Corn	<0.1	<0.2
Soybean	<0.1	<0.2
Red bean	<0.1	<0.2
Potato	<0.1	<0.1
Sweet potato	<0.1	<0.1
Chinese cabbage	<0.2	<0.3
Radish	<0.1	<0.1
Green onion	<0.05	<0.1
Spinach	<0.2	<0.3

Table 9. Heavy metal standards of agricultural products by EU and FAO/WHO

	Heavy metals	Agricultural Products	mg/kg
EU	Cd	vegetables and fruits1 (except leaf vegetables, herbs[fresh], mushrooms, stem vegetables, root vegetables, potatoes)	0.05
	Cd	cereals(except wheat bran, buds, wheat, rice), stem vegetables, root vegetables, potatoes (except celeries, potatoes : apply to ML2 of peeled potatoes)	0.1
	Cd	wheat bran, buds, wheat, rice, soybeans, leaf vegetables, herbs[fresh], celery, mushroom	0.2
	Pb	vegetables(except brassica, leaf vegetables, herbs[fresh], mushrooms, potatoes : apply to ML of peeled potatoes), fruits(except berries and small berries)	0.1
	Pb	cereals(included buckwheats), beans, berries and small berries	0.2
	Pb	brassica, leaf vegetables, cultivated mushrooms	0.3
FAO/ WHO	Cd	leguminous plants, beans(except dried soybeans), cereals(except buckwheats, quinoa)	0.1
	Pb	tropical fruits, citrus fruits, pome fruits, stone fruits, bulbous plant, fruit vegetables, root vegetables	0.1
	Pb	berries and small berries, leguminous plants, beans	0.2
	Pb	Brassica vegetables, leaf vegetables	0.3

[1] Defined Article 1 of Directive 90/642/EEC; [2] ML: Maximum Level

III. Risk Assessment of Heavy Metals in the Abandoned Metalliferous Mines

A. Introduction

Risk analysis is comprised of risk assessment, risk management and risk communication. Risk assessment can be defined as the process of estimating both the probability that an event will occur, and the probable magnitude of its adverse effects – economic, health/safety related, or ecological – over a specified time period (Pepper et al., 2006). There are two types of risk assessment: health-based risk and ecological risk. Whatever its focus, the risk assessment process consists of four basic steps; hazard identification, exposure assessment, dose-response assessment and risk characterization (Paustenbach, 2002) (Figure 1). Once the risks are characterized, various regulatory options are evaluated in a process called risk management, which includes consideration of social, political and economic issues, as well as the engineering problems inherent in a proposed solution. One important component of risk management is risk communication, which is the interactive process of information and opinion exchange among individuals, groups, and institutions.

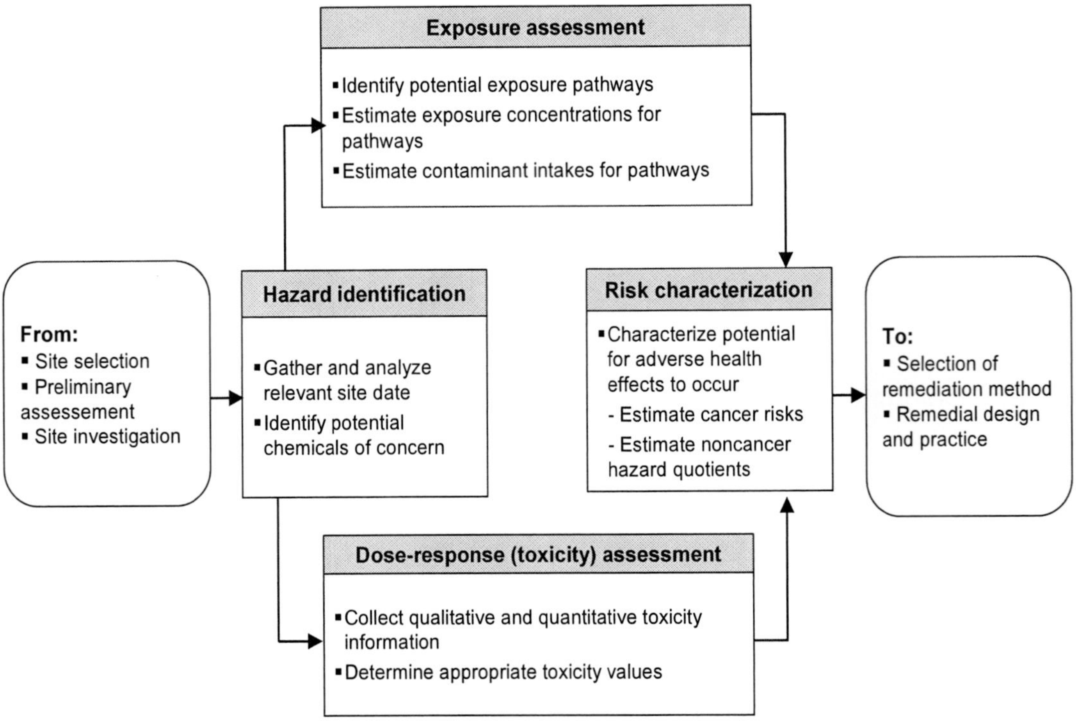

Figure 1. The principal components of the risk assessment process (EPA, 1989).

As stated in the previous sections, metal pollution in soils and crops in Korea is closed connected to the mining activities. The abandoned mines produced the solid wastes and wastewater into streams and agricultural fields. In the typical metalliferous mines, the enormous amounts of sulfide ores in the mine tailings, waste rocks and waste water are weathered and oxidized at an accelerated rate due to an atmospheric exposure. Thus, elevated levels of heavy metals discharged from the mine wastes are found in nearby streams, agricultural soils and food crops. They may pose a potential health risk to the residents in the vicinity of the mines (Davies and Ballinger, 1990; Merrington and Alloway, 1994).

Accurate prediction and quantification of the toxicological risk to the residents in the toxic element-contaminated environments are necessary. The risk assessment models incorporate toxic element data for a range of important exposure pathways; drinking water, food consumption, dust inhalation and hand-to-mouth soil ingestion. From these pathways, a total human intake is derived (Paustenbach, 2002). Recent works on the health based risk and ecological risk assessments have been conducted in the abandoned metalliferous mines in Korea by Lee and Chon (2005), and Lee et al. (2005a,b).

In this section, we discuss about the case study on the risk assessment of the adverse human health effects in the abandoned mine areas through the geochemical analyses and the exposure assessment to heavy metals.

B. Approaches

Three closed metalliferous mimes, Okdong, Dokok and Hwacheon, are selected as the experimental sites (Table 10). Large amounts of mine wastes such as tailings have been discharged into the slope terrain without a proper management being subject to erosion. Samples of soil, rice and the drinking groundwater were collected nearby the mining areas and analyzed for heavy metals (As, Cd, Cu, Pb and Zn) using ICP-AES and ICP-MS. The analytical data was used for the toxic risk assessment. The preliminary step in an exposure assessment is the construction of a conceptual model that represents the exposure pathways for the specific case study. The conceptual model was based on the exposure of a typical Korean farmer, who has lived within the abandoned mine territory for a long term, to the toxic metals through the direct or indirect pathways; ingestion of groundwater as drinking water, soil ingestion through bad hygiene and the ingestion of contaminated rice grains (Figure 2; Lee and Chon., 2005). Risk assessment was coupled with chemical analytical data for these exposure components (Table 11). The risk assessment could not consider Cu and Pb due to the lack of toxicity information such as reference doses and slope factors. Figure 3 shows the risk assessment modeling parameters for each principal component (Lee et al., 2005b).

Table 10. Site descriptions for the Okdong, the Dokok and the Hwacheon metalliferous mines (Lee et al., 2005a)

Mines	Type of ore deposits	Major metals	Main geology	Sulfide minerals
Okdong	Hydrothermal vein type	Cu, Pb, Zn	Sandstone, Black shale	Chalcopyrite, Galena, Sphalerite
Dokok	Hydrothermal vein type	Au, Ag, Cu	Quartz porphyry, Sedimentary rocks	Chalcopyrite, Pyrite, Sphalerite
Hwacheon	Hydrothermal vein type	Au, Ag, Pb, Zn	Conglomerate, Sand-stone, Black shale	Arsenopyrite, Galena, Pyrite, Sphalerite

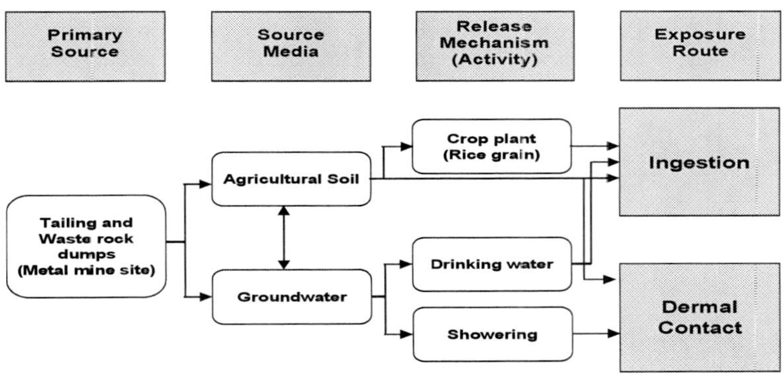

Figure 2. Conceptual site model (CSM) for the risk assessment in the abandoned metal mine areas (Lee and Chon, 2005).

C. Heavy Metals in Soil, Rice and Groundwater

The geochemical data of heavy metals for tailings, soils, rice and groundwater in three mines are shown in Table 11. Heavy metal concentrations in tailing and soils were relatively high exceeding the Korea soil pollution criteria (Table 7). These high levels can affect the metal uptake by rice and other crops grown on the agricultural fields.

Mean concentrations of heavy metals in rice grains in the Okdong and Hwacheon mines were also high and exceeded the natural abundance (Tables 2 and 5). Arsenic concentration in groundwater used as drinking water from the Okdong mine was higher than the permissible level for drinking water for WHO and USA (0.01 mg/L) and Cd concentration from the Dokok mine exceeded the permissible level for drinking water in Korea and WHO (0.01 mg/L) (Table 11). The chronic consumption of rice and groundwater by the local residents could pose a potential health problem due to a long-term As and Cd exposures in these mines.

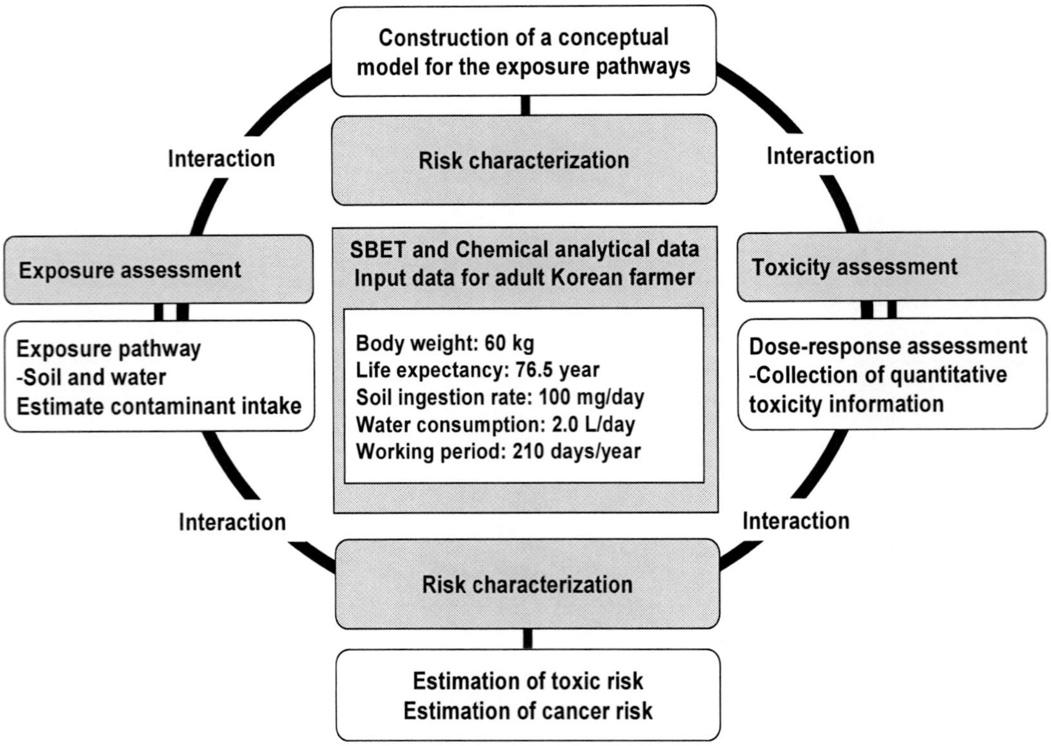

Figure 3. The risk assessment modeling for heavy metals in the abandoned metalliferous mines (Lee et al., 2005b).

Table 11. Mean concentrations of As, Cd, Cu, Pb and Zn in tailings and soils from the abandoned metal mines (mg/kg)

Mine	Sample type	As	Cd	Cu	Pb	Zn
Okdong	Tailings(mg/kg)	72	53.6	910	1,590	5,720
	Soils(mg/kg)	14	3.5	57	44	104
	Rice(mg/kg)	0.24	0.12	3.37	0.20	21.2
	Groundwater(mg/L)	0.038	0.006	0.029	0.058	0.098
Dokok	Tailings(mg/kg)	254	98.2	2,550	4,200	18,020
	Soils(mg/kg)	10	2.9	37	52	156
	Groundwater (mg/L)	0.001	0.054	0.005	0.001	0.317
Hwacheon	Tailings(mg/kg)	72	12.4	34	580	1,300
	Soils(mg/kg)	15	2.8	19	173	178
	Rice(mg/kg)	0.23	0.16	2.24	0.08	28.3
	Groundwater(mg/L)	0.007	n.d.	0.011	0.007	0.046

D. Exposure Assessment

The dose assessment is carried out by estimating the total environmental exposure to a specific heavy metal identified in the source. In order to assess the exposure, the average daily dose (ADD) of the contaminant via the three identified pathways, such as soil, groundwater and rice grain, was needed to be estimated. The formula for the ADD is as follows (Lee et al., 2005a):

$$\text{ADD} = \frac{C \times IR \times ED \times EF}{BW \times AT \times 365}$$

where C is the concentration of the metal in the soil, water and rice (ppm)

IR is the ingestion rate per unit time (mg/day, L/day)

ED is the exposure duration (years)

EF is the exposure frequency (days/year)

BW is the body weight of the receptor (kg)

AT is the averaging time (years)

The estimated ADD values for As, Cd and Zn with three exposure pathways in three mines are shown in Table 12. In the groundwater exposure pathway, the daily intake of As in the Okdong mine site was highest, as in the cases of Cd and Zn in the Dokok mine. In the soil pathway, doses of As and Cd in the Okdong mine were highest. There was no significant difference in the ADD values of heavy metals between the Okdong and the Hwacheon mines through the rice grain pathway.

Table 12. ADD values of As, Cd and Zn with the various exposure pathways (mg/kg-day)

Exposure pathways	Mines	As	Cd	Zn
Soil	Okdong	2.16E-06	1.73E-06	3.77E-05
	Dokok	1.44E-06	1.67E-06	4.49E-05
	Hwacheon	2.11E-06	1.61E-06	5.12E-05
Rice	Okdong	1.45E-03	7.24E-04	1.28E-01
	Hwacheon	1.39E-03	9.65E-04	1.71E-01
Groundwater	Okdong	1.21E-03	1.92E-04	3.13E-03
	Dokok	3.20E-05	1.73E-03	1.01E-02
	Hwacheon	2.24E-04	n.d.	1.47E-03

Table 13. Hazard indices for As, Cd and Zn in the abandoned metal mine sites

Mines	As	Cd	Zn
Okdong	8.88	1.11	0.44
Dokok	0.11	3.45	0.03
Hwacheon	5.38	0.97	0.57

E. Toxic Risk

Toxic risks are the non-carcinogenic exposures being described in terms of a hazard quotient (HQ). The HQ formula is ADD/RfD, where the RfD is the reference dose which can be obtained from the US-EPA IRIS (Integrated Risk Information System) database (http://www.epa.gov/iriswebp/ iris/index.html). The hazard index (HI) was obtained by summing the ADD/RfD ratios of three exposure pathway values as follows (Lee et al., 2005a).

$$HI = \Sigma HQs = \{ADD(\text{rice grain})/RfD + ADD(\text{groundwater})/RfD + ADD(\text{soil})/RfD\}$$

The HI values for As, Cd and Zn were site specific showing the differences with the abandoned mining sites (Table 13). In the Okdong and the Hwacheon mine sites, the HI values for As were higher than 1.0 representing a toxic risk for As via exposure (ingestion) to three exposure pathways. A toxic risk for Cd was higher in the Dokok mine than in the Okdong. The toxic risk in the Okdong mine was relatively higher than the other two mines. A toxic risk for Zn was relatively low as compared to the other two metals.

F. Risk Characterization: Cancer Risk

The final phase pf risk assessment is risk characterization. In this phase, exposure and dose-response assessments are integrated to yield probabilities of effects occurring in human under specific exposure conditions. Quantitative risks are calculated for appropriate media and pathways. In other word, the cancer risks are statements of probability. An individual excess risk is an estimate of the probability that an individual will get cancer from an exposure, not the probability of dying from it. If the dose-response curve is assumed to be linear at low doses for a carcinogen, then the cancer risk is estimated as follows with the combinations of the chronic daily intake and the potency factor (PF) or the slope factor (SF) of the dose-response curve. The ADD is the average daily dose averaged over a lifetime and the SF is the slope factor, i.e. the gradient of the laboratory determined dose-response curve in the low-dose region, usually assumed to be linear. The SF and RfD are compound-specific and may be obtained from the US-EPA IRIS database (Table 14).

Cancer risk = ADD (from the exposure assessment) x SF (from IRIS database)

Table 14. Reference doses (RfD) and slope factors (SF) of As, Cd and Zn (US-EPA database)

Metals	RfD (mg/kg-d)	SF (mg/kg-d)
As	3×10^{-4}	1.5
Cd	5×10^{-4} /1×10^{-3}	n.a.
Zn	3×10^{-1}	n.a.

Based on the RfD and SF values, the cancer risk for As was assessed in this case study (Table 15). The cancer risk for As via the consumption of groundwater was in the orders of Okdong>Hwacheon>Dokok mines. The probabilities of the cancer risk for As in the Okdong and Hwacheon mines were 7*E*-04 and 1*E*-04, respectively. Also, in these sites, those values for As through the exposure to rice grain consumption were 8*E*-04, indicating the cancer risk exceeded one cancer case in ten thousand. The US-EPA considers the risks of greater than one in one hundred thousand (1 in 100,000) as significant.

Table 15. The cancer risk assessment of As for three exposure pathways in the three abandoned mines

Mines	Rice grain pathway	Soil pathway	Groundwater pathway
Okdong	8.5×10^{-4}	1.3×10^{-6}	7.1×10^{-4}
Dokok	n.a.	8.5×10^{-7}	1.9×10^{-5}
Hwacheon	8.2×10^{-4}	1.2×10^{-6}	1.3×10^{-4}

G. Discussion

In this case study, the geochemical data showed that heavy metal contents in tailings, soil, groundwater and rice were greater than the natural abundance and exceeded the soil and crop safety standards of the Korea, indicating those exposure pathways were affected by the mining activities. The risk assessment revealed that the HI (hazard index) values for As and Cd exceeded the value of 1, depending on the mine sites, representing a potential toxic risk of As and Cd to the human health through the three exposure pathways. The cancer risk for As via the pathways of rice and groundwater consumptions in the Okdong and the Hwacheon mine sites exceeded one cancer case in ten thousand. These results demonstrated that a significant human risk could be present from the pollutions of soil, groundwater and crops in the abandoned mining sites. However, the uncertainty is inherent in every step of the risk assessment process. Thus, before we can begin to characterize the risk, we need some idea of the nature and magnitude of the uncertainty in this type of risk estimate, which requires a further study such as sensitivity analysis and the Conte Calro simulation (Pepper et al., 2006).

IV. SOIL POLLUTION COUNTERMEASURES: REMEDIATION

A. Introduction

Once the risk is characterized, various regulatory options should be followed to evaluate the pollution cleanup through the risk management process, which includes consideration of social, political and economic issues, as well as the engineering problems inherent in a proposed solution. There are many ways to remediate the polluted soils (Sparks, 2003) (Table 16). In most cases, these methods have been applied to the industrial and mining sites, where the soil pollutions with various inorganic and organic pollution sources are extremely high, such as the superfund sites in US. However, under paddy fields conditions for rice production in the Southeast and East Asia, uses of these technologies are not economically feasible. For this reason, in Korea, many researchers have tried to find alternative remedial methods which are economically feasible and environmentally sound. Those candidates are classified into chemical, physical and biological remediation (SGIS, 2007).

Table 16. In situ and non-in-situ techniques used in soil decontamination (Sparks, 2003)

Technologies	Advantages	Limitations	Relative costs
In situ			
Volatilization	Can remove some compounds resistant to biodegradation	Volatile organic compounds only	Low
Biodegradation	Effective on some nonvolatile compounds	Long-term timeframe	Moderate

Technologies	Advantages	Limitations	Relative costs
Phytoremediation	Effective with a number of inorganic and organic compounds	Plants are often specific for particular contaminants	Low to medium
Leaching	Could be applicable to wide variety of compounds	Not commonly practiced	Moderate
Passive	Lowest cost and simplest to implement	Varying degrees of removal	Low
Isolation/containment	Physically prevents or impedes migration	Compounds not destroyed	Low to moderate
Non-in-situ			
Land treatment	Uses natural degradation	Some residuals remain	Moderate
Thermal treatment	Complete destruction possible	Usually requires special features	High
Asphalt incorporation	Use of existing facilities	Incomplete removal of heavier compounds	Moderate
Solidification	Immobilizes compounds	Not commonly practiced for soils	Moderate
Excavation	Removal of soils from site	Long-term liability	Moderate

Physical methods are soil covering/dressing, soil removing/dressing, reversing, soil layer mixing, isolation of contaminated soils, and burial of the contaminated soil into impermeable layer sheet. Chemical methods are to precipitate metals to form insoluble hydroxide, carbonate or phosphate compounds by increasing soil pH with additions of silicate, calcium carbonate, or calcium hydroxide etc. or to make metals less soluble by promoting reducing condition in a soil by adding organic materials such as compost, grasses etc. or S-containing compounds. Biological methods are to grow specific plants that absorb these metals from the contaminated soil (phytoremediation) using ornamental or wild plants.

Recently in Korea, two approaches have been considered to prevent the heavy metal transfer to rice from the contaminated paddy soils: the soil and the plant management option. The soil management options include the uses of soil ameliorants, fertilizers and irrigation control to inhibit the metal uptake by rice. Also, soil covering/dressing, reversing and soil layer mixing methods are introduced and under evaluation to set the guidelines in the mining affected soils. In the plant management options, the rice cultivars have been screened to find the accumulating or excluding species. This approach intends to find cultivars that accumulate a specific heavy metal at a low enough for the safe consumption even when growing in the contaminated soil environment. Yu et al. (2006) screened the 43 cultivars of rice (*Oryza sativa* L.) for Cd uptake for the evaluation of the pollution safe cultivars of rice. In this section, we will introduce the case studies on the remediation of the contaminated paddy soils using the soil and plant management options.

B. Mine Hazard Prevention Law (Mhpl) and Remediation Initiative

The Mine Hazard Prevention Law (MHPL) was promulgated in May of 2005 by the Ministry of Commerce, Industry and Energy (MOCIE) for the integrated control of the remediation of the polluted environments caused by the mining activities. This law also covers the remediation of the contaminated agricultural soils that are located in the territories of the mining activities. According to this Law, the MIRECO, Mine Reclamation Cooperation, as a government subsidiary was installed in 2006 to manage the remedial projects.

There exist 2006 metalliferous mines in Korea, of which 730 mines are active and 1,276 mines are inactive/abandoned/closed. Among them, 936 mines have been assessed for the pollution levels, of which 388 mines have produced mine hazards; such as erosion of solid wastes and tailings, soil pollution, waste water discharges from the abandoned tailings and waste rocks, portal water discharge, subsidence, and abandoned facilities (MIRECO, 2007). The MHPL deals with prevention of environments from the mining activities by the forestry revegetation, wastewater treatment, soil remediation, erosion control and subsidence prevention etc.

In Korea, the remediation of the contaminated agricultural soils has been shared to some extents by the roles of MOE with the SECL and MAF by the Arable Land Law. The KFDA is in charge of food safety guidelines. With the MHPL promulgation, the integrated remediation is initiated for the contaminated soils by various sources of contaminants.

C. Prioritization of the Abandoned Mines for Remediation: Case Study by MOE (2006)

The MOE conducted the corrective action assessment through the detailed survey for soil pollution in the abandoned mining areas following the SECL guideline. Among of the 936 inactive and closed metal mines, 529 mines are classified to the corrective action assessments. Until 2005, 191 mines were finished the survey and 28 mines were surveyed in 2006. From 2007, the rest of mines will be under assessment to be finished by 2009.

For 28 mines, the survey protocol was consisted of the general environmental conditions, site characterization and the detailed survey for soil pollution. Land use, mine history, mining wastes such as tailings and rocks, surface and groundwater uses and heavy metal contents in soils were investigated. Based on survey results, the MOE determined the priority for the mining sites remediation and grouped them into I, II and III.

- I : Heavily contaminated and need the urgent remediation: 5 mines
- II: Moderately contaminated and need to remediate later: 8 mines
- III: Less contaminated and need a further pollution dispersion prevention: 13 mines

In those mines, most of the soils had been used for paddy and upland fields and surface water was used as irrigation source. Among 28 mines, 23 mines were selected as the corrective action assessment, of which heavy metals in soils from 19 mimes were exceeded the corrective action level (Table 7), and from 14 mines were higher than the threshold of danger level. Soils were mostly contaminated with As, Cd, and Cu. This protocol was useful in deciding the priority to select the remediation project site.

D. Fertilization and Irrigation Practices to Inhibit Cd Uptake by Rice

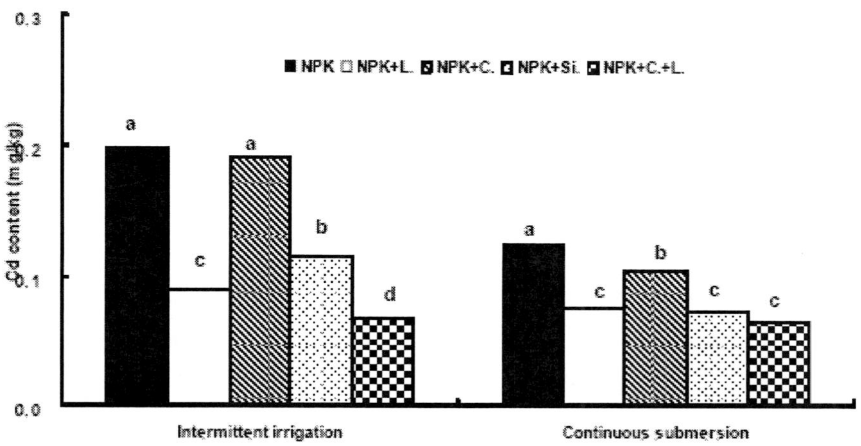

Figure 4. The interaction of fertilizer and water management practices on the Cd uptake by rice. The different letters above the bar indicate the significant difference among treatments at $P<0.05$ level.

Figure 4 shows the combined treatment effects of fertilizer and irrigation practices on the Cd uptake by rice. The continuous submersion of the soil was interacted better with fertilizer than the intermittent irrigation to retard the Cd uptake by rice. The combination of NPK fertilizer with lime and compost was effective to control the Cd transfer to the rice. The effect of the silicate fertilizer was moderate. In addition to these, the soil reaction can affect the Cd availability to rice. When the soil pH of the contaminated soil was 5.2, the transfer factor percentage was 16% but when the pH was increased to 7.2 with lime treatment, only 1.3% of soil Cd was transferred to rice. Also, phosphate fertilizers with 30 kg/10a reduced the Cd in the brown rice from 0.9 to 0.5 mg/kg. Deep plowing of the soil to 60 cm decreased the Cd contents in the brown rice from 0.8 to 0.3 mg/kg. These management practices also increased the rise yield. Results show that with such conventional soil management practices in the metal contaminated soils, soils can be reclaimed to support the yield and prevent the metal contamination for the food crops.

E. Rice Cultivars Screening for Cd Accumulation

As the plant management option, 24 cultivars of rice, commonly cultivated in Korea, were screened to compare the Cd accumulation. The Indicana cultivars of Hanyang and IR 72 rice and other cultivars such as M23 and IRI326 were classified as a relatively high accumulating rice, even though the Cd contents were lower than the food safety guideline of the KFDA. On the other hand, the Japonica cultivars were considerably low accumulating rice for Cd (Figure 8). Yu et al. (2006) compared 43 rice cultivars in China to screen the pollution safe cultivars of rice for the purpose to select the rice that can take up Cd even in the contaminated soils at a level low enough for the safe consumption. They reported that, at a low level of Cd exposure, 30 out of the 43 cultivars screened were found to be Cd pollution-safe cultivars (PSC). Grain Cd concentrations were highly correlated between the experiments, suggesting that Cd accumulation in rice is genotype dependent and that the selection of PSC is possible, at least at a certain level of soil contamination. No Cd-PSCs were found under the high level of Cd exposure. They concluded that the screening for PSCs and the PSC breeding programs are feasible to effectively reduce the risk of human exposure to soil pollutants such as Cd through crop consumption.

F. Dilemma on Soil Remediation vs. the Metal Standard Criteria

Based on the standard criteria for soil pollution and food safety, the quantity of Cd which should be remediate at most can be estimated (Table 17). Based on these estimations, the remediation should be taking action in order to remove the only 0.04% of Cd in the contaminated soils. At this point, we should be considerate in evaluating which criteria between soil and rice is correctly designated: Are those criteria the risk-based or the concentration-based? There should be a concrete discrimination in setting the standards.

Table 18 estimated the duration required to remove the metals using a plant taking up a certain level of metals. Those estimations are based on the soil pollution criteria (threshold of danger level) in Korea. This plant can be classified as a non-resistant based on the metal accumulation level, but considering the amounts of biomass production, this can be used for phytoremediation source. Using this plant, removal of Cd, Pb, Cu and Zn will take 450, 14,000, 10,000 and 2000 years, respectively. Phytoremediation in these days is considered as the promising remediation tools, but selection of plants based on the level of accumulation and the biomass yield should be considered, along with the clean-up goal. Again, is the soil standard correctly set?

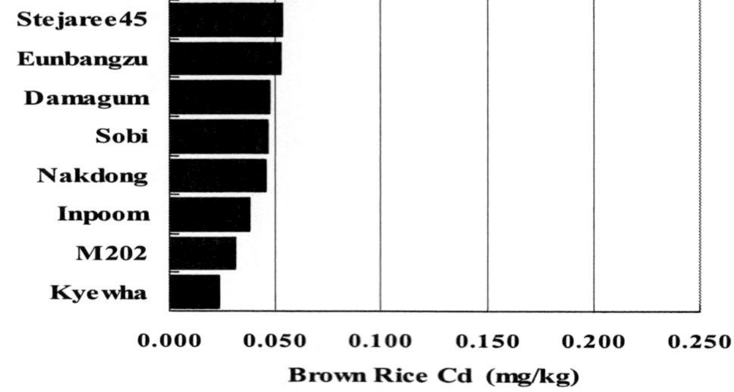

Figure 5. Comparison of the rice cultivars for the Cd accumulations.

Table 17. The estimation of the Cd removal ration by the remediation based on the standard criteria for soil and food safety in Korea

Parameters to be considered	Estimation
The corrective action level of the Korean soil pollution criteria for Cd	4 mg/kg
Quantity of Cd in surface soil (10 cm depth) with 1.2 g/cm^3 bulk density	4,800 g/ha
The safety guideline of the KFDA for the polished rice	0.2 mg/kg
Total biomass of rice (grain, straw and root)	10,000 kg/ha
Total quantity of Cd taken up by rice at the safety guideline level	2 g/ha
Ratio of Cd in rice over soils at both standard levels	(2/4800)*100=0.04%

Table 18. Estimation of time required to phytoremediate the heavy metal contaminated soil to the level of the soil pollution standard using the high biomass producing plant

Parameter Estimations	Metal Conc (mg/kg)			
	Cd	Pb	Cu	Zn
Field Soil	3.0	150.0	92.0	700
Korean Soil Standard (threshold of danger level): (A-level)	1.5	100	50	300
Plant (Biomass production of 8 ton/ha)	2.5	2.8	3.0	156
Mass balance				
Metal (kg) in plowing layer (6,000ton) (50 cm depth, bulk density of 1.2 g/cm^3)	18	900	552	4,200
Metal (kg) in clean soil	9.0	600	300	1,800
X: to be removed from field soil (kg)	9.0	300	252	2,400
Y: annual removal (kg) by plant with biomass of 8ton/ha	0.02	0.022	0.024	1.248
X/Y: Period (yrs) for biological cleanup to achieve A-level	450	13636	10500	1923

REFERENCES

Adriano DC 1986. *Trace elements in the terrestrial environment*, pp18-20. Springer-Verlag, New York, USA.

Albasel N, Cottenie A 1985: Heavy metals uptake from contaminated soils as affected by peat, lime, and chelates. *Soil Sci. Soc. Am.* J. 49, 386-390.

Chang AC, Page AL, Warnke JE, Grguevic E 1984: Sequential extraction of soil heavy metals following a sludge application. *J. Environ. Qual.* 13, 33-38.

Chandrajith RLR, Okumura M, Hashitani H 1995: Human influence on the Hg pollution in lake Jinzai, Japan. *Appl. Geochem.* 10, 229-235.

Chen HM, Zheng CR, Tu C, Shen ZG 2000: Chemical methods and phytoremediation of soil contaminated with heavy metals. *Chemosphere* 41, 229-234.

Chen ZS, Lee GJ, Liu JC 2000. The effects of chemical remediation treatments on the extractability and speciation of cadmium and lead in contaminated soils. *Chemosphere* 41, 235-242.

Davies BD, Ballinger RC 1990. Heavy metals in soils in north Somerset, England, with special reference to contamination from base metal mining in the Mendips. *Environ. Geochem. Health* Vol. 12, pp. 291-300.

Farago, ME 1994. *Plants and the chemical elements: Biochemistry, uptake, tolerance and toxicity*. VCH, New York, USA.

Im, IG 1994. Influence of heavy metals on soil urease activity. M.S. Thesis, Graduate School, Kangwon National University, Chuncheon, Korea. (Korean)

Jung MC, Thornton I 1997. Environmental contamination and seasonal variation of metals in soils, plants and water in the paddy fields around a Pb-Zn mine in Korea. *Sci. Tot. Environ.* 198, 105-121.

Kabata-Pendias A 2001. *Trace elements in soils and plants*. CRC Press, Boca Raton, USA.

Kang MJ 2007. Bioavailability of arsenic and cadmium to crops in the mining-affected soils. MS Thesis, Graduate School, Kangwon National University, Chunchon, Korea. (Korean)

Kim BY, Kim KS, Lee CS, Woo SH 1993. Survey on the natural heavy metal contents in fruits and orchard soils in Korea. RDA. *J. Agri. Sci.* 35, 280-290.

Kim BY 1993. Soil pollution status and its improvement countermeasures. *Symp. Soil Management for Sustainable Agric*. pp. 68-98, Kor. Soc. Soil Fert.

Kim JJ 1989. Soil pollution. In: K. Han et al. (eds.), *Agricultural Environmental chemistry*. pp.169-214. Dong-Hwa Technol. Pub. Co., Seoul.

Lee CG, Chon HT, Jung MC 2001. Heavy metal contamination in the vicinity of the Daduk Au-Ag-Pb-Zn mine in Korea. *Appli. Geochem.* 16, 1377-1386.

Lee JS, Chon HT 2005. Risk assessment of arsenic by human exposure of contaminated soil, groundwater and rice grain. *Econ. Environ. Geol.* 38:535-545.

Lee JS, Chon HT, Jung MC 2005a. Toxic risk assessment and environmental contamination of heavy metals around abandoned metal mine sites in Korea. *Key Engineering Materials,* Vol. 277-279:542-547.

Lee JS, Chon HT, Kim KW 2005b. Human risk assessment of As, Cd, Cu and Zn in the abandoned metal mine site. *Environ. Geochem. Health* 27:185-191.

Lee KW 1995. Heavy metal-induced changes of ionic distribution and species between soil and soil solution. M.S. thesis. Kangwon National University. Chuncheon, Korea.

Lim SK 1994. Survey on the establishment of the soil quality standard. Korea Council *Environ. Sci.. Seoul.*

Lindim C, de Varennes A, Toress MO, Mota AM 2001. Remediation of sandy soil artificially contaminated with cadmium using a polyacrylate polymer. *Commum. Soil Sci. Plant* Anal. 32, 1567-1574.

Lothenbach B, Furrer G, Scharli H, Schulin R 1999. Immobiization of zinc and cadmium by montmorillonite compounds: effects of aging and subsequent acidification. *Environ. Sci. Technol.* 32, 2945-2952.

Mench MJ, Manceau A, Vangronsveld, J, Clijsters H, Mocquot B 2000. Capacity of soil amendments in lowering the phytoavailability of sludge-borne zinc. *Agronomie* 20, 383-397.

Merrington G, Alloway BJ. 1994. The transfer and fate of Cd, Cu, Pb and Zn from two historic metalliferous mine sites in the U.K. *Applied Geochem. Vol.* 9, pp. 677-687.

Ministry of Environment. 1993. Korea Environmental Yearbook. Vol. 6.

MIRECO (Mine Reclamation Corporation). 2007. http://www.mireco.or.kr

Nargal RP, Singh BR 1998. Effect of organic materials on partitioning, extractability and plant uptake of metals in an alum shale soil. *Water Air Soil Pollu.* 103, 405-421.

Parker DR, Zelazny LW, Kinraide TB 1987. Improvement to the program GEOCHEM. Soil *Sci. Soc. Am. J.* 51, 488-491.

Paustenbach DJ 2002. *Human and Ecological Risk Assessment: Theory and Practice* (John Wiley and Sons, New York), p.1556.

Pepper IL, Gerba CP, Brusseau ML 2006. *Environmental and pollution science*. Academic Press, New York, USA.

Salomons W, Forstner U 1984. Metals in the hydrocycle. Springer-Verlag, Berlin, Germany.

Soil Groundwater Information System (SGIS). 2007. http://www.sgis.or.kr (Korean)

Sparks DL. 2003. *Soil environmental chemistry*, Academic Press, New York, USA.

US EPA. 1989. Risk assessment guidance for superfund: Volume I. Human health evaluation manual (Part A). EPA/540/1-89/002. Washington, DC, USA

Wilkin RT, McNeil MS 2003. Laboratory evaluation of zero-valent iron to treat water impacted by acid mine drainage. *Chemosphere* 53, 715-725.

Yang JE, Skogley EO 1989. Influence of copper or cadmium on soil K availability properties. *Soil Sci. Soc. Am. J.* 53, 1019-1023.

Yang JE, Skogley EO 1990. Effects of copper or cadmium on potassium adsorption and buffering capacity. *Soil Sci. Soc. Am. J.* 54, 739-744.

Yang JE, Kim YK, Kim JH, Park YH 2000. Environmental impacts and management strategies of trace metals in soil and groundwater in the Republic of Korea. In: Huang PM and Iskandar IK (eds) *Soils and Groundwater Pollution and Remediation: Asia, Africa, and Oceania*, pp 270-289. Lewis Publishers, Boca Raton, FL, USA.

Yang JE, Skousen GS, Ok YS, Yoo KY, Kim HJ. 2006. Reclamation of abandoned coal mine waste in Korea using lime cake by-product. *Mine Water and the Environment*, 25:227-232.

Yang JE 2007. Remediation of Tailings and Leachate in the Samkwang Mine. *Mine Reclamation Corporation* (MIRECO), Seoul, Korea.

Yoo SH, Lee CY 1990. Contents of Cd and Zn in paddy soil and brown rice in Zn mining areas. *Kor. Nat'l Acad. Sci. J.* 19, 255-266.

Yoo SH 1990. Investigation of the Agricultural Environment pollution status. Kor. Rural Develop Agency, 3-44.

Yu H, Wnag J, Fang W, Yuan J, Yang Z 2006. Cadmium accumulation in different rice cultivars and screening for pollution-safe cultivars of rice. *Science of the Total Environment*, 370:302-306.

INDEX

A

B

C

E

F

G

H

I

J

K

L

N

O

P

Q

R

S

T

U

V